ANNALS OF THE NEW YORK ACADEMY OF SCIENCES

Volume 514

EDITORIAL STAFF
Executive Editor
BILL BOLAND
Managing Editor
JUSTINE CULLINAN
Associate Editor
KEVIN M. MAYER

The New York Academy of Sciences
2 East 63rd Street
New York, New York 10021

MECHANISMS OF CHEMICAL-INDUCED PORPHYRINOPATHIES

ANNALS OF THE NEW YORK ACADEMY OF SCIENCES
Volume 514

MECHANISMS OF CHEMICAL-INDUCED PORPHYRINOPATHIES

Edited by Ellen K. Silbergeld and Bruce A. Fowler

The New York Academy of Sciences
New York, New York
1987

The electron micrograph on the cover of the paperback volume shows a kidney proximal tubule cell from a rat exposed to lead-contaminated drinking water for 6 months. Note the swollen mitochondria. See the paper by Fowler et al. *(pp. 172–182) for more details.*

Library of Congress Cataloging-in-Publication Data

Mechanisms of chemical-induced porphyrinopathies.

(Annals of the New York Academy of Sciences, ISSN 0077-8923; v. 514)
Based on a conference held in Rye Brook, N.Y. on Oct. 27–29, 1986 by the New York Academy of Sciences.
Bibliography: p.
Includes index.
1. Porphyria—Environmental aspects—Congresses. 2. Porphyrin and porphyrin compounds—Metabolism—Congresses. 3. Metals—Toxicology—Congresses. 4. Organohalogen compounds—Toxicology—Congresses. 5. Lead—Toxicology—Congresses. I. Silbergeld, Ellen K. II. Fowler, Bruce A. III. New York Academy of Sciences. IV. Series.
Q11.N5 vol. 514 500 s 87-34795
[RC632.P6] [616.3'99]
ISBN 0-89766-423-X
ISBN 0-89766-424-8 (pbk.)

SP
Printed in the United States of America
ISBN 0-89766-423-X (cloth)
ISBN 0-89766-424-8 (paper)
ISSN 0077-8923

ANNALS OF THE NEW YORK ACADEMY OF SCIENCES

Volume 514
December 31, 1987

MECHANISMS OF CHEMICAL-INDUCED PORPHYRINOPATHIES[a]

Editors and Conference Organizers

ELLEN K. SILBERGELD and BRUCE A. FOWLER

CONTENTS

Preface. *By* ELLEN K. SILBERGELD and BRUCE A. FOWLER ix

Mechanisms of Chemical-Induced Porphyrinopathies: Introductory Comments. *By* ATTALLAH KAPPAS 1

Part I. Regulation of Heme and Porphyrin Metabolism

Zinc Protoporphyrin: Past, Present, and Future. *By* ROBERT F. LABBÉ, REBECCA L. RETTMER, ALKA G. SHAH, and JUDITH R. TURNLUND 7

Genetic Regulation of the Heme Pathway. *By* SHIGERU SASSA, HIROYOSHI FUJITA, and OSAMU SUGITA 15

δ-Aminolevulinic Acid Dehydratase Isozymes and Lead Toxicity. *By* K. H. ASTRIN, D. F. BISHOP, J. G. WETMUR, B. KAUL, B. DAVIDOW, and R. J. DESNICK 23

Part II. Mechanisms of Metal-Induced Porphyrinopathies

Occurrence and Biological Properties of *N*-Methyl Protoporphyrin. *By* FRANCESCO DE MATTEIS, ANTHONY H. GIBBS, and ANN E. HOLLEY 30

Metal-Induced Alterations of δ-Aminolevulinic Acid Dehydratase. *By* ALFRED BERNARD and ROBERT LAUWERYS 41

Effect of Lead on Uroporphyrin and Heme Content in the Bone Marrow of Rats Exposed to Nitrous Oxide. *By* WALTER N. PIPER and W. RUSS CHRISTENSON 48

Metal Alteration of Uroporphyrinogen Decarboxylase and Coproporphyrinogen Oxidase. *By* JAMES S. WOODS and BRUCE A. FOWLER 55

Metal Induction of Heme Oxygenase. *By* F. WILLIAM SUNDERMAN, JR. 65

Metal Inhibition of Ferrochelatase. *By* HARRY A. DAILEY 81

Control of Heme Metabolism by Synthetic Metalloporphyrins. *By* GEORGE S. DRUMMOND 87

[a]This volume is the result of a conference entitled Mechanisms of Chemical-Induced Porphyrinopathies, which was held in Rye Brook, New York on October 27–29, 1986 by the New York Academy of Sciences.

Part III. Mechanisms of Halogenated Organic-Induced Porphyrinopathies

Role of Cytochrome P-450 in Porphyria Caused by Halogenated Aromatic Compounds. *By* HERBERT L. BONKOVSKY, PETER R. SINCLAIR, W. JAY BEMENT, RICHARD W. LAMBRECHT, and JACQUELINE F. SINCLAIR 96

Patterns of Porphyrin Accumulation in Response to Xenobiotics: Parallels between Results in Chick Embryo and Rodents. *By* GERALD S. MARKS, JANE POWLES, MARTHA LYON, STUART MCCLUSKEY, ELAINE SUTHERLAND, and DAVID ZELT 113

Effects of Chlorinated Organics on Intermediates in the Heme Pathway and on Uroporphyrinogen Decarboxylase. *By* LAVINIA CANTONI, MILENA RIZZARDINI, ANDREA GRAZIANI, CARMEN CARUGO, and SILVIO GARATTINI 128

Alterations of Uroporphyrinogen Decarboxylase by Chlorinated Organics. *By* G. H. ELDER, A. G. ROBERTS, and A. J. URQUHART 141

Inhibitory Effect of Tetrachloro-*p*-Hydroquinone and Other Metabolites of Hexachlorobenzene on Hepatic Uroporphyrinogen Decarboxylase Activity with Reference to the Role of Glutathione. *By* GÜNTER KOSS, MARGA LOSEKAM, JUTTA SEIDEL, KLAUS STEINBACH, and WOLFGANG KORANSKY . 148

Biochemical Studies on Griseofulvin-Induced Protoporphyria. *By* TOKIO SHIMOYAMA and SHIGEO NONAKA 160

Part IV. Porphyrins as Indicators of Chemical Injury and Exposure

Porphyrins as Indicators of Chemical Injury and Exposure. *By* ELLEN K. SILBERGELD 170

Metal- and Metalloid-Induced Porphyrinurias: Relationships to Cell Injury. *By* BRUCE A. FOWLER, AGNETA OSKARSSON, and JAMES S. WOODS 172

Turkish Epidemic Hexachlorobenzene Porphyria: A 30-Year Study. *By* HENRY PETERS, DEREK CRIPPS, AYHAN GÖCMEN, GEORGE BRYAN, ERDOGAN ERTÜRK, and CARL MORRIS 183

Lead Effects on the Heme Biosynthetic Pathway: Relationship to Toxicity. *By* MICHAEL R. MOORE, ABRAHAM GOLDBERG, and ALBERT A. C. YEUNG-LAIWAH 191

Porphyrinurias and Occupational Disease. *By* MANFRED O. DOSS 204

Porphyrins in Urine as an Indication of Exposure to Chlorinated Hydrocarbons. *By* J. J. T. W. A. STRIK 219

Mechanism of PCB-Induced Porphyria and Yusho Disease. *By* YOSHIHIKO SEKI, SHOSUKE KAWANISHI, and SEIYO SANO 222

Part V. Factors Influencing Alterations in Heme Metabolism and Function

Regulatory Roles of High-Affinity Metal-Binding Proteins in Mediating Lead Effects on δ-Aminolevulinic Acid Dehydratase. *By* PETER L. GOERING and BRUCE A. FOWLER 235

Drug-Induced Modulations of Hepatic Heme Metabolism: Neurological Consequences. *By* MARIA ALMIRA CORREIA, DAVID A. LITMAN, and JENNINE M. LUNETTA 248

Hormonal Regulation of the Heme Pathway. *By* RICHARD A. GALBRAITH 256

Part VI. Porphyrinopathies in Chemical-Induced Disease States Involving Specific Organ Systems

The Dermatologic Manifestations of Human Porphyria. *By* DAVID R. BICKERS . 261

Alterations in Renal Heme Biosynthesis during Metal Nephrotoxicity. *By* AGNETA OSKARSSON and BRUCE A. FOWLER 268

Lead-Induced Abnormalities of Porphyrin Metabolism: The Relationship with Iron Deficiency. *By* SERGIO PIOMELLI, CAROL SEAMAN, and SHIV KAPOOR . 278

Porphyrins and Hepatotoxicity. *By* RENATE D. KIMBROUGH 289

Role of Altered Heme Synthesis in Chemical Injury to the Nervous System. *By* ELLEN K. SILBERGELD 297

Mechanisms of Toxicity of Photoactivated Artificial Porphyrins: Role of Porphyrin–Protein Interactions. *By* ANN SMITH 309

Principles for Application of Porphyrin Markers in Epidemiological Studies. *By* PHILIP J. LANDRIGAN 323

Comments on Porphyrin. *By* NORTON NELSON 327

Poster Papers

Hematoporphyrin Derivative Uptake and Tumor Cell Density. *By* GRETA M. GARBO, RICK W. KECK, STEVEN H. SELMAN, and MARTHA KREIMER-BIRNBAUM 328

Mechanism of Urinary Excretion of δ-Aminolevulinic Acid after Intratracheal Instillation of Gallium Arsenide. *By* PETER L. GOERING and BRUCE A. FOWLER 330

Studies on the Role of the *Ah* Receptor in Hexachlorobenzene-Induced Porphyria. *By* MARK E. HAHN, THOMAS A. GASIEWICZ, PATRICIA LINKO, and JOYCE A. GOLDSTEIN 333

Porphyria Cutanea Tarda: Manifestation and Therapy. *By* VÁCLAV KORDAČ, PAVEL MARTÁSEK, VLASTIMIL ZOUBEK, and MILAN KALÁB 335

Comparison of Uroporphyrinogen Decarboxylase Activities and Uroporphyrin Accumulation in Mice, Japanese Quail, and Cultured Chick Embryo Hepatocytes Treated with Polyhalogenated Aromatic Compounds. *By* R. LAMBRECHT, P. SINCLAIR, W. J. BEMENT, J. SINCLAIR, and H. CARPENTER 337

Mathematical Models for Effects of Xenobiotics on Heme Biosynthesis: Application to Lead. *By* ALLAN H. MARCUS 339

Induction of Acute Attack in Acute Hepatic Porphyrias: Experiences with Diagnosis and Therapy over Thirty Years. *By* PAVEL MARTÁSEK, VÁCLAV KORDAČ, MICHAELA KOZÁKOVÁ, MILAN JIRSA, PETR KOTAL, and JIŘÍ HORÁK 342

Overcoming the Limitations of Hematofluorometry for Assaying Zinc Protoporphyrin. *By* REBECCA L. RETTMER, ELAINE W. GUNTER, and ROBERT F. LABBÉ 345

Evidence for 3,4,3′,4′-Tetrachlorobiphenyl-Mediated Oxidation of Uroporphyrinogen by Chick Embryo Liver Supernatants. *By* PETER R. SINCLAIR and RICHARD W. LAMBRECHT 347

An Association between Zinc Protoporphyrin Elevation and Recovery from Acute Hematopoietic Damage. *By* THOMAS L. WALDEN, JR., HUDA M. AL-ANSARI, and WALTER R. FARKAS 349

Index of Contributors 351

Financial assistance was received from:

- ENVIRONMENTAL PROTECTION AGENCY
- HOECHST-ROUSSEL PHARMACEUTICALS, INC.
- NATIONAL INSTITUTE OF ENVIRONMENTAL HEALTH SCIENCES
- SANDOZ CORPORATION
- SCHERING-PLOUGH CORPORATION

Preface

This volume follows upon a tradition of landmark publications in the *Annals of the New York Academy of Sciences* to examine the progress in the knowledge of the chemistry, biology, and physiology of the porphyrin pathway. But for the first time, the significance of this major cellular pathway has been considered as a source of information of relevance to toxicology, in the context of papers presented on the endogenous and exogenous factors that regulate this complex system and the physiological consequences of dysfunction in its biochemistry.

Since the last porphyrin publication in the *Annals,* in 1973,[a] much has been learned of the genetic and molecular factors that control heme biosynthesis in cell systems. Considerable attention has been focused upon the critical importance of metalloporphyrins and the functional relationships among early- and late-stage enzymes in the pathway; these issues have been critically reviewed by many contributors to this volume. The intracellular compartmentation of the pathway, within and outside mitochondria, and the variations in porphyrin biochemistry across different organ systems attest to the complexities of understanding signals from this pathway.

But in addition, and of critical importance in the organization of this volume, a substantial biomedical literature has developed on alterations in porphyrin biochemistry associated with relatively low-level exposure of experimental animals and humans to several environmentally and occupationally significant toxins. These toxins include toxic metals, such as lead, arsenic, gallium, manganese, cobalt, nickel, and copper, and polycyclic aromatic hydrocarbons, such as polychlorinated biphenyls (PCBs), hexachlorobenzene, and 2,3,7,8-tetrachlorodibenzo-*p*-dioxin (TCDD). Although our understanding of the mechanisms of action of these agents ranges in completeness, it is clear from the many papers in this volume, which draw upon a large literature of clinical and experimental findings, that the porphyrin pathway is clearly a target for the actions of these agents. Thus, like the exemplary pharmacologic compounds griseofulvin and allylisopropylacetamide, these agents serve to dissect the intricate interrelationships within this highly regulated system.

In addition, these agents are of considerable real-world importance. Exposure to them has been the subject of considerable controversy. Lead and PCBs are some of the most extensively regulated substances worldwide; TCDD remains one of the most controversial chemicals in modern society. The promise of porphyrin toxicology is hence twofold: as a source of "embedded signals of toxicity," to quote a journalist who attended the meeting upon which this volume is based; and as a set of critical mechanisms that may help to explain the multiple manifestations of toxicity that have been described in populations exposed to these agents. This volume thus considers chemically acquired porphyrinopathy at two levels: as a set of phenomena that might yield clinically useful indicators, or biological markers, of exposure and the progress of intoxication; and as causative factors in organ-specific toxicity involving the skin, liver, kidney, and nervous system.

As biological markers, the enzymes, intermediates, and products of the porphyrin pathway are highly promising because of their detectability in available compartments. One porphyrin marker—zinc protoporphyrin, or erythrocyte protoporphyrin—has an accepted status and widespread use as a clinical screen for excess lead exposure. Other porphyrin signals have not been as thoroughly exploited, or have been less well

[a] See volume 206, entitled The Chemical and Physical Behavior of Porphyrin Compounds and Related Structures, edited by Alan D. Adler.

correlated with exposure—several papers in this volume, however, do demonstrate the potential of evaluating the expression of porphyrin signals for specific toxicant exposure, particularly of certain halogenated organics and metals.

The remarkable similarities between inherited diseases of heme synthesis and the acquired porphyrinopathies, including hepatic porphyria and coproporphyrinuria, have long intrigued biologists with the possibility that similar mechanisms of organ-level damage might be occurring. These mechanisms presumably involve alterations in porphyrin synthesis because these alterations are common to the inherited and acquired disorders. The causative role of porphyrins and heme in dermatotoxicity, hepatotoxicity, renal dysfunction, and central and peripheral neurotoxicity is discussed in several papers. Heme depletion, photocatalyzed reactions, toxicity of heme precursors and substituted metalloporphyrins, and effects of altered heme metabolism on related amino acid metabolism remain the hypotheses under discussion. Although considerable progress has been made in understanding the role of these factors in cell function, it remains a challenge to elucidate why porphyrias—both inherited and acquired—appear to affect the specific organ systems as they do.

This volume provides the first critical discussion of the significance of altered heme synthesis in chemical intoxication states involving TCDD and related compounds. Clearly, altered biochemical patterns are observed in experimental animals exposed chronically to TCDD, although the step(s) at which this extremely toxic agent acts remain unclear. The effects of TCDD in humans are less clear: genetically predisposed individuals, as at Seveso, can be precipitated into overt disease by exposure to porphyrinopathic agents like TCDD, but its effects on nonsusceptible persons are less clear. Until better correlations with tissue-level dosimetry can be made, it is not clear how to interpret the available clinical findings. Several contributors discuss the importance of measuring comprehensive patterns of porphyrin signals in urine. Although the significance of these signals as predictors of clinical disease is as yet undefined, it has been noted that alterations in biochemistry of sufficient magnitude to be reflected in patterns of excreted products are invariably accompanied by morphologic correlates of organ damage.

Porphyrins and related compounds have pharmacologic properties that, in addition to playing an hypothesized causative role in toxicity in porphyrinopathic conditions, may be exploited in pharmacotherapies of oncogenic disease. Two papers report on exciting developments in this area and point to the power of this system, which can be exploited in certain conditions but in others requires extensive evaluation and intervention to prevent toxicity.

This volume provides an example of an interdisciplinary exchange of information among basic biologists, pharmacologists, biochemists, and toxicologists, whose interests have brought them into the common arena of studying this remarkable cellular system. Heme synthesis is essential to cell function; its development, external and internal regulation, and response to genetic and environmental conditions make it a uniquely rich area to explore critical concepts of genetically determined responsiveness to toxicants, the effects of combined exposures, and the consequences of long-term biochemical alterations on cell morphology and organ function. It is hoped that this arena will continue to promote fruitful collaborations among these disciplines and that our growth in knowledge will permit rational exploitation of this pathway for both signals and mechanistic hypotheses.

ELLEN K. SILBERGELD
BRUCE A. FOWLER

Mechanisms of Chemical-Induced Porphyrinopathies

Introductory Comments

ATTALLAH KAPPAS

The Rockefeller University Hospital
New York, New York 10021

This volume has as its principal focus a consideration of the mechanisms by which chemicals induce porphyrinopathies.

It is a timely subject for review. We have recently passed the two-decade mark 4ince the first demonstration was made by Granick and Urata[1] that the rate-limiting enzyme in the metabolic sequence through which porphyrins and heme are produced could be induced by foreign chemicals. Within this same time frame heme b was shown by Omura and Sato[2] to be the prosthetic group of the hepatic CO-binding microsomal protein, opening the way to the remarkable series of discoveries that have defined the various functions and gene classes of those liver hemoproteins collectively known as cytochrome P-450. And finally it has been within the same two decades that mitochondrial P-450 was shown by Estabrook and colleagues[3] to catalyze a crucial oxygenation step in the biosynthesis of adrenocortical hormones. This observation concerning the physiological role of P-450 in synthetic reactions was a seminal discovery—one that may at times seem to be obscured by the mass of literature defining P-450 as a central enzyme in catabolic or detoxification processes, or as a hemoprotein upon which a myriad of foreign chemicals exert a biological impact.

The opening lecture at the conference upon which this volume is based was originally entitled "The Basic Biochemistry of Cellular Heme Metabolism." Ultimately, however, the lecture did not emphasize the biosynthetic sequence and metabolic fate of heme (and neither will this paper, which is based on the lecture) because there was no one present at the conference who was not knowledgeable about the biosynthetic sequence of this pigment; its role as the prosthetic group of a variety of hemoproteins; the metabolic route by which it is degraded to bile pigments; or the fact that it represents the point of impact, as well as the active site of biotransformation, of a majority of the organic chemicals with which we come into contact in our environment. Heme (FIGURE 1), which is the subject—directly (heme itself) or indirectly (in the form of its precursors or protein complexes)—of most of the papers in this volume, will therefore be examined from a broad perspective in this paper.

FIGURE 2, although it may remind us of the complex intermediates and metabolic events that are involved in heme formation and function, also draws attention to types of problems in heme biology that could perhaps be more intensively addressed by those who are concerned with the impact of environmental chemicals on porphyrin, that is, heme metabolism in the broadest sense. These problems include those related to the compartmentalization of heme and its intermediates in the cell; the transit of these compounds between cytosol and various cellular organelles; the manner in which this metabolic traffic is regulated normally and how it may be perturbed by such agents as the foreign chemicals that will be discussed in this volume; the mechanism and sites of assembly of heme with its apohemoproteins and the impact of environmental agents on these processes; and how the cell signals that the life span of a given molecule of heme has been concluded and that it is time for the pigment to end its functional existence and yield up its iron atom for reutilization elsewhere.

The interactions shown in FIGURE 2 seem almost chaotic when one attempts to schematize them in such a fashion; but complex as they are, they also proceed in an orderly and closely coupled manner so that there is little "wastage" of intermediate compounds; and scant evidence of inefficiency of metabolic interconnections can normally be detected in this complex of processes in the whole animal. It is primarily through the perturbation of key steps in this array of metabolic events, which are produced by genetic or environmental factors, that one can identify the impact of such factors on cellular heme metabolism, as well as learn, through the study of the mechanisms of such perturbations, how porphyrin, heme, and hemoprotein metabolism is normally controlled—or can be made to be controllable—by pharmacological means.

FIGURE 3 represents a great simplification of all of the pathways depicted in FIGURE 2 and has the advantage of defining not only the principal metabolic axes of heme in the cell but probably represents a reasonable, schematic summary of the main themes of most of the papers included in this volume.

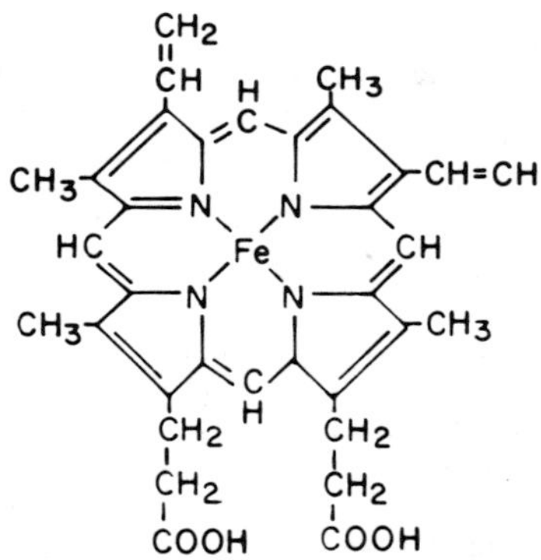

FIGURE 1. Structure of heme.

The pathway of synthesis of heme involves a sequence of eight enzymatic steps, and a number of investigators in this audience have conducted major, sustained programs of study related to these enzymes. Such studies have included systematic efforts to isolate, purify, and characterize the mode of action of these enzymes and to define their regulatory mechanisms. To date, a number of these enzymes have been purified to homogeneity; and some of them have had their gene structures partially or fully defined. The interest in such research is driven not only by the basic importance of such work but by the desire to understand the nature of the genetically determined biochemical lesions of the human porphyrias—and the manner in which environmental factors interact with the genetic lesions in these disorders to produce overt clinical syndromes.

Although in most instances the biochemical lesions underlying the porphyrias can be identified by appropriate study, it is also true that most of these hereditary disorders remain in a latent state for long periods of time in most afflicted subjects, and erupt only when some detrimental environmental agent or circumstance imposes itself on the patient. The hereditary porphyrias must therefore also be considered diseases of the

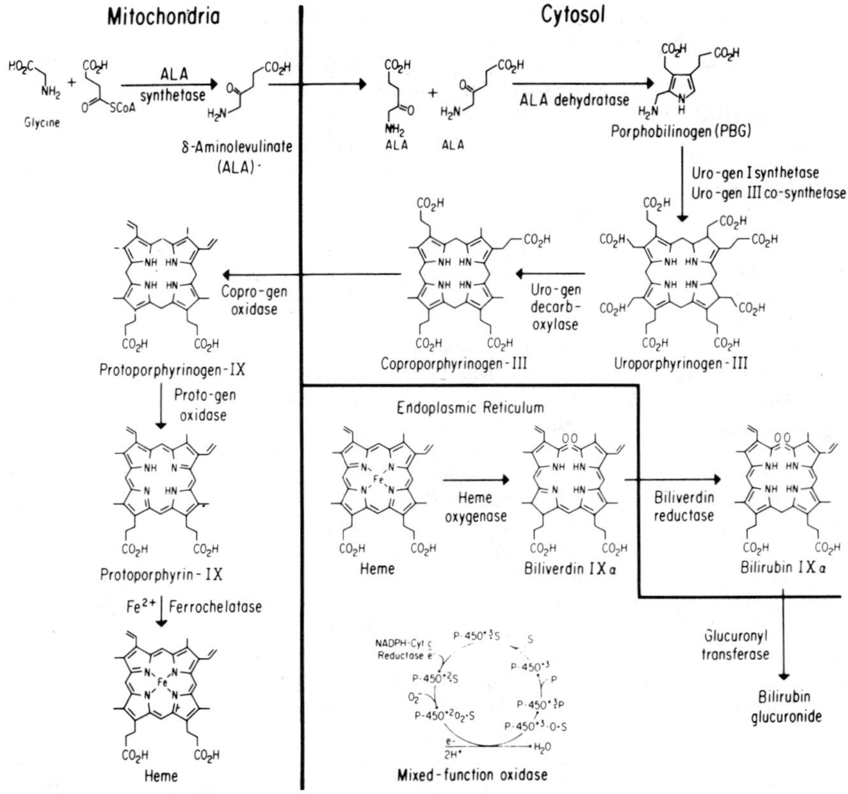

FIGURE 2. Interrelationship of porphyrins, heme, and cytochrome P-450 in various cellular sites. Reproduced from reference 4.

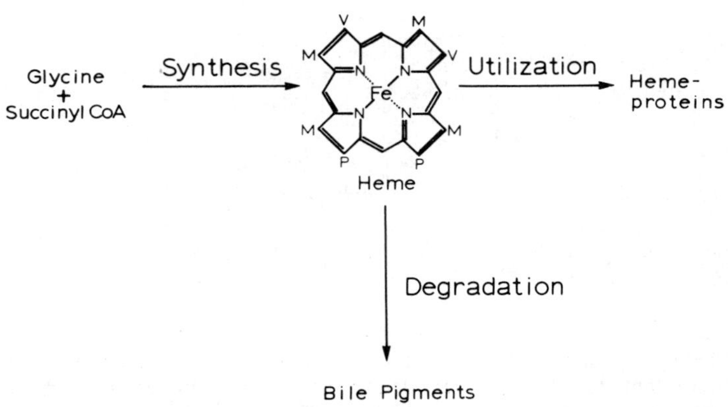

FIGURE 3. Schematic relating the synthesis, utilization, and degradation of heme in cells.

environment, and they represent in this sense a paradigm of the manner in which environmental and genetic factors can interact to produce serious and even fatal illnesses in man. For the environmental biologist the enzymes involved in the biosynthetic sequence of heme constitute a rich field of study because at least one-half of these enzymes are direct targets of attack by environmental pollutants and because the aberrations in porphyrin-heme synthesis that such agents produce are readily detectable—and even made vivid—by the striking chromatic properties of the intermediates in heme formation.

Most of the contributors to this volume are concerned with problems relating to those proteins that utilize heme as a prosthetic group and that are involved in chemical oxidations: the hemoproteins that are known collectively as cytochrome P-450. The first paper on the schedule of the conference upon which this volume is based[a] dealt with the subject of nutrition in this field; one could wish there was a stronger representation of the relation of dietary composition to various aspects of heme–P-450 biology at these meetings. A distinguished gourmandizer and litterateur, in an essay on the pleasures of eating, once referred to food as "the sex of the middle-aged." If we consider our diets in less sardonic terms and, for example, view a thick filet mignon with a heavy layering of sauce bearnaise, accompanied by a large platter of vegetables and fried potatoes, followed by a rich pastry for dessert, as nothing more than a collection of complex organic molecules containing carbon, nitrogen, oxygen, and other elements, it seems evident that as is the case with other organic molecules there could readily be a significant impact of such dietary molecules on the metabolism of other types of chemicals, such as hormones, drugs, and environmental pollutants, *in vivo;* the importance of studying the effects of diet components on heme metabolism and P-450-mediated chemical biotransformations in humans thus becomes evident. When one considers that the exposure of humans to environmental chemicals such as polycyclic aromatic hydrocarbons from dietary constituents greatly exceeds that from any other source, except for selective exposures associated with specific occupational settings, then what seems evident, in fact, becomes imperative.[5–7]

It is perhaps disenchanting to think of the foods we eat in somewhat the same terms as we choose to view more classically defined environmental chemicals, but surely if the estrogen receptor is prepared to recognize, to some extent, a polycyclic aromatic hydrocarbon,[8] as well as estradiol itself, then we have as much reason to examine the impact of the human diet on heme–P-450 metabolism, and the metabolic consequences thereof, as we do the impact on these systems of many of the foreign chemicals discussed in this volume.

Of the three metabolic axes shown in FIGURE 3, the one relating to heme utilization, specifically for P-450 formation, has in the past two decades seen the greatest advances made in both physiological terms as well as in relation to the biochemistry and the molecular genetics of these crucial proteins. It is not imagining too much to suppose that in the next decade the bulk of the constitutive and inducible P-450s will have been physically and functionally characterized; that the structures of their genes will have been defined; and that we might, at a later time, even be able to provide some semblance of the "P-450 profile" of a given individual. As a physician, I believe such metabolic profiles would be valuable in individualizing the administration

[a]The paper, which was entitled "Nutritional Factors Regulating the Heme Pathway," was presented by K. R. Mahaffey at the conference but unfortunately was not submitted for publication—*Ed.*

and enhancing the effectiveness of therapeutics in man and also in diminishing the high frequency of toxic reactions induced by drugs used in the treatment of disease. Such profiles would also have great importance clinically in helping to identify population groups that may be especially susceptible (or resistant) to the deleterious biological impacts of various types of environmental chemicals.

Studies along the pathway of heme catabolism also interest environmental biologists because the degradative pathway of heme is as sensitive to the impact of environmental pollutants and other foreign chemicals as is the biosynthetic pathway of this pigment. The impact of such agents on heme degradation is nearly always to enhance this process by inducing synthesis of the rate-limiting enzyme of heme catabolism, heme oxygenase.[9,10]

The term "metal" is mentioned in the titles of more than one-third of the papers in this volume; clearly, metals continue to be of singular interest to those concerned with the regulation of heme catabolism. Metal compounds—inorganic metal compounds; organic metal compounds (such as the organometals widely utilized in manufacturing and in agriculture); metal compounds incorporated into or used as drugs (such as gold compounds and the parasiticidal metal complexes)—are perhaps the broadest category known of consistently active and very potent chemical inducers of heme oxygenase and thus of enhanced heme degradation, and consequent impairment of cellular cytochrome P-450 content and function.[11]

We have learned a great deal about how metals influence all of the three axes of heme metabolism depicted in FIGURE 3, but I will mention only one recent advance in the area of research involving metals and heme degradation because this example gives clear substance to an old dictum that one might discover in studies of what is now called environmental toxicology, medical uses for substances that upon first examination appear to have only detrimental biological effects in animals and man.

Metals powerfully degrade heme in part by the mechanism of heme oxygenase induction as I have just noted. But we have in our own studies learned that insertion of certain metals other than the iron atom into several types of porphyrin macrocycles leads to the formation of competitive inhibitors of heme oxygenase.[12,13] Such metals include the elements tin and chromium, which in their inorganic forms greatly induce heme oxygenase activity. A chemical agent that can act in this manner, that is, as an inhibitor of heme oxidation, not only has the potential of preserving P-450 against degradation by other chemicals that enhance the catabolism of the heme moiety of this hemoprotein,[11] but can greatly inhibit bilirubin formation in circumstances in which this heme metabolite may have a toxic potential. The neurotoxicity of bilirubin in severely jaundiced human newborns is one such circumstance—and one inorganic metal species, tin chelated with an appropriate porphyrin macrocycle, has, because of its especially potent heme oxygenase inhibitory properties, proved highly effective in blocking heme degradation to bilirubin, thus suppressing hyperbilirubinemia in various forms of experimentally induced and naturally occurring jaundice in experimental animals and man.[12,14] This unique capacity of tin chelated within a porphyrin macrocycle, to inhibit heme catabolism *in vivo* and to suppress hyperbilirubinemia, stands in striking contrast to the potent ability of the inorganic form of the metal to induce heme oxygenase and to thereby enhance heme degradation and bile pigment formation in the whole animal.[11]

Toxicology then, like the Roman god Janus, has its alternative face; perhaps, then, we can not only devote attention to the pathologic or toxic actions elicited by certain chemicals, which is described in the individual papers, but also consider the possible ways in which such actions produced by noxious chemicals might find, if they can be suitably moderated, some useful biological or clinical purpose.

REFERENCES

1. GRANICK, S. & G. URATA. 1963. J. Biol. Chem. **238:** 821–827.
2. OMURA, T. & R. SATO. 1964. J. Biol. Chem. **239:** 2370–2378.
3. ESTABROOK, R. W., D. Y. COOPER & O. ROSENTHAL. 1963. Biochem. J. **338:** 741–755.
4. MAINES, M. D. & A. KAPPAS. 1977. Science **198:** 1215.
5. ANDERSON, K. E., A. H. CONNEY & A. KAPPAS. 1986. *In* Ethnic Differences in Reactions of Drugs and Xenobiotics. W. Kalow, W. Goedde & D. P. Agarwal, Eds. Vol. **214:** 39–54. Alan R. Liss. New York, NY.
6. CONNEY, A. H. & A. KAPPAS. 1985. *In* Carcinogenesis. E. Huberman & S. H. Barr, Eds. Vol. **10:** 147–166. Raven Press. New York, NY.
7. KAPPAS, A. 1983. *In* Polycyclic Aromatic Hydrocarbons: Evaluation of Sources and Effects. Chapter 6: 1–76. National Academy Press. Washington, DC.
8. BERTHOIS, Y., J. A. KATZENELLENBOGEN & B. S. KATZENELLENBOGEN. 1986. Proc. Natl. Acad. Sci. USA **83:** 2496–2500.
9. TENHUNEN, R., H. S. MARVER & R. SCHMID. 1968. Proc. Natl. Acad. Sci. USA **61:** 748–755.
10. TENHUNEN, R., H. S. MARVER & R. SCHMID. 1968. J. Biol. Chem. **244:** 6388–6394.
11. KAPPAS, A. & G. S. DRUMMOND. 1984. Environ. Health Perspect. **57:** 301–306.
12. DRUMMOND, G. S. & A. KAPPAS. 1981. Proc. Natl. Acad. Sci. USA **78:** 6466–6470.
13. YOSHINAGA, T., S. SASSA & A. KAPPAS. 1982. J. Biol. Chem. **257:** 7778–7785.
14. KAPPAS, A. & G. S. DRUMMOND. 1986. J. Clin. Invest. **77:** 335–339.

Zinc Protoporphyrin

Past, Present, and Future[a]

ROBERT F. LABBÉ AND REBECCA L. RETTMER

Department of Laboratory Medicine
University of Washington
Seattle, Washington 98195

ALKA G. SHAH AND JUDITH R. TURNLUND

United States Department of Agriculture
Western Human Nutrition Research Center
Albany, California 94710

INTRODUCTION

The objective of this report is threefold: 1) to review the course of zinc protoporphyrin biomedical research, 2) to consider the mechanism of zinc protoporphyrin formation together with *in vivo* effects of lead, and 3) to redefine the relationship of zinc protoporphyrin to certain other iron parameters.

Fischer and his school first identified and named protoporphyrin in 1924, and it now appears, with the benefit of hindsight, that his receiving the Nobel prize for protoporphyrin research foreshadowed the future importance of this metabolite in medicine. In 1933, Hijmans von den Bergh and Grotepass established the presence of protoporphyrin in erythrocytes. Soon after (1936), Vigliani and Angeleri found increased erythrocyte protoporphyrin to accompany lead poisoning, but not until 1948 did Vanotti first point out that lead interferes with the utilization of iron during hemoglobin synthesis. Years later (1966), after the heme biosynthetic pathway had been described, Ludwig Heilmeyer wrote the following: "The union of protoporphyrin and iron appears to be particularly sensitive to disturbance. This enzymatic reaction is the Achilles heel of heme biosynthesis." These remarks represent another instance of the foreshadowing of future developments—in Heilmeyer's case, of developments only two decades into the future. It is now well established that protoporphyrin and iron form the two substrates in the terminal enzymic reaction of heme biosynthesis, a process through which protoporphyrin chelates a ferrous ion with the liberation of two protons.

PROTOPORPHYRIN FLUORESCENCE IN BLOOD

Until a decade ago, virtually all blood porphyrin assays involved an initial acid solvent extraction of the porphyrins from proteins and other blood constituents. After further purification and separation, uroporphyrin, coproporphyrin, and protoporphyrin were measured fluorometrically. In 1971, it was pointed out that the only blood

[a]This investigation was supported in part by U.S. Public Health Service Grant AM35816 and by U.S. Department of Agriculture Competitive Grant 85-CRCR-1-1581.

porphyrin of widespread clinical interest was protoporphyrin.[1] Since that time, numerous assays have been developed specifically for protoporphyrin; the very simple and effective Piomelli assay[2] has perhaps become the most widely used, though other procedures have been common. Because these procedures have routinely resulted in the isolation and measurement of free-base protoporphyrin, the term "free erythrocyte protoporphyrin," or "FEP," came into being and has been used frequently in the clinical literature. Of course, it is now known that the analyte for which the term is used is neither free nor protoporphyrin, strictly speaking, but was the zinc chelate before acid treatment.

As interest in blood porphyrin analysis grew, studies were initiated for the direct measurement of porphyrin in blood by fluorometry as a new approach to porphyrin assay.[3] It fell, however, to Lamola and his colleagues at Bell Laboratories, who were working independently on a similar approach, to identify the principal porphyrin fluorescence in blood as being due to zinc protoporphyrin.[4] The same group at Bell Laboratories then developed the hematofluorometer,[5] a simple instrument designed specifically to measure zinc protoporphyrin in intact erythrocytes.

About this time, a few researchers began to express results as μg protoporphyrin/g hemoglobin rather than in concentration units. In 1979, use of the protoporphyrin/heme molar ratio was proposed as a concept that reflected a simplified equilibrium constant for the terminal reaction in heme biosynthesis.[6] Viewed in this way, the ratio can be considered a functional test for iron status, or availability, during hemoglobin formation. Experience has now demonstrated that the ratio provides a measure of relative iron-deficient erythropoiesis more sensitive than any expression of porphyrin concentration.

Having experienced the inherent problems of hematofluorometry for studying marginal iron deficiency, we subsequently modified a commercial instrument to increase its sensitivity and to display results as μmoles zinc protoporphyrin/mole heme, which is a more accurate method of achieving results.[7] This ratio has proven to be an extremely sensitive indicator of iron status, especially when newly developed reagents are utilized in conjunction with the hematofluorometer; additional details of this analytical procedure are described elsewhere in this volume.[8] Suffice it to be noted here that further benefits of this functional test include a common reference range that applies to both sexes and all age groups, except neonates. Further, precision and accuracy are improved while the use of hematofluorometry can become more cost effective through simplification of the assay procedure.

IRON–ZINC–LEAD INTERACTIONS

At several metabolic sites, these three metals appear to exhibit interactions. Though mechanistic details probably differ, they presumably compete for common binding sites. Porphobilinogen synthase requires zinc for activity. The enzyme operates at a branch point in metabolism, namely, the site at which δ-aminolevulinate can proceed either through the porphyrin pathway or be oxidized to succinyl coenzyme A and reenter the tricarboxyclic cycle. In 1977, we discovered, somewhat serendipitously, that iron deficiency leads to an increase in porphobilinogen synthase activity,[9] suggesting an apparent interaction or competition of iron and zinc at a specific binding site. We proposed at the time that the iron and zinc interaction might provide a regulatory mechanism through which porphobilinogen synthase could serve to direct δ-aminolevulinate into one pathway or the other. Hypothetically, a site for iron–zinc interaction may be the same site that is so highly sensitive to lead inhibition of porphobilinogen synthase.

Reference to these particular metal–metal interactions is made because in the terminal reaction of heme biosynthesis catalyzed by ferrochelatase, a possibly similar relationship among iron, zinc, and lead may occur. Ferrochelatase was first described in the late 1950s. The enzyme is now known to be located in the inner mitochondrial membrane. It has a rather high degree of substrate specificity for both the metal and the porphyrin. By utilizing existing knowledge of ferrochelatase, one can hypothesize a mechanism by which the porphyrin attaches to the enzyme through the 7 and 8 carboxyl groups while the metal is attached to an associated separate chelate site; finally, the metal is chelated by the porphyrin with the liberation of two protons. During early studies of metal specificity of ferrochelatase, enzymic zinc incorporation into protoporphyrin was not detected. In recent years, however, enzyme preparations from both mammalian and microbial sources have been shown to catalyze the incorporation of zinc into porphyrins.[10] Hence, one can seriously question whether the long-assumed nonenzymic incorporation of zinc really occurs *in vivo*. The fact that zinc chelation by porphyrins does occur *in vitro* certainly does not establish that the same reaction occurs in a developing red cell, especially when a model enzyme system as

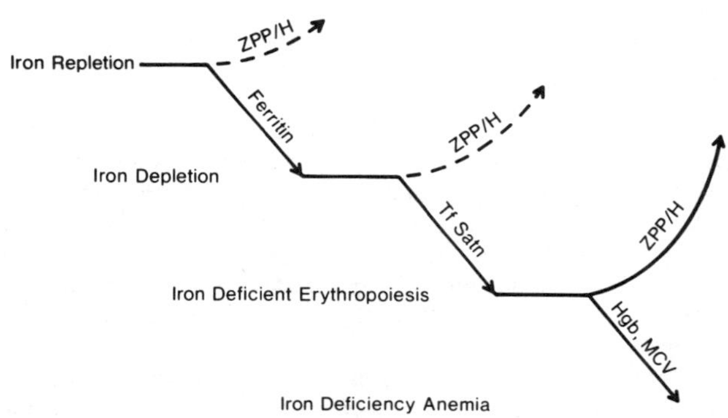

FIGURE 1. The sequence of biochemical changes that occur during the development of iron deficiency.

described years ago by Lowe and Phillips[11] consisting of a surfactant and a chelating agent is created in the reaction matrix found with most experimental conditions. A more plausible explanation has been proposed[10] that is based upon differences in metal–metal interactions with the enzyme; that is, ferrous ion is an effective inhibitor of zinc utilization. Given that iron substrate is used preferentially, it is reasonable to conclude that zinc is utilized as a ferrochelatase substrate when the concentration of iron decreases to a suboptimal level.

Where, then, does this leave the action of lead? Again, the repeatedly stated presumption is that lead reacts directly with ferrochelatase to inhibit the enzyme—after all, this is a sulfhydryl enzyme, and, in purified preparations, it is inhibited by metals that react readily with sulfhydryl groups. These other heavy metals, however, are not known to produce a clinical increase in zinc protoporphyrin in the way that lead induces this response, implying that some other mechanism should be considered.

Schwartz *et al.*[12] have shown that acute lead toxicity in rabbits initially induces an increase in free-base protoporphyrin, followed some time later by an increasing accumulation of the zinc chelate. This response to acute lead exposure suggests the possibility of two modes of action. Thus, the response to acute lead poisoning may be, in fact, a direct inhibition of ferrochelatase, followed by a second effect relating to interference with intracellular iron transfer. In the far more common case of low-level, chronic exposure to lead, this latter interference with iron transfer becomes the most likely toxic reaction. The induced iron deficiency state then allows zinc to become the metal substrate for ferrochelatase.

Finally, if one is to presume that zinc is incorporated nonenzymically because lead inhibits ferrochelatase, then one must explain the great increase in free-base protoporphyrin, but not zinc protoporphyrin, in the inherited disorder of ferrochelatase deficiency, namely, protoporphyria. Indeed, Sandberg *et al.*[13] have demonstrated clearly that the administration of large amounts of zinc cannot by mass action alter the formation of zinc protoporphyrin in this particular metabolic disorder, as might be expected for a nonenzymic reaction.

It was concluded several years ago[14] that when both clinical and biochemical data are considered *in toto,* the case for lead inhibiting ferrochelatase finds less supportive experimental and clinical evidence than does the case for lead interfering with iron transfer and creating a condition resembling iron deficiency. More recent investigations have clarified a distinction between acute and chronic effects while substantiating the earlier conclusion.

CLINICAL STUDIES

Before zinc protoporphyrin had been identified in erythrocytes, Dagg *et al.*[15] noted the potential diagnostic value of protoporphyrin in identifying latent iron deficiency, which they have described as "the stage of iron deficiency before frank anemia has developed." Today, zinc protoporphyrin formation is established to be closely linked to iron status. The general assumptions about the development of iron deficiency and some supporting data suggest the following chain of events (FIGURE 1): First, iron stores (in ferritin) are diminished to a depletion level; subsequently, a decrease in the iron saturation of transferrin occurs and iron-deficient erythropoiesis develops; finally, as less iron is delivered to the marrow by transferrin, zinc protoporphyrin increases and iron deficiency anemia develops. Our several studies with states of marginal iron deficiency, especially in athletes and blood donors where uncomplicated iron depletion can occur, suggested that these parameters might vary in concert rather than in a step-wise fashion, as just described. We were further led to this conclusion by finding that the zinc protoporphyrin/heme ratio was actually a better predictor of iron deficiency anemia than was the serum ferritin concentration.[16]

In the course of these marginal deficiency studies, we initiated another study in which adult men were placed on a metabolic ward over a 12-week period. The decrease in iron stores observed in this study may have been caused by repeated blood draws. The zinc protoporphyrin/heme ratio was determined with a customized hematofluorometer (described briefly above and elsewhere in this volume[8]). Measurements were made using ProtoFluor reagent (Helena Laboratories, Beaumont, TX). Ferritin was assayed with Ferrizyme reagents (Abbott Laboratories, North Chicago, IL). Iron and iron-binding capacity were assayed using the Gemeni Serum Iron and TIBC kit (Electro-Nucleonics, Fairfield, NJ).

As illustrated in FIGURE 2, the serum ferritin concentration in all subjects decreased from the beginning to the end of the experimental period whereas the zinc protoporphyrin/heme ratio uniformly increased. By contrast, the changes in percent transferrin saturation (not shown) were inconsistent. The hematocrit results were also inconsistent, but usually showed little change. The temporal relationship between changes in ferritin concentration and the zinc protoporphyrin/heme ratio were particularly significant; these clearly do change in a concerted pattern as shown in FIGURE 3. One can obtain a truly dramatic view of changing iron status by dividing the zinc protoporphyrin/heme ratio by the ferritin concentration (FIG. 4). This presentation of the data may be of primarily academic interest because the zinc protoporphyrin/heme ratio in itself is so sensitive to changes in iron status that one need not

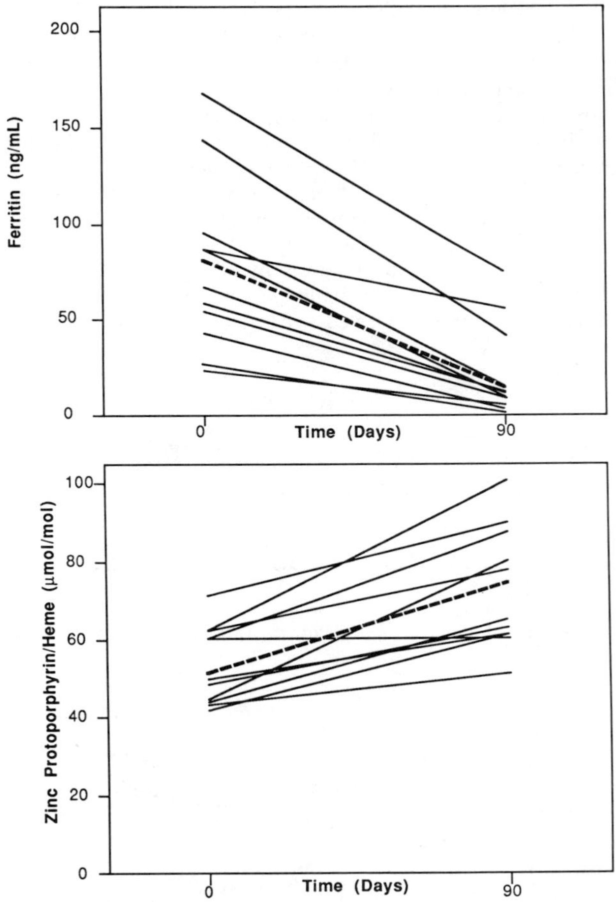

FIGURE 2. The decrease in the serum ferritin concentration and the increase in the whole blood zinc protoporphyrin/heme ratio in individual young men before and after becoming marginally iron deficient.

manipulate the data further to improve its capacity to detect relative iron-deficient erythropoiesis or to gain clinical significance.

In 1975, McLaren *et al.*[17] demonstrated the association of erythrocyte protoporphyrin with marrow iron levels. This observation is consistent with the close metabolic association between the zinc protoporphyrin/heme ratio and the serum ferritin concentration. The fact that zinc protoporphyrin increases concomitantly with ferritin decreases suggests that iron-deficient erythropoiesis should be reconsidered as an early rather than a late manifestation of iron deficiency. The very high sensitivity of the zinc protoporphyrin/heme ratio as an indicator of iron status is especially noteworthy. The test has been evaluated extensively in many subject groups and has never failed to identify known cases of iron-deficient erythropoiesis; it is highly sensitive in this regard. False-negative results have never knowingly been found; false-positive results have been limited principally to the interference from bilirubin that occurs with present hematofluorometers. Of course, many etiologies (inadequate diet, blood loss, inflammation) can produce iron-deficient erythropoiesis. These etiologies, however, are differentiated by appropriate secondary procedures.

Although the main use of hematofluorometry has been to quantify iron deficiency, the interference of lead with iron utilization indicates that the hematofluorometer may be equally effective in identifying the extremely low levels of lead that are now commonly accepted as being toxic, at least in young children. Technological improvements allow hematofluorometry to emerge from the status of a simple, yet effective, screening test into an accurate, precise analytical tool. Potential applications lie not only in lead surveillance and the diagnosis of chronic toxicity, but also in blood banking, pediatrics, obstetrics, sports medicine, and other clinical situations where a high level of sensitivity combined with cost effectiveness is required.

SUMMARY

The course of zinc protoporphyrin research has progressed at an increasingly rapid pace on several fronts. A variety of biochemical and clinical evidence viewed *in toto*

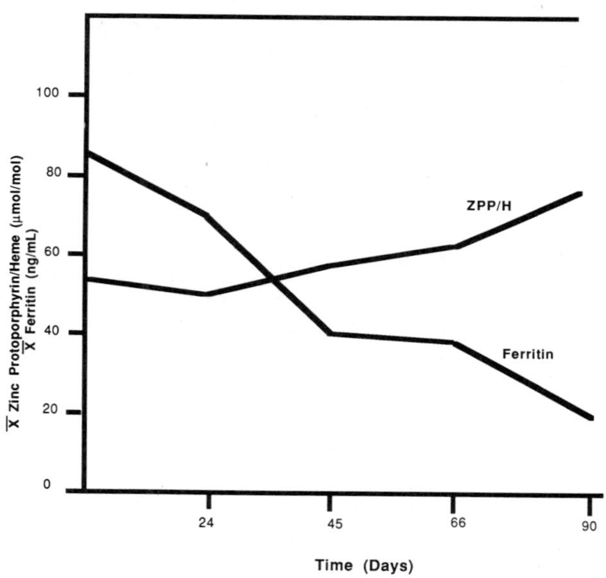

FIGURE 3. The decrease in the serum ferritin concentration and the increase in the whole blood zinc protoporphyrin/heme ratio during a course of iron loss in young men.

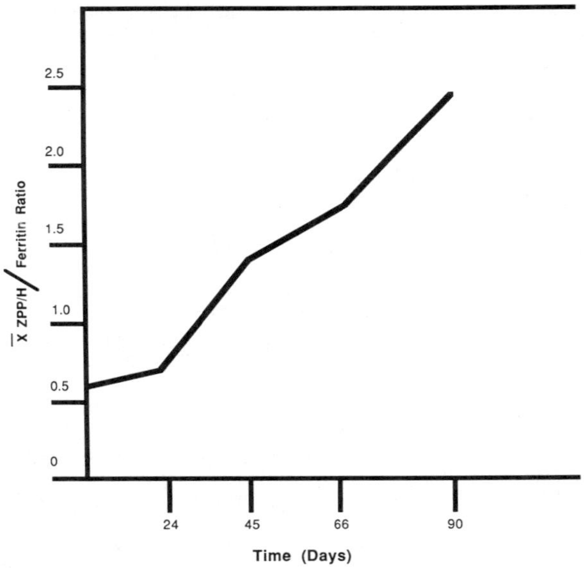

FIGURE 4. Combining two measures of iron status, the whole blood zinc protoporphyrin/heme ratio and the serum ferritin concentration, during a course of iron loss in young men.

now suggests that ferrochelatase catalyzes zinc protoporphyrin formation in states of relative iron-deficient erythropoiesis and in lead-inhibited iron metabolism. Furthermore, a redefinition of the relationship of zinc protoporphyrin to certain other parameters of iron status has been made based upon changes during the earliest states of iron depletion. These clinical studies show that the zinc protoporphyrin level and the ferritin level vary in concert but that changes in the percent transferrin saturation and in the hematocrit results are less consistent. Thus zinc protoporphyrin and ferritin are closely linked metabolically such that iron-deficient erythropoiesis becomes an initial manifestation of iron depletion. The measurement and expression of results as μmoles zinc protoporphyrin/mole heme have improved the quality of results, partly by the elimination of the assumed hematocrit designed into existing instruments. Other refinements in hematofluorometry technology have permitted exploration of the potentially extensive applications of zinc protoporphyrin measurements for lead surveillance and diagnosis, blood banking, pediatrics, obstetrics, sports medicine, and other clinical situations where a very sensitive, cost-effective indication of iron status is required.

REFERENCES

1. Heller, S. R., R. F. Labbé & J. Nutter. 1971. A simplified assay for porphyrins in whole blood. Clin. Chem. **17:** 525–528.
2. Piomelli, S., D. Davidow, V. F. Guinee, P. Young & G. Gay. 1973. The FEP (free erythrocyte porphyrins) test: A screening micromethod for lead poisoning. Pediatrics **51:** 254.
3. Sayer, P., R. F. Labbé & M. Gouterman. 1974. Fluorescence of blood protoporphyrin in the presence of hemoglobin. Biochem. Med. **10:** 24–35.

4. Lamola, A. A. & T. Yamane. 1974. Zinc protoporphyrin in the erythrocytes of patients with lead intoxication and iron deficiency anemia. Science **186:** 936–938.
5. Blumberg, W. E., J. Eisinger, A. A. Lamola & D. Zuckerman. 1977. The hematofluorometer. Clin. Chem. **23:** 270–274.
6. Labbé, R. F., C. A. Finch, N. J. Smith, R. N. Doan, S. K. Sood & N. Madan. 1979. Erythrocyte protoporphyrin/heme ratio in the assessment of iron status. Clin. Chem. **25:** 87–92.
7. Lamola, A. A., J. Eisinger & W. E. Blumberg. 1980. Erythrocyte protoporphyrin/heme ratio by hematofluorometry. Clin. Chem. **26:** 677–678.
8. Rettmer, R. L., E. W. Gunter & R. F. Labbé. 1987. Overcoming the limitations of hematofluorometry for assaying zinc protoporphyrin. Ann. N. Y. Acad. Sci. This volume.
9. Labbé, R. F. & C. A. Finch. 1977. Effects of iron status on δ-aminolevulinic acid dehydratase activity. Biochem. Med. **18:** 323–329.
10. Camadro, J.-M., N. G. Ibraham & R. D. Levere. 1984. Kinetic studies of human liver ferrochelatase: Role of endogenous metals. J. Biol. Chem. **259:** 5678–5682.
11. Lowe, M. B. & J. N. Phillips. 1961. Catalysis of metalloporphyrin formation: A possible enzyme model for haem iron incorporation. Nature **190:** 262–264.
12. Schwartz, S., B. Stephenson, D. Sarkah, H. Freyholtz & G. Ruth. 1980. Quantitative assay of erythrocyte "free" and zinc-protoporphyrin: Clinical and genetic studies. Int. J. Biochem. **12:** 1053–1057.
13. Sandberg, S., A. Brun, G. Hovding, M. Bjordol & I. Romslo. 1980. Effect of zinc on protoporphyrin-induced photohaemolysis. Scand. J. Clin. Lab. Invest. **40:** 185–189.
14. Labbé, R. F. 1980. History and background of protoporphyrin testing. Clin. Chem. **23:** 256–259.
15. Dagg, J. H., A. Goldberg & A. Lochhead. 1966. Value of erythrocyte protoporphyrin in the diagnosis of latent iron deficiency (sideropenia). Br. J. Haematol. **12:** 326–330.
16. Rettmer, R. L., P. Fernandez-Cano, M. L. Sayers, E. R. Ashwood & R. F. Labbé. 1985. The zinc protoporphyrin/heme ratio in laboratory diagnosis. Clin. Chem. **31:** 1026.
17. McLaren, G. D., J. T. Carpenter, Jr. & H. P. Nino. 1975. Erythrocyte protoporphyrin in the detection of iron deficiency. Clin. Chem. **21:** 1121–1127.

Genetic Regulation of the Heme Pathway[a]

SHIGERU SASSA, HIROYOSHI FUJITA,
AND OSAMU SUGITA

The Rockefeller University
New York, New York 10021

INTRODUCTION

Genetic regulation of the heme pathway can best be illustrated by the porphyrias. The porphyrias are inherited and acquired disorders characterized by overproduction of porphyrins and their precursors, and are invariably associated with abnormalities in the activities of specific enzymes of the heme biosynthetic pathway. These diseases are classified as either hepatic or erythropoietic depending on the principal site of expression of the enzymatic defect in each disorder. Because the bone marrow and liver are the two major organs involved in heme biosynthesis, it is perhaps not surprising that porphyrias primarily express themselves in these organs.

In almost all porphyrias, environmental factors, such as lead, polyhalogenated aromatic hydrocarbons, and nonheritable metabolic factors, play a significant role in determining the clinical expression of the primary gene defect. In this respect, porphyrias are the paradigm of disorders in which the interactions between genetic factors and environmental chemicals are critically involved in the pathogenesis of human disease. Because it is not possible to discuss the genetic regulation of all enzymes in the heme pathway within the limited space available, we will describe the recent findings from our laboratory on the genetic and chemical modulation of uroporphyrinogen decarboxylase, and its implications in porphyria cutanea tarda (PCT). In this study, we examined both the enzyme activity and the enzyme protein content in erythrocytes from normal subjects, patients with PCT, and patients with hepatoerythropoietic porphyria (HEP) using a monospecific antibody against normal uroporphyrinogen decarboxylase. In addition, we examined a spectrum of polychlorinated biphenyls (PCBs), a group of toxic environmental chemicals that cause experimental PCT in animals and in isolated chick embryo hepatocyte cultures, and analyzed the structure–activity relationships of PCB and uroporphyrin accumulation.

MATERIALS AND METHODS

Enzyme Purification and Preparation of Antibodies

Uroporphyrinogen decarboxylase was purified from normal human erythrocyte lysates as described previously.[1] A monospecific polyclonal antiserum against the purified uroporphyrinogen decarboxylase was prepared in rabbits. An immunoglobulin G fraction was prepared from the antiserum according to the method of Masters *et al.*[2]

[a]This study was supported in part by Grants ES-01055 and DK-32890 from the U.S. Public Health Service.

Blood Samples

Blood samples were obtained at the Rockefeller University Hospital from 23 normal subjects, 23 patients with sporadic PCT, and 6 patients with a familial history of PCT in four families. All patients were characterized by typical skin lesions, and the diagnosis of PCT was confirmed by quantitation of porphyrins in urine and feces by high-performance liquid chromatography (HPLC). In addition, samples from two male HEP children (one was 9 years old and the other was 7 years old) from a single family[3] and a sample from a 31-year-old male with HEP[4] were examined.

Enzyme Assay and Immunoquantitation

Uroporphyrinogen decarboxylase activity was measured using an HPLC assay previously described.[1] The enzyme protein was quantitated by rocket immunoelectrophoresis on hemoglobin-free lysates prepared by DEAE cellulose chromatography.[5] In addition, radioimmunoassay (RIA) was carried out for the quantitation of uroporphyrinogen decarboxylase protein in the lysate from the adult HEP patient who had an extremely low immunologically detectable enzyme protein. The specific activity was calculated based on the enzyme activity and the enzyme concentration determined in the same specimen.

Cell Cultures

Monolayer cultures of chick embryo liver cells were prepared from 17-day-old chick embryos and incubated in a serum-free medium supplemented with insulin (1 μg/ml), hydrocortisone (0.05 μg/ml), and triiodothyronine (1 μg/ml) in 24-well plates in a humidified CO_2 incubator at 37 °C with a 5:95 mixture (v/v) of CO_2 and air as

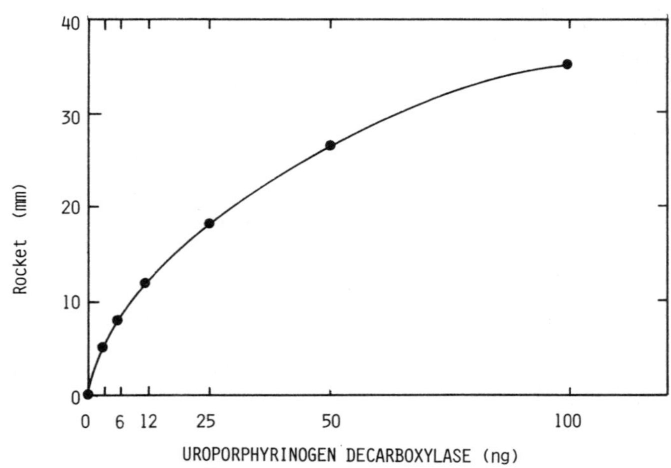

FIGURE 1. Rocket immunoelectrophoresis of uroporphyrinogen decarboxylase. One hundred microliters of rabbit immunoglobulin G versus human uroporphyrinogen decarboxylase were used in 6 ml of 1% agarose solution for the preparation of an electrophoretic gel film.

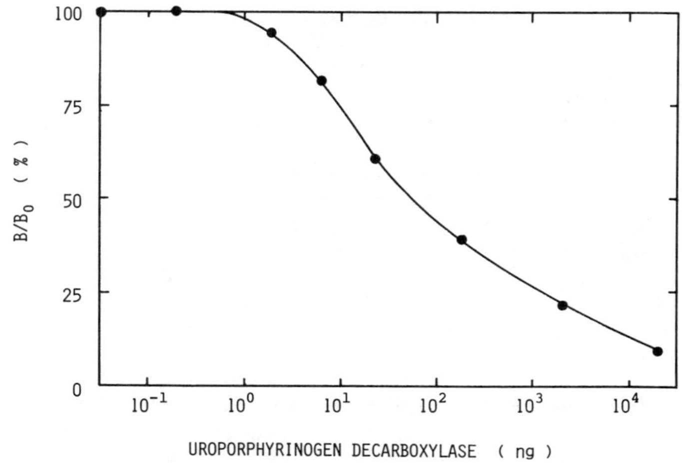

FIGURE 2. Radioimmunoassay of uroporphyrinogen decarboxylase. Each point represents the mean of triplicate assays.

described previously.[6] After 24 hr of incubation, the medium was replaced and the addition of chemicals was made. Cells were then incubated for another 24 hr with the chemicals.

Fluorometry of Porphyrins

Porphyrins were extracted into a 1:1 mixture (v/v) of 1 N $HClO_4$ and methanol, and their concentrations were determined in a Hitachi–Perkin Elmer MPF IV fluorescence spectrophotometer equipped with an R928 photomultiplier. The percentages of uro-, copro-, and protoporphyrin were quantitated using the method of Grandchamp *et al.*[7]

Molecular Orbital Calculations

Molecular orbital calculations for PCBs were performed using the semiempirical molecular orbital method (MNDO) as described previously.[8]

RESULTS

Immunochemical Assay

Typical calibration curves for rocket immunoelectrophoresis and RIA on uroporphyrinogen decarboxylase are shown in FIGURES 1 & 2. Enzyme concentrations as low as 3 ng and 1.6 ng per assay can be determined using rocket analysis and RIA, respectively.

Erythrocyte Uroporphyrinogen Decarboxylase Activity in Normal Subjects, Patients with PCT, and Patients with HEP

The mean uroporphyrinogen decarboxylase activity for adult patients with familial PCT (who, for the purpose of this study, are defined as those having a positive family history of the disease) was approximately one-half of the mean value for the normal subjects (TABLE 1). These adult subjects with familial PCT, who ranged from 26 to 69 years of age, can be considered to be heterozygous for uroporphyrinogen decarboxylase deficiency. Uroporphyrinogen decarboxylase activities for the sporadic PCT patients (defined as those devoid of a family history of PCT) were between those for the normal subjects and those for the heterozygous PCT patients. A much wider distribution of enzyme activities was noted in the sporadic group than in the familial group or in normal subjects, suggesting that the sporadic PCT group may consist of patients with a

Congeners		Uroporphyrin (pmol/mg prot.)
BP	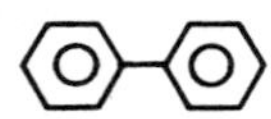	10
4-BP	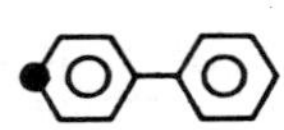	10
3,4-BP	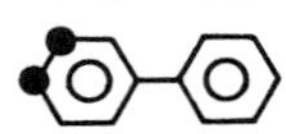	10
4,4'-BP	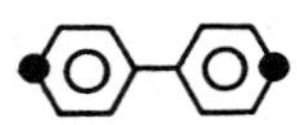	10
2,6,2',6'-BP	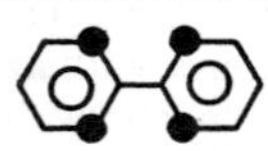	10
2,3,2',3'-BP	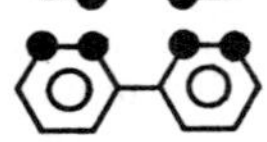	10
2,4,3',4'-BP	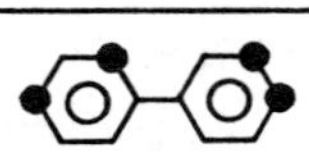	1000
3,4,3',4'-BP	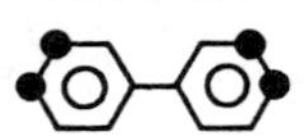	2200
3,4,5,3',4',5'-BP	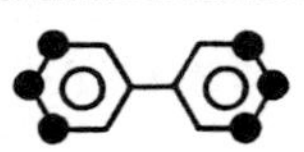	1500

FIGURE 3. Structure–activity relationship between polyhalogenated biphenyls and uroporphyrin accumulation in chick embryo liver cultures. Chick embryo liver cultures were prepared as described in Materials and Methods. After 24 hr of incubation, the medium was changed and chemicals (0.5 μg/ml) were added. Cultures were incubated for another 24 hr. Porphyrins were quantitated fluorometrically as described in Materials and Methods. Data are shown for representative congeners for each class.

TABLE 1. Activity and Specific Activity for Erythrocyte Uroporphyrinogen Decarboxylase

Subject	Age (years)	Uroporphyrinogen Decarboxylase[a] Activity (unit/ml RBC)	Specific Activity (unit/μg protein)
Normal	22–75	80.0 ± 15.0 (23)	0.85 ± 0.32 (8)
PCT			
Sporadic	22–77	65.1 ± 23.8 (23)	1.01 ± 0.12 (11)
Familial	26–69	40.8 ± 8.6 (6)	0.96 ± 0.04 (3)
HEP			
Case 1	7	7.3	0.42
Case 2	9	11.3	0.38
Case 3	31	22.3	>12[b]

[a]Each value for the normal subjects and PCT subjects is a mean ± SD; following each of these values, in parentheses, is the number of subjects studied. RBC: red blood cells; unit: 1 nmole coproporphyrin formed/hr at 37 °C.

[b]Value obtained by RIA. Normal range for RIA: 3.0 ± 0.3 ($N = 5$).

heterogeneous enzyme defect. In contrast to the half-normal enzyme activity found in familial PCT patients, the enzyme activities for three HEP patients were substantially lower than normal activities, that is 9, 11, and 28% of the normal mean.

Specific Activity of Uroporphyrinogen Decarboxylase

In order to define the nature of the uroporphyrinogen decarboxylase defect in PCT and in HEP, rocket immunoelectrophoresis was performed using a monospecific rabbit antibody against normal uroporphyrinogen decarboxylase. No significant differences in the specific activity of the enzyme, which was expressed in unit/μg protein, were found for normal subjects or for patients with sporadic or familial PCT (TABLE 1). On the other hand, the two children with HEP had abnormally decreased specific activities, indicating the presence of catalytically inactive protein. In contrast, the adult case of HEP showed a high specific activity (>12) due to an unusual combination of enzyme activity but no detectable enzyme protein.

Relationship between the Conformational Probability of PCB and Uroporphyrin Induction

Hepatic uroporphyrinogen decarboxylase activity is known to be inhibited in experimental animals exposed to polyhalogenated aromatic hydrocarbons such as PCBs, hexachlorobenzene, and 2,3,7,8-tetrachlorodibenzo-*p*-dioxin (TCDD). In order to elucidate the possible structural requirements of environmental chemicals for uroporphyrinogen decarboxylase inhibition, we examined the influence of several PCBs on uroporphyrin accumulation in chick embryo hepatocyte cultures. Addition of 3,4,3′,4′-tetrachlorobiphenyl (3,4,3′,4′-tetra-CB), 3,4,5,3′,4′,5′-hexa-CB, or 3,4,5,3′,4′,5′-hexabromobiphenyl (HBB) caused a marked induction of uroporphyrin, whereas biphenyl, 4′-mono-CB, 3,4-di-CB, 4,4′-di-CB, 2,6,2′,6′-tetra-CB, and 2,3,2′,3′-tetra-CB were ineffective (FIGURE 3). 2,4,5,2′,4′,5′-Hexa-CB and HBB, 2,4,6,2′,4′,6′-hexa-CB, 2,3,4,5,2′,3′,4′,5′-octa-CB, and 2,3,4,5,6,2′,3′,4′,5′,6′-deca-CB were ineffective as well (results not shown).[9] The probability of the coplanar conformation of tetrachlorobiphenyl congeners calculated by MNDO was linearly correlated to their ability to increase uroporphyrin formation ($r = .99$).[9]

DISCUSSION

PCT patients are commonly classified into three categories: 1) sporadic cases in which there is no evidence of family history for the disease, 2) familial cases in which there is evidence of autosomal dominant inheritance, and 3) cases due to exposure to polyhalogenated aromatic hydrocarbon compounds. It is, however, often difficult to classify patients with PCT on these grounds alone, because complete family pedigree studies for enzyme deficiencies are often difficult to obtain and chemical exposure is often overlooked.

In familial PCT there is a deficiency of uroporphyrinogen decarboxylase activity of approximately 50% in the liver, in erythrocytes (TABLE 1), and presumably in other tissues, and this enzyme activity is inherited in an autosomal dominant fashion.[10] In contrast, sporadic PCT probably represents a nonhomogeneous population of patients, comprising those who have normal erythrocyte uroporphyrinogen decarboxylase activities and those with deficient uroporphyrinogen decarboxylase activities in their red blood cells as well as in liver; the latter patients would most likely have a genetic determinant of the disease. Although clinical symptomatology of sporadic and familial PCT is almost indistinguishable, the factors responsible for the clinical expression of sporadic PCT are often not associated with familial PCT. For example, in childhood cases of familial PCT, there are indications that liver dysfunction is often absent and serum iron levels are frequently normal.[10] Phlebotomy was not required in the three HEP cases described in TABLE 1 and there was no indication of iron deposition in the liver biopsy from HEP case 3.[4] Thus, although liver damage from alcohol, iron deposition, estrogens, or other causes are believed to be important contributing factors in PCT, they may play little role in the pathogenesis of familial PCT or HEP. It also appears that HEP may be a homozygous form of familial PCT, but the deficiency of uroporphyrinogen decarboxylase in sporadic and familial PCT (or HEP) is brought about by different factors.

It is important to note that all reported cases of familial PCT are cross-reactive material (CRM) negative, including the three subjects in whom the specific activity of the enzyme was examined in this study (TABLE 1), whereas both CRM-negative and CRM-positive mutations have been reported for HEP. Case 3 of HEP in our study represents the third new mutation, that is, abnormally decreased immunoreactive enzyme.[11] The three different types of mutations of uroporphyrinogen decarboxylase in HEP patients is remarkable in that only 16 cases of HEP have been reported and immunochemical studies on uroporphyrinogen decarboxylase were carried out in only 5 HEP patients. These findings indicate that the pathogenesis of uroporphyrinogen decarboxylase deficiency is quite complex and that abnormal CRM can be definitively characterized only in homozygous enzyme deficiency, but not necessarily in heterozygous enzyme deficiency such as PCT.

A disorder similar to PCT occurs when humans or animals are exposed to hexachlorobenzene or other polyhalogenated aromatic hydrocarbons.[12] Porphyrin accumulation in chick embryo hepatocytes in culture occurs within 8 hr after treatment with PCB, whereas porphyria development *in vivo* in animals occurs several weeks after treatment with these chemicals.[12,13] The reason for the discrepancy in the time course of porphyria development *in vivo* and *in vitro* is not understood. It is certainly not due to the difference between mammals and birds, since the chick embryos *in ovo* are highly resistant to development of porphyria in response to treatment with 3,4,3′,4′-tetra-CB, although the characteristic toxicity and induction of cytochrome P-450 take place.[14] Pretreatment of chick embryo liver cultures with inducers of cytochrome P-450, however, significantly augments the effect of PCB on porphyrin induction,[13] suggesting that an inhibitor of uroporphyrinogen decarboxylase may be generated from the parent compound by metabolic activation involving the induction of cytochrome P-450.

It is also evident that there is a clear structure–activity relationship between PCB and uroporphyrin accumulation in cultured hepatocytes. Certain substituent sites for halogens in PCBs play a key role in determining uroporphyrin accumulation, such that 1) at least two, but no more than four, halogens must be present in each phenyl ring in the biphenyl molecule; 2) halogens must be present at the lateral adjacent site in both phenyl rings; and 3) halogens must not occupy 2,2′ or 6,6′ positions. These structural requirements are very similar to those for the induction of arylhydrocarbon hydroxylase whose expression is under the control of the *Ah* locus. It is possible that the metabolic transformation of the parent PCB to an inhibitor of uroporphyrinogen decarboxylase may also be a function of *Ah* gene regulation. A conformational probability index for coplanarity obtained by molecular orbital calculations was highly correlated with uroporphyrin accumulation, indicating that the structure–activity relationship within the same PCB congener series, for example, tetra-CB, is largely determined by the dihedral angle of these congeners.

These findings demonstrate that uroporphyrinogen decarboxylase activity is regulated not only by genetic factors, but also by chemical agents such as PCB. Genetic influences on the enzyme are also complex in that they may lead to loss of the normal enzyme, production of catalytically inactive enzyme protein, or a catalytically active but immunologically altered enzyme protein. The existence of a CRM-negative mutation, a CRM-positive mutation, and an abnormally decreased CRM mutation in HEP patients suggests that the pathogenesis of the defect of uroporphyrinogen decarboxylase is heterogenous in HEP, and perhaps also in PCT. To what extent the clinical expression of uroporphyrinogen decarboxylase deficiency can be precipitated in latent gene carriers by exposure to environmental chemicals such as polyhalogenated aromatic hydrocarbons remains to be determined.

ACKNOWLEDGMENTS

We are grateful to Dr. A. Kappas and Dr. T. Noguchi for their support in this research. We also thank Dr. Karl E. Anderson for providing us with blood samples from patients with PCT; Dr. Maureen B. Poh-Fitzpatrick for blood samples from HEP cases 1 and 2; Dr. Arnold Toback for a blood sample from HEP case 3; and Dr. Richard A. Galbraith for his helpful discussions. The excellent technical assistance of Ms. Luba Garbaczewski and Mr. Thomas Ferraro is gratefully acknowledged.

REFERENCES

1. DE VERNEUIL, H., S. SASSA & A. KAPPAS. 1983. J. Biol. Chem. **254:** 2454–2460.
2. MASTERS, B. S. S., J. BARON, W. E. TAYLOR, E. L. ISAACSON & J. LOSPALLUTO, 1971. J. Biol. Chem. **246:** 4143–4150.
3. LIM, H. W. & M. B. POH-FITZPATRICK. 1984. J. Am. Acad. Dermatol. **11:** 1103–1111.
4. TOBACK, A. C., S. SASSA, M. B. POH-FITZPATRICK, J. SCHECHTER, E. ZAIDER, L. C. HARBER & A. KAPPAS. 1987. N. Engl. J. Med. **316:** 645–650.

5. SASSA, S., H. DE VERNEUIL, K. E. ANDERSON & A. KAPPAS, 1983. Trans. Assoc. Am. Phys. **96:** 65–75.
6. SASSA, S. & A. KAPPAS. 1977. J. Biol. Chem. **252:** 2428–2436.
7. GRANDCHAMP, B., J. C. DEYBACH, M. GRELIER, H. DE VERNEUIL & Y. NORDMANN. 1980. Biochim. Biophys. Acta **629:** 577–586.
8. SASSA, S., O. SUGITA, N. OHNUMA, S. IMAJO, T. OKUMURA, T. NOGUCHI & A. KAPPAS. 1986. Biochem. J. **235:** 291–296.
9. SASSA, S., O. SUGITA, A. KAPPAS, N. OHNUMA, S. IMAJO, T. OKUMURA & T. NOGUCHI. 1986. Biochem. J. **240:** 622–623.
10. KAPPAS, A., S. SASSA & K. E. ANDERSON. 1982. *In* Metabolic Basis of Inherited Disease. 5th edit. J. B. Stanbury, J. B. Wyngaarden, D. S. Frederickson, J. L. Goldstein & M. S. Brown, Eds.: 1301–1384. McGraw-Hill. New York, NY.
11. FUJITA, H., S. SASSA, A. C. TOBACK & A. KAPPAS. 1987. J. Clin. Invest. **71:** 1533–1537.
12. ELDER, G. H. 1978. *In* Handbook of Experimental Pharmacology. F. de Matteis & W. N. Aldridge, Eds. Springer-Verlag. Berlin.
13. DE VERNEUIL, H., S. SASSA & A. KAPPAS, 1983. Biochem. J. **214:** 145–151.
14. RIFKIND, A. B., S. SASSA, J. REYERS & H. MUSCHICK. 1985. Toxicol. Appl. Pharmacol. **78:** 268–279.

δ-Aminolevulinic Acid Dehydratase Isozymes and Lead Toxicity[a]

K. H. ASTRIN,[b] D. F. BISHOP,[b] J. G. WETMUR,[c] B. KAUL,[d]
B. DAVIDOW,[d] AND R. J. DESNICK[b]

[b]*Division of Medical Genetics*
Department of Pediatrics
[c]*Department of Microbiology*
Mount Sinai School of Medicine
New York, New York 10029

[d]*Public Health Laboratories*
New York City Department of Health
New York, New York 10016

INTRODUCTION AND RATIONALE

ALAD Activity Is Exquisitely Sensitive to Lead

δ-Aminolevulinic acid dehydratase (ALAD, EC 4.2.1.24), the second enzyme in the heme biosynthetic pathway, catalyzes the asymmetric condensation of two molecules of δ-aminolevulinic acid (ALA) to form the monopyrrole, porphobilinogen (PBG), the precursor of heme, chlorophylls, cytochromes, and cobalamins. The mammalian enzyme has been purified to homogeneity from bovine liver[1,2] and human erythrocytes.[3] The enzymes from both species are metalloenzymes composed of eight identical subunits and eight zinc atoms.[1–7] The zinc atoms are essential for enzymatic activity. Jaffe and co-workers[7,8] have suggested that four zinc atoms are required to prevent oxidation of the sulfhydryl groups in the active site whereas the other four are presumed to bridge subunits and maintain the homooctameric structure.

ALAD activity is inhibited by lead[9–11] as well as by the oxidation of critical thiol groups.[4,12–14] Lead atoms replace the zinc atoms that are required to maintain ALAD activity.[5] The inhibitory effect of lead on ALAD activity is reversible *in vitro* by incubation with zinc atoms and dithiothreitol (DTT).[7–10,13,14] The inhibition of erythrocytic ALAD activity has been used as a sensitive diagnostic indicator of lead exposure.[15] The inhibition is stoichiometric; for example, 15 and 30 μg of Pb/dl of blood results in 50 and 75% ALAD inhibition, respectively. More recently, the ratio of ALAD present before and after reactivation with zinc and DTT has been shown to correlate best to blood lead levels.[16] The inhibition of ALAD activity results in a proportionate accumulation of ALA in blood and urine.

Recently, a cDNA for human ALAD has been cloned and sequenced in our laboratory.[17,18] Characterization of the predicted amino acid sequence revealed a region that was distinctly similar to the recently identified consensus sequence for a zinc-binding domain.[19] FIGURE 1 shows the zinc-binding site of ALAD as well as the consensus sequence for zinc binding. In addition to the required amino acids represented by cysteines at 119, 122, and 132 and by the histidine at 129, the zinc-binding site of ALAD had a fourth cysteine at 124 and a second histidine at 131. This unique structure may be responsible for the strong affinity of ALAD for lead as well as zinc.

[a]This work was supported in part by Grant DK 26824 from the National Institutes of Health.

Neurological Symptoms of ALAD Porphyria Resemble Lead Poisoning

Doss and co-workers have described two unrelated German patients who had less than 2% of normal ALAD activity, excessive urinary excretion of ALA, and severe neurological manifestations.[20,21] Lead poisoning was ruled out because the deficient enzymatic activity could not be restored with zinc and DTT. The parents and some family members were asymptomatic and had half-normal levels of ALAD activity, consistent with autosomal recessive inheritance. More recently, Thunell *et al.*[22] described a third family in which an affected infant had severe neurological impairment. The discovery of this hepatic porphyria is of particular interest because it implicates ALA accumulation as a cause of the neurological manifestations in lead poisoning, which often resemble those of acute intermittent porphyria.[23,24] Although lead itself is neurotoxic *in vitro,*[25] the role of ALA as a neurotoxic agent also has been supported by a variety of studies (for reviews, see references 26 and 27). Furthermore, Doss *et al.*[28] recently suggested that heterozygotes for ALAD deficiency are predisposed to lead poisoning. In fact, such heterozygous individuals who were exposed to low levels of lead developed acute lead poisoning.[27–29]

RESIDUE:	119 120 121 122 123 124 125 126 127 128 129 130 131 132
ALAD:	CYS asp val CYS leu CYS pro tyr thr ser HIS gly HIS CYS
CONSENSUS:	CYS..X_{2-4}...CYS..........X_{4-12}..........HIS..X_{3-5}...CYS

FIGURE 1. The human ALAD amino acid sequence at the zinc-binding site. Residues 119–132 are shown with the potential zinc-chelating residues, cysteine and histidine, capitalized. A consensus sequence for zinc-binding sites in many proteins[19] is also shown (X: any amino acid).

Human ALAD Is a Polymorphic Enzyme

Human ALAD was first shown to be a polymorphic enzyme in 1981 by Petrucci and co-workers.[30,31] The ALAD gene, localized to chromosome 9q34,[32] has two common alleles, $ALAD_1$ and $ALAD_2$, with gene frequencies of about 0.9 and 0.1, respectively. The expression of these alleles results in a polymorphic enzyme system with three distinct charge isozyme phenotypes, designated ALAD 1-1, 1-2, and 2-2. The erythrocytic isozymes are readily separated by starch gel or cellulose acetate electrophoresis.[31] In a normal Italian population, the frequencies of the isozyme phenotypes were 1-1 (81%), 2-1 (17%), and 2-2 (2%), consistent with gene frequencies of 0.90 and 0.10 for the $ALAD_1$ and $ALAD_2$ alleles, respectively. Similar results were obtained in other European populations, whereas occurrence of the $ALAD_2$ allele was not observed in a large sample of Black individuals from Liberia.[33] At the biochemical level, all three ALAD isozymes reportedly have essentially the same activity in erythrocyte lysates.[31] These isozymes, however, have not been purified, nor have any of their individual physical and kinetic properties been determined, especially the effects of lead on their inhibition, binding, and stability.

ALAD Polymorphism and Susceptibility to Lead Toxicity

Polymorphisms at certain genetic loci are known to be related to differentially inherited responses to environmental challenges.[34–36] The existence of this common

ALAD polymorphism and the fact that ALAD is markedly inhibited by lead suggested that there could be a physiologic relationship between the frequency of the $ALAD_2$ allele and lead poisoning. For example, individuals with the $ALAD_2$ allele could be more susceptible to lead exposure if the ALAD 2 subunit bound lead more tightly than the ALAD 1 subunit. They would have higher blood and bone lead concentrations as well as higher total body lead stores, making them even more likely to express subclinical and clinical manifestations of chronic low-level or acute lead exposure. To test this hypothesis, we determined the ALAD isozyme phenotypes in over 1000 blood samples obtained from the New York City Lead Screening Program. The blood lead level and ALAD isozyme determinations were performed in a double-blind fashion. These data support a relationship between the $ALAD_2$ allele and the accumulation of lead in the blood.

METHODS

Collection and Preparation of Blood Specimens

Clinical samples were obtained by the Toxicology Laboratories of the Department of Health, City of New York, primarily from potentially lead-poisoned individuals. Heparinized blood samples were collected from individuals who had high free erythrocytic protoporphyrin (FEP) levels that could be attributable to lead intoxication.[37,38] An aliquot of each sample was analyzed for lead content by standard methods[37–40] and separately (blindly) for the ALAD isozyme phenotype. For determination of ALAD activity and isozyme phenotype, heparinized whole blood was centrifuged at 500 g for 30 min at 4 °C, and the plasma and buffy coat were removed. The erythrocytes were washed twice with an equal volume of 0.9% sodium chloride followed by centrifugation at 500 g for 30 min at 4 °C. For enzyme assay, the washed, pelleted erythrocytes were lysed by the addition of 1 mM potassium phosphate buffer, pH 6.8, containing 1 mM magnesium chloride, 1 mM DTT, and 0.5% Triton X-100. The mixture was subjected to a vortex for 2 min and centrifuged at 15 000 g for 40 min, and the supernatant (lysate) was used to determine ALAD activity as previously described.[41–43] For determination of isozyme phenotype, the washed, pelleted erythrocytes were lysed with an equal volume of 2.5 mM zinc acetate and 20 mM DTT and equalized for lysate protein.

Electrophoretic Discrimination of the ALAD Isozyme Phenotypes

The ALAD isozyme phenotypes were determined by starch gel electrophoresis as previously described.[30,31] Briefly, the starch gels were 10% with a gel buffer of 0.01 M sodium phosphate, pH 6.8. The bridge buffer was 0.1 M sodium phosphate, pH 6.8.[31] Sample slots contained 50 to 75 μl of lysate. Electrophoresis was carried out at 5 V/cm for 18 hr at 4 °C. The gel was stained first with ALA and DTT, and then with modified Ehrlich's reagent.[31] When the activities of the 1-1, 1-2, and 2-2 isozymes in normal erythrocytes were inactivated by incubation with 40 to 200 mM lead acetate, and then electrophoresed, under the above conditions, no differences were observed in their respective migrations. Lysates also were analyzed by cellulose acetate gel electrophoresis, which improved the resolution of the isozyme phenotypes. Lysates (0.5 to 4 μl) were applied to the gel and subjected to electrophoresis for 2 hr at 200 V and 4 °C with a running buffer of 0.03 M phosphate–citrate, pH 7.0. The activity was detected as described above for the starch gels.

RESULTS

TABLE 1 shows the frequencies of the ALAD isozyme phenotypes in erythrocytes for a normal population of 1074 individuals of various ethnic and demographic backgrounds from New York City. The population was 87% Caucasian, 10% American Black, 2% Hispanic, and 0.5% Asian. The $ALAD_1$ allele frequency in the total population was 0.89, a frequency similar to that (0.90) reported in a large Italian population.[30,31] It was notable that the $ALAD_2$ allele frequency in the 115 American Black individuals studied was 0.03, whereas the frequencies found in Hispanic, Asian, and Ashkenazi Jewish individuals ranged from 0.08 to 0.13 (TABLE 1). This finding was consistent with the reported absence of the $ALAD_2$ allele in a Black population from Liberia.[33]

Because blood specimens from lead-poisoned individuals were unavailable, efforts were directed to obtain blood specimens from lead-exposed individuals in order to determine the ALAD allele frequencies in those with elevated blood lead levels. In collaboration with the New York City Public Health Toxicology Laboratories, six sets of blood specimens submitted for lead toxicity studies were tested independently in a double-blind fashion for blood lead levels and the ALAD isozyme phenotypes. TABLE 2 shows the frequency of the isozyme phenotypes in individuals with blood lead levels ≥ 30 $\mu g/dl$, a level previously designated as evidence of lead intoxication. For each of the four sets of blood samples containing adequate numbers of lead-poisoned subjects ($N > 20$), the results demonstrated that a higher than expected proportion of lead-exposed individuals had the $ALAD_2$ allele. The presence of the $ALAD_2$ allele was associated with a fourfold increase in an individual's ability to retain lead at blood levels ≥ 30 $\mu g/dl$. Among the total 1051 individuals studied, 30% of the individuals with blood lead levels ≥ 30 $\mu g/dl$ were heterozygous or homozygous for the $ALAD_2$ allele whereas only 8% of those with blood levels <30 $\mu g/dl$ had an $ALAD_2$ allele. Similarly, in the subset of 343 known American Blacks in this study, 30% of those with blood lead levels ≥ 30 $\mu g/dl$ had the $ALAD_2$ allele compared with only 3% of those with blood levels <30 $\mu g/dl$.

DISCUSSION

These results indicate that individuals carrying the $ALAD_2$ allele are more likely to retain lead when exposed to the metal in the workplace or environment. The blood

TABLE 1. Human ALAD Polymorphism: Phenotype and Gene Frequencies

Sample	N	Phenotype Frequency 1-1	Phenotype Frequency 2-1	Phenotype Frequency 2-2	Gene Frequency $ALAD_1$	Gene Frequency $ALAD_2$
Italy[a]	762	0.81	0.17	0.02	0.90	0.10
New York						
Ashkenazi Jewish	872	0.76	0.22	0.02	0.87	0.13
Caucasian, other	62	0.82	0.18	—	0.91	0.09
Black	115	0.95	0.05	—	0.97	0.03
Asian	5	0.80	0.20	—	0.89	0.11
Hispanic	20	0.85	0.15	—	0.92	0.08
Total	1074	0.786	0.198	0.016	0.889	0.11

[a]See reference 30.

TABLE 2. Human ALAD Polymorphism: Association with Lead Poisoning

Sample Set	Blood Lead Level (μg/dl)	ALAD Polymorphism[a] Total	1-1	1-2	2-2
1	<30	157	138 (88)	18 (11)	1
	≥30	49	25 (51)	22 (45)	2
2	<30	119	106 (89)	13 (11)	0
	≥30	30	24 (80)	5 (17)	1
3	<30	121	117 (97)	2 (2)	2
	≥30	21	17 (81)	4 (19)	0
4	<30	152	138 (91)	10 (7)	4
	≥30	28	22 (79)	4 (14)	2
5	<30	205	194 (95)	11 (5)	0
	≥30	3	2 (67)	1 (33)	0
6	<30	161	148 (92)	12 (7)	1
	≥30	5	5 (100)	0 (0)	0
Totals	<30	915	841 (92)	66 (7)	8
	≥30	136	95 (70)	36 (26)	5
Black	<30	292	282 (97)	9 (3)	1
	≥30	53	37 (70)	15 (28)	1
Hispanic	<30	226	200 (89)	25 (11)	1
	≥30	33	26 (79)	7 (21)	0

[a]Percentages are shown in parentheses.

samples tested were primarily obtained from children living in neighborhoods at high risk for lead exposure. These individuals were initially screened for lead exposure by determination of their FEP levels. Individuals with FEP levels greater than 50 μg/dl or blood lead levels greater than 30 μg/dl or both were recalled for collection of the venous blood samples used in these studies. Thus, the ascertainment bias in these data was skewed to reflect the ALAD frequencies in individuals with elevated FEP levels. In addition to lead exposure, FEP levels can be increased by iron deficiency. The fact that the $ALAD_2$ allele frequency was found to be higher among individuals with blood lead levels ≥30 μg/dl than among the general population, however, may have particular relevance to the susceptibility to lead poisoning in high-risk groups.

At about the same time, Ziemsen *et al.*[44] measured blood lead levels, ALAD activities (without activation with zinc and DTT), urinary ALA concentrations, and ALAD isozyme types in lead-exposed workers in Germany. When their data for lead-poisoned individuals with the 2-2 phenotype (N = 10) were paired with individuals with the 1-1 phenotype (N = 20) on the basis of blood lead levels, no significant differences were seen in the levels of ALAD activity or urinary ALA concentration. These results, however, may not reflect the role of the ALAD polymorphism: Ziemsen *et al.* did not determine the percentage of lead inactivation of ALAD activity by assaying enzymatic activity in the absence and presence of 2.5 mM zinc chloride and DTT. Even more relevant is the fact that they did not analyze the distribution of isozyme phenotypes as a function of blood lead levels. Instead, and in support of our hypothesis, they found that the average blood lead levels increased in individuals heterozygous or homozygous for the $ALAD_2$ allele: isozyme phenotype 1-1, 38 ± 17 μg/dl (N = 160); isozyme phenotype 1-2, 44 ± 17 μg/dl (N = 32); and isozyme phenotype 2-2, 56 ± 18 μg/dl (N = 10). Although the sample sizes are small, the percentage of individuals with isozyme phenotype 2-2 was 5%, an unusually high value. Moreover, their data are consistent with our results, which suggest that the presence of the $ALAD_2$ allele may lead to increased lead accumulation.

Based on the findings reported here, we hypothesize that the $ALAD_2$ allele encodes an enzyme subunit that binds lead more tightly than the subunit expressed by the $ALAD_1$ allele. The increased binding would lead to a slower lead clearance; a greater total body lead concentration; and, as a consequence, a greater susceptibility to the subclinical and clinical effects of lead exposure. Support for this hypothesis will require additional molecular biological and biochemical studies to characterize the $ALAD_1$ and $ALAD_2$ alleles and the physical and kinetic properties of each of the purified ALAD isozymes. Analysis of the nucleotide sequence of each allele will identify their molecular difference and provide insight into their relative abilities to bind lead and zinc. Large amounts of the ALAD 1-1 and 2-2 isozymes could then be produced in expression vectors, and characterization of the purified recombinant isozymes would permit (for the first time) careful studies of their relative (and competitive) binding of zinc and lead as well as their respective physical and kinetic properties. We anticipate that understanding of the pathophysiological interaction of the ALAD isozymes and lead will provide insights for the treatment and prevention of lead poisoning.

SUMMARY

ALAD is a zinc metalloenzyme whose inhibition by lead is the first and most sensitive indicator of lead exposure and whose decreased activity has been implicated in the pathogenesis of lead poisoning. This heme biosynthetic enzyme is encoded by a gene located at chromosome 9q34, which has two codominant alleles, $ALAD_1$ and $ALAD_2$. The occurrence of two frequent alleles for ALAD stimulated an investigation into the possible pharmacogenetic role of the enzyme polymorphism in lead poisoning. In a New York City population at high risk for lead exposure, individuals heterozygous or homozygous for the less common allele, $ALAD_2$, had blood lead levels $\geq 30\ \mu g/dl$ more frequently than expected. These findings suggest a potential genetic susceptibility to lead poisoning in individuals with the ALAD 1-2 and 2-2 phenotypes.

ACKNOWLEDGMENTS

The authors thank Dr. Nain-hu Sun, Dr. Xian-Zhang Hu, and Mr. Constantine Zamfirescu for their valuable contributions to this study.

REFERENCES

1. Wu, W. H., D. Shemin, K. E. Richards & R. C. Williams. 1974. Proc. Natl. Acad. Sci. USA **71:** 1767–1770.
2. Gurne, D., J. Chen & D. Shemin. 1977. Proc. Natl. Acad. Sci. USA **74:** 1383–1387.
3. Anderson, P. M. & R. J. Desnick. 1979. J. Biol. Chem. **254:** 6924–6930.
4. Bevan, D. R., P. Bodlaender & D. Shemin. 1980. J. Biol. Chem. **255:** 2030–2035.
5. Tsukamoto, I., T. Yoshinaga & S. Sano. 1979. Biochim. Biophys. Acta **570:** 167–178.
6. Tsukamoto, I., T. Yoshinaga & S. Sano. 1980. Int. J. Biochem. **12:** 751–756.
7. Jaffe, E. K., S. P. Salowe, N. T. Chen & P. A. DeHaven. 1984. J. Biol. Chem. **259:** 5032–5036.
8. Jaffe, E. K. & D. Hanes. 1986. J. Biol. Chem. **261:** 9348–1353.
9. Finelli, V. N., D. S. Klauder, M. A. Karaffa & H. G. Petering. 1975. Biochem. Biophys. Res. Commun. **65:** 303–312.

10. GIBBS, P. N. B., M. G. GORE & P. M. JORDAN. 1985. Biochem. J. **225:** 573–580.
11. GIBBS, P. N. B. & P. M. JORDAN. 1986. Biochem. J. **236:** 447–451.
12. JORDAN, P. M. & P. N. B. GIBBS. 1985. Biochem. J. **227:** 1015–1020.
13. BARNARD, G. F., R. ITOH, L. H. HOHBERGER & D. SHEMIN. 1977. J. Biol. Chem. **252:** 8965–8974.
14. GIBBS, P. N. B. & P. M. JORDAN. 1981. Biochem. Soc. Trans. **9:** 232–233.
15. MORGAN, J. M. & H. B. BURCH. 1972. Arch. Intern. Med. **130:** 335–341.
16. CHISHOLM, J. J., JR., D. J. THOMAS & T. J. HAMILL. 1985. Clin. Chem. **31:** 662–668.
17. WETMUR, J. G., D. F. BISHOP, L. OSTASIEWICZ & R. J. DESNICK. 1986. Gene **43:** 123–130.
18. WETMUR, J. G., D. F. BISHOP, C. CANTELMO & R. J. DESNICK. 1986. Proc. Natl. Acad. Sci. USA **83:** 7703–7707.
19. BERG, J. M. 1986. Nature **319:** 264–265.
20. DOSS, M., R. VON TIEPERMANN, J. SCHNEIDER & H. SCHMID. 1979. Klin. Wochenschr. **57:** 1123–1127.
21. DOSS, M., R. VON TIEPERMANN & J. SCHNEIDER. 1982. Clin. Biochem. **15:** 52–55.
22. THUNELL, S., L. HOLMBERG & J. LUNDGREN. 1987. J. Clin. Chem. Clin. Biochem. **25:** 5–14.
23. DAGG, J. H., A. GOLDBERG, A. LOCHHEAD & J. A. SMITH. 1965. Q. J. Med. **34:** 163–175.
24. BERK, P. D., D. P. TSCHUDY, L. A. SHEPLEY, J. G. WAGGONER & N. I. BERLIN. 1970. Am. J. Med. **48:** 137–144.
25. AUDESIRK, G. 1985. Prog. Neurobiol. **24:** 199–231.
26. KAPPAS, S. SASSA & K. E. ANDERSON. *In* The Metabolic Basis of Inherited Disease. J. B. Stanbury, J. B. Wyngaarden, D. F. Fredrickson, J. L. Goldstein & M. S. Brown, Eds.: 1340–1341. McGraw-Hill. New York, NY.
27. CUTLER, M. G., M. MCLAUGHLIN, E. MCNEIL & M. R. MOORE. 1985. Neuropharmacology **24:** 1005–1009.
28. DOSS, M., U. BECKER, F. SIXEL, S. GEISSE, H. SOLCHER & J. SCHNEIDER. 1982. Klin. Wochenschr. **60:** 599–606.
29. DOSS, M. & W. A. MULLER. 1982. Blut **45:** 131–139.
30. PETRUCCI, R., A. LEONARDI & G. BATTISTUZZI. 1982. Hum. Genet. **60:** 289–290.
31. BATTISTUZZI, G., R. PETRUCCI, L. SILVAGNI, F. R. URBANI & S. CAIOLA. 1981. Ann. Hum. Genet. **45:** 223–229.
32. POTLURI, V. R., K. ASTRIN, J. G. WETMUR, D. F. BISHOP & R. J. DESNICK. 1987. Hum. Genet. **76:** 236–239.
33. BENKMANN, H. G., P. BOGDANSKI & H. W. GOEDDE. 1983. Hum. Hered. **33:** 62–64.
34. PASVAL, G., D. J. WEATHERALL & J. M. WILSON. 1978. Nature **274:** 7801–7803.
35. PROPPING, P. 1978. J. Physiol. Biochem. Pharmacol. **83:** 124–173.
36. KALOW, W., H. W. GOEDDE & D. P. AGARWAL. 1986. *In* Ethnic Differences in Reactions to Drugs and Xenobiotics. Vol. 214: 583. Alan R. Liss. New York, NY.
37. KAUL, B., G. SLAVIN & B. DAVIDOW. 1983. Clin. Chem. **29:** 1467–1470.
38. PIOMELLI, S. & J. GRAZIANO. 1980. Pediatr. Clin. North Am. **27:** 843–853.
39. PIOMELLI, S., C. SEAMAN, D. ZULLOW, A. CURRAN & B. DAVIDOW. 1982. Proc. Natl. Acad. Sci. USA **79:** 3335–3339.
40. KAUL, B., B. DAVIDOW, Y. M. ENG & M. H. GEWIRTZ. 1983. Arch. Environ. Health **38:** 296–300.
41. GIAMPIETRO, P. F. & R. J. DESNICK. 1982. Enzyme **28:** 144–145.
42. GIAMPIETRO, P. F. & R. J. DESNICK. 1983. Anal. Biochem. **131:** 83–92.
43. DESNICK, R. J. & D. F. BISHOP. 1985. Methods Enzymol. **123:** 339–345.
44. ZIEMSEN, B., J. ANGERER, G. LEHNERT, H.-G. BENKMANN & H. W. GOEDDE. 1986. Int. Arch. Occup. Environ. Health **58:** 245–247.

Occurrence and Biological Properties of *N*-Methyl Protoporphyrin

FRANCESCO DE MATTEIS, ANTHONY H. GIBBS, AND ANN E. HOLLEY

Toxicology Unit
Medical Research Council
Carshalton, Surrey SM5 4EF, England

INTRODUCTION

Because *N*-methylprotoporphyrin (*N*-MePP) was first isolated from the liver of animals made protoporphyric by drugs and is intimately involved in the pathogenesis of this type of experimental porphyria, a short introduction on drug-induced hepatic protoporphyria will first be given to set the more recent work in its appropriate historical context.

In mice, 3,5-diethoxycarbonyl-1,4-dihydrocollidine (DDC) and griseofulvin (GF) (FIGURE 1) both cause hepatic protoporphyria,[1] which is characterized by a very large accumulation of protoporphyrin. Onisawa and Labbe[2] first showed that the activity of ferrochelatase (protoheme ferrolyase, EC 4.99.1.1) was markedly reduced in the liver of mice made porphyric by treatment with DDC, and a similar depression of liver ferrochelatase in mice after they were fed GF was reported by Lockhead *et al.*[3] The depression of ferrochelatase provided an explanation for the marked accumulation of protoporphyrin (the substrate of the inhibited enzyme) and also for the compensatory stimulation of 5-aminolevulinate synthase (the first enzyme of the pathway)—a response secondary to the inhibition of heme biosynthesis and consequent failure of the negative feedback control that heme exercises on the synthase. Onisawa and Labbe[2] also reported that when the cytosolic fraction from the liver of mice made porphyric with DDC was preincubated with normal mitochondria, the ferrochelatase activity of the mitochondria became slightly inhibited, suggesting the presence of an inhibitor of ferrochelatase in the porphyric livers. Many years were to pass before the significance of this important observation could be fully appreciated.

Two additional findings were reported in animals treated with DDC, and, as both of these have a bearing on the mechanism by which inhibition of ferrochelatase is brought about by this drug, they will now be briefly discussed. First, it was shown that an inhibitior of cytochrome P-450-dependent drug metabolism, compound SKF 525-A, could prevent the loss of ferrochelatase activity caused by DDC and also that DDC was not very effective at inhibiting the ferrochelatase in newborn animals, where the concentration of cytochrome P-450 in the liver is known to be low. So it was suggested that DDC had to be metabolized before it could lead to loss of ferrochelatase activity.[4] In addition, it was reported that both DDC and GF caused a rapid loss of cytochrome P-450 in the liver,[5] an effect reminiscent of that seen with other compounds (for example, 2-allyl-2-isopropylacetamide), which are metabolized to reactive derivatives and are then responsible for modifying the heme prosthetic group into abnormal porphyrins, or green pigments.[6] Although the loss of cytochrome P-450 seen after DDC had originally been interpreted as resulting from inhibition of ferrochelatase and blocking of heme biosynthesis, further work showed that increased heme degradation was also involved with loss of radioactivity from prelabeled liver heme (mostly from microsomal heme) and accumulation of nonheme products, particularly in the cytosolic fraction.[7]

As a result of these studies the following main points could be considered established: 1) DDC and GF caused a marked inhibition of ferrochelatase when given to rodents *in vivo.* 2) The drugs probably had to be metabolized in order to produce this effect. 3) A transient decline in the concentration of microsomal cytochrome P-450 and heme took place at the same time that the enzyme ferrochelatase became inhibited, and this decline was due—at least in part—to drug-induced conversion of heme into unidentified products. It is only in the last few years that these three main points have been brought together in a coherent molecular mechanism with the discovery and characterization of *N*-MePP, as will now be discussed.

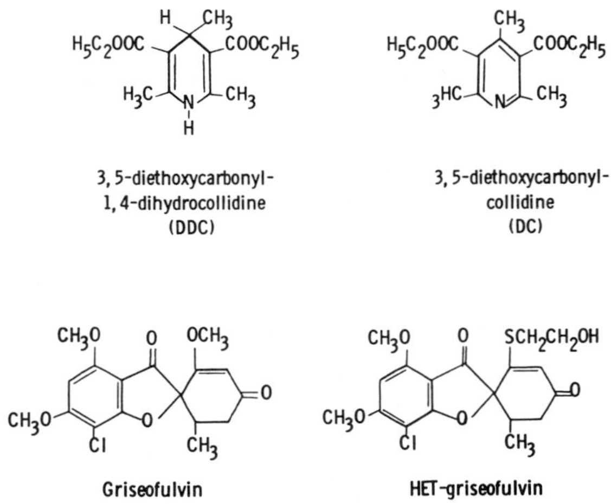

FIGURE 1. Two classes of compounds that induce hepatic protoporphyria in rodents. For each porphyrogenic compound, the structure of a closely related analogue that does not cause porphyria is also shown.

ISOLATION OF A PORPHYRIN INHIBITOR OF FERROCHELATASE

Tephly *et al.*[8] extracted from the liver of mice made porphyric with DDC a porphyrin with strong *in vitro* inhibitory properties toward the ferrochelatase of normal mitochondria. A similar inhibitory porphyrin was later isolated from the liver of mice treated with GF.[9] Thus both these porphyria-inducing drugs not only inhibit the activity of hepatic ferrochelatase as measured in the liver of treated animals, but also cause accumulation of an inhibitor capable of inhibiting the enzyme *in vitro.* When the liver accumulation of the inhibitory porphyrin was related to the inhibition of the ferrochelatase seen directly in the liver mitochondria from porphyric animals, a close correlation was found not only when the mouse (a responsive species) was compared to the rat (a relatively nonresponsive species),[8] but also when DDC and GF were compared to their inactive analogues, 3,5-diethoxycarbonyl collidine (DC) and the thioether analogue of GF (HET-GF),[8,9] respectively (FIGURE 1). In addition, in the case of DDC where this was also studied, a good correlation between liver levels of the inhibitor and inhibition of hepatic ferrochelatase was found at different times after administration of a single dose of either 2 or 10 mg/kg body wt.[10] These findings showed convincingly that DDC and GF inhibit the hepatic ferrochelatase activity by causing accumulation in the liver of the porphyrin inhibitor of the enzyme.

In further experiments, the inhibitory porphyrin (which could be isolated not only from the mitochondrial fraction where the enzyme ferrochelatase is located, but also—in larger amounts—from the microsomal fraction) was shown to originate from prelabeled heme.[8] Its production after DDC administration could be decreased by pretreatment with compound SKF 525-A, an inhibitor of cytochrome P-450-dependent drug metabolism.[10] These findings therefore suggested that DDC metabolism was in some way involved in converting preexisting microsomal heme into the inhibitory porphyrin, which could then be released from the membranes of the endoplasmic reticulum and migrate within the cell to reach its selective target—the ferrochelatase—in the inner membrane of the mitochondrion.

IDENTIFICATION OF THE INHIBITORY PORPHYRIN AS *N*-MePP

Abnormal porphyrins originating from the heme prosthetic group of cytochrome P-450 had previously been described after treatment with 2-allyl-2-isopropylacetamide,[6] other alkenes, or alkynes,[11] and here too metabolism of the drug responsible to reactive derivatives had been implicated. The mechanism of production of these porphyrins (also named green pigments) and their chemical structure had recently been clarified: they were found to originate through covalent binding of an oxygenated metabolite of the unsaturated drug to the protoporphyrin nucleus of heme,[12] and their electronic absorption spectrum and titration behavior with acid indicated for all of them the structure of *N*-monoalkylated protoporphyrin,[13] the site of attachment of the reactive metabolite being in all cases one of the pyrrole nitrogens of protoporphyrin IX. When the inhibitory porphyrins isolated from the liver of mice made porphyric by treatment with DDC or GF were similarly examined, these were also found to exhibit the electronic spectrum and titration behavior with acid characteristic of *N*-alkylated protoporphyrins[9]; however, the spectral properties of the divalent cation and zinc-

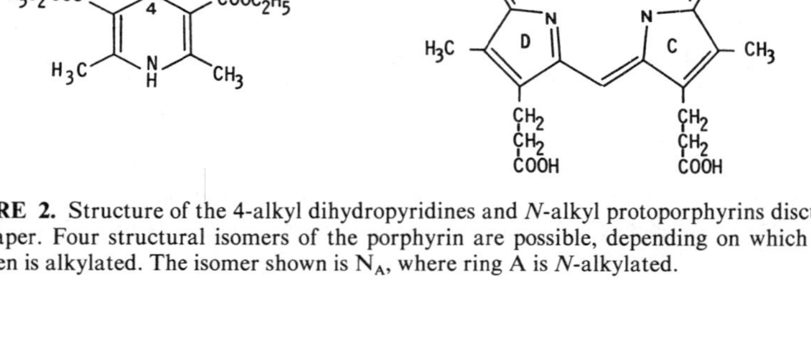

FIGURE 2. Structure of the 4-alkyl dihydropyridines and *N*-alkyl protoporphyrins discussed in this paper. Four structural isomers of the porphyrin are possible, depending on which pyrrole nitrogen is alkylated. The isomer shown is N_A, where ring A is *N*-alkylated.

TABLE 1. HPLC Absorption and Mass Spectral Properites of the DDC and GF Inhibitory Pigments and of Authentic Synthetic *N*-MePP

Compound[a]	HPLC Retention Times (min)	Absorption Maxima (nm) Soret	IV	III	II	Ia	I	Molecular Mass Ion (m/e)
DDC pigment	2.85, 3.66	419	513	546	595	625	652	604
GF pigment	2.85, 3.70	419	513	546	596		653	
N-MePP	2.9, 3.72	419	514	547	594	625	653	604

NOTE: The HPLC system employed allowed separation of two isomeric pairs (isomers N_A and N_B being eluted first, followed by isomers N_C and N_D) and involved a mobile phase of CH_2Cl_2:MeOH:ammonia (s.g.: 0.88) 50:50:0.1, a Nucleosil 5 column, and a flow rate of 2 ml/min. The absorption spectrum was that of the neutral porphyrin (unresolved mixture of isomers) and was obtained in $CHCl_3$. Data from references 9 and 14.

[a] All compounds were methyl esters.

complex derivatives suggested that in the case of the DDC and GF pigments, the substituent on the pyrrole nitrogen must be small.[9] In agreement with this, the DDC pigment was identified as *N*-MePP (FIGURE 2) by comparison of its chromatographic, absorption, and mass spectral properties with those of authentic synthetic *N*-MePP (TABLE 1), and also by showing that the synthetic porphyrin was, like the biological product, a powerful inhibitor of ferrochelatase.[14] These conclusions were confirmed by Tephly *et al.*[15] and Ortiz de Montellano *et al.*,[16] who also resolved and fully characterized by elegant NMR studies all four structural isomers of *N*-MePP. The inhibitory porphyrin isolated after GF treatment has so far been studied less extensively; nevertheless, its spectral characteristics are indistinguishable from those of *N*-MePP[9] (TABLE 1), as are the HPLC chromatographic properties of its four isomers, which are present in a similar proportion to that seen in the DDC pigment (A. Holley, L. J. King, and F. De Matteis, unpublished results).

MECHANISM OF PRODUCTION OF *N*-MePP

The source of the *N*-methyl group and of the tetrapyrrole macrocycle of N-MePP has more recently been clarified. Two findings had suggested that the methyl group on the pyrrole nitrogen might originate from the drug itself. 1) Loev and Snader[17] had reported that, during the oxidative process leading to aromatization, a number of dihydropyridines could lose their 4-alkyl group in a reactive form and transfer this to suitable nucleophiles. If the alkyl group could be similarly transferred to heme, this would explain the inability of the oxidized analogue of DDC, DC (FIGURE 1), to promote formation of *N*-MePP *in vivo,* as this drug already possesses the aromatic structure and as, on account of this, the 4-methyl substituent is not readily eliminated. 2) The second line of evidence compatible with this mechanism had come from Marks and collaborators: they had reproduced the inhibition of ferrochelatase by DDC in chick embryo livers and also reported that a number of dihydropyridines (all differing from DDC in the nature of the 4-alkyl group) showed considerable variation in inhibitory activity on ferrochelatese, 4-desmethyl DDC (the analogue lacking a 4-alkyl group) being almost completely inactive.[18] To investigate whether the 4-alkyl group of a dihydropyridine (FIGURE 2) could in fact be donated to a tetrapyrrole, mice were given a series of analogues differing in the 4-alkyl group and the corresponding *N*-alkyl protoporphyrins produced *in vivo* were isolated from their liver and characterized by mass spectrometry (TABLE 2). The molecular ion found after giving DDC and its 4-ethyl analogue was compatible with a structure of *N*-MePP and *N*-ethylprotoporphyrin, respectively,[19] quite in agreement with the anticipated mechanism of 4-alkyl transfer. Conclusive evidence for this mechanism was obtained by treating mice with the deuterated analogue, 4-CD_3-dihydropyridine, and demonstrating an increase of 3 mass units (TABLE 2) in the molecular ion of the corresponding *N*-alkylated porphyrin (compared with the molecular ion obtained after unlabeled DDC) compatible with the 4-alkyl group being donated intact, with its full complement of hydrogen atoms.[19] Similar conclusions as to the source of the *N*-methyl group were reached independently by Tephly *et al.*[15] and by Ortiz de Montellano *et al.*[20]

The possibility that cytochrome P-450 might be involved in the reaction by contributing its own prosthetic group as the tetrapyrrolic macrocycle of *N*-MePP was suggested by early findings showing rapid loss of the cytochrome after administration of DDC[5] and concurrent conversion of prelabeled microsomal heme into unidentified products.[7] A precedent for drug-induced conversion of cytochrome P-450 heme to *N*-alkylated porphyrins was already known because several olefinic and acetylenic compounds had been shown to produce this effect. In this case, however, no inhibition of the ferrochelatase had been found in the liver of treated animals, and the corresponding alkylated porphyrins also failed to exhibit inhibitory activity on the enzyme *in vitro*.[21] In order to ascertain whether liver cytochrome P-450 was in fact the source of the tetrapyrrole moiety of *N*-MePP, DDC and certain of its analogues were incubated with isolated rat hepatocytes, and accumulation of *N*-alkyl porphyrins was related to the loss of cytochrome P-450 (TABLE 3). The analogue lacking a 4-alkyl group (4-desmethyl DDC) was inactive in both respects; when the size of the 4-substituent of the dihydropyridine was increased from a methyl to an ethyl or propyl residue, the production of *N*-alkylated porphyrins increased and a corresponding greater loss of cytochrome P-450 was observed. With the 4-isopropyl and 4-benzyl analogues, however, an even greater loss of cytochrome P-450 heme was observed but no alkylated product could be demonstrated. It was therefore concluded that the heme of cytochrome P-450 was the target of the transalkylation reaction leading to the *N*-alkyl porphyrin products; however, for a *N*-alkylated porphyrin to accumulate, the 4-alkyl group needed to be of special size and shape, possibly because bulky residues would be sterically hindered in their access to the pyrrole nitrogen atoms and would attack instead more accessible sites in the molecule of heme.[22] In more recent experiments, Marks *et al.* have tested the same analogues *in vitro* for destruction of cytochrome P-450 heme in chick liver microsomes and found that they ranked in the same order of potency.[23]

TABLE 2. The 4-Alkyl Group of a Dihydropyridine Is Transferred Intact to a Pyrrole Nitrogen of Protoporphyrin IX

4-Alkyl Group of the Dihydropyridine Administered	Molecular Ion of the Resulting Pigment (m/e)	*N*-Alkyl Group Present in the Isolated *N*-Alkyl Protoporphyrin IX
CH_3–	604	CH_3–
C_2H_5–	618	C_2H_5–
CD_3–	607	CD_3–

NOTE: The *N*-alkylated porphyrins produced *in vivo* by several 4-alkyl dihydropyridines were purified as the methyl ester derivatives and characterized by mass spectrometry. The mass of the molecular ion is given, as is the structure of the corresponding *N*-alkylated pigment. Data from reference 19.

TABLE 3. Loss of Cytochrome P-450 and Accumulation of *N*-Alkylporphyrins Caused by Several 4-Substituted Dihydropyridines in Isolated Hepatocytes

4-Substituent of the Dihydropyridine Administered[a]	Cytochrome P-450 (% of unincubated)	*N*-Alkylated Porphyrins (nmoles/100 × 10^6 cells)
Control (ethanol alone)	89.5	<0.17
Desmethyl (H)	91.5	<0.17
Methyl	77.0	0.79
Ethyl	31.5	1.69
n-Propyl	27.0	2.54
Isopropyl	25.0	<0.17
Benzyl	50.7	<0.17

NOTE: Ethanolic solutions of various 4-substituted dihydropyridines were added to isolated rat hepatocyte suspensions and incubated for 2 hr. Data from reference 22.

[a]Final concentration: 50 μM.

The following additional evidence also implicated cytochrome P-450: 1) When isolated liver microsomes were incubated with 4-alkyl dihydropyridines *in vitro,* a destruction of cytochrome P-450 was only seen when NADPH, the cofactor essential for cytochrome P-450-dependent drug metabolism, was also present.[20,22,23] 2) The specific radioactivity of the liver heme pool, from which the *N*-alkyl porphyrin originated, decayed with a half-life[22] similar to that reported for the rapidly turning over component of liver cytochrome P-450. 3) The alkylated porphyrins all exhibited chiral properties, as expected from the *N*-alkylation reaction taking place on an enzymic template,[16,25] and, in addition, the proportion of the four structural isomers of *N*-ethyl protoporphyrin (in each isomer a different pyrrole nitrogen is ethylated) was found to depend on the particular cytochrome P-450 enzyme predominating in the liver.[26] 4) Finally, although exchangeable heme (including exogenous heme analogues, such as mesoheme) could also be utilized in the dihydropyridine-dependent *N*-alkylation pathway, preliminary uptake by the apoprotein of cytochrome P-450 was probably required, as in this case too the isomeric profile of the *N*-ethylated products was influenced by the type of cytochrome P-450 that predominated in the cell.[24]

It can therefore be concluded that DDC and other 4-alkyl dihydropyridines inactivate cytochrome P-450 by donating their intact 4-alkyl group to one pyrrole nitrogen of the prosthetic heme, giving rise in this way to *N*-MePP and other *N*-alkylated protoporphyrins. The molecular mechanism suggested for this suicidal transalkylation reaction[19,20,23,27] is based on that proposed by Loev and Snader[17] for the oxidative aromatization of dihydropyridines in chemical systems. An initial oxidative attack at the dihydropyridine nitrogen can lead to aromatization by one of two competing pathways: 1) loss of the 4-alkyl group or 2) loss of the 4-hydrogen atom, the first pathway being favored with leaving 4-alkyl groups of increased stability and under the influence of steric factors, such as when the substituents in the 3 and 5 positions of the dihydropyridine are increased in size, as discussed.[22] In both the chemical and biological systems, the oxidation reaction is mediated by electron transfer without oxygen insertion, so in this case the cytochrome appears to operate as an oxidase rather than in its more usual role as an monooxygenase. Augusto *et al.*[27] have suggested that with cytochrome P-450 the first step is a one-electron oxidation of the dihydropyridine to the corresponding radical cation, leading to elimination of the 4-ethyl group as an alkyl radical. By analogy with the finding discussed above for DDC, the *N*-methyl group of the GF pigment could conceivably originate from the drug, but this has not yet been established. The possibility that the drug may merely accelerate an alternative pathway of production of *N*-MePP is less likely but cannot yet be discounted, as a similar *N*-alkylated pigment with inhibitory properties on ferrochelatase is present in trace amounts in the liver of control mice that have received no drug treatment.[8,15]

MECHANISM OF INHIBITION OF FERROCHELATASE

Early studies conducted with the DDC-inhibitory pigment showed it to be very powerful, a 50% loss of enzyme ferrochelatase activity being obtained under standard assay conditions with a concentration as low as 8 nM.[21] Since the identification of the inhibitor as *N*-MePP, most of the work on ferrochelatase has been carried with synthetic *N*-MePP, which has been found to be inhibitory toward the enzyme from a number of sources, including liver, spleen, and microbial systems.[28,29] Several analogues of *N*-MePP have also been found to be active, and from the structural requirements for the inhibitory activity certain important points have emerged that may have a bearing on the stereochemistry of the active site of ferrochelatase, as discussed below.

The following evidence indicates that in order to inhibit ferrochelatase, *N*-MePP must bind the porphyrin-binding site of the enzyme: 1) Preincubation of mitochondria with porphyrin substrate affords protection against subsequent challenge by *N*-MePP, indicating that substrate and inhibitor compete for the same binding site.[21] 2) Only *N*-alkylated porphyrins with two carboxylic acid sidechains can inhibit ferrochelatase, and these are the very porphyrins that are substrates of the enzyme, when not *N*-alkylated. 3) The two propionate side chains of *N*-MEPP (in positions 6 and 7) must be free for inhibition to occur, as shown by inactivity of the corresponding methyl ester derivatives, unless they are first hydrolyzed by endogenous carboxylesterases.[30] 4) Using the enzyme purified from bovine liver, Dailey and Fleming have found the kinetics of ferrochelatase inhibition to be competitive with respect to the porphyrin substrate, with a K_i value of approximately 7 nM, indicating that *N*-MePP is a high-affinity, tight-binding inhibitor of the enzyme.[31]

TABLE 4. Comparison of the Specific Inhibitory Activities of the N_A/N_B and N_C/N_D Alkylated Isomers for Several *N*-Alkyl Protoporphyrin Derivaties

N-Alkyl Porphyrin Derivative	Specific Inhibitory Activity (units/nmole) N_A/N_B	N_C/N_D
N-Methylprotoporphyrin IX	9.2	10.7
N-Ethylprotoporhyrin IX	6.7	6.9
N-*n*-Propylprotoporphyrin IX	6.1	0.1
N-Methylprotoporphyrin IX (zinc complex)	15.7	2.1

NOTE: Units of specific inhibitory activity are defined in Tephly *et al.*[8] Data from reference 30.

Differences in inhibitory activity between different isomers of the same *N*-alkylated porphyrin have been reported, particularly when zinc is chelated in the center of the tetrapyrrole macrocycle or the *N*-alkyl group is rather large. Thus the N_C and N_D isomers of *N*-MePP, which have the *N*-methyl group on propionic acid-substituted rings, are less inhibitory than the corresponding N_A and N_B isomers, where vinyl-substituted rings are *N*-methylated (see FIGURE 2 for the nomenclature of each

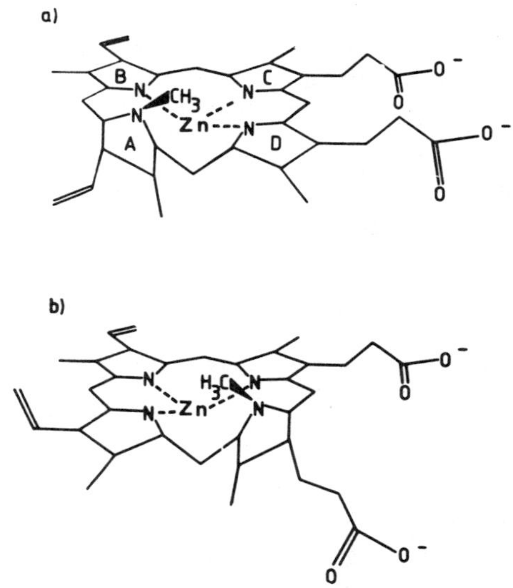

FIGURE 3. Two structural isomers of *N*-MePP zinc complex, where a vinyl-substituted ring (**a**) or a propionic acid-substituted ring (**b**) is *N*-methylated. Note distortion of the *N*-methylated ring out of the porphyrin plane and, in **b**, change in alignment of two propionate side chains. Modified from reference 25.

pyrrole ring). With the free *N*-MePP the difference is small,[14,25] and has not been found consistently in all studies (TABLE 4); however, when zinc is incorporated in *N*-MePP, the inhibitory activity of the N_C and N_D isomers (as compared to the N_A and N_B isomers) markedly declines[25] (TABLE 4). From X-ray crystallographic studies of *N*-alkylated porphyrins (cited in reference 25), it is known that the alkylated ring is tilted out of the porphyrin plane, particularly when a metal is chelated in the center of the tetrapyrrole system. Because—as discussed above—the two propionate side chains are required for binding of the inhibitor to the enzyme, it is possible that in the isomers where the propionate-substituted rings are *N*-methylated, the alignment of the two propionate side chains with respect to each other may be significantly altered (FIGURE 3) and the affinity of the inhibitor for the ferrochelatase active site correspondingly reduced.[25] Similar steric factors may be at play in the reduction of inhibitory activity seen when the size of the *N*-substituent is increased from one carbon unit to two or three units,[25] as in this case too the N_C and N_D isomers are those that show the most pronounced loss of inhibitory activity (see reference 32 and TABLE 4). These steric

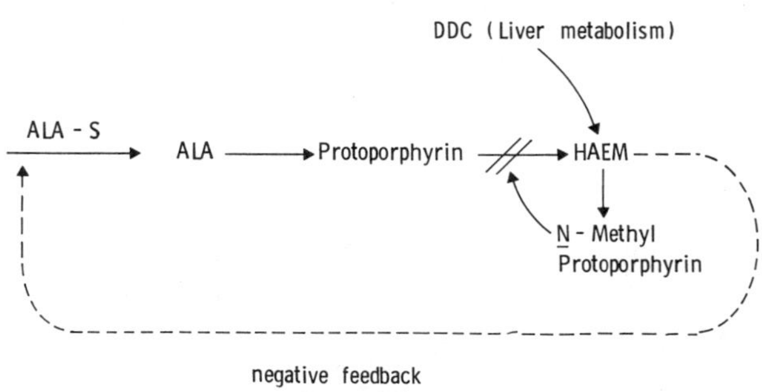

FIGURE 4. Main steps involved in the induction of hepatic protoporphyrin by DDC.

factors, which reduce the affinity of the inhibitor for the active site of ferrochelatase, may explain why the *N*-alkylated protoporphyrin produced by ethylene does not inhibit ferrochelatase,[33] as in this case the *N*-alkyl group is relatively large[14] and is also present on a propionate-substituted ring.[34]

Another factor involved in modifying the affinity of the inhibitor for ferrochelatase is probably the nature of the substituents at the 2 and 4 positions at the periphery of a dicarboxylic porphyrin, as differences in inhibitory activity between *N*-alkylated deuteroporphyrin, mesoporphyrin, and protoporphyrin[30] were found, particularly for the N_C and N_D isomers. So these substituents in the 2 and 4 positions (vinyl groups in protoporphyrin) are probably involved in binding.[30] Accordingly, the porphyrin-binding site at the active center (shared by the porphyrin substrate and the *N*-alkylated inhibitor) must include, in addition to appropriately spaced positive charges for binding of the two propionates, an area of interaction for the vinyl side chains of rings A and B. The binding site may then hold the porphyrin in a plane with the portion of the enzyme concerned with the insertion of iron, possibly extending over the porphyrin plane to allow insertion of the metal in the center of the tetrapyrrole. Some of the features described above for the binding site of ferrochelatase may be shared by heme oxygenase, where two propionate side chains are also required for binding of the substrate[35] and where the zinc complexes of several *N*-alkylated porphyrins are also inhibitory with similar steric factors involved in modifying the inhibitory activity.[30]

CONCLUSION

The sequence of events leading to formation of *N*-MePP and development of hepatic protoporphyria after treatment with DDC is summarized in FIGURE 4. A powerful antimetabolite, *N*-MePP originates in the endoplasmic reticulum of the hepatocytes from metabolic transfer of the 4-methyl group of DDC to a pyrrole nitrogen of the heme prosthetic group of cytochrome P-450. The *N*-MePP, which then migrates within the cell, reaches the ferrochelatase in the inner membrane of the mitochondrion, where its presence leads to enzyme inhibition. A compensatory stimulation of 5-aminolevulinate synthase follows, and this, combined with the inhibition of ferrochelatase, is then responsible for the marked accumulation of protoporphyrin. Direct confirmation of the involvement of *N*-MePP in the induction of prophyria has been obtained in mice[33] and chick embryos[36] by treating them with an *N*-methylated dicarboxylic porphyrin and showing inhibition of ferrochelatase, as well as stimulation of 5-aminolevulinate synthase and accumulation of protoporphyrin.

REFERENCES

1. SMITH, A. G. & F. DE MATTEIS. 1980. Drugs and the hepatic porphyrias. Clin. Haematol. **9:** 399–425.
2. ONISAWA, J. & R. F. LABBE. 1963. Effects of diethyl-1,4-dihydro-2,4,6-trimethylpyridine-3,5-dicarboxylate on the metabolism of porphyrins and iron. J. Biol. Chem. **238:** 724–727.
3. LOCKHEAD, A. C., J. H. DAGG & A. GOLDBERG. 1967. Experimental griseofulvin porphyria in adult and foetal mice. Br. J. Dermatol. **79:** 96–102.
4. DE MATTEIS, F., G. ABBRITTI & A. H. GIBBS. 1973. Decreased liver activity of porphyrin-metal chelatase in hepatic porphyria caused by 3,5-diethyoxycarbonyl-1,4-dihydrocollidine: Studies in rats and mice. Biochem. J. **134:** 717–727.
5. WADA, O., Y. YANO, G. URATA & K. NAKAO. 1968. Behaviour of hepatic microsomal cytochromes after treatment of mice with drugs known to disturb porphyrin metabolism in liver. Biochem. Pharmacol. **17:** 595–603.

6. DE MATTEIS, F. 1971. Loss of haem in rat liver caused by the porphyrogenic agent 2-allyl-2-isopropylacetamide. Biochem. J. **124:** 767–777.
7. ABBRITTI, G. & F. DE MATTEIS. 1973. Effect of 3,5-diethoxycarbonyl-1,4-dihydrocollidine on degradation of liver heme. Enzyme **16:** 196–202.
8. TEPHLY, T. R., A. H. GIBBS & F. DE MATTEIS. 1979. Studies on the mechanism of experimental porphyria produced by 3,5-diethoxycarbonyl-1,4-dihydrocollidine: Role of a porphyrin-like inhibitor of protohaem ferro-lyase. Biochem. J. **180:** 241–244.
9. DE MATTEIS, F. & A. H. GIBBS. 1980. Drug-induced conversion of liver haem into modified porphyrins. Biochem. J. **187:** 285–288.
10. TEPHLY, T. R., A. H. GIBBS, G. INGALL & F. DE MATTEIS. 1980. Studies on the mechanism of experimental porphyria and ferrochelatase inhibition produced by 3,5-diethoxycarbonyl-1,4-dihydrocollidine. Int. J. Biochem. **12:** 993–998.
11. WHITE, I. 1984. Suicidal destruction of cytochrome P-450 by ethynyl-substituted compounds. Pharm. Res.: 141–188.
12. ORTIZ DE MONTELLANO, P. R., K. L. KUNZE, G. S. YOST & B. A. MICO. 1979. Self-catalyzed destruction of cytochrome P-450: Covalent binding of ethynyl sterols to prosthetic heme. Proc. Natl. Acad. Sci. USA **76:** 746–749.
13. DE MATTEIS, F. & L. CANTONI. 1979. Alteration of the porphyrin nucleus of cytochrome P-450 caused in the liver by treatment with allyl-containing drugs: Is the modified porphyrin *N*-substituted? Biochem. J. **183:** 99–103.
14. DE MATTEIS, F., A. H. GIBBS, A. H. JACKSON & S. WEERASINGHE. 1980. Conversion of liver haem into *N*-substituted porphyrins or green pigments: Nature of the substituent at the pyrrole nitrogen atom. FEBS Lett. **119:** 109–112.
15. TEPHLY, T. R., B. L. COFFMAN, G. INGALL, M. S. ABOU ZEIT-HAR, H. M. GOFF, H. D. TABBA & K. M. SMITH. 1981. Identification of *N*-methylprotoporphyrin IX in livers of untreated mice and mice treated with 3,5-diethoxycarbonyl-1,4-dihydrocollidine: Source of the methyl group. Arch. Biochem. Biophys. **212:** 120–126.
16. ORTIZ DE MONTELLANO, P. R., H. S. BEILAN & K. L. KUNZE. 1981. *N*-Methylprotoporphyrin IX: Chemical synthesis and identification as the green pigment produced by 3,5-diethoxycarbonyl-1,4-dihydrocollidine treatment. Proc. Natl. Acad. Sci. USA **78:** 1490–1494.
17. LOEV, B. & K. M. SNADER. 1965. The Hantzsch reaction. I. Oxidative dealkylation of certain dihydropyridines. J. Org. Chem. **30:** 1914–1916.
18. COLE, S. F. C. & G. S. MARKS. 1980. Structural requirements in dihydropyridines for ferrochelatase inhibition and 5-aminolevulinic acid synthetase induction. Int. J. Biochem. **12:** 989–992.
19. DE MATTEIS, F., A. H. GIBBS, P. B. FARMER & J. H. LAMB. 1981. Liver production of *N*-alkylated porphyrins caused in mice by treatment with substituted dihydropyridines: Evidence that the alkyl group on the pyrrole nitrogen atom originates from the drug. FEBS Lett. **129:** 328–331.
20. ORTIZ DE MONTELLANO, P. R., P. R. BEILAN & K. L. KUNZE. 1981. *N*-Alkylprotoporphyrin IX formation in 3,5-dicarbethoxy-1,4-dihydrocollidine-treated rats: Transfer of the alkyl group from the substrate to the porphyrin. J. Biol. Chem. **256:** 6708–6713.
21. DE MATTEIS, F., A. H. GIBBS & T. R. TEPHLY. 1980. Inhibition of protohaem ferro-lyase in experimental porphyria: Isolation and partial characterization of a modified porphyrin inhibitor. Biochem. J. **188:** 145–152.
22. DE MATTEIS, F., C. HOLLANDS, A. H. GIBBS, N. DE SA & M. RIZZARDINI. 1982. Inactivation of cytochrome P-450 and production of *N*-alkylated porphyrins caused in isolated hepatocytes by substituted dihydropyridines. FEBS Lett. **145:** 87–92.
23. MARKS, G. S., D. T. ALLEN, C. T. JOHNSTON, E. P. SUTHERLAND, K. NAKATSU & R. A. WHITNEY. 1985. Suicidal destruction of cytochrome P-450 and reduction of ferrochelatase activity by 3,5-diethoxycarbonyl-1,4-dihydro-2,4,6-trimethylpyridine and its analogues in chick embryo liver cells. Mol. Pharmacol. **27:** 459–465.
24. DE MATTEIS, F., C. HARVEY & S. R. MARTIN. 1986. *N*-Alkylation of exogenous haem analogues caused by drugs in isolated hepatocytes: Structural isomerism and chirality of the resulting porphyrins. Biochem. J. **238:** 263–268.
25. DE MATTEIS, F., A. H. JACKSON, A. H. GIBBS, K. R. N. RAO, J. ATTON, S. WEERASINGHE & C. HOLLANDS. 1982. Structural isomerism and chirality of *N*-mono-substituted

protoporphyrins: Possible relevance to binding of *N*-methyl protoporphyrin by ferrochelatase and to orientation of haem in cytochrome P-450. FEBS Lett. **142:** 44–48.
26. De Matteis, F., A. H. Gibbs & C. Hollands. 1983. *N*-Alkylation of the haem moiety of cytochrome P-450 caused by substituted dihydropyridines: Preferential attack of different pyrrole nitrogen atoms after induction of various cytochrome P-450 isoenzymes. Biochem. J. **211:** 455–461.
27. Augusto, O., H. S. Beilan & P. R. Ortiz de Montellano. 1982. The catalytic mechanism of cytochrome P-450: Spin-trapping evidence for one-electron substrate oxidation. J. Biol. Chem. **257:** 11288–11295.
28. Brown, S. B., A. Holroyd, D. I. Vernon, R. F. Troxler & K. M. Smith. 1982. The effect of *N*-methyl protoporphyrin IX on the synthesis of photosynthetic pigments in *Cassidium caldarium*. Biochem. J. **208:** 487–491.
29. Houghton, J. D., C. L. Honeybourne, K. M. Smith, H. D. Tabba & O. T. G. Jones. 1982. The use of *N*-methyl protoporphyrin dimethyl ester to inhibit ferrochelatase in *Rhodopseudomones spheroides* and its effect in promoting biosynthesis of magnesium tetrapyrroles. Biochem. J. **208:** 479–486.
30. De Matteis, F., A. H. Gibbs & C. Harvey. 1985. Studies on the inhibition of ferrochelatase by *N*-alkylated dicarboxylic porphyrins: Steric factors involved and evidence that the inhibition is reversible. Biochem. J. **226:** 537–544.
31. Daily, H. A. & J. E. Fleming. 1983. Bovine ferrochelatase: Kinetic analysis of inhibition by *N*-methyl protoporphyrin, manganese and heme. J. Biol. Chem. **258:** 11453–11459.
32. Ortiz de Montellano, P. R., K. L. Kunze, P. C. Cole & G. S. Marks. 1981. Differential inhibition of hepatic ferrochelatase by the isomers of *N*-ethyl protoporphyrin IX. Biochem. Biophys. Res. Commun. **103:** 581–586.
33. De Matteis, F., A. H. Gibbs & A. G. Smith. 1980. Inhibition of protohaem ferro-lyase by *N*-substituted porphyrins: Structural requirements for the inhibitory effect. Biochem. J. **189:** 645–648.
34. Kunze, K. L., B. L. K. Mangold, C. Wheeler, H. S. Beilan & P. R. Ortiz de Montellano. 1983. The cytochrome P-450 active site: Regiospecificity of prosthetic heme alkylation by olefins and acetylenes. J. Biol. Chem. **258:** 4202–4206.
35. Tomaro, M. L., R. B. Frydman, B. Frydman, R. K. Pandey & K. M. Smith. 1984. The oxidation of hemins by microsomal heme oxygenase. Biochim. Biophys. Acta **791:** 342–349.
36. De Matteis, F. & G. S. Marks. 1983. The effect of *N*-methyl protoporphyrin and succinyl-acetone on the regulation of heme biosynthesis in chicken hepatocytes in culture. FEBS Lett. **159:** 127–131.

Metal-Induced Alterations of δ-Aminolevulinic Acid Dehydratase

ALFRED BERNARD[a] AND ROBERT LAUWERYS

Industrial Toxicology and
Occupational Medicine Unit
University of Louvain
B-1200 Brussels, Belgium

INTRODUCTION

δ-Aminolevulinic acid dehydratase (ALAD) or porphobilinogen synthetase (EC 4.2.1.24) is a cytoplasmic enzyme that catalyzes the second step of the heme biosynthesis pathway, that is, the condensation of two molecules of δ-aminolevulinic acid to a pyrrole, porphobilinogen. The mechanism of this unusual and complex reaction involves the intermediate formation of a Schiff's base between the keto group of one molecule of δ-aminolevulinic acid and an ϵ-amino group of a lysine residue of the enzyme.[1] The functional integrity of ALAD is essential not only for hemoglobin formation but also for the synthesis of all heme-containing enzymes such as the cytochromes or catalase.

ALAD has been found in many organisms and tissues. In animals, the highest activities are observed in liver, kidneys, and bone marrow. In all animal tissues examined so far, ALAD has about the same structure.[2–5] It has a molecular weight between 250 000 and 290 000 and is composed of eight apparently identical subunits of molecular weight of about 30 000. The optimum pH of ALAD is between 6.3 and 6.7. Lysine, histidine, and cysteine have been identified in the catalytic site of the enzyme. ALAD is a zinc-dependent enzyme. The isolated enzyme protein contains two to six zinc atoms per octamer, and it binds a maximum number of eight zinc atoms per octamer.[6,7] Zinc chelators such as EDTA or 1,10-phenanthroline reversibly inhibit the enzyme[4,6] whereas the *in vivo* or *in vitro* supplementation of zinc protects the enzyme from the inhibitory effect of lead and increases enzyme activity.[7,8] At very high concentrations, however, zinc inhibits ALAD *in vitro*.[4] Zinc does not bind to the enzyme unless the essential thiol groups of the enzyme are in the reduced state (hence the need for a reductant such as dithiothreitol to restore the enzyme activity). The function of zinc in ALAD is not yet well defined. Some findings suggest that zinc plays a catalytic role,[9,10] but, according to a recent investigation of ALAD structure by X-ray absorption[11] the role of zinc would be structural, enabling the enzyme to adopt the active conformation. Zinc would thus not bind or activate the substrate directly. The latter study[11] also showed that in the native enzyme, zinc is coordinated to three sulfur atoms and to one nitrogen or oxygen atom. A related and also unresolved question is whether ALAD behaves as an allosteric enzyme. The quaternary structure of ALAD is compatible with the hypothesis of an allosteric enzyme, but the existence of cooperativity has not been consistently observed.[4,12] There is also some suggestion that pyridoxal 5′-phosphate (the cofactor of δ-aminolevulinic acid synthetase) could bind to ALAD.[4]

[a]Chercheur Qualifié du Fonds National Belge de la Recherche Scientifique.

The aim of this review is to summarize our present knowledge on the *in vitro* and *in vivo* interactions between metals and ALAD.

LEAD

The inhibitory effect of lead on ALAD has been known for a long time and has been the object of many investigations. *In vitro* studies with ALAD purified from human[4] or guinea pig[3] erythrocytes show that lead alters both the K_m and V_{max} of the enzyme; these alterations suggest both a competitive and noncompetitive inhibition. The bovine hepatic enzyme, however, has been reported to be inhibited by lead in a noncompetitive manner.[13] This difference could be due to either tissue-specific or species differences. The inhibition of ALAD by lead, as observed in erythrocytes from humans or animals exposed to this heavy metal, also occurs *in vivo*. The *in vivo* inhibition is not an artifact resulting from hemolysis, which would, for instance, bring the intracellular enzyme in contact with the membrane-bound lead.[14]

ALAD activity in erythrocytes is currently the most sensitive indicator of human exposure to lead. The degree of inhibition closely correlates with the lead concentration in whole blood, the log of ALAD activity being linearly related to the blood lead concentration.[15] This type of relationship, initially established in occupationally exposed workers,[15–18] has been subsequently demonstrated in normal urban populations and in populations living in the vicinity of lead smelters.[19–22]

This relationship has also been demonstrated in children living in lead-polluted environments and ingesting large amounts of lead through pica.[23–25] The high sensitivity of ALAD to lead inhibition makes its assay impractical for the biological monitoring of occupationally exposed workers. For that purpose, the determination of urinary δ-aminolevulinic acid, together with that of free erythrocyte protoporphyrin and lead in blood, is usually preferred. The urinary excretion of δ-aminolevulinic acid starts to increase only after a 70–80% ALAD activity reduction, which occurs at a blood lead concentration of about 400 μg/l.[26] This value is considered close to the threshold effect level for occupationally exposed populations.

Recent studies suggest that the accumulation of δ-aminolevulinic acid due to ALAD inhibition is responsible for the increased synthesis of ALAD observed in animals[27,28] or humans[29,30] exposed to lead. As shown by these studies, the decrease of ALAD activity is accompanied by a parallel increase in the total amount of ALAD; however, this compensatory mechanism, which takes place in the precursors of red blood cells, seems to operate only at blood lead levels below 400–500 μg/l. At higher exposure levels, the amount of ALAD reaches a plateau corresponding to about twice the normal value. This ratio probably corresponds to the maximum capacity of ALAD synthesis. A practical consequence of this adaptive phenomenon is that measurement of ALAD activity alone does not accurately reflect the degree of enzyme inhibition. When a precise assessment of ALAD inhibition is desirable, it is recommended to express the ALAD activity as a percentage of the activity obtained after reactivation of the enzyme with zinc, dithiothreitol, glutathione, heat, or a combination of these treatments (for example, zinc and dithiothreitol). The ALAD activity restored under these conditions reflects the total amount of enzyme, this amount being determined by an immunochemical method.[29,30]

The inhibition of erythrocyte ALAD has no pathological significance per se, and can be considered as a nonadverse biological effect reflecting a concomitant impairment of ALAD in tissues such as bone marrow, liver, brain, or kidneys. The response of tissue ALAD to lead, however, depends on the distribution of the metal in the organism and on specific defense mechanisms of the tissue. The effect of lead on renal ALAD illustrates this point. In lead-treated rats, the extent of ALAD inhibition correlates with the accumulation of lead: the kidney is the main site of lead deposition and also the first organ where ALAD inhibition is detected.[31] This is true at least during the first weeks of treatment. Renal ALAD inhibition is indeed transient and disappears when lead exposure is prolonged for more than 6 months.[31,32] This reversal of lead-induced inhibition of renal ALAD results from the synthesis in kidney of a high-affinity cytoplasmic lead-binding protein (M_r: 11 500) that attenuates ALAD inhibition, as demonstrated by Goering and Fowler,[33,34] by chelating lead and also by serving as a zinc donor.

The inhibition of ALAD in the precursors of red blood cells is certainly involved in the development of the lead anemia, in combination with the other effects of lead on the heme synthesis pathway and on the membranes of circulating erythrocytes. The mechanism by which lead causes neuropathy has not yet been fully elucidated. It could be partly related to the accumulation of δ-aminolevulinic acid. Neuropathy occurs in diseases in which δ-aminolevulinic acid alone accumulates (ALAD deficiency). In addition, δ-aminolevulinic acid has been shown to competitively inhibit the binding of the central nervous system neurotransmitter γ-aminobutyric acid to synaptic membranes in brain tissue.[35]

ALUMINUM

Data on the effects of aluminum on ALAD are contradictory and difficult to interpret. According to Abdulla *et al.*,[36] aluminum inhibits ALAD *in vitro* but activates the enzyme *in vivo*. On the other hand, Meredith *et al.*[37] found an activation of ALAD *in vitro* and *in vivo*. In another study,[38] the latter authors showed a biphasic effect of aluminum on ALAD: an activation at low concentrations followed by an inhibition at high concentrations. This observation was confirmed by Despaux *et al.*[12]

In man, the only evidence of an interaction between aluminum and ALAD comes from studies in patients who have end-stage renal disease and are undergoing chronic hemodialysis. The aluminum accumulated by these patients is an important detrimental factor in the development of the microcytic anemia due to renal insufficiency (that is, uremic anemia).[39] The mechanism by which aluminum causes this anemia is unknown. As most of the serum aluminum is bound to transferrin, it has been hypothetized that aluminum could displace or prevent iron from binding to this protein and thus reduce the iron availability for heme synthesis.[32] Another mechanism that has been postulated is that aluminum would inhibit some steps in the heme synthesis pathway, particularly the step catalyzed by ALAD. Meredith *et al.*[40] reported a significant decrease of ALAD activity in hemodialyzed patients, but this effect was not correlated with plasma aluminum concentration. Buchet *et al.*[41] recently also found a depressed ALAD activity in hemodialysis patients, but, in contrast to the study by Meredith *et al.*,[40] the decrease of ALAD activity was significantly correlated with the aluminum concentration in plasma ($r = .55$; $p < .001$; $N = 45$). This correlation suggests a causal association between both variables.

OTHER METALS

In addition to lead and aluminum, several other metals have been found to alter the ALAD activity *in vitro*. FIGURE 1 summarizes those observations for which there is some consistency in the literature.[12,42–44] The values on the abscissa are given just to

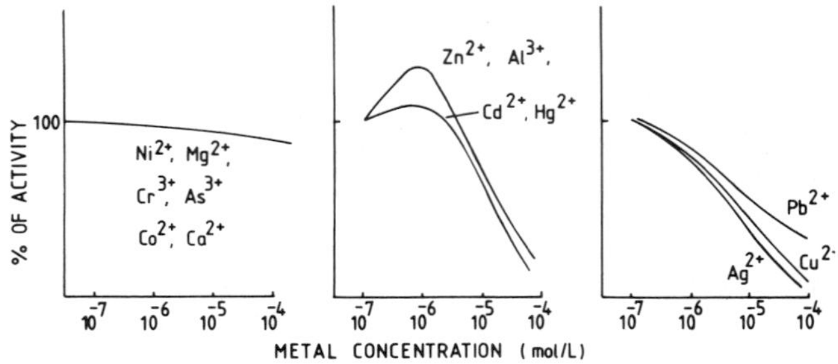

FIGURE 1. *In vitro* effects of metals on the activity of ALAD.

indicate the magnitude of metal concentrations used *in vitro*. The precise pattern of inhibition or activation varies with the source of enzyme, its degree of purity, its oxidation state, or its zinc content. Metals can be classified in three categories according to their action on ALAD: metals that exert little or no effect on the enzyme (such as Ni^{2+}, Mg^{2+}, Cr^{3+}, and As^{3+}); metals that are activators (weak: Cd^{2+} and Hg^{2+}; strong: Zn^{2+} and Al^{3+}) at low concentrations but become inhibitors at high concentrations; and metals that are potent inhibitors of ALAD (Ag^{2+}, Cu^{2+}, and Pb^{2+}). To explain this mode of action of metals on ALAD, Despaux *et al.*[12] have put forward the hypothesis of an allosteric regulation: at low concentrations, metal ions would bind to thiol groups of allosteric sites and, according to their structure, would provoke allosteric transitions to the active or inactive form of the enzyme. At high concentrations ($>10^{-5}$ mol/l), however, metals would be inhibitors, binding to all thiol groups of the enzyme, including those at the active site.

In man, there is no suggestion that metals other than lead, zinc, and, possibly, aluminum can affect ALAD activity *in vivo*. Studies performed in the sixties and seventies found no evidence of decreased ALAD activity in workers exposed to cadmium[45] or elemental mercury.[46,47] TABLE 1, which shows data collected recently in workers exposed to lead, mercury vapor, or cadmium, confirms these earlier observations. When ALAD activity is standardized for lead exposure, the only significant effects observed are with lead. A slight decrease of ALAD activity, however, has been reported in subjects highly exposed to mercury in the environment[48] or in the

TABLE 1. Effect of Occupational Exposure to Cadmium, Elemental Mercury, or Lead on Erythrocyte ALAD Activity

Group	N	Metal Concentration in Blood[a] (μg/l) Cadmium	Mercury	Lead	ALAD[a] Activity (U/ml RBC)	ALAD[a,b] Inhibition (%)
Control	100	1.3 ± 1	2.4 ± 1.4	143 ± 67	24.5 ± 9.3	64 ± 7
Cadmium	42	13 ± 13[c]	3 ± 2	206 ± 56[c]	24.6 ± 7.7[d]	63 ± 1[d]
Elemental mercury	69	1 ± 0.7	29 ± 29[c]	100 ± 41[c]	25.9 ± 7.2[d]	63 ± 10[d]
Lead	38	—	—	485 ± 91[c]	7.6 ± 10.9[c]	91 ± 3[c]

[a]Each value is a mean ± SD.
[b]ALAD inhibition = [1 − (nonrestored activity/restored activity)] × 100.
[c]$p < .05$ (one-way variance analysis followed by Dunett's test).
[d]Standardized to a lead concentration of 143 μg/l on the basis of the regression lines calculated in control subjects.

industry.[49] The lack of effect of cadmium and mercury on ALAD under current conditions of human exposure can be explained by both the low concentrations of these two metals in the human organism compared to lead levels and the ability of storage tissues (mainly kidney and liver) to inactivate them through binding to metallothionein.

SUMMARY

ALAD is a cytoplasmic enzyme that catalyzes the second step of the heme biosynthesis pathway, that is, the condensation of two molecules of δ-aminolevulinic acid into porphobilinogen. ALAD is a zinc-dependent enzyme; thiol groups are essential for its activity; and *in vitro* experiments show that ALAD can be activated or inhibited by several metal ions including Al^{3+}, Pb^{2+}, Cd^{2+}, Hg^{2+}, Ag^{2+}, and Cu^{2+}. To explain these effects, it has been postulated that metals bind to thiol groups of allosteric sites and, according to their structure, provoke allosteric transitions to the active or inactive form of the enzyme. Under current environmental and occupational exposure levels, lead is practically the only metal that can affect ALAD activity. Erythrocyte ALAD is the most sensitive indicator of lead exposure: effects of exposure are detectable even when blood lead levels are within the "normal" range. Zinc protects ALAD *in vitro* and *in vivo* from the inhibitory effect of lead. There is also some suggestion that aluminum could be responsible for the decreased erythrocyte ALAD activity observed in patients on chronic hemodialysis.

REFERENCES

1. Shemin, D. 1976. δ-Aminolevulinic acid dehydratase: Structure, function, and mechanism. Philos. Trans. R. Soc. London Ser. B. **273:** 109–115.
2. Coleman, D. L. 1966. Purification and properties of δ-aminolevulinate dehydratase from the tissues of two strains of mice. J. Biol. Chem. **241:** 5511–5517.
3. Weinberg, J. P. & P. E. Voytek. 1974. Liver and red cell porphobilinogen synthase in the adult and fetal guinea pig. Biochem. Biophys. Acta **364:** 304–319.
4. Anderson, P. M. & R. J. Desnick. 1979. Purification and properties of δ-aminolevulinate dehydratase from human erythrocytes. J. Biol. Chem. **254**(15): 6924–6930.
5. Chang, C. S., S. Sassa & D. Doyle. 1984. An immunological study of δ-aminolevulinic acid dehydratase specificity consistent with the phylogeny of species. Biochim. Biophys. Acta **797:** 297–301.
6. Bevan, D. R., P. Bodlaender & D. Shemin. 1980. Mechanism of porphobilinogen synthetase: Requirement of Zn^{2+} for enzyme activity. J. Biol. Chem. **255:** 2030–2035.
7. Tsykamoto, I., T. Yoshinaga & S. Sano. 1979. The role of zinc with special references to the essential thiol groups in δ-aminolevulinic acid dehydratase of bovine. Biochim. Biophys. Acta **570:** 167–178.
8. Abdulla, M. & S. Svensson. 1979. Effect of oral zinc intake on δ-aminolevulinic acid dehydratase in red blood cells. Scand. J. Clin. Invest. **39:** 31–36.
9. Sommer, R. & D. Beyersman. 1984. Zinc and cadmium in 5-aminolevulinic acid dehydratase: Equilibrium, kinetic and ^{113}Cd NMR studies. J. Inorg. Biochem. **20:** 131–145.
10. Beyersmann, D. & M. Cox 1984. Affinity labeling of 5-aminolevulinic acid dehydratase with 2-bromo-3-(5-imidazolyl)propionic acid. Biochim. Biophys. Acta **788:** 162–166.
11. Hasnain, S. S., C. Wardell, D. Garner, M. Schlösser & D. Beyerman. Extended X-ray absorption fine structure investigations of zinc in 5-aminolevulinate dehydratase. Biochem. J. **230:** 625–633.
12. Despaux, N., C. Bohuon, E. Comoy & C. Boudène. 1977. Postulated mode of action of metals on purified human ALAD dehydratase (EC 4.2.1.24). Biomedicine **27:** 358–361.

13. WILSON, E. L., P. E. BURGER & E. B. DOWALLE. 1972. Beef liver δ-aminolevulinic acid dehydratase: Purification and properties. Eur. J. Biochem. **29:** 563–571.
14. ROELS, H., J. P. BUCHET & R. LAUWERYS. 1974. Inhibition of human erythrocyte δ-aminolevulinate dehydratase by lead: *In vitro* artifact or real phenomenon *in vivo*. Int. Arch. Arbeitsmed. **33:** 277–284.
15. BONSIGNORE, D. 1966. L'attivata ALA deidratasica eritrocitara quale test diagnostico del saturnismo professionale. Med. Lav. **57:** 647–654.
16. DE BRUIN, A. 1968. Effect of lead exposure on the level of δ-aminolevulinic dehydratase activity. Med. Lav. **59:** 411–418.
17. HERNBERG, S., J. NIKKANEN, G. MELLIN & H. LILIUS. 1970. δ-Aminolevulinic acid dehydratase as a measure of lead exposure. Arch. Environ. Health **21:** 140–145.
18. HAEGER-ARONSEN, B., M. ABDULLA & B. I. FRISTEDT. 1971. Effect of lead on δ-aminolevulinic acid dehydratase activity in red blood cells. Arch. Environ. Health **23:** 440–445.
19. SARTOR, F. & D. RONDIA. 1981. δ-Aminolevulinate dehydratase and blood lead levels in urban male adults. Int. Arch. Occup. Environ. Health **49:** 45–51.
20. HERNBERG, S. & J. NIKKANEN. 1970. Enzyme inhibition by lead under normal urban conditions. Lancet **i:** 63–64.
21. SECCHI, G. C., L. ALESSIO, G. GAMBIOCHI & F. ANDREOLETTI. 1973. ALA dehydratase activity of erythrocytes and blood lead levels in "critical" population groups. *In* Environmental Health Aspects of Lead. D. Barth, A. Berlin, R. Engel, P. Recht & J. Smeet, Eds.: 595–602. EEC-EPA. Luxembourg.
22. NORDMAN, C. H., S. HERNBERG, J. NIKKANEN & A. RYKANEN. 1973. Blood lead levels and erythrocyte δ-aminolevulinic acid dehydratase activity in people living around a secondary lead smelter. Work Environ. Health **10:** 19–25.
23. MILLAR, J. A., R. L. CUMMING, V. BATTISTINI, F. CABSWELL & A. GOLDBERG. 1970. Lead and δ-aminolevulinic acid dehydratase levels in mentally retarded children and in lead-poisoned suckling rats. Lancet **ii:** 695–698.
24. WEISBERG, J. B., F. LIPSCHUTZ & F. A. OSKO. 1971. δ-Aminolevulinic acid dehydratase activity in circulating blood cells: A sensitive laboratory test for the detection of childhood lead poisoning. N. Engl. J. Med. **284:** 565–569.
25. ROELS, H., J. P. BUCHET, R. LAUWERYS, G. HUBERMONT, P. BRUAUX, F. CLAEYS-THOREAU, A. LAFONTAINE & J. VAN OVERSCHELDE. 1976. Impact of air pollution by lead on the heme biosynthesis pathway in school-age children. Arch. Environ. Health **31:** 310–316.
26. LAUWERYS, R., J. P. BUCHET, H. ROELS & D. MATERNE. 1974. Relationship between urinary δ-aminolevulinic acid excretion and the inhibition of red cell δ-aminolevulinate dehydratase by lead. Clin. Toxicol. **7**(4): 383–388.
27. MAES, J. & G. B. GERBER. 1978. Increased ALA dehydratase activity and spleen weight in lead-intoxicated rats: A consequence of increased blood cell destruction. Experientia **34:** 381–382.
28. FUJITA, H., Y. ORII & S. SANO 1981. Evidence of increased synthesis of δ-aminolevulinic acid dehydratase in experimental lead-poisoned rats. Biochim. Biophys. Acta **678:** 39–50.
29. FUJITA, H., K. SATO & S. SANO. 1982. Increase in the amount of erythrocyte δ-aminolevulinic acid dehydratase in workers with moderate lead exposure. Int. Arch. Occup. Environ. Health **50:** 287–297.
30. BOUDÈNE, C., N. DESPAUX-PAGES, E. COMOY & C. BOHUON. 1984. Immunological and enzymatic studies of erythrocytic δ-aminolevulinate dehydratase. Int. Arch. Occup. Environ. Health **55:** 87–96.
31. BUCHET, J. P., H. ROELS, G. HUBERMONT & R. LAUWERYS. 1976. Effect of lead on some parameters of the heme biosynthetic pathway in rat tissues *in vivo*. Toxicology **6:** 21–34.
32. FOWLER, B. A., C. A. KIMMEL, J. S. WOODS, E. E. MCCONNELL & L. D. GRANT. 1980. Chronic low-level lead toxicity in the rat. III. An integrated assessment of long-term toxicity with special reference to the kidney. Toxicol. Appl. Pharmacol. **56:** 59–77.
33. GOERING, P. L. & B. A. FOWLER. 1984. Regulation of lead inhibition of δ-aminolevulinic acid dehydratase by a low molecular weight, high-affinity renal lead-binding protein. J. Pharmacol. Exp. Ther. **231:** 66–71.

34. GOERING, P. L. & B. A. FOWLER. 1985. Mechanisms of renal lead-binding protein reversal of δ-aminolevulinic acid dehydratase inhibition by lead. J. Pharmacol. Exp. Ther. **234**(2): 365–371.
35. BRESMAN, M. J. W., R. C. CANTRILL & S. KRAMER. 1980. Effect of δ-aminolevulinic acid on GABA receptor binding in synaptic plasma membranes. Int. J. Biochem. **12:** 833–835.
36. ABDULLA, M., S. SVENSSON & B. HAEGER-ARONSON. 1979. Antagonistic effects of zinc and aluminum on lead inhibition of δ-aminolevulinic acid dehydratase. Arch. Environ. Health **34:** 464–469.
37. MEREDITH, P. A., M. R. MOORE & A. GOLDBERG. 1977. Effects of aluminum, lead and zinc on δ-aminolevulinic acid dehydratase. Enzyme **22:** 22–27.
38. MEREDITH, P. A., M. R. MOORE & A. GOLDBERG 1974. The effects of aluminum, lead and zinc on δ-aminolevulinic acid dehydratase. Biochim. Soc. Trans. **2:** 1243–1244.
39. KAISER, L. & V. A. SCHWARTZ. 1985. Aluminum-induced anemia. Am. J. Kidney Dis. **6**(5): 348–352.
40. MEREDITH, P. A., H. L. ELLIOTT & B. C. CAMPBELL. 1979. Changes in serum aluminum, blood zinc, blood lead and erythrocyte δ-aminolevulinic acid dehydratase activity during hemolysis. Toxicol. Lett. **4:** 419–424.
41. BUCHET, J. P., R. LAUWERYS, A. HASSOUN, M. DRATWA, R. WENS, F. COLLART & C. TIELEMANS. 1987. Effect of aluminum on porphyrin metabolism in hemodialyzed patients. Nephron (in press).
42. ALLIAN, P. & A. PREMEL-CABIC. 1977. Influence de quelques métaux et de quelques anti-inflammatoires non-stéroïdiens sur l'activité de l'ALA déhydratase érythrocytaire humaine *in vitro*. Pathol. Biol. **25**(8): 535–540.
43. TREVIRAN, A., G. P. GORI, A. ZANGIROLAMI, C. BENEVENTO, A. ROSA & P. CHIESURA. 1980. Site of action of metals on the δ-aminolevulinic acid dehydratase of human erythrocytes. Enzyme **25:** 33–36.
44. THOMPSON, J., D. D. JONES & W. H. BEASLEY. 1977. The effect of metal ions on the activity of δ-aminolevulinic acid dehydratase. Br. J. Ind. Med. **34:** 32–36.
45. LAUWERYS, R. R., J. P. BUCHET & H. ROELS. 1973. Comparative study of the effect of inorganic lead and cadmium on blood δ-aminolevulinate dehydratase in man. Br. J. Ind. Med. **30:** 359–364.
46. LAUWERYS, R. & J. P. BUCHET. 1973. Occupational exposure to mercury vapors and biological action. Arch. Environ. Health **27:** 65–68.
47. GOLDWATER, L. J. & M. M. JOSELOW. 1967. Absorption and excretion of mercury in man. XIII. Effects of mercury exposure on urinary excretion of coproporphyrin and δ-aminolevulinic acid. Arch. Environ. Health **15:** 327–331.
48. SCHÜTZ, A. & S. SKERFVING. 1975. Blood cell δ-aminolevulinic acid dehydratase activity in humans exposed to methylmercury. Scand J. Work Environ. **1:** 54–59.
49. WADA, O., K. TOYOKOWA, T. SUZUKI, Y. YANO & K. NAKAO. 1969. Response to a low concentration of mercury vapor. Arch. Environ. Health **19:** 485–488.

Effect of Lead on Uroporphyrin and Heme Content in the Bone Marrow of Rats Exposed to Nitrous Oxide[a]

WALTER N. PIPER[b] AND W. RUSS CHRISTENSON

Department of Pharmacology
College of Medicine
University of Nebraska Medical Center
Omaha, Nebraska

INTRODUCTION

The synthesis of the porphyrinogen ring begins with four molecules of the pyrrole porphobilinogen (PBG) combining to form uroporphyrinogen (URO). This reaction, when mediated by uroporphyrinogen I synthase and uroporphyrinogen III cosynthase acting in a concerted mechanism, forms uroporphyrinogen III and ultimately leads to the biosynthesis of heme. Uroporphyrinogen I, an inappropriate substrate for heme biosynthesis, is formed when cosynthase is absent. Recently, evidence has been presented that cosynthase has an attached polyglutamate folate derivative that may act as a coenzyme to ensure that uroporphyrinogen III is formed.[1]

Because uroporphyrinogen synthase has low activity in liver and lower than normal activity in various organs in individuals with certain porphyric diseases, it is thought that the synthesis of uroporphyrinogen may be potentially rate-limiting for the biosynthesis of heme. Thus, it is anticipated that an understanding of the regulation of uroporphyrinogen formation may be important for understanding the effects of exposure to environmental toxins such as heavy metals.

Several heavy metals of potential toxicological significance (cadmium, copper, iron, lead, manganese, mercury, and zinc) have been reported to be inhibitors of uroporphyrinogen I synthase or cosynthase purified from a variety of sources such as avian and human erythrocytes, spinach, *Rhodopseudomonas spheroides,* cow, and rat liver.[2–7] Furthermore, it has been reported that exposure to lead results in excretion of abnormally high amounts of PBG in the urine of rabbits and humans,[8,9] increased bone marrow levels of uroporphyrin I in rabbits and rats,[10,11] and inhibition of human and rat erythrocytic uroporphyrinogen I synthase activity.[12,13] Lead has been shown to inhibit erythrocytic uroporphyrinogen I synthase activity but to have no effect on activity in rat hepatic cytosol.[13] Subsequently, a factor was isolated from hepatic cytosol that protects against lead inhibition of uroporphyrinogen I synthase activity, which has been identified as polyglutamate folic acid.[14]

Such findings suggest a possible susceptibility of uroporphyrinogen formation in

[a]This research was supported by Grant ES-02424 from the National Institutes of Health.
[b]Present address: Toxicology Division, Department of Environmental and Industrial Health, SPH-1, 109 South Observatory Street, University of Michigan, Ann Arbor, Michigan 48109-2029.

erythropoietic tissue to the effects of lead that may be regulated by folic acid. Evidence for this concept is reported herein.

MATERIALS AND METHODS

Animals

Male Sprague–Dawley rats (180–200 g) were obtained from Sasco (Omaha, NE). Rats were housed in hanging cages in groups of at least three and were exposed to a 12-hr light–dark cycle (0600 lights on; 1800 lights off). Food (Purina Laboratory Rodent Chow, Ralston Purina, St. Louis, MO) and water were provided *ad libitum*. Lead acetate (15 mg/kg) was administered i.p. in 0.9% NaCl at days 0 and 2. The rats were sacrificed at day 7. Controls received equal volumes of 0.9% NaCl (4 ml/kg). For experiments involving nitrous oxide, rats were placed in a 30-l metabolic chamber through which a mixture of nitrous oxide and oxygen (1:1) flowed at 8 l/min for 4 hr/day (0900–1300) beginning at day 0 of the 7-day experimental period. Controls were housed in a similar chamber and exposed to room air. Rats were sacrificed 24 hr after the last exposure. Bone marrow (pooled from the femurs of at least three rats) was flushed with 0.9% NaCl at 5° C and centrifuged at 3000 × g for 10 min at 5° C. Supernatants were discarded and cells suspended twice with 0.9% NaCl and centrifuged at 1000 × g for 10 min at 5° C after each suspension. Tissue samples of 1.0–1.5 g wet weight were used for subsequent analyses.

Porphyrin Extraction and Analysis

Uroporphyrin was extracted from bone marrow samples and analyzed by HPLC by the method of Reddy *et al.*[15] Analyses were performed on a Waters Associates liquid chromatograph (Model 6000A solvent delivery system and a Rheodyne 7125 injector) using a Schoeffel FS-970 fluorometer and a C18-bonded column (spherical 5 mm, 4.5 mm inner diameter × 25 cm) from IBM Instruments (Danbury, CT). A C18-bonded guard column (spherical 10 mm, 4.6 mm inner diameter × 3 cm) from Brownlee Labs (Santa Clara, CA) was placed between the pump and the analytical column. Recovery of porphyrins from spiked tissue samples was greater than 90.0 ± 1.0%.

Polyglutamate Folate Extraction and Analysis

Tissue samples were homogenized with three volumes of 6.7% (w/v) trichloracetic acid and centrifuged at 20 000 × g for 20 min. Supernatants were extracted with 24 ml of diethylether three times for removal of trichloroacetic acid. The method of Foo *et al.*[16] was used to cleave the C-9–N-10 bond to yield a pteridine residue and the *p*-aminobenzoylpolyglutamate derivatives; these derivatives were diazotized and converted to the azo dye of *N*-(1-naphthyl)ethylene diamine by the method of Brody *et al.*[17] The azo dyes were purified from the cleaved pteridine residues by chromatography on Bio Gel P2 columns and were subjected to HPLC analysis by the method of Shane.[18] All HPLC analyses were performed with a Perkin Elmer Series 3 liquid chromatograph, using a Hitachi 204-A fluorescence detector with excitation and emission wavelengths of 310 and 360 nm, respectively, and a Whatman Partisil 10 SAX analytical column (4.5 mm × 25 cm).

TABLE 1. The Effect of Lead on Uroporphyrin Isomer Composition in Rat Bone Marrow

Dosage (mg/kg)	Uroporphyrin Content (pmol/g)	
	I	III
0	74 ± 9	9.0 ± 1.4
5	73 ± 10	9.2 ± 1.5
15	94 ± 11	9.3 ± 1.7
30	118 ± 9[a]	9.0 ± 1.8

NOTE: Lead acetate was administered i.p. at days 0 and 2. Rats were sacrificed at day 7. Bone marrow was assayed for uroporphyrin as described in the Methods section. Each value represents the mean ± SEM for six determinations.

[a]Significantly different from the corresponding control value ($p < .05$).

Microsomal Heme

Levels of microsomal heme were determined in bone marrow microsomes from the difference spectrum of the oxidized/reduced pyridine hemochromagen between 541 and 557 nm using an extinction coefficient of 20.7 $mM^{-1} \cdot cm^{-1}$. For elimination of any spectral interference from hemoglobin, microsomal suspensions were gassed with carbon monoxide as described by Estabrook *et al.*[19]

Microsomal Heme Oxygenase

The activity of microsomal bone marrow heme oxygenase was determined by the method of Ibrahim and Levere[20] using a partially purified hepatic biliverdin reductase preparation as described by Tenhunen *et al.*[21]

Protein

The concentration of protein was determined by the method of Bradford[22] with bovine serum albumin type V as the standard.

Statistics

Data were analyzed by Student's *t* test for determination of the significance of differences between means.

TABLE 2. The Effect of Nitrous Oxide on Uroporphyrin Isomer Composition in Rat Bone Marrow

Treatment	Uroporphyrin Content (pmol/g)	
	I	III
Control	74 ± 8	9.4 ± 1.6
Nitrous oxide	102 ± 10[a]	8.7 ± 1.3

NOTE: A nitrous oxide:oxygen mixture (1:1) was administered 4 hr/day for 7 days, and control animals were exposed to air. Bone marrow was assayed for uroporphyrin as described in the Methods section. Each value represents the mean ± SEM for three determinations.

[a]Significantly different from the corresponding control value ($p < .05$).

TABLE 3. The Effect of Nitrous Oxide on Polyglutamated Folate Content in Rat Bone Marrow

Number of Residues in the Glutamate Chain	Polyglutamated Folate Content (nmol/g)	
	Control	Nitrous Oxide
2	ND	ND
3	ND	ND
4	ND	ND
5	0.26 ± 0.02	0.12 ± 0.02[a]
6	0.20 ± 0.01	0.12 ± 0.05[a]
7	0.08 ± 0.01	0.05 ± 0.01[a]
Total	0.54 ± 0.04	0.29 ± 0.03[a]

NOTE: A nitrous oxide:oxygen mixture (1:1) was administered 4 hr/day for 7 days, and control animals were exposed to air. Bone marrow was assayed for polyglutamate folate as described in the Methods section. Each value represents the mean ± SEM for three determinations. ND: not detectable.

[a]Significantly different from the corresponding control value.

RESULTS

The Effect of Lead on Uroporphyrin Isomer Composition in Rat Bone Marrow

Previous findings from this laboratory have shown that uroporphyrin I levels are increased in rat bone marrow after lead exposure at a dosage having no effect on liver, kidney, or erythrocyte uroporphyrin content.[11] A dose-response study was conducted in order to determine the threshold for uroporphyrin I increases in rat bone marrow (TABLE 1). It was anticipated that such information would be useful for determining whether parameters associated with the formation of uroporphyrinogen and heme biosynthesis in bone marrow might be potentiated in folate-deficient animals exposed to lead.

The Effect of Nitrous Oxide on Uroporphyrin Isomer Composition in Rat Bone Marrow

The findings that polyglutamate folate is associated with purified rat hepatic cosynthase[1] and that polyglutamate folates protect against lead inhibition of uropor-

TABLE 4. The Effect of Lead and Nitrous Oxide on Microsomal Heme Content in Rat Bone Marrow

Treatment	Microsomal Heme Content (pmol/mg protein)
Control	86 ± 10
Lead	92 ± 18
Nitrous oxide	61 ± 10[a]
Nitrous oxide and lead	37 ± 12[b]

Lead acetate (15 mg/kg) was administered i.p. at days 0 and 2. Rats were sacrificed at day 7. A nitrous oxide:oxygen mixture (1:1) was administered 4 hr/day for 7 days. Bone marrow was assayed for heme as described in the Methods section. Each value represents the mean ± SEM for three determinations.

[a]Significantly different from the control value ($p < .05$).

[b]Significantly different from the nitrous oxide value ($p < .05$).

phyrinogen synthase activity[14] suggested that reduction of tissue folate might alter the isomeric composition of uroporphyrinogen in bone marrow. Thus, rats were exposed to nitrous oxide at a dosage regimen (4 hr/day) known to rapidly inhibit methionine synthetase activity and to decrease tissue levels of tetrahydrofolate.[23,24] As shown in TABLE 2, exposure to nitrous oxide for 7 days resulted in a 38% increase of uroporphyrin content in the bone marrow. Levels of uroporphyrin III in bone marrow were not altered by exposure to nitrous oxide.

The Effect of Nitrous Oxide on Polyglutamated Folate Content in Rat Bone Marrow

Total polyglutamate folate content in bone marrow was decreased 46% after 7 days of exposure to nitrous oxide (TABLE 3). This depression of total polyglutamate folate content was associated with significant decreases in the levels of folates with glutamate chains five to seven residues long (TABLE 3).

TABLE 5. The Effect of Lead and Nitrous Oxide on Microsomal Heme Oxygenase Activity in Rat Bone Marrow

Treatment	Microsomal Heme Oxygenase Activity (pmol bilirubin/mg protein/hr)
Control	509 ± 24
Lead	482 ± 19
Nitrous oxide	532 ± 34
Nitrous oxide and lead	512 ± 29

NOTE: Lead acetate (15 mg/kg) was administered i.p. at days 0 and 2. Rats were sacrificed at day 7. A nitrous oxide:oxygen mixture (1:1) was administered for 4 hr/day for 7 days. Bone marrow was assayed for heme oxygenase as described in the Methods section. Each value represents the mean ± SEM for three determinations.

The Effect of Lead and Nitrous Oxide on Microsomal Heme Content in Rat Bone Marrow

Increased levels of uroporphyrin I in bone marrow of rats exposed to nitrous oxide indicates that synthesis of uroporphyrinogen might be impaired, leading to depressed levels of heme. As shown in TABLE 4, levels of microsomal heme in bone marrow were depressed after exposure to nitrous oxide. Lead acetate (15 mg/kg) had no effect on levels of microsomal heme in bone marrow. The previous report of a polyglutamate folate attached to rat hepatic cosynthase[1] and increased levels of uroporphyrin I in bone marrow of rats exposed to lead[11] suggested that the formation of uroporphyrinogen and the subsequent synthesis of heme might be affected by the folate status of this tissue. Thus, rats exposed to nitrous oxide (4 hr/day for 7 days) were also administered lead acetate (15 mg/kg at days 0 and 2; sacrifice at day 7). Lead had no effect, and nitrous oxide caused a 29% decrease in microsomal heme (TABLE 4). The administration of lead to rats exposed to nitrous oxide resulted in a 57% decrease in the level of microsomal heme in bone marrow. The microsomal heme content observed in rats exposed to both nitrous oxide and lead was 39% lower than that observed in rats exposed to nitrous oxide.

The Effect of Lead and Nitrous Oxide on Microsomal Heme Oxygenase Activity in Rat Bone Marrow

The lower microsomal heme content in the bone marrow of rats exposed to both lead and nitrous oxide could be explained by an increased rate of breakdown of heme rather than an impaired biosynthesis. To test this possibility, microsomal heme oxygenase was measured. As shown in TABLE 5, the activity of heme oxygenase in rat bone marrow was not altered by lead, nitrous oxide, or their combination.

DISCUSSION

The results herein indicate that decreased polyglutamate folate content in bone marrow produced by exposure to nitrous oxide is associated with elevated tissue content of uroporphyrin I. These findings are consistent with the report of a polyglutamate folate bound to purified rat hepatic cosynthase, and the ability of tetrahydrofolates to form uroporphyrinogen III in a cosynthase-free preparation of uroporphyrinogen I synthase.[1] It is possible that the effects of exposure to nitrous oxide—depression of tetrahydrofolate and associated impairment of folylpolyglutamate formation—decrease the amount of reduced polyglutamate folate available to cosynthase in bone marrow.

Administration of lead at a dosage regimen (15 mg/kg) that had no effect on bone marrow uroporphyrin I content to nitrous oxide-exposed rats caused a greater depression of microsomal heme than that produced by nitrous oxide alone. These findings suggest that the ability of lead to impair the formation of uroporphyrinogen III and the subsequent biosynthesis of heme is enhanced in the folate-deficient state. The lack of any change in microsomal heme oxygenase activity in the bone marrow of rats exposed to lead, nitrous oxide, or both indicates that an altered rate of heme catabolism is not responsible for decreased heme content.

REFERENCES

1. KOHASHI, M., R. P. CLEMENT, J. TSE & W. N. PIPER. 1984. Biochem. J. **220:** 755–765.
2. BOGORAD, L. 1958. J. Biol. Chem. **233:** 501–509.
3. LOCKWOOD, W. H. & A. BENSON. 1960. Biochem. J. **75:** 372–381.
4. JORDAN, P. M. & D. SHEMIN. 1973. J. Biol. Chem. **248:** 1019–1024.
5. SANCOVICH, H. A., A. M. FERRAMOLA, A. M. DEL C. BATLLE, A. KIVILEVICH & M. GRINSTEIN. 1976. Acta Physiol. Lat. Am. **26:** 379–386.
6. ANDERSON, P. M. & R. J. DESNICK. 1980. J. Biol. Chem. **255:** 1993–1999.
7. FUMAGALLI, S. A., M. L. KOTLER, M. V. ROSSETTI & A. M. DEL C. BATLLE. 1985. Int. J. Biochem. **17:** 485–494.
8. DEKRETSER, A. J. & H. A. WALDRON. 1963. Br. J. Ind. Med. **20:** 35–40.
9. GIBSON, S. L. M. & A. GOLDBERG. 1970. Clin. Sci. **38:** 63–72.
10. SCHMID, R., B. HANSON & S. SCHWARTZ. 1952. Proc. Soc. Exp. Biol. Med. **79:** 459–462.
11. CHRISTENSON, W. R., V. R. REDDY & W. N. PIPER. 1985. Biochem. Pharmacol. **34:** 4345–4347.
12. PETERSON, L. R., P. HAMERNYIK, T. D. BIRD & R. F. LABBE. 1976. Clin. Chem. **22:** 1835–1840.
13. PIPER, W. N. & T. R. TEPHLY. 1974. Life Sci. **14:** 873–876.
14. PIPER, W. N. & R. B. L. VAN LIER. 1977. Mol. Pharmacol. **13:** 1126–1135.
15. REDDY, R. V., W. R. CHRISTENSON & W. N. PIPER. 1987. J. Pharmacol. Methods **17:** 51–57.

16. Foo, S. K., D. J. Cichowicz & B. Shane. 1980. Anal. Biochem. **107:** 109–115.
17. Brody, J., B. Shane & E. L. R. Stokstad. 1979. Anal. Biochem. **92:** 501–509.
18. Shane, B. 1982. Am. J. Clin. Nutr. **35:** 599–608.
19. Estabrook, R. W., J. Peterson, J. Baron & A. Hildebrandt. 1972. *In* Methods of Pharmacology. C. F. Chignell, Ed. Vol. **2:** 303–350. Elsevier. New York, NY.
20. Ibrahim, N. G. & R. D. Levere. 1980. Life Sci. **26:** 525–531.
21. Tenhunen, R., H. S. Marver & R. Schmid. 1969. J. Biol. Chem. **244:** 6388–6394.
22. Bradford, M. 1976. Anal. Biochem. **72:** 248–254.
23. Black, K. A. & T. R. Tephly. 1983. Mol. Pharmacol. **23:** 724–730.
24. Lumb, M., J. Perry, R. Deacon & I. Chanarin. 1981. Am. J. Clin. Nutr. **34:** 2412–2417.

Metal Alteration of Uroporphyrinogen Decarboxylase and Coproporphyrinogen Oxidase[a]

JAMES S. WOODS

Battelle Seattle Research Center
Seattle, Washington 98105

BRUCE A. FOWLER

National Institute of Environmental Health Sciences
Research Triangle Park, North Carolina 27709

INTRODUCTION

Uroporphyrinogen decarboxylase and coproporphyrinogen oxidase are the principal porphyrin oxidative enzymes of the heme biosynthetic pathway. Preparations of both enzymes from various sources have been found to contain SH groups and to be inhibited or inactivated *in vitro* by various trace metals as well as by other sulfhydryl-directed binding agents. The elicitation of porphyrinurias with specific porphyrin patterns during prolonged exposure to subtoxic doses of certain metals suggests metal-specific inhibition of uroporphyrinogen decarboxylase or coproporphyrinogen oxidase *in vivo*. This paper describes results of experimental studies that were undertaken to investigate the nature of metal interactions with these enzymes in various sources and to define the effects metals have on the development of porphyrinurias that characterize exposure to them.

UROPORPHYRINOGEN DECARBOXYLASE

Uroporphyrinogen decarboxylase (UD) catalyzes the stepwise decarboxylation of 8- to 4-carboxyl porphyrinogen (CP) through the 7-, 6-, and 5-carboxylated intermediates. The decarboxylation of the four acetic acid sidechains of 8-CP to methyl groups is considered to be mediated by a single enzyme and, at least for the physiologically relevant III isomers, to proceed through the clockwise sequence beginning with the D ring of 8-CP as shown in FIGURE 1. There is no evidence that a cofactor of any sort is required for this reaction, although the enzyme must be assayed anaerobically in order to prevent oxidation of porphyrinogen substrates and intermediates. The enzyme from all tissues thus far examined decarboxylates both the I and III isomers of 8-CP but not the II or IV isomer forms. Typically, the physiologically relevant III isomer of 8-CP is decarboxylated more rapidly than the I form, possibly because of a higher binding affinity of the III form for the enzyme's active site.

Studies of the purified decarboxylase from bovine liver have indicated that the

[a]This work was supported in part by Grant R01 ES03628 from the U.S. Public Health Service.

decarboxylation of 8-CP to 7-CP III proceeds rapidly, whereas the rate-limiting step in the conversion of 8-CP to 4-CP III is the decarboxylation of 7-CP to 6-CP III.[1] Studies with the purified enzyme from human red cells[2,3] have demonstrated that porphyrinogens of both I and III isomeric series with the same numbers of carboxylic groups are decarboxylated at the same active center. The sequential decarboxylation of 8-CP to 4-CP, however, occurs at four different active centers of the enzyme. Quantitative immunochemical studies conducted by Elder and colleagues[4,5] have shown that the erythrocyte and liver enzymes are very similar, although not identical, differing only with respect to possible slight variations in secondary enzyme structure. Tissue-specific isozymes of UD, however, do not seem likely.

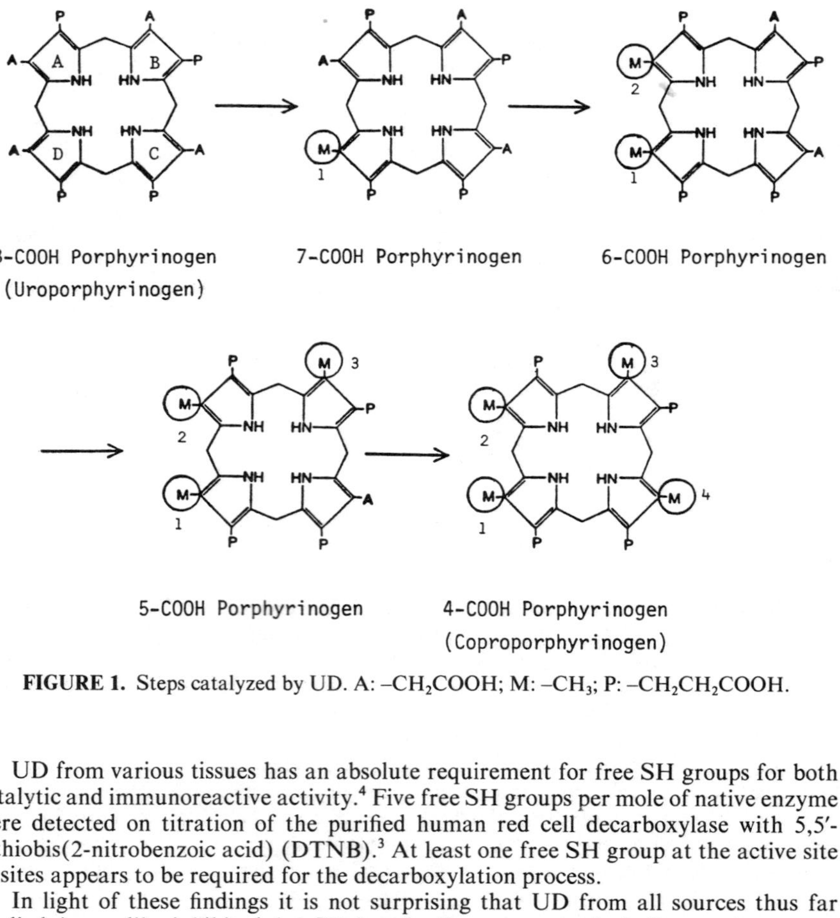

FIGURE 1. Steps catalyzed by UD. A: $-CH_2COOH$; M: $-CH_3$; P: $-CH_2CH_2COOH$.

UD from various tissues has an absolute requirement for free SH groups for both catalytic and immunoreactive activity.[4] Five free SH groups per mole of native enzyme were detected on titration of the purified human red cell decarboxylase with 5,5′-dithiobis(2-nitrobenzoic acid) (DTNB).[3] At least one free SH group at the active site or sites appears to be required for the decarboxylation process.

In light of these findings it is not surprising that UD from all sources thus far studied is readily inhibited by SH-inactivating reagents, including various trace metals.

TABLE 1 describes the effects of various SH-binding agents on the activity of UD derived from different sources.[1,3,4] It can be seen that all agents inhibit enzyme activity in a dose-dependent manner, suggesting direct binding with SH groups of the enzyme.

TABLE 1. Effects of SH-Binding Agents on UD Activity

Reagent	Concentration (mM)	Activity (% Control)	Enzyme Source	Reference
p-Hydroxymercuribenzoate	0.1	8	Human erythrocyte	3
	0.2	0	Human erythrocyte	3
p-Chloromercuribenzoate	0.01	0	Chicken erythrocyte	6
N-Ethylmaleimide	2.0	25	Human erythrocyte	3
	0.01	0	Chicken erythrocyte	6
Iodoacetamide	5.0	5	Bovine liver	1
	5.0	86	Human erythrocyte	3
	10.0	47	Human erythrocyte	3

These effects are prevented or reversed by glutathione (GSH) or by other SH reagents.

Comparably, some but not all metals inhibit or inactivate UD in various tissue preparations. As shown in TABLE 2, the most potent of the metals evaluated is Hg^{2+}, which effectively eliminates enzyme activity *in vitro* in relatively low concentrations. Catalytic activity was completely abolished when the human red cell enzyme was incubated with 0.5 mM $HgCl_2$. Interestingly, UD from rat kidney appears to be much more readily inhibited by Hg^{2+} than does the hepatic enzyme by a factor of at least 10 when measured in crude enzyme preparations.[7,8] Moreover, prolonged treatment of animals with subtoxic levels of Hg compounds elicits a uroporphyrinuria indicative of inhibition of UD by Hg *in vivo*.[9]

In contrast to Hg^{2+}, Pb does not effectively inhibit UD *in vivo* or *in vitro* except at fairly high concentrations. No inhibition of UD from human erythrocytes or from rat liver or kidney was observed by Pb at concentrations as high as 1 mM. This observation is consistent with the property of Pb to readily inhibit only those enzymes that possess vicinal dithiols at the catalytic site (for example, δ-aminolevulinate dehydratase[10]) but

TABLE 2. Effects of Metals on UD *in Vitro*

Metal Ion	Concentration (mM)	Activity (% Control)	Enzyme Source	Reference
Hg^{2+}	0.1	28	Human erythrocyte	3
	0.5	0	Human erythrocyte	3
	0.1	61	Rat liver	7
	1.0	5	Rat liver	7
	0.1	6	Rat kidney	8
Pb^{2+}	0.5	96	Human erythrocyte	3
	1.0	94	Rat liver	7
	1.0	100	Rat kidney	8
	10.0	14	Bovine liver	1
Fe^{2+}	0.1	98	Human erythrocyte	3
	1.0	84	Human erythrocyte	3
	1.0	91	Rat liver	7
	0.5	88	Bovine liver	1
	10.0	55	Bovine liver	1

not those possessing only a single thiol group at the active site, as may be the case for UD.[1,3,4]

Considerable controversy has surrounded the question as to whether UD is inhibited by Fe^{2+}, as has been postulated to occur in porphyria cutanea tarda. Ferrous iron was ineffective in inhibiting the purified enzyme from human erythrocytes and had no effect in our laboratory on partially purified UD from rat liver or kidney at concentrations as high as 1 mM. In contrast, Straka and Kushner[1] have demonstrated significant inhibition of the purified bovine liver enzyme by Fe^{2+} in reaction mixtures containing buffer systems that do not complex Fe^{2+}. Inhibition of rat liver decarboxylase by Fe^{2+} and a direct interaction of Fe^{2+} with cysteinyl residues at the active site of the enzyme have recently been reported by Mukerji and Pimstone.[11]

The observation that Hg^{2+} is a potent inhibitor of UD *in vitro* prompted us to undertake studies to evaluate the potential consequences of exposure to Hg *in vivo* on UD in several target organ tissues, specifically liver and kidney. In initial studies,[9] rats were exposed to Hg as methyl mercury hydroxide at various concentrations in their drinking water for 6 weeks. Animals so exposed developed an extensive porphyrinuria characterized by a significant dose-related elevation in the urinary concentration of uroporphyrins, principally 8- and 7-CP. Exposure to mercury methyl hydroxide at 10 ppm, for example, resulted in a 12.5-fold increase in the concentration of uroporphyrin during the course of exposure. Subsequent studies on the etiology of this observation[8] have indicated that this effect is principally of renal origin and involves inhibition of renal UD by Hg^{2+}.

The basis of this conclusion is that, whereas no changes in heme biosynthetic enzyme activites were observed in *livers* of Hg-exposed animals, substantial alterations in *renal* enzyme activities were seen, as depicted in FIGURE 2. The most interesting observations here are an initial decrease, followed by a secondary increase, in renal δ-aminolevulinate synthetase activity to at least 2.5 times controls levels; concomitantly, a substantial and continual decline in UD activity during the course of Hg exposure is observed. These observations suggest that the uroporphyrinuria observed during Hg

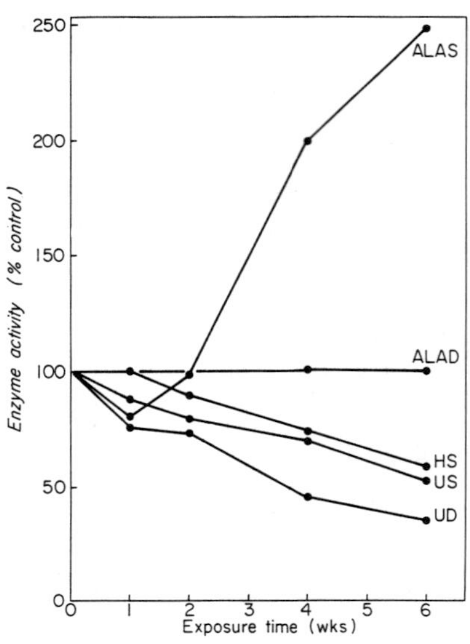

FIGURE 2. Changes in renal heme biosynthetic pathway enzymes during prolonged exposure to Hg as methyl mercury hydroxide (10 ppm in drinking water) in rats. ALAS: δ-aminolevulinic acid synthetase; ALAD: δ–aminolevulinic acid dehydratase; US: uroporphyrinogen I synthetase; UD: uroporphyrinogen decarboxylase; HS: heme synthetase (ferrochelatase).

TABLE 3. Percentage of Porphyrinogen Isomers Formed by Renal UD at Various Concentrations of GSH in the Reaction Mixture

GSH Concentration (mM)	Percentage of Reaction Product[a]			
	7-CP	6-CP	5-CP	4-CP
0	32	4	16	48
2.5	28	5	16	51
5	23	6	18	53
10	16	8	21	55
20	12	4	17	67

[a]Assays were conducted and porphyrins were separated, using reversed-phase HPLC, and quantitated spectrofluorometrically, as previously described.[8]

exposure may be linked to impaired porphyrin decarboxylation at the level of UD in the kidney. Further evaluation of this question has shown that when tissue extracts containing UD from either liver or kidney are assayed under identical conditions, using enzymatically generated 8-CP I as the substrate, the specific activity of the renal enzyme, measured as the production of 4-CP, is approximately one-third that of the enzyme from the liver. Thus, renal and hepatic UD activities are 97.1 and 296.1 pmol of 4-CP/mg protein/hr, respectively. Additionally, substantially greater proportions of porphyrinogen intermediates with 7, 6, and 5 carboxyl groups are seen to accumulate in the renal enzyme assay mixture than is observed in the case of the liver enzyme. In particular, 7-CP constitutes over 30% of reaction products of the renal enzyme, whereas 7-CP constitutes only 7% of reaction products of the hepatic enzyme.

Reduced GSH added to the reaction mixture greatly facilitates stoichiometric conversion of 8-CP to 4-CP by the renal, but not by the hepatic, enzyme. Data presented in TABLE 3 indicate that, as the concentration of GSH in the reaction mixture of the renal decarboxylase is increased, the proportion of 7-CP remaining in the assay medium decreases substantially, whereas the proportion of 4-CP increases. The pattern of porphyrinogen isomers produced at higher GSH concentrations approaches that produced by the liver enzyme, where 4-CP constitutes nearly 80% of total reaction products.

Further studies on the effects of GSH on renal UD activity indicated that GSH acts principally by direct stimulation of renal enzyme activity rather than by stabilizing or protecting porphyrinogen products from degradation in the reaction mixture. Thus, enzyme activity, measured as the production of 4-CP, increases linerarly over a 2-hr period, both in the presence and absence of GSH. The rate of increase, however, is substantially greater in the presence of GSH. Thus, approximately 200 pmol of 4-CP per mg of protein is generated over a 2-hr period in the absence of added GSH, whereas more than 400 pmol of 4-CP is generated over the same time period in the presence of 5 mM GSH in the reaction mixture. These observations indicated that GSH directly stimulates renal enzyme activity, rather than protects reaction products from degradation in the assay medium.

The observation that the concentration of GSH in the reaction mixture moderates the conversion of 8-CP to 4-CP in kidney but not in liver tissue extracts is of interest with regard to porphyrin metabolism in the intact cell, because it suggests that tissues such as the kidney, in which endogenous GSH levels are relatively quite low in comparison with those of the liver, may have increased sensitivity to impairment of porphyrin decarboxylation by low levels of SH-directed binding agents, including trace metals *in vivo*.Thus, although the kidney ordinarily engages in low rates of porphyrin decarboxylation and contributes relatively little to total excreted porphyrin levels, a substantial increase in the contribution of this organ to total urinary porphyrins during metal exposure might be expected. This concept is supported by further observations made *in vitro* when $HgCl_2$ is added to extracts of the renal decarboxylase in the absence of added GSH. As indicated in TABLE 4, the production of 6-, 5-, and 4-CP is significantly impaired at Hg^{2+} concentrations as low as 50 μM, and the production of 4-CP is effectively eliminated at 100 μM. In contrast, the decarboxylation of 8-CP to 7-CP by renal UD did not begin to be substantially inhibited until the Hg^{2+} concentration in the reaction mixture equalled or exceeded 100 μM. This observation demonstrates differential effects of Hg^{2+} at catalytic sites that convert 8-CP to 7-CP versus 7-CP to lesser CPs. This observation also is consistent with the selectively increased urinary concentrations of 8-CP and 7-CP observed in animals exposed to low levels of Hg compounds for prolonged periods.

The inhibitory effects of Hg^{2+} ion on the decarboxylation of 7-CP to lesser CPs are substantially reduced by the addition of GSH to the reaction mixture (TABLE 4). Interestingly, this effect appears to involve more than mere complex formation between Hg and GSH per se, since overall stimulation of enzyme activity, as well as a shifting of the reaction toward more complete production of 4-CP by GSH, occur even in the presence of inhibiting concentrations of Hg^{2+}.[8]

TABLE 4. Effects of $HgCl_2$ on Renal UD *in Vitro*

$HgCl_2$ Concentration (μM)	Enzyme Activity as Isomer Formed[a]			
	7-CP	6-CP	5-CP	4-CP
0	100	100	100	100
10	93	95	90	80
50	86	17	48	40
100	79	8	7	6
500	0	0	0	0
50[b]	91	111	94	85

[a]Values are percentages of control values.
[b]With 5 mM GSH in the reaction mixture.

From the observations described above it is apparent that UD from various tissues requires free SH groups for catalytic activity and is susceptible to inhibition or inactivation by agents such as trace metals that bind SH groups. It is also apparent that the extent of 8-CP decarboxylated by UD may vary in different tissues *in vivo* owing to the influence of endogenous factors such as GSH concentration on enzyme activity. In this regard, the kidney probably contributes very little to overall urinary porphyrin concentrations under ordinary nonclinical conditions, but may assume increased importance in the underlying etiology of porphyrinurias that develop during exposure to trace metals for which the kidney is a principal target organ.

COPROPORPHYRINOGEN OXIDASE

Coproporphyrinogen oxidase (CO) catalyzes the oxidative decarboxylation of propionic acid groups of the A and B rings of coproporphyrinogen III (and IV), leaving vinyl groups at these positions (FIGURE 3). The resultant molecule is protoporphyrin-

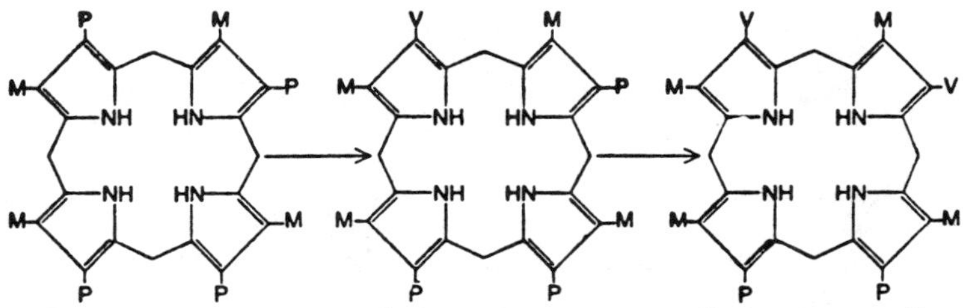

FIGURE 3. Steps catalyzed by CO. M: $-CH_3$; P: $-CH_2CH_2COOH$; V: $-C{=}CH_2$.

ogen IX. An intermediate 3-CP with a single vinyl group in the A ring, harderoporphyrinogen, is formed in the process. CO is unusual as an oxidase in that metals do not appear to be involved in its oxidative function, nor is its activity affected by treatment with metal chelators.[12]

Studies on preparations of CO from various sources have clearly characterized it as an SH-containing protein,[13] although the extent to which functional SH groups are required for oxidase activity is not clear. A preparation of CO from bovine liver[13] was found to contain three SH groups in the intact enzyme and seven more when the protein was unfolded with sodium dodecyl sulfate. Camadro and colleagues[14] have recently reported purification of a CO from yeast (*Saccharomyces cerevisiae*); this CO is partially inhibited by thiol-directed reagents, indicating that at least one SH group is required for functional activity in that preparation.

Relatively few reports have described the effects of metals per se on CO activity, although coproporphyrinuria is a well-characterized clinical manifestation of intoxication by Pb[15] and several other metals of occupational or environmental importance. As indicated in TABLE 5, Batlle and colleagues[16] reported inhibition *in vitro* of a purified preparation of CO from rat liver by a variety of metal ions including As^{3+}, Cd^{2+}, and Pb^{2+}. Enzyme activity was not affected by monovalent cations, however. We have observed findings comparable to those of Batlle in our laboratory with respect to inhibition of activity in partially purified preparations of CO from rat liver. Thus, As^{5+} produced inhibition similar to that reported of As^{3+}. Cadmium and Pb also inhibit the enzyme *in vitro*, although at relatively high concentrations.

Interestingly, higher concentrations of trace metals are required *in vitro* than *in*

TABLE 5. Effects of Metals on CO *in Vitro*

Metal Ion	Concentration (mM)	Activity (% Control)	Enzyme Source	Reference
As^{3+}	40	65	Rat liver	16
As^{5+}	1	86	Rat liver	This laboratory
	10	71	Rat liver	This laboratory
Cd^{2+}	20	60	Rat liver	16
	1	80	Rat liver	This laboratory
	10	59	Rat liver	This laboratory
Pb^{2+}	20	70	Rat liver	16
	1	86	Rat liver	This laboratory
	10	61	Rat liver	This laboratory

vivo to inactivate CO. As demonstrated in TABLE 6, CO activity was depressed to 30% of control activity in renal tissues of rabbits fed lead acetate in drinking water in doses sufficient to achieve a renal Pb concentration of 0.20 mM.[15] In contrast, as previously presented, 50 or 100 times that concentration *in vivo* inhibited cell-free extracts of the enzyme to only 70% of control levels.

There is no evidence that free SH groups are directly involved in the catalytic function of CO. Therefore, it seems reasonable, in light of the information presented here, that metals such as Pb compromise enzyme activity through processes that do not involve direct binding at the active site. One likely alternative site of interaction would be the SH groups of the enzyme that are required to maintain the structural integrity of the protein. In addition, metals that compromise CO activity *in vivo* probably do so by disrupting the structure of mitochondrial membranes with which CO is associated in the intact cell. Previously reported studies from our laboratories[17–19] have demonstrated that many metals indirectly impair membrane-associated biochemical functions at much lower doses *in vivo* than *in vitro* by causing degradative changes in the structural integrity of membranes of subcellular organelles with which such biochemical functions are associated. Inasmuch as CO is considered to be associated in the intact cell with membrane surfaces surrounding the mitochondrial inner membrane (an association that may be required for optimal activity *in vivo*),[20] it is reasonable to

TABLE 6. Inactivation of CO by Pb^{2+} *in Vivo* and *in Vitro*

	Pb^{+2} Concentration (mM)	Activity (% Control)	Reference
In vivo	0.20[a]	30	15
In vitro	10.0[b]	71	This laboratory
	20.0[b]	70	16

[a]Concentration in tissue.
[b]Concentration in reaction medium.

postulate that disruption of this association would potentiate any direct effect of the metal on enzyme activity. Lead is a particularly likely candidate for such an effect, inasmuch as Pb causes mitochondrial membrane distortion at tissue concentrations as low as 0.01 mM when fed to animals for prolonged periods.[18] This question is considered in further detail in the paper by Fowler *et al.* in this volume.[21]

SUMMARY

Both UD and CO are susceptible to alteration by sulfhydryl-directed binding agents including a variety of trace metals. UD apparently requires a functional SH group or groups for catalytic activity, and the various steps of decarboxylation catalyzed by the enzyme can be differentially inhibited by divalent cations such as Hg^{2+} at very low concentrations. There is evidence that tissue-specific factors such as the endogenous GSH concentration may influence the susceptibility of UD in some tissues to metal inhibition, and this circumstance could be highly relevant to the etiology of porphyrinopathies or porphyrinurias that arise during prolonged metal exposures.

CO does not appear to have a requirement for functional SH groups at the active site, but several SH groups on the enzyme appear to be involved in maintaining the protein's noncovalent structural characteristics. CO appears to be substantially more readily inhibited by metals *in vivo* than *in vitro*. This observation may reflect effects of metals on both the structural integrity of the enzyme per se, as well as on the mitochondrial membranal structures with which the enzyme is functionally associated in the intact cell.

Finally, it seems reasonable to suggest that tissues, such as the kidney, that ordinarily contribute only sparingly to total excreted porphyrin levels may assume increased importance in this regard when challenged by specific porphyrinogenic chemicals such as trace metals. Advantage might be taken of such chemical- and organ-specific changes in porphyrin metabolism and porphyrin excretion patterns in monitoring prolonged, subclinical exposure to such chemicals in human populations.

REFERENCES

1. Straka, J. G. & J. P. Kushner. 1983. Purification and characterization of bovine hepatic uroporphyrinogen decarboxylase. Biochemistry **22:** 4664–4672.
2. de Verneuil, H., B. Grandchamp & Y. Nordmann. 1980. Some kinetic properties of human red cell uroporphyrinogen decarboxylase. Biochim. Biophys. Acta **611:** 174–186.
3. de Verneuil, H., S. Sassa & A. Kappas. 1983. Purification and properties of uroporphyrinogen decarboxylase from human erythrocytes. J. Biol. Chem. **258:** 2454–2460.
4. Elder, G. H., J. A. Tovey & D. M. Sheppard. 1983. Purification of uroporphyrinogen decarboxylase from human erythrocytes. Biochem. J. **215:** 45–55.
5. Elder, G. H. & A. J. Urquhart. 1984. Human uroporphyrinogen decarboxylase: Do tissue-specific isoenzymes exist? Biochem. Soc. Trans. **12:** 663–664.
6. Kawanishi, S., S. Yoshihiko & S. Sano. 1983. Uroporphyrinogen decarboxylase: Purification, properties, and inhibition by polychlorinated biphenyl isomers. J. Biol. Chem. **258:** 4285–4292.
7. Woods, J. S., R. M. Kardish & B. A. Fowler. 1981. Studies on the action of porphyrinogenic trace metals on the activity of hepatic uroporphyrinogen decarboxylase. Biochim. Biophys. Res. Commun. **103:** 264–271.
8. Woods, J. S., D. L. Eaton & C. B. Lukens. 1984. Studies on porphyrin metabolism in the kidney: Effects of trace metals and glutathione on renal uroporphyrinogen decarboxylase. Mol. Pharmacol. **26:** 336–341.
9. Woods, J. S. & B. A. Fowler. 1977. Renal porphyrinuria during chronic methyl mercury exposure. J. Lab. Clin. Med. **90:** 266–272.
10. Sheehra, J. S., M. G. Gore, A. G. Chaudhry & P. M. Jordan. 1981. 5-Aminolevulinic acid dehydratase: The role of sulphydryl groups in 5-aminolevulinic acid dehydratase from bovine liver. Eur. J. Biochem. **114:** 263–269.
11. Mukerji, S. K. & N. R. Pimstone. 1986. *In vitro* studies of the mechanism of inhibition of rat liver uroporphyrinogen decarboxylase activity by ferrous iron under anaerobic conditions. Arch. Biochem. Biophys. **244:** 619–629.
12. Yoshinaga, T. & S. Sano. 1980. Coproporphyrinogen oxidase. I. Purification and properties and activation by phospholipids. J. Biol. Chem. **255:** 4722–4726.
13. Yoshinaga, T. & S. Sano. 1980. Coproporphyrinogen oxidase. II. Reaction mechanism and role of tyrosine residues on the activity. J. Biol. Chem. **255:** 4727–4731.
14. Camadro, J. M., H. Chambon, J. Jolles & P. Labbe. 1986. Purification and properties of coproporphyrinogen oxidase from the yeast *Saccharomyces cerevisiae*. Eur. J. Biochem. **156:** 579–587.
15. Henderson, M. J. & C. Toothill. 1983. Urinary coproporphyrin in lead intoxication: A study in the rabbit. Clin. Sci. **65:** 527–532.
16. Batlle, A. M. del C., A. Benson & C. Rimington. 1965. Purification and properties of coproporphyrinogenase. Biochem. J. **97:** 731–740.

17. WOODS, J. S. & B. A. FOWLER. 1986. Alterations of hepatocellular structure and function by thallium chloride: Ultrastructural, morphometric and biochemical studies. Toxicol. Appl. Pharmacol. **83:** 218–229.
18. FOWLER, B. A., R. M. KARDISH & J. S. WOODS. 1983. Alteration of hepatic microsomal structure and function by indium chloride. Lab. Invest. **48:** 471–478.
19. FOWLER, B. A., C. A. KIMMEL, J. S. WOODS, E. E. MCCONNEL & L. D. GRANT. 1980. Chronic low-level toxicity of lead in the rat. III. An integrated assessment of long-term toxicity with special reference to the kidney. Toxicol. Appl. Pharmacol. **56:** 59–77.
20. ELDER, G. H. & J. O. EVANS. 1978. Evidence that the coproporphyrinogen oxidase activity of rat liver is situated in the intermembrane space of mitochondria. Biochem. J. **172:** 345–351.
21. FOWLER, B. A., A. OSKARSSON & J. S. WOODS. 1987. Metal- and metalloid-induced porphyrinurias: Relationship to cell injury. Ann. N.Y. Acad. Sci. This volume.

Metal Induction of Heme Oxygenase[a]

F. WILLIAM SUNDERMAN, JR.

Departments of Laboratory Medicine and Pharmacology
University of Connecticut School of Medicine
Farmington, Connecticut 06032

INTRODUCTION

In 1974, Maines and Kappas[1] discovered that treatment of rats with $CoCl_2$ greatly stimulates heme oxidation by hepatic microsomes, but not by splenic microsomes. From 1975 to 1977, Maines and Kappas[2–7] showed that numerous metal ions cause induction of microsomal heme oxygenase activity in rat liver, kidney, heart, lung, and gut mucosa, albeit to varying extents and with differing tissue selectivities. During the past decade, these pioneering studies have been confirmed and extended by many investigations, but the literature on metal induction of heme oxygenase activity has not hitherto been comprehensively reviewed. The objectives of this paper are 1) to collate the published experiments on induction of microsomal heme oxygenase activity after administration of inorganic metal compounds to rodents, and 2) to summarize the studies on $NiCl_2$ induction of heme oxygenase activity that have been performed in the author's laboratory. The important effects of organometallic compounds, and especially metalloporphyrins, on heme oxygenase activity are excluded from consideration, as this topic is discussed in another paper in this volume.[8]

MOLECULAR MECHANISMS

Metal induction of microsomal heme oxygenase activity involves fresh synthesis of mRNA and protein because induction of the enzyme by $CoCl_2$, $CdCl_2$, and $NiCl_2$ is inhibited by administration of actinomycin D, puromycin, and cycloheximide.[2,9–13] Moreover, Kikuchi *et al.*[14] and Ishizawa *et al.*[15] demonstrated that translatable mRNA specific for heme oxygenase is increased in liver or kidney of Co-, Cd-, and Sn-treated rats, in approximate proportion to the induced heme oxygenase activity of the respective organs. Maines *et al.*[16] showed that microsomal heme oxygenase activity in rat liver is present in two constitutive forms, designated "HO-1" and "HO-2." The hepatic concentration of the HO-1 isoenzyme becomes increased after administration of certain metal compounds (for example, $CdCl_2$ and $CoCl_2$) or organic inducers (for example, bromobenzene and hematin).[16] Knowledge is scanty regarding the specific mechanisms whereby metal ions stimulate the synthesis of mRNA for HO-1 in target tissues. The hypothesis[17] that metal induction of heme oxygenase is mediated by heme derived for accelerated degradation of hemoproteins, such as cytochrome P-450, has been contradicted by several studies.[14,18–20] Administration of certain thiol compounds (for example, cysteine and glutathione) to rodents inhibits metal induction of heme oxygenase.[4,7,13,21] Conversely, depletion of tissue SH levels (for example, by treatment with diethyl maleate) enhances metal induction of heme oxygenase.[4,7,22,23] These

[a]Supported by grants from the Northeast Utilities Company, the U.S. Department of Energy (EV-01340), and the National Institute of Environmental Health Sciences (ES-01337).

findings furnish indirect evidence that binding of metal ions to SH groups in unidentified regulatory molecules may be involved in the induction of heme oxygenase activity.[4]

EXPERIMENTAL MODELS

The acute effects of metals on heme oxygenase activities in tissues of rodents are listed in TABLES 1–4 and 8. The entries for 1) dosages of metal compounds, 2) intervals between injection and sacrifice, and 3) heme oxygenase activities (expressed as a multiple of the mean activity in controls) designate the experimental conditions that yielded the maximal alterations of microsomal heme oxygenase activity in the respective studies. The abbreviation "NS" (that is, not significant) in the tables indicates that the mean heme oxygenase activity in the specified tissue of metal-treated rodents did not differ significantly from the corresponding mean value in the controls. The experiments on heme oxygenase induction by Co, Cd, Sn, and Ni compounds are considered first (TABLES 1–4) because the effects of these metals have been studied much more extensively than the remaining metals, which are summarized in TABLE 8.

TABLE 1. Acute Effects of $CoCl_2$ Administration on Heme Oxygenase Activity

Species and Route	Dose (μmol/kg)	Time before Sacrifice (hr)	Organ	Activity[a]	Authors
Rat, sc	100 (×2)	48, 24	Liver	8	Maines & Kappas[1,2]
			Spleen	NS	
Rat, sc	100 (×2)	42, 18	Liver	28	Correia & Schmid[24]
			Gut	5	
Rat, sc	256	17	Liver	10	DeMatteis & Gibbs[17]
Rat, sc	250	24	Liver	9	Maines & Kappas[4–7]
			Kidney	5	
			Heart	2	
			Lung	1.5	
			Gut	4	
			Brain	NS	
Rat, ip	75 (×2)	36, 12	Liver	3	Eaton *et al.*[19]
Rat, sc	250	24	Ovary	4	Maines & Kutty[25]
			Testis	NS	
Rat, sc	250	24	Brain	NS	Maines *et al.*[26]
Rat, sc	250	24	Adrenal	2	Veltman & Maines[27]

[a]Each value for heme oxygenase activity below is a multiple of the mean activity in controls. NS: not significant.

Cobalt

Administration of $CoCl_2$ to rats by sc or ip injection, at dosages from 150 to 250 μmol/kg body wt, caused pronounced induction of microsomal heme oxygenase activity in liver, and less marked induction in heart, lung, kidney, ovary, adrenal, and

TABLE 2. Acute Effects of Cd on Heme Oxygenase Activity

Compound	Species and Route	Dose (μmol/kg)	Time before Sacrifice (hr)	Organ	Activity[a]	Authors
$CdCl_2$	Rat, sc	25	24	Liver	9	Maines & Kappas[4,5]
				Kidney	NS	
				Heart	NS	
	Rat, ip	11	24	Liver	14	Yoshida *et al.*[32]
	Mouse, ip	22	16	Liver	20	Yoshida *et al.*[33]
	Rat, ip	20 (×2)	36, 12	Liver	7	Eaton *et al.*[19]
	Rat, sc	25	16	Liver	11	Yoshida & Ishizawa[34]
$Cd(C_2H_3O_2)_2$	Rat, ip	7.5	32	Liver	4	Krasny & Holbrook[10]
	Rat, sc	20	24	Testis	0.3	Maines *et al.*[35]
	Rat, sc	20	24	Adrenal	1.4	Veltman & Maines[27]

[a]Each value for heme oxygenase activity below is a multiple of the mean activity in controls. NS: not significant.

TABLE 3. Acute Effects of Sn on Heme Oxygenase Activity

Compound	Species and Route	Dose (μmol/kg)	Time before Sacrifice (hr)	Organ	Activity[a]	Authors
$SnCl_2$	Rat, sc	250	16	Liver	3	Kappas & Maines[38]
				Kidney	23	
				Heart	NS	
				Lung	NS	
				Gut	NS	
	Rat, sc	250	16	Kidney	28	Drummond & Kappas[39]
	Rat, sc	250	16	Kidney	48	Yoshida & Ishizawa[34]
	Rat, sc	125	24	Testis	NS	Maines & Kutty[25]
	Rat, sc	125	44	Liver	4	Kutty & Maines[40]
				Kidney	19	
$SnC_4H_4O_6$	Rat, ip	168	24	Liver	1.2	Dwivedi *et al.*[41]

[a]Each value for heme oxygenase activity below is a multiple of the mean activity in controls. NS: not significant.

gut mucosa (TABLE 1). Heme oxygenase activity was not significantly increased in spleen, brain, or testis of $CoCl_2$-treated rats. Hino *et al.*[28] showed that administration of $CoCl_2$ to rats (250 μmol/kg body wt, sc, at 42 and 18 hr before sacrifice) produced a sevenfold increase of heme oxygenase activity in the smooth-surfaced fraction of hepatic microsomes, whereas the activity was not increased in the rough-surfaced microsomal fraction. A small increase (twofold) was noted in the Golgi fraction. According to Maines and Kappas,[3] $CoCl_2$ induction of heme oxygenase activity of rat kidney occurs primarily in the renal cortex, and to a relatively minor degree in the renal medulla. Sardana *et al.*[29] reported that the susceptibility of pregnant rats to $CoCl_2$ induction of hepatic heme oxygenase activity during late gestation was equal to that in nonpregnant controls. Maines and Kappas[30] found that administration of $CoCl_2$ (250 μmol/kg body wt, sc, 24 hr before sacrifice) to pregnant dams during late gestation did not induce hepatic heme oxygenase activity in the fetuses, nor did administration of $CoCl_2$ to lactating dams induce hepatic heme oxygenase activity in their neonatal pups. Woods and Carver[31] showed that administration of a higher dose of $CoCl_2$ (400 μmol/kg body wt, ip, 24 hr before sacrifice) to pregant rats during late gestation yielded a threefold increase of microsomal heme oxygenase activity in fetal liver. According to Maines and Kappas,[30] 4-day-old neonatal rats exhibited a small (less than twofold) induction of hepatic heme oxygenase activity when treated directly with $CoCl_2$ (100 μmol/kg body wt, sc, 24 hr before sacrifice); the inducibility of hepatic heme oxygenase activity in $CoCl_2$-treated pups increased to fully adult levels by 4 weeks postpartum.

Cadmium

Induction of microsomal heme oxygenase activity by sc or ip administration of $CdCl_2$ or $Cd(C_2H_3O_2)_2$ to rats occurs primarily in the liver and to a minor extent in the adrenal gland; kidney and heart are evidently refractory (TABLE 2). According to Yoshida *et al.*,[32] $CdCl_2$ induction of hepatic heme oxygenase activity occurs more rapidly in female rats than in males. Induction of hepatic heme oxygenase activity by $CdCl_2$ is more pronounced in mice than in rats.[32,33] Krasny and Holbrook[10] noted a twofold increase of heme oxygenase activity in the smooth-surfaced microsomal fraction and a threefold increase in the rough-surfaced microsomal fraction of livers from $Cd(C_2H_3O_2)_2$-treated rats. The Cd^{2+} ion causes prompt inhibition of microsomal heme oxygenase in rat testis, with marked diminution of activity in both the smooth and rough microsomal fractions.[35] In rats killed 24 hr after injection of $Cd(C_2H_3O_2)_2$ (20 μmol/kg body wt, sc), heme oxygenase activity in testicular Leydig cells was inhibited by 80%, whereas heme oxygenase activity in Sertoli cells was not significantly affected.[36] Maines *et al.*[35] speculated that heme oxygenase activity in rat testis is regulated by factors that are totally different from those that control the enzyme activity in liver. Trakshel *et al.*[37] showed by electrophoretic, chromatographic, and immunochemical techniques that the predominant isoenzyme of heme oxygenase in rat testis is HO-1, with only minute amounts of the HO-2 isoform; in contrast, rat liver has an approximately 1:2 ratio of HO-1 to HO-2. Although heme oxygenase activity was inhibited when measured in intact testicular microsomes from $CdCl_2$-treated rats,[35–37] when the testicular isoforms were chromatographically separated, the activity of HO-1 was slightly decreased and the activity of HO-1 was increased twofold.[37] To explain this paradox, Trakshel *et al.*[37] speculated that Cd^{2+}-mediated inhibition of heme oxygenase activity in intact testicular microsomes may reflect a disturbance of molecular interactions in the microsomal membranes (for example, altered spatial relationships between the molecules of heme oxygenase and NADPH–cytochrome P-450 reductase).

Tin

Stannous chloride is exceptionally specific and potent as an inducer of heme oxygenase activity in renal microsomes; it yields slight stimulation of hepatic heme oxygenase activity, and has little or no effect on the enzyme activity in other tissues, including heart, lung, testis, and gut mucosa (TABLE 3). The induction of renal heme oxygenase activity after sc administration of $SnCl_2$ to rats is greater than has been observed in tissues of rodents treated with other metals. For this reason, Sn^{2+} induction

TABLE 4. Acute Effects of $NiCl_2$ Administration on Heme Oxygenase Activity

Species and Route	Dose (μmol/kg)	Time before Sacrifice (hr)	Organ	Activity[a]	Authors
Rat, sc	250	24	Liver	4	Maines & Kappas[4-6]
			Kidney	11	
			Heart	1.5	
Rat, sc	250	16	Kidney	7	Drummond & Kappas[39,42]
Rat, sc	250	16	Kidney	12	Maines & Kappas[43,44]
Rat, ip	120 (×2)	36, 12	Liver	4	Eaton *et al.*[19]
Rat, sc	250	16	Liver	2	Yoshida & Ishizawa[34]
			Kidney	5	
Rat, sc	750	17	Liver	4	Sunderman *et al.*[13,45]
			Kidney	11	
Rat, sc	250	17	Liver	2	Sunderman *et al.*[45,46]
			Kidney	5	
			Lung	1.6	
			Brain	1.3	
			Spleen	NS	
			Gut	NS	
Mouse, sc	250	17	Kidney	5	
Hamster, sc	250	17	Kidney	1.4	
Guinea pig, sc	250	17	Kidney	4	

[a]Each value for heme oxygenase activity below is a multiple of the mean activity in controls. NS: not significant.

of renal heme oxygenase activity was selected by Drummond and Kappas[39,42] as an experimental model to investigate metal interactions in enzyme induction. They found that $MnCl_2$, when administered simultaneously with $SnCl_2$ to rats in equimolar dosages (250 μmol/kg body wt, sc, 16 hr before sacrifice), produced a 98% inhibition of renal heme oxygenase induction.[39] Under similar experimental conditions, $MnCl_2$ produced 80% inhibition of $NiCl_2$ induction of renal heme oxygenase.[39] In rats pretreated up to 8 hr with Mn^{2+}, the level of induction of renal heme oxygenase by Sn^{2+} or Ni^{2+} was markedly depressed in a time-dependent fashion. Conversely, administration of Mn^{2+} after Sn^{2+} or Ni^{2+} was ineffective in preventing the induction of renal heme oxygenase.[39] As will be noted subsequently, $MnCl_2$ alone does not significantly affect renal heme oxygenase activity at the specified dosage (250 μmol/kg body wt, sc, 16 hr before sacrifice).[39] Drummond and Kappas[42] also found that $ZnCl_2$ treatment

TABLE 5. Effects of Various Treatments on Ni and Hb Induction of Heme Oxygenase in Rat Kidney[a]

			Heme Oxygenase Activity (nmol/hr/mg protein)		
Treatment	Dose and Route	Time before Sacrifice (hr)	Vehicle Controls	Induced by $NiCl_2$[b]	Induced by Hb[c]
Vehicle			13 ± 3	75 ± 26	96 ± 14
Actinomycin D[d]	1.25 mg/kg, ip	18, 15, 12, 10	16 ± 3	25 ± 5[e]	26 ± 7[e]
Cycloheximide	2 mg/kg, ip	18	15 ± 4	22 ± 4[e]	42 ± 9[e]
Cysteine	5 mmol/kg, po	17.75	19 ± 7	57 ± 21[e]	87 ± 18
DDC	1.33 mmol/kg, po	17.2	21 ± 5	163 ± 36[e]	110 ± 16

[a]Adapted from Sunderman *et al.*[13]
[b]Dose of $NiCl_2$: 250 μmol/kg body wt, sc, 17 hr before sacrifice.
[c]Dose of Hb: 0.3 g/kg body wt, iv, 17 hr before sacrifice.
[d]Administered in divided doses.
[e]$p < .05$ versus Ni or Hb alone (4–7 rats/group).

blocked $SnCl_2$ or $NiCl_2$ induction of heme oxygenase activity in rat kidney. The blocking effect of Zn^{2+} was greatest on simultaneous administration, time dependent when administered up to 8 hr before Sn^{2+} or Ni^{2+}, and ineffective when administered as soon as 10 min after the inducing metal ions. Zinc chloride was also able to block the induction of hepatic heme oxygenase activity by $CoCl_2$, $NiCl_2$, or $CdCl_2$.[42] Yoshida and Ishizawa[34] questioned the biological significance of such metal interactions in the regulation of microsomal heme oxygenase activity. They confirmed that Sn^{2+}-mediated induction of renal heme oxygenase in rats is substantially prevented by Mn^{2+} when the mixture of Sn^{2+} and Mn^{2+} is administered to rats as a single injection, but they failed to detect inhibition of Sn^{2+}-mediated induction when Mn^{2+} and Sn^{2+} (250 μmol/kg body wt, sc, 16 hr before sacrifice) were injected simultaneously at separate sites. Similar results were noted when the combination of 1) Sn^{2+} and Zn^{2+}, 2) Ni^{2+} and Mn^{2+}, and 3) Cd^{2+} and Zn^{2+} were tested for induction of heme oxygenase activity

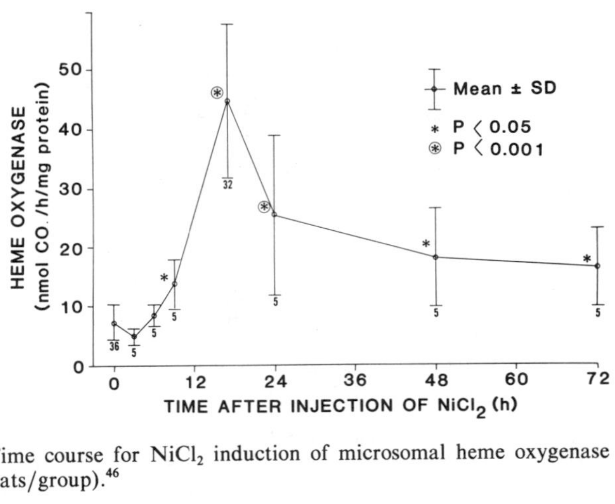

FIGURE 1. Time course for $NiCl_2$ induction of microsomal heme oxygenase activity in rat kidney (5–36 rats/group).[46]

TABLE 6. Heme Oxygenase Activity in Rat Tissues during Short-term $NiCl_2$ Administration[a]

Treatment	Dose and Route	Number of Rats	Number of Days	Heme Oxygenase Activity (nmol/hr/mg protein)	
				Kidney	Liver
Vehicle		20	1–9	18 ± 9	28 ± 8
$NiCl_2$	125 μmol/kg, sc (daily)	3–7/group	1	93 ± 27[b]	41 ± 13[b]
			2	110 ± 30[b]	51 ± 16[b]
			3	92 ± 31[b]	49 ± 12[b]
			5	73 ± 26[b]	51 ± 16[b]
			7	41 ± 17[b]	46 ± 10[b]
			9	27 ± 9[b]	41 ± 7[b]
	1 mmol/1, po (in drinking water)	3–7/group	1	23 ± 8	33 ± 13
			2	21 ± 4	33 ± 12
			3	18 ± 5	27 ± 6
			5	23 ± 6	31 ± 4
			7	20 ± 8	28 ± 5
			9	20 ± 8	28 ± 5

[a] Adapted from reference 47.
[b] $p < .05$.

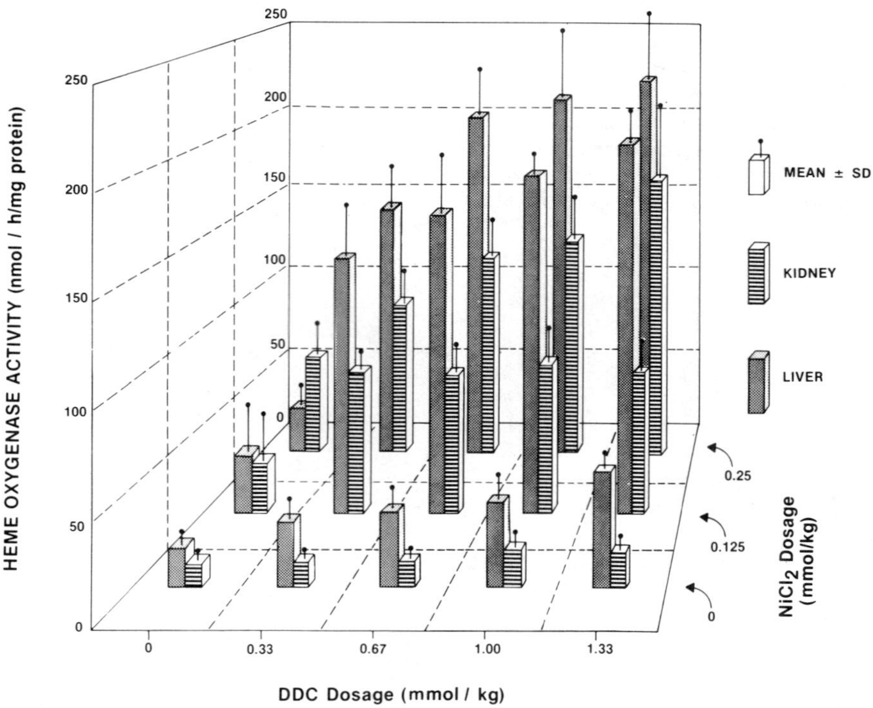

FIGURE 2. Microsomal heme oxygenase activities in liver and kidney of rats treated with DDC and $NiCl_2$, singly and in various dosage combinations, 17 hr before sacrifice.[13]

in rat kidney or liver.[34] Further research is needed to resolve these discordant results, since the matter may by critically important in elucidating the molecular mechanisms for metal induction of heme oxygenase activity.

Nickel

Acute effects of parenteral injections of $NiCl_2$ on microsomal heme oxygenase activity in rodent tissues are summarized in TABLE 4. In $NiCl_2$-treated rats, the greatest induction of heme oxygenase activity occurred in the kidney, but significant induction was also observed in liver, heart, lung, and brain.[46] No significant induction of heme oxygenase activity in spleen or gut mucosa was detected in $NiCl_2$-treated rats.[46] Addition of $NiCl_2$ to drinking water (3.3 mmol/l) rats during the 3 days before sacrifice did not yield any significant induction of heme oxygenase activity in duodenal mucosa.[46] Likewise, intrarenal implantation of Ni_3S_2, a potent carcinogen, did not affect heme oxygenase activity in kidneys of rats sacrificed 1, 2, 3, or 4 weeks after implantation.[46] As indicated in TABLE 5, administration of $NiCl_2$ (250 μmol/kg body

TABLE 7. Effects of Ni and DDC on Constituents of Rat Liver[a]

$NiCl_2$[b]	DDC[c]	HO (nmol/hr/ mg protein)	Ni (μg/kg)	MT (mg/kg)	GSH (mol/kg)	GSSG (% of total)
0	0	21 ± 4	<20	39 ± 10	4.5 ± 0.5	2.6 ± 0.8
+	0	26 ± 5	53 ± 26[d]	156 ± 38[d]	4.5 ± 0.5	1.7 ± 0.4[d]
0	+	65 ± 7[d]	<20	296 ± 55[d]	5.0 ± 0.5	2.5 ± 1.1
+	+	221 ± 37[d]	308 ± 63[d]	358 ± 37[d]	6.7 ± 1.3[d]	1.8 ± 0.6[d]

NOTE: HO: heme oxygenase; MT: metallothionein; GSH: glutathione; GSSG: oxidized glutathione.

[a]Adapted from references 48 and 49.

[b]Dose of $NiCl_2$: 250 μmol/kg body wt, sc, 17 hr before sacrifice.

[c]Dose of DDC: 1.33 mmol/kg body wt, im, 17.2 hr before sacrifice.

[d]$p < .05$ versus vehicle controls (4–7 rats/group).

wt, sc, 17 hr before sacrifice) yielded 4- to 5-fold increases of renal heme oxygenase activity in rats, mice, and guinea pigs, but only a 1.4-fold increase in hamsters.[46] The time course for $NiCl_2$ induction of renal heme oxygenase activity in rats is illustrated in FIGURE 1. Induction of renal heme oxygenase activity was not maintained during sustained $NiCl_2$ injections (125 μmol/kg body wt, daily, for 1 to 9 days, with the last injection 24 hr before sacrifice). As shown in TABLE 6, the peak activity in kidney microsomes occurred after two injections of $NiCl_2$ and gradually diminished thereafter.[47] The adaptive mechanism for acquired tolerance to $NiCl_2$ treatments has not been elucidated. Heme oxygenase activity was not significantly increased in rat kidney or liver during 9 days of oral exposure to $NiCl_2$ (1 mmol/l, in drinking water).[47]

Effects of actinomycin D, cycloheximide, cysteine, and sodium diethyldithiocarbamate (DDC) on the induction of renal heme oxygenase activity are listed in TABLE 5, based on experiments in which $NiCl_2$ or hemoglobin were used as the inducing agents.[13] Actinomycin D and cycloheximide significantly inhibited the induction of renal heme oxygenase activity by either $NiCl_2$ or hemoglobin, whereas oral cysteine treatment blocked $NiCl_2$ induction of the enzyme without affecting hemoglobin induction. Administration of DDC, a Ni^{2+} chelator, produced remarkable synergistic induction of renal heme oxygenase activity when administered to rats in combination with $NiCl_2$;

TABLE 8. Effects of Other Metals on Heme Oxygenase Activity in Rats

Metal	Compound	Dose (μmol/kg)	Time before Sacrifice (hr)	Organ	Activity[a]	Authors
Ag	$AgNO_3$	65 (×2), ip	36, 12	Liver	4	Eaton *et al.*[19]
As	$NaAsO_2$	100, sc	16	Liver	11	Sardana *et al.*[51]
				Kidney	15	
	Na_2HAsO_4	100, sc	16	Liver	NS	
				Kidney	6	
Au	$AuCl_4$	125, sc	16	Liver	2	Maines[52]
	$AuNaO_4SC_4H_4$	150, ip	24	Liver	2	Eiseman & Alvares[53]
				Kidney	8	
	$AuNaO_4SC_4H_4$	12 (×3), ip	72, 48, 24	Liver	1.4	Leonard *et al.*[54]
				Kidney	NS	
Cr	K_2CrO_4	125, sc	24	Liver	2.5	Maines & Kappas[3,4]
				Kidney	NS	
				Heart	NS	
Cu	$CuCl_2$	125, sc	24	Liver	6	Maines & Kappas[3–5]
				Kidney	2	
				Heart	1.5	
	$CuSO_4$	75 (×2), ip	36, 12	Liver	2	Eaton *et al.*[19]
		125, sc	24	Testis	NS	Maines & Kutty[25]
Fe	$FeCl_2$	250, sc	24	Liver	4	Maines & Kappas[4,5]
				Kidney	NS	
				Heart	1.5	
	$FeSO_4$	300 (×2), ip	36, 12	Liver	3	Eaton *et al.*[19]
		1000, ip	24	Liver	6.5	Ariyoshi *et al.*[55]
Hg	$Hg(C_2H_3O_2)_2$	250, sc	24	Liver	5	Maines & Kappas[3–5]
				Kidney	4	
				Heart	4	
	$HgCl_2$	5 (×2), ip	36, 12	Liver	NS	Eaton *et al.*[19]
		30, sc	24	Testis	NS	Maines & Kutty[25]
		10, sc	24	Brain	NS	Maines *et al.*[26]
		30, sc	24	Adrenal	7	Veltman & Maines[27]

In	$InCl_3$	180, ip	16	Liver	5	Woods *et al.*[56]
Mg	$MgCl_2$	250, sc	15	Kidney	NS	Drummond & Kappas[39]
	$MgSO_4$	250, sc	24	Brain	NS	Maines *et al.*[26]
Mn	$MnCl_2$	250, sc	24	Liver	3	Maines & Kappas[4]
		250, sc	16	Kidney	NS	Drummond & Kappas[39, 42]
		125 (×2), ip	12	Liver	3	Eaton *et al.*[19]
		500, sc	16	Kidney	2	Yoshida & Ishizawa[34]
	$MnSO_4$	250, sc	24	Brain	NS	Maines *et al.*[26]
Pb	$Pb(C_2H_3O_2)_2$	250, sc	24	Liver	4	Maines & Kappas[3–5]
				Kidney	3	
				Heart	NS	
		200 (×2), ip	36, 12	Liver	3	Eaton *et al.*[19]
		125, ip	24	Testis	NS	Maines & Kutty[25]
Pt	PtK_2Cl_2	250, sc	24	Liver	6	Maines & Kappas[6]
				Kidney	5	
				Heart	1.5	
	$PtCl_2H_6N_2$	30, ip	24	Liver	1.5	Jollie & Maines[57]; Maines[58]
				Kidney	2	
Sb	$SbCl_3$	125, sc	16	Liver	10	Drummond & Kappas[59]
				Kidney	11	
	$SbCl_5$	125, sc	16	Liver	3	
				Kidney	1.5	
Se	$NaSeO_3$	25 (×2), ip	36, 12	Liver	3	Eaton *et al.*[19]
Zn	$ZnCl_2$	250, sc	24	Liver	5	Maines & Kappas[3–5]
				Kidney	NS	
				Heart	NS	
	$ZnCl_2$	250, sc	16	Liver	NS	Drummond & Kappas[39,42]
				Kidney	NS	
	$ZnCl_2$	200 (×2), ip	36, 12	Liver	NS	Eaton *et al.*[19]

[a]Each value for heme oxygenase activity is a multiple of the mean activity in controls. NS: not significant.

combined administration of DDC and hemoglobin did not significantly increase the enzyme activity above that observed in rats that received only hemoglobin. The synergistic effect of DDC and $NiCl_2$ on heme oxygenase activity was even more pronounced in liver than in kidney, as illustrated by FIGURE 2. In an attempt to identify the factors that contribute to the synergistic effect, measurements of hepatic Ni^{2+}, metallothionein, glutathione, and oxidized glutathione levels were performed in rats that received $NiCl_2$ and DDC, alone or in combination (TABLE 7).[48–50] The combined administration of $NiCl_2$ and DDC greatly increased the hepatic concentrations of nickel, evidently as a result of enhanced hepatic uptake of the lipophilic Ni–DDC complex.[49,50] DDC and $NiCl_2$, administered separately, were both potent inducers of hepatic metallothionein,[48] but the increase of hepatic metallothionein concentration was only slightly greater in rats that received $NiCl_2$ and DDC, in combination, than in rats that received DDC alone. A modest synergistic effect of $NiCl_2$ and DDC was noted on hepatic concentrations of gluthathione. These findings suggest that 1) increased hepatic uptake of Ni^{2+} is a major factor that contributes to the synergistic induction of hepatic heme oxygenase by $NiCl_2$ and DDC, and 2) disturbances of cellular constituents containing SH residues may also a play a role.[50] The possibilities that $NiCl_2$ and DDC may exert synergistic effects on enzymes in the heme biosynthetic pathway, on cytochrome P-450 concentrations, and on intracellular heme pools have not yet been investigated.

Other Metals

Soluble salts of 15 other metals have been tested in rodents for acute effects on heme oxygenase activity (TABLE 8). These metal compounds have been administered only by ip or sc injections; there is a dearth of published studies on heme oxygenase activities after exposures of animals to metal compounds by oral or respiratory routes. Arsenic (administered as As^{3+} in $NaAsO_2$) and antimony (administered as Sb^{3+} in $SnCl_3$) are clearly the most potent of the metals listed in TABLE 8 for induction of heme oxygenase activity in rat liver and kidney.[51] Treatments of rats with As^{5+} (as Na_2HAsO_4) or Sb^{5+} (as $SbCl_5$) were less effective for induction of heme oxygenase activity in liver or kidney than treatments with As^{3+} or Sb^{3+} at equivalent dosages.[51] Drummond and Kappas[59] noted that Sb^{3+} induction of renal heme oxygenase activity was not prevented by pretreatment with oral cysteine or parenteral $ZnCl_2$, whereas Co^{2+} induction of the enzyme was blocked by cysteine or Zn^{2+} under identical conditions. Mercury (administered as Hg^{2+} in $HgCl_2$) produced greater induction of heme oxygenase in adrenal than in the other tissues that have been tested.[27] Moderate induction of heme oxygenase activity in liver, kidney, or heart has been reported in rats after parenteral injections of Ag, Au, Cr, Cu, Fe, Hg, In, Mn, Pb, Pt, Se, and Zn salts (TABLE 8). Magnesium (administered to rats as Mg^{2+} in $MgCl_2$ or $MgSO_4$) failed to stimulate microsomal heme oxygenase activity in kidney or brain, respectively, at the specified dosages. Meredith and Moore[60] reported that repeated daily injections of lead acetate (10 μmol/day, ip, for 14 days) caused progressive induction of hepatic heme oxygenase activity in rats, reaching a peak increase of 1.5-fold on day 15. Acute administration of gold sodium thiomalate to rats caused greater induction of heme oxygenase in kidney than in liver.[53] Eiseman and Alvares[61] observed that heme oxygenase activity was increased 1.4-fold in fetal livers, but not in maternal livers, after repeated daily administration of gold sodium thiomalate (50 μmol/kg body wt, ip, for 3 days) to pregnant rats during late gestation. Administration of gold sodium thiomalate at the same dosage for 3 days to lactating rats increased hepatic heme oxygenase activity 1.6-fold in the dams, but did not significantly affect the enzyme activity in livers of suckling pups.[61]

SUMMARY AND CONCLUSIONS

The experimental data that have been reviewed support the following conclusions regarding metal induction of microsomal heme oxygenase activity:

1. Induction of heme oxygenase in liver, kidney, and other organs of rodents is a nonspecific, toxic response to parenteral administration of numerous metal compounds.
2. The Co^{2+} and Cd^{2+} ions are especially potent for induction of heme oxygenase in rat liver; Sn^{2+}, Ni^{2+}, and As^{3+} are especially potent for induction of the enzyme in rat kidney; Hg^{2+} is especially potent for induction of the enzyme in rat adrenal.
3. Rat spleen, testis, and brain are relatively refractory to metal induction of heme oxygenase activity; in testicular microsomes from Cd^{2+}-treated rats, heme oxygenase activity is markedly inhibited.
4. Metal induction of heme oxygenase requires *de novo* synthesis of mRNA and protein, based on 1) experiments with metabolic inhibitors (actinomycin D, puromycin, and cycloheximide) and 2) translation assays of heme oxygenase mRNA.
5. Heme oxygenase induction by metals is generally suppressed by treatments with SH compounds (for example, cysteine and glutathione) and enhanced by agents that deplete tissue SH levels (for example, diethyl maleate), suggesting that the induction mechanism may involve binding of metal ions to SH-containing regulatory molecules.
6. Administration of DDC exerts a pronounced synergistic effect on Ni^{2+} induction of heme oxygenase activity in rat tissues, attributable in part to enhanced cellular uptake of nickel.
7. Induction of heme oxygenase is not sustained during repeated daily treatments of rats with $NiCl_2$, pointing to an adaptive mechanism for tolerance to the enzyme inducer.

Investigations of metal induction of heme oxygenase activity have commonly involved sc or ip administration of the test compounds to rats. The paucity of studies in other species and the dearth of experiments with oral or respiratory exposures hinder extrapolations of the animal data to human environmental or occupational exposures to metal compounds.

REFERENCES

1. MAINES, M. D. & A. KAPPAS. 1974. Cobalt induction of hepatic heme oxygenase, with evidence that cytochrome P-450 is not essential for this enzyme activity. Proc. Natl. Acad. Sci. USA **71:** 4293–4297.
2. MAINES, M.D. & A. KAPPAS. 1975. Cobalt stimulation of heme degradation in the liver. J. Biol. Chem. **250:** 1471–1477.
3. MAINES, M. D. & A. KAPPAS. 1976. The induction of heme oxidation in various tissues by trace metals: Evidence for the catabolism of endogenous heme by hepatic heme oxygenase. Ann. Clin. Res. **9**(Suppl. 17):39–46.
4. MAINES, M. D. & A. KAPPAS. 1976. Studies on the mechanisms of induction of haem oxygenase by cobalt and other metals. Biochem. J. **154:** 125–131.
5. MAINES, M. D. & A. KAPPAS. 1977. Metals as regulators of heme metabolism. Science **198:** 1215–1221.

6. Maines, M. D. & A. Kappas. 1977. Enzymes of heme metabolism in the kidney: Regulation by trace metals which do not form heme complexes. J. Exp. Med. **146:** 1286–1293.
7. Maines, M. D. & A. Kappas. 1977. Regulation of cytochrome P-450-dependent microsomal drug-metabolizing enzymes by nickel, cobalt, and iron. Clin. Pharmacol. Ther. **22:** 780–790.
8. Drummond, G. S. 1987. Control of heme metabolism by synthetic metalloporphyrins. Ann. N.Y. Acad. Sci. This volume.
9. Maines, M. D. & P. Sinclair. 1977. Cobalt regulation of heme synthesis and degradation in avian embryo liver cell culture. J. Biol. Chem. **252:** 219–223.
10. Krasny, H. C. & D. J. Holbrook, Jr. 1978. Effects of cadmium on heme oxygenase and hemoproteins in smooth and rough endoplasmic reticulum of rat liver. Biochem. Pharmacol. **27:** 364–366.
11. Guzelian, P. S. & N. A. Elshourbagy. 1979. Induction of hepatic heme oxygenase activity by bromobenzene. Arch. Biochem. Biophys. **196:** 178–185.
12. Sardana, M. D., S. Sassa & A. Kappas. 1982. Metal ion mediated regulation of heme oxygenase induction in cultured avian liver cells. J. Biol. Chem. **257:** 4806–4811.
13. Sunderman, F. W., Jr., L. M. Bibeau & M. C. Reid. 1983. Synergistic induction of microsomal heme oxygenase activity in rat liver and kidney by diethyldithiocarbamate and nickel chloride. Toxicol. Appl. Pharmacol. **71:** 436–444.
14. Kikuchi, G., T. Yoshida & S. Ishizawa. 1982. Effects of drugs and metals on heme degradation by the heme oxygenase systems. Adv. Pharmacol. Ther. **5:** 121–130.
15. Ishizawa, S., T. Yoshida & G. Kikuchi. 1983. Induction of heme oxygenase in liver. J. Biol. Chem. **258:** 4220–4225.
16. Maines, M. D., G. M. Trakshel & R. K. Kutty. 1986. Characterization of two constitutive forms of rat liver microsomal heme oxygenase. J. Biol. Chem. **261:** 411–419.
17. De Matteis, F. & A. H. Gibbs. 1976. The effect of cobaltous chloride on liver haem metabolism in the rat. Ann. Clin. Res. **8**(Suppl. 17): 193–197.
18. Paine, A. J. & R. F. Legg. 1978. Apparent lack of correlation between the loss of cytochrome P-450 in hepatic parenchymal cell culture and the stimulation of haem oxygenase activity. Biochem. Biophys. Res. Commun. **81:** 672–679.
19. Eaton, D. L., N. H. Stacey, K. L. Wong & C. D. Klaassen. 1980. Dose-response effects of various metal ions on rat liver metallothionein, glutathione, heme oxygenase, and cytochrome P-450. Toxicol. Appl. Pharmacol. **55:** 393–402.
20. Maines, M. D. 1984. New developments in the regulation of heme metabolism and their implications. CRC Crit. Rev. Toxicol. **12:** 241–314.
21. Maines, M. D. & A. Kappas. 1977. Regulation of heme pathway enzymes and cellular glutathione content by metals that do not chelate with tetrapyrroles: Blockade of metal effects of thiols. Proc. Natl. Acad. Sci. USA **74:** 1875–1878.
22. Burk, R. F. & M. A. Correia. 1979. Stimulation of rat hepatic microsomal heme oxygenase by diethyl maleate. Res. Commun. Chem. Pathol. Pharmacol. **24:** 205–207.
23. Maines, M. D. 1982. Enhancement and inhibition of enzymes of heme metabolism by diethyl maleate in the rat kidney. Arch. Biochem. Biophys. **216:** 17–26.
24. Correia, M. A. & R. Schmid. 1975. Effect of cobalt on microsomal cytochrome P-450: Differences between liver and intestinal mucosa. Biochem. Biophys. Res. Commun. **65:** 1378–1384.
25. Maines, M. D. & R. K. Kutty. 1983. Differential response of testicular and ovarian heme oxygenase activity to metal ions. Arch. Biochem. Biophys. **226:** 134–144.
26. Maines, M. D., M. Qato & C. Carranza. 1984. Alteration of heme and cytochrome metabolism in birth by metal ions. *In* Cellular and Molecular Neurotoxicology. T. Narahashi, Ed.: 241–256. Raven Press. New York, NY.
27. Veltman, J. C. & M. D. Maines. 1986. Alterations of heme, cytochrome P-450, and steroid metabolism by mercury in rat adrenal. Arch. Biochem. Biophys. **248:** 467–478.
28. Hino, Y., H. Asagami & S. Minakami. 1979. Topological arrangement in microsomal membranes of hepatic haem oxygenase induced by cobalt dichloride. Biochem. J. **178:** 331–337.
29. Sardana, M. K., S. Sassa & A. Kappas. 1981. Differential responses to inducers of

δ-aminolaevulinic synthase and haem oxygenase during pregnancy. Biochem. J. **198:** 403–408.

30. MAINES, M. D. & A. KAPPAS. 1975. Study of the developmental pattern of heme catabolism in liver and the effects of cobalt on cytochrome P-450 and the rate of heme oxidation during the neonatal period. J. Exp. Med. **141:** 1400–1410.
31. WOODS, J. S. & G. T. CARVER. 1977. Action of cobalt chloride on the biosynthesis, degradation, and utilization of heme in fetal rat liver. Metab. Disp. **5:** 487–492.
32. YOSHIDA, T., Y. SUZUKI & Y. HASHIMOTO. 1979. Sex-related effect of cadmium on hepatic cytochrome P-450, drug-metabolizing enzymes, δ-aminolevulinic acid synthetase and heme oxygenase in the rat. Toxicol. Lett **4:** 97–102.
33. YOSHIDA, T., M. OKAMOTO, Y. SUZUKI & Y. HASHIMOTO. 1979. Cadmium-induced alterations in the activities of hepatic δ-aminolevulinic acid synthetase and heme oxygenase in mice. J. Pharmacol. Dyn. **2:** 84–91.
34. YOSHIDA, T. & S. ISHIZAWA. 1981. Question on the biological significance of metal ion interactions in the control of heme oxygenase induction. Biochem. Int. **3:** 181–187.
35. MAINES, M. D., A. S. CHUNG & R. K. KUTTY. 1982. The inhibition of testicular heme oxygenase by cadmium. J. Biol. Chem. **257:** 14116–14121.
36. MAINES, M. D. 1984. Characterization of heme oxygenase activity in Leydig and Sertoli cells of the rat testes. Biochem. Pharmacol. **33:** 1493–1502.
37. TRAKSHEL, G. M., R. K. KUTTY & M. D. MAINES. 1986. Purification and characterization of the major constitutive form of testicular heme oxygenase. J. Biol. Chem. **261:** 11131–11137.
38. KAPPAS, A. & M. D. MAINES. 1976. Tin: A potent inducer of heme oxygenase in kidney. Science **192:** 60–62.
39. DRUMMOND, G. S. & A. KAPPAS. 1980. Metal ion interactions in the control of haem oxygenase induction in liver and kidney. Biochem. J. **192:** 637–648.
40. KUTTY, R. K. & M. D. MAINES. 1984. Effects of induction of heme oxygenase by cobalt and tin on the *in vivo* degradation of myoglobin. Biochem. Pharmacol. **33:** 2924–2926.
41. DWIVEDI, R. S., G. KAUR, R. C. SRIVASTAVA & C. R. K. MURTI. 1985. Role of tin on heme and drug biotransformation mechanism of partially hepatectomized rats. Ind. Health **23:** 1–7.
42. DRUMMOND, G. S. & A. KAPPAS. 1979. Manganese and zinc blockade of enzyme induction: Studies with microsomal heme oxygenase. Proc. Natl. Acad. Sci. USA **76:** 5331–5335.
43. MAINES, M. D. & A. KAPPAS. 1977. Nickel-mediated alterations in the activity of hepatic and renal enzymes of heme metabolism and heme-dependent cellular activities. *In* Clinical Chemistry and Chemical Toxicology of Metals. S. S. Brown, Ed.: 75–81. Elsevier. Amsterdam.
44. MAINES, M. E. 1981. Nickel alterations of heme biosynthesis and degradation: Implications for the oxidative degradation of drugs and carcinogens. *In* Nickel in the Environment. J. O. Nriagu, Ed.: 547–568. John Wiley & Sons. New York, NY.
45. SUNDERMAN, F. W., JR., J. R. DOWNS, M. C. REID & L. M. BIBEAU. 1982. Gas chromatographic assay for heme oxygenase activity. Clin. Chem. **28:** 2026–2032.
46. SUNDERMAN, F. W., JR., M. C. REID, L. M. BIBEAU & J. V. LINDEN. 1983. Nickel induction of microsomal heme oxygenase activity in rodents. Toxicol. Appl. Pharmacol. **68:** 87–95.
47. SUNDERMAN, F. W., JR., C. PELZ & O. ZAHARIA. Effects of short-term $NiCl_2$ treatments on microsomal heme oxygenase activity in kidney and liver of Fischer rats. Paper presented at the IUPHAR 9th International Congress of Pharmacology. Abstract 1371-P. August 1984. London.
48. SUNDERMAN, F. W., JR. & C. B. FRASER. 1983. Effects of nickel chloride and diethyldithiocarbamate on metallothionein in rat liver and kidney. Ann. Clin. Lab. Sci. **13:** 489–495.
49. SUNDERMAN, F. W., JR., O. ZAHARIA, M. C. REID, J. F. BELLIVEAU, G. P. O'LEARY, JR. & H. GRIFFIN. 1984. Effects of diethyldithiocarbamate and nickle chloride on glutathione and trace metal concentrations in rat liver. Toxicology **32:** 11–21.
50. SUNDERMAN, F. W., JR. 1985. Effects of $NiCl_2$ and diethyldithiocarbamate on heme oxygenase, metallothionein, glutathione, and trace metals in rat liver and kidney. *In* Progress in Nickel Toxicology. S. S. Brown & F. W. Sunderman, Jr., Eds.: 157–160. Blackwells. Oxford.

51. Sardana, M. K., G. S. Drummond, S. Sassa & A. Kappas. 1981. The potent heme oxygenase inducing action of arsenic and parasiticidal arsenicals. Pharmacology **23:** 247–253.
52. Maines, M. D. 1979. Role of trace metals in regulation of cellular heme and hemoprotein metabolism: Sensitizing effects of chronic iron treatment on acute gold toxicity. Drug Metab. Rev. **9:** 237–255.
53. Eiseman, J. L. & A. P. Alvares. 1978. Alterations induced in heme pathway enzymes and monooxygenases by gold. Mol. Pharmacol. **14:** 1176–1188.
54. Leonard, T. B., M. E. Graichen, L. J. Dahm & J. G. Dent. 1986. Effects of the crysotherapeutic agents aurofin and gold sodium thiomalate on hepatic and renal drug metabolism and heme metabolism. Biochem. Pharmacol. **35:** 3057–3063.
55. Ariyoshi, T., T. Nakai, N. Terada, K. Hamasaki & K. Arizono. 1981. Effect of acute iron overload on heme metabolism. J. Pharmacol. Dyn. **4:** S75.
56. Woods, J. S., G. T. Carver & B. A. Fowler. 1979. Altered regulation of hepatic heme metabolism by indium chloride. Toxicol. Appl. Pharmacol. **49:** 455–461.
57. Jollie, D. R. & M. D. Maines. 1985. Effect of *cis*-platinum on kidney cytochrome P-450 and heme metabolism: Evidence for the regulatory role of the pituitary hormones. Arch. Biochem. Biophys. **240:** 51–59.
58. Maines, M. D. 1986. Differential effect of *cis*-platinum (*cis*-diamminedichloroplatinum) on regulation of liver and kidney haem and haemoprotein metabolism. Biochem. J. **237:** 713–721.
59. Drummond, G. S. & A. Kappas. 1981. Potent heme-degrading action of antimony and antimony-containing parasiticidal agents. J. Exp. Med. **153:** 245–256.
60. Meredith, P. A. & M. R. Moore. 1979. The influence of lead on haem biosynthesis and biodegradation in the rat. Biochem. Soc. Trans. **7:** 637–639.
61. Eiseman, J. L. & A. P. Alvares. 1980. Heme metabolism in the fetoplacental unit and neonatal livers during lactation: Effect of gold sodium thiomalate. J. Pharmacol. Exp. Ther. **214:** 250–256.

Metal Inhibition of Ferrochelatase

HARRY A. DAILEY

Department of Microbiology
University of Georgia
Athens, Georgia 30602

Ferrochelatase (protoheme ferrolyase EC 4.99.1.1), the terminal enzyme of the heme biosynthetic pathway, catalyzes the insertion of iron into protoporphyrin IX to form protoheme. The enzyme is bound to the inner mitochondrial membrane with its active site oriented toward the matrix side of the membrane.[1] Ferrochelatase has been purified from a variety of sources, and details of its substrate specificity and kinetic parameters have been reported.[2-4] The enzyme utilizes ferrous, but not ferric, iron and the divalent cations of cobalt and zinc. Based upon chemical inactivation and kinetic studies it has been proposed that the first event in catalysis is the binding of ferrous iron via vicinal sulfhydryl groups[5] on ferrochelatase. Below we report on the effects that a variety of heavy metals have on ferrochelatase activity in cultured cells and on the inhibitory activity of the same metal ions on purified ferrochelatase.

MATERIALS AND METHODS

Bovine and mouse ferrochelatase were purified as described previously[2,3] and were assayed with either the pyridine hemochromogen[6] or ^{59}Fe[3] method. Protoporphyrinogen oxidase was assayed fluorometrically by a modification of the procedure of Brenner and Bloomer.[7]

MEL Cultures

Friend virus-transformed murine erythroleukemia (MEL) cells were maintained in culture at 37° C in 5% CO_2 in Dulbecco's Modified Eagle Medium supplemented with 15% heat-inactivated fetal bovine serum. Cell lines (DS19) were maintained in culture as described previously.[8] Cultures (10^5 to 10^6 cells/ml) were induced to differentiate with 2% dimethylsulfoxide (DMSO) when appropriate. Upon DMSO induction, metal solutions were added to cell cultures (TABLE 1). Metal stock solutions (1 mM) were freshly prepared in cell culture media for each experiment.

Determination of Heme Content

The fluorometric assay for determination of heme content has been described by Sassa *et al.*[8] Endogenous porphyrins in the cells were measured by preparing a tissue blank that contained the same amount of cells, boiled for 30 sec in the presence of 1.0% sodium dodecylsulfate. Fluorescence was determined with a Perkin–Elmer 650-40 fluorescence spectrophotometer (405 nm excitation, 635 nm emission).

TABLE 1. Effects of Metal Ions on Purified Ferrochelatase[a]

Metal Solution[b]	Relative Activity[c]
NaCl	100
KCl	98
$NiCl_2$	100
$HgCl_2$	15
$MnCl_2$	44
$MgSO_4$	96
$CuSO_4$	104
$Pb(NO_3)_2$	37
$CdCl_2$	70

[a]Assays were conducted as described in Materials and Methods. All metal solutions were freshly prepared the day of the experiment.
[b]Final concentration for each metal ion was 10 μM.
[c]Each value is a percentage of the control activity.

Source of Compounds

Dulbecco's Modified Eagle Medium was purchased from Grand Island Biological Co. (Grand Island, NY). Protoporphyrin IX, mesoporphyrin IX, and hemin were obtained from Porphyrin Products (Logan, UH). All other reagents were of the highest purity available.

RESULTS

Effect of Divalent Cations on Ferrochelatase and Protoporphyrinogen Oxidase

The effect of a variety of divalent cations on purified bovine ferrochelatase activity was examined, and the results of these assays are shown in TABLE 1. In all cases, except with Pb^{2+}, the chloride salts of the metals were used. These same metals were tested with purified mouse ferrochelatase, and essentially identical results were found (except for Pb^{2+}, where 100 μM Pb^{2+} caused only about 50% inhibition). Cadmium chloride was also tested with the mouse enzyme, and it was found that 10 μM $CdCl_2$ caused about 30% inhibition. Purified mouse protoporphyrinogen oxidase activity was also determined in the presence of these divalent cations. It was found that none of these metals had any effect, either inhibitory or stimulatory, on protoporphyrinogen oxidase.

Inhibition of Purified Ferrochelatase by Mn^{2+} and Cd^{2+}

Manganese and cadmium ions, which were found to inhibit ferrochelatase activity, were examined to determine their mode of inhibition. It was found that both Mn^{2+} and Cd^{2+} are competitive inhibitors of ferrochelatase with respect to Fe^{2+} (FIGURE 1). Secondary plots produced calculated K_i values of 15 μM for $MnCl_2$ and 50 μM for $CdCl_2$. Although Co^{2+} and Zn^{2+} were not tested, it would be expected that both of these would also be found to be competitive inhibitors with respect to Fe^{2+} because both serve as substrates for the enzyme.

Inhibition by Hg^{2+} and Arsenite

Ferrochelatase is strongly inhibited by $HgCl_2$ (TABLE 1), with 85% inhibition at 10 μM Hg^{2+}. One mechanism of inhibition by Hg^{2+} is via interactions with protein sulfhydryl residues. Arsenite is also known to bind to some proteins via vicinyl sulfhydryl residues. Because of this binding ability, arsenite was tested to see if it had any effect on purified ferrochelatase activity. In these assays dithiothreitol was omitted from the reactions because it would bind with all of the added arsenite.[5] The data from this experiment showed that inclusion of 1 mM sodium arsenite inhibited activity 92% and that addition of 5 mM dithiothreitol to the assay mixture restored activity.

Effect of Metals on Differentiating MEL Cells

MEL cells were induced to differentiate in the presence of added $CoCl_2$, $CdCl_2$, $MnCl_2$, or $Pb(NO_3)_2$, and the effects of these various metals at a variety of concentrations on the ability of the cells to produce heme was examined. Cobaltous chloride above 200 μM was cytotoxic, but below that concentration cell growth appeared unaffected. The effect of 15–180 μM $CoCl_2$ on heme production is shown in FIGURE 2. Addition of $MnCl_2$ at over 15 μM proved toxic to cells; concentrations below 15 μM caused decreased heme production (FIGURE 3). Cadium chloride above 100 μM was also cytotoxic, but, in the range of 0–50 μM, it inhibited heme production without

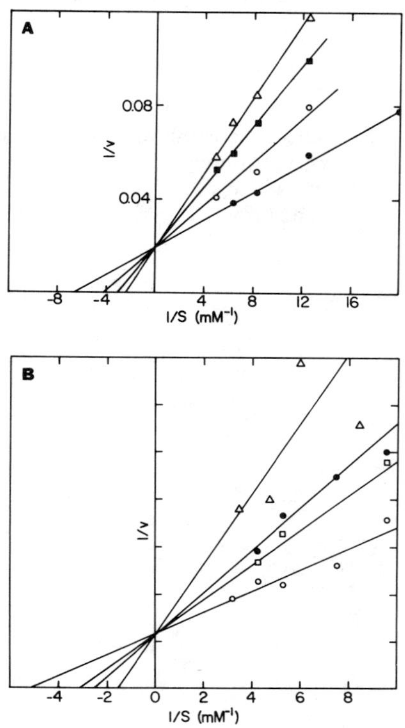

FIGURE 1. Lineweaver–Burke plots for ferrochelatase. (**A**) Inhibition of activity by Mn^{2+} with respect to Fe^{2+}. ●: no added $MnCl_2$; ○: 24 μM $MnCl_2$; ■: 50 μM $MnCl_2$; and △: 100 μM $MnCl_2$. (**B**) Inhibition of activity by Cd^{2+}. ○: no added $CdCl_2$; □: 10 μM $CdCl_2$; ●: 25 μM $CdCl_2$; and △: 100 μM $CdCl_2$.

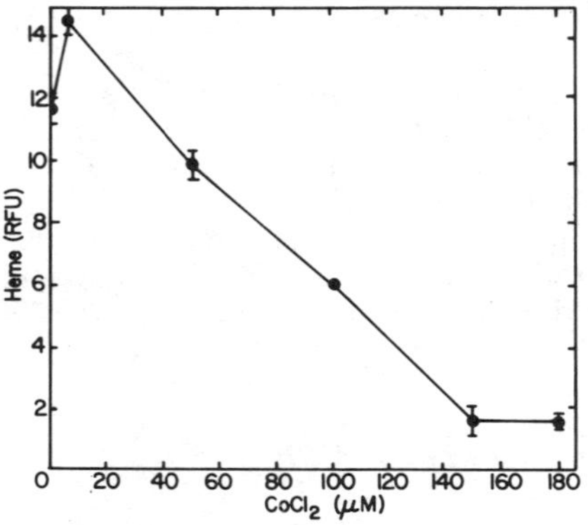

FIGURE 2. The effects of $CoCl_2$ on heme production in MEL cells induced with 2% DMSO. The points plotted on the graph represent the means of the duplicates. RFU: relative fluorescence units.

cell destruction (FIGURE 4). In MEL cells, $Pb(NO_3)_2$ above 25 μM was cytotoxic, but below that concentration there was no inhibition of heme formation (data not shown). For all of these metal ions examined, there was no detectable increase in cellular free porphyrins.

The effect of these same metals, when added in culture, on ferrochelatase activity is shown in TABLE 2. Whenever heme production *in vivo* was decreased, *in vitro* ferrochelatase activity of these cells was also decreased.

DISCUSSION

Ferrochelatase catalyzes the insertion of Fe^{2+} into protoporphyrin to form protoheme. Kinetic studies have suggested that the binding of iron occurs before binding of the porphyrin substrate.[2] Chemical modification studies indicate that the binding or orientation of the porphyrin substrate is mediated in part via charge pair interactions

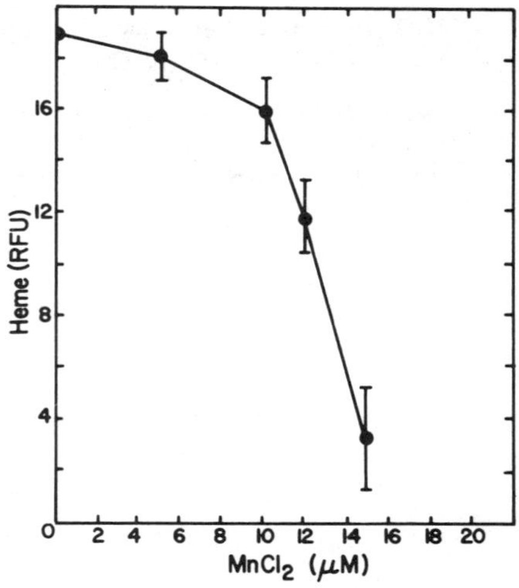

FIGURE 3. The effect of $MnCl_2$ on heme production in MEL cells 72 hr after induction with 2% DMSO. The points plotted on the graph represent the means of duplicates. RFU: relative fluorescence units.

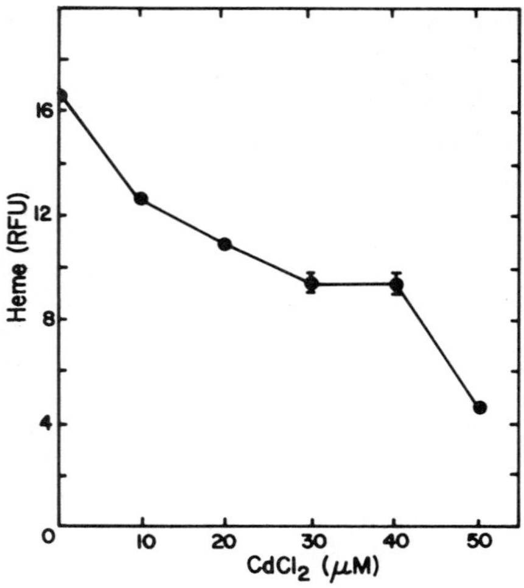

FIGURE 4. The effect of $CdCl_2$ on the heme production in MEL 72 hr after induction with 2% DMSO. The points plotted on the graph represent the means of the duplicates. RFU: relative fluorescence units.

that involve an enzyme arginyl residue.[9] The presence of porphyrin prevents the inactivation of the enzyme by arginyl reagents, but Fe^{2+} has no protective effect.

Conversely, the binding of Fe^{2+} appears to be mediated via vicinyl sulfhydryl residues.[5] Data presented above and published previously show that a variety of sulfhydryl reagents can inactivate ferrochelatase. For the reagent *N*-ethylmaleimide it has been shown that reaction of a single sulfhydryl residue is sufficient to inactivate the enzyme. Ferrous iron, but not porphyrin, protects ferrochelatase from inactivation by *N*-ethylmaleimide as well as from inactivation by a variety of other sulfhydryl reactive reagents.

Ferrochelatase catalyzes the insertion of Co^{2+} and Zn^{2+} in addition to that of Fe^{2+}, so these two metals can serve as competitive inhibitors.[10] Manganese and cadmium ions at micromolar concentrations were also found to inhibit ferrochelatase activity. The kinetics of inhibition by these two metal ions have been examined, and the data obtained suggest that both are competitive inhibitors of the enzyme with respect to Fe^{2+}. The calculated K_i values are 15 μM ($MnCl_2$) and 50 μM ($CdCl_2$).

A variety of cations that were shown to effect ferrochelatase activity were also examined for the effect upon heme biosynthesis in differentiating MEL cells. It was found that Mn^{2+}, Cd^{2+}, and Co^{2+} at micromolar concentrations strongly affected heme biosynthesis, and it is apparent that these divalent cations were exerting their effect on

TABLE 2. Effects of $CoCl_2$, $CdCl_2$, and $MnCl_2$ on Ferrochelatase Activity[a]

Addition to Culture	Relative Activity[b]
$CoCl_2$, 10 μM	100
$CoCl_2$, 50 μM	48
$MnCl_2$, 15 μM	56
$CdCl_2$, 8.6 μM	64

[a]Cells of clone DS19 were induced with 2% DMSO, and either $CoCl_2$, $MnCl_2$, or $CdCl_2$ was added at the concentration indicated. An aliquot from each was analyzed 72 hr after induction for ferrochelatase activity as described in Materials and Methods.

[b]Each value is a percentage of the control activity.

heme biosynthesis because of their effect on ferrochelatase activity. Although it might be argued that these divalent cations have their effect on iron uptake or on some other system with inhibition of ferrochelatase activity being of only secondary importance, the fact that similar concentrations of the metals affect both cellular heme production and ferrochelatase activity suggests that these cations can express their effect by inhibition of ferrochelatase.

Although it is of interest to note that the concentrations at which Mn^{2+} and Cd^{2+} inhibit heme production in MEL cells are both approximately the same concentration as the K_i for purified ferrochelatase, it is impossible to correlate the two because the intracellular concentration of either ion may be considerably different from the culture concentration.

Under no conditions did we find significant cellular accumulation of protoporphyrin. Although this might have been expected from inhibition of ferrochelatase, it is possible that any excess protoporphyrin produced was excreted into the medium. Because of the large amount of serum in the medium (15%) and the relatively low levels of porphyrin expected to be produced, we did not attempt to quantitate media protoporphyrin.

In conclusion, it can be stated that ferrochelatase, but not protoporphyrinogen oxidase, is sensitive to inhibition by a variety of divalent cations. In particular, the heavy metals Hg^{2+}, Mn^{2+}, Cd^{2+}, and Co^{2+} were shown to inhibit ferrochelatase activity *in vitro*. These metals, when added to the culture media of differentiating MEL cells, also caused the decreased production of cellular heme.

SUMMARY

Ferrochelatase activity was examined both in growing MEL cells and in *in vitro* assays of the purified enzyme to determine what effect a variety of divalent cations would have. Data obtained with the purified enzyme demonstrated that Mn^{2+} strongly inhibits the activity in a competitive fashion with respect to Fe^{2+} with a calculated K_i of 15 μM. Cadmium ion is also inhibitory (K_i: 50 μM), as is Hg^{2+} and arsenite, but Pb^{2+} is a poor inhibitor. All other metals tested had no effect. When these same metal ions were tested on differentiating MEL cells it was found that their ability to inhibit both heme formation and ferrochelatase activity mimicked their *in vitro* effect on purified ferrochelatase.

REFERENCES

1. HARBIN, B. M. & H. A. DAILEY. 1985. Biochemistry **24:** 366–370.
2. DAILEY, H. A. & J. E. FLEMING. 1983. J. Biol. Chem. **258:** 11453–11459.
3. DAILEY, H. A., J. E. FLEMING & B. M. HARBIN. 1986. Methods Enzymol. **123:** 401–408.
4. DAILEY, H. A. & A. SMITH. 1984. Biochem. J. **223:** 441–445.
5. DAILEY, H. A. 1984. J. Biol. Chem. **259:** 2711–2715.
6. DAILEY, H. A. 1986. Methods Enzymol. **123:** 408–415.
7. BRENNER, D. A. & J. R. BLOOMER. 1980. Clin. Chim. Acta **100:** 259–266.
8. SASSA, S., J. L. GRANICK, H. EISEN & W. OSTERTAG. 1978. In *In Vitro* Aspects of Erythropoiesis. M. J. Murphy, Ed.: 135–142. Springer-Verlag. New York, NY.
9. DAILEY, H. A. & J. E. FLEMING. 1986. J. Biol. Chem. **261:** 7902–7905.
10. JONES, M. S. & O. T. G. JONES. 1969. Biochem. J. **113:** 507–514.

Control of Heme Metabolism by Synthetic Metalloporphyrins[a]

GEORGE S. DRUMMOND

The Rockefeller University Hospital
New York, New York

INTRODUCTION

The red pigment heme consists of four nonidentical subunits called pyrroles that are joined together to form a single large tetrapyrrole (porphyrin) ring structure. Iron is the central metal atom of heme, and the ability of iron to undergo reversible changes in oxidation state is essential to the involvement of the heme molecule in a number of physiological systems. In these systems heme is the prosthetic group of both circulating proteins (such as hemoglobin) and membrane-bound proteins (such as the cytochromes) that are concerned with cellular respiration, energy generation, oxygen transport, and chemical oxidations and biotransformations. Inorganic iron can mimic some of the catalytic properties of heme; however, incorporation of iron into the protoporphyrin macrocycle and binding of this metalloporphyrin complex to specific apoproteins can enhance the catalytic activity of iron by several hundred fold.

Chemical perturbations of δ-aminolevulinic acid (ALA) synthase and heme oxygenase, the rate-limiting enzymes of heme synthesis and degradation, respectively, have been extensively studied in experiments on heme biology: the chemicals utilized for this purpose have included metals, drugs, and a variety of organic compounds including organometals. However, apart from the use of heme itself to suppress exacerbations of the genetic disease acute intermittent porphyria (AIP), an action that is attributed to heme repression of hepatic ALA synthase formation, there have been few investigations of the potential clinical use of the biological properties of metalloporphyrins in which metals other than iron have figured as the central metal atom.

We have studied in detail over the last several years the ability of certain synthetic metalloporphyrins to both inhibit and induce heme catabolism *in vitro* and *in vivo*. The results of these studies have defined novel biological properties of these synthetic heme analogues that are both quantitatively and qualitatively different from those exhibited by heme. In addition, it may be possible to envisage potentially useful clinical and experimental roles for these compounds.

METHODS

Male (160–200 g) and 15-day-synchronized pregnant rats were purchased from Taconic Farms (Germantown, NY) and Holtzman (Madison, WI), respectively. The animals were maintained on standard rat chow and drinking water *ad libitum*, housed

[a]These studies were supported in part by the U.S. Public Health Service (Grant ES01055), by the New York Community Trust, by the March of Dimes Birth Defects Foundation (Grant I-804), and by gifts from the Dyson Foundation, the Ogden Corporation, and the Renfield Foundation.

in the Rockefeller University Laboratory Animal Research Facility, and maintained in a controlled environment with a 12 hr:12 hr light/dark cycle. Metalloporphyrins were purchased from Porphyrin Products (Logan, UT), and solutions were prepared in subdued light by dissolving the metalloporphyrin in a small volume of 0.5 M NaOH (0.2 ml/1.0 ml of final volume of metalloporphyrin solution) adjusted to pH 7.4 with 1 N HCl and made up to final volume with 0.9% NaCl (saline). All other chemicals were of the highest grade obtainable from either Fisher or Sigma. The dose of metalloporphyrin administered is indicated in the legends to the figures and tables; microsomes were prepared and assays performed as previously described.[1]

RESULTS AND DISCUSSION

A series of synthetic metalloporphyrins were examined for their ability to inhibit heme oxygenase activity *in vitro*. These studies were carried out using human spleen as a source of microsomal heme oxygenase. Five of the compounds examined—Sn(tin)-protoporphyrin, Cr(chromium)-protoporphyrin, Mn(manganese)-protoporphyrin,

TABLE 1. Inhibition of Heme Oxygenase *in Vitro* by a Series of Metalloporphyrins

Metalloporphyrin	K_i (μM)
Sn-protoporphyrin	0.018 ± 0.001
Cr-protoporphyrin	0.033 ± 0.009
Mn-protoporphyrin	0.530 ± 0.140
Co-protoporphyrin	0.670 ± 0.190
Zn-protoporphyrin	2.523 ± 0.720

NOTE: The K_i were determined using double reciprocal plots of substrate concentration versus velocity of human splenic microsomal heme oxygenase in the presence of the appropriate metalloporphyrin.

Co(cobalt)-protoporphyrin, and Zn(zinc)-protoporphyrin—proved to be competitive inhibitors of heme oxygenase *in vitro;* Sn-protoporphyrin was by far the most potent of the compounds examined (TABLE 1). The K_i (inhibition coefficient) of Sn-protoporphyrin was 1.8×10^{-8} M whereas the K_m for human spleen heme oxygenase was reported to be 1.64×10^{-5} M.[2]

The five synthetic metalloporphyrins that were inhibitors of heme oxygenase activity *in vitro* (TABLE 1) were administered to adult male rats and the activity of the enzyme in liver measured 16 hr after the administration of the metalloporphyrin. The administration of Sn-, Cr-, Mn-, or Zn-protoporphyrin resulted in a significant decrease in the activity of hepatic heme oxygenase; in contrast, Co-protoporphyrin administration resulted in a marked and significant increase in the activity of hepatic heme oxygenase (TABLE 2).

On the basis of these findings (TABLES 1 & 2), which indicate that within this group of compounds (metalloporphyrins) there exist both inducers and inhibitors of the synthesis or the catalytic function of heme oxygenase, and on the basis of the well-documented effects of the inorganic elements on heme metabolism,[3] the protoporphyrin chelates of tin and cobalt were extensively studied with respect to their ability to affect heme homeostasis. This includes the synthesis and function of heme in the form of cytochrome P-450.

TABLE 2. Effect of Metalloporphyrins on Heme Oxygenase in Liver

Metalloporphyrin	Heme Oxygenase Activity (nmol bilirubin/mg/hr)
Control	2.37 ± 0.12
Sn-protoporphyrin	1.08 ± 0.13[a]
Cr-protoporphyrin	0.29 ± 0.04[a]
Mn-protoporphyrin	1.05 ± 0.16[a]
Co-protoporphyrin	10.92 ± 0.79[a]
Zn-protoporphyrin	1.24 ± 0.14[a]

NOTE: Metalloporphyrins were administered subcutaneously in a single dose (50 μmol/kg body wt) to rats that were sacrificed 16 hr later. Three to six animals were used for each determination (each determination is a mean ± SEM).

[a] $p < .01$.

Sn-protoporphyrin: Inhibitor of Heme Oxygenase

Sn-protoporphyrin (FIGURE 1) inhibits heme oxygenase activity both *in vitro* (TABLE 1) and *in vivo* (TABLE 2). This is in direct contrast to inorganic tin and organotin compounds, which are powerful inducers of this enzyme in kidney[4] and liver,[5] respectively. Thus the ability of this synthetic metalloporphyrin to inhibit heme degradation considerably diminishes the rate of heme catabolism to bile pigments. It should be noted that Sn-protoporphyrin does not itself bind molecular oxygen and is thus *not* degraded to bile pigments. In addition, the administration of small amounts of Sn-protoporphyrin to mammals results in 1) an immediate (30–60 min) complete heme saturation of the heme-dependent enzyme tryptophan pyrrolase in liver, an effect that declines with time over a 72-hr period[6]; 2) a substantial increase in heme output in bile,[7] an effect in mammals that can last for 4–5 days (the heme that is excreted can be derived from both endogenous and exogenous sources); 3) a marked decrease in biliary bilirubin output[8] that is rapid in onset and can last for 4–5 days (FIGURE 2); 4) a diminution that lasts for 12–24 hr of the production of carbon monoxide derived from the catabolism of endogenous and exogenous heme[9]; 5) an elimination of the increased output in biliary bilirubin that occurs after the infusion of heme and a reduction in the increased output of this bile pigment produced by administration of heat-damaged red blood cells[8]; and 6) a prompt excretion of Sn-protoporphyrin (19–35% of the administered dose) in bile.[8]

In the experimental model of the bile duct cannulated rat, the infusion of heme, which is taken up primarily in hepatocytes, or of heat-damaged red blood cells, which are thought to be taken up mainly by the spleen, resulted in a significant increase in the levels of biliary bilirubin. Sn-protoporphyrin treatment either prevented (heme) or reduced (heat-damaged red blood cells) these increases.[8] It is possible that some

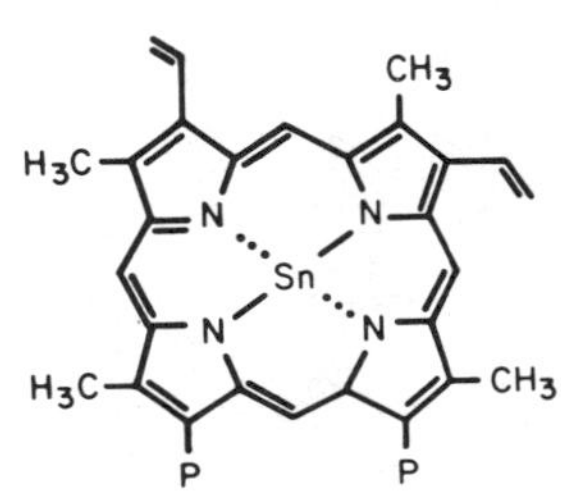

FIGURE 1. The structure of Sn-protoporphyrin.

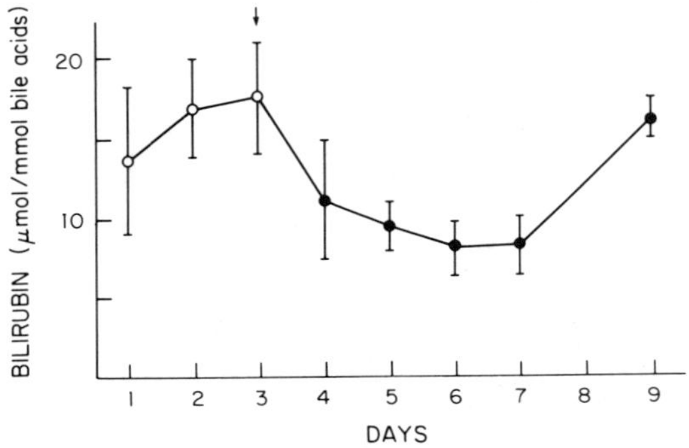

FIGURE 2. The effect of Sn-protoporphyrin on bilirubin concentration in bile. The values are the mean values for 10 healthy human volunteers and are expressed as bilirubin:bile acid ratios. The time of administration of Sn-protoporphyrin is indicated by the arrow.

fraction of the heme of the heat-damaged red cells exits the spleen and reaches the liver undegraded. Thus the decrease in biliary bilirubin output noted after infusion of heat-damaged red cells may reflect both hepatic and splenic inhibition of heme oxidation by Sn-protoporphyrin. Therefore the question of whether Sn-protoporphyrin will inhibit the normal extent of heme degradation that results from red cell senescence, which takes place primarily in the spleen, remains unresolved. However, the ability of Sn-protoporphyrin to reduce the *in vivo* production of bilirubin in both animals[8] and man,[10] together with its ability to significantly reduce carbon monoxide production in animals,[9] clearly establishes that this metalloporphyrin can decrease bilirubin levels in mammals by inhibiting bilirubin production *in vivo*.

Extensive toxicology studies have been completed in animals, and Sn-protoporphyrin has proved to be quite innocuous. No acute toxicity has been noted in adult or neonatal rats or mice in studies involving over 5000 animals; the therapeutic/toxicity ratio is at least 150:1 in newborn rats. In addition, long-term weekly treatment of genetically anemic mice with hemolytic anemia (sph^{ha}/sph^{ha}) with Sn-protoporphyrin (100 μmol/kg body wt for 32 weeks, a cumulative dose of 3200 μmol/kg body wt) did not alter hematological indices, histological findings, or enzyme activities related to heme biosynthesis, even though it resulted in sustained decreases in hepatic, renal, and splenic heme oxygenase activity and a prolonged decrease in serum bilirubin levels.[11] In humans a transient mild photosensitivity has been noted in some subjects after administration of the metalloporphyrin at doses exceeding 1.0 μmol/kg body wt.[10] This has been the only adverse action of this compound so far identified in over 150 studies in adults. Sn-protoporphyrin also suppresses both naturally occurring and experimentally induced jaundice in newborn (FIGURE 3) and adult mammals, including

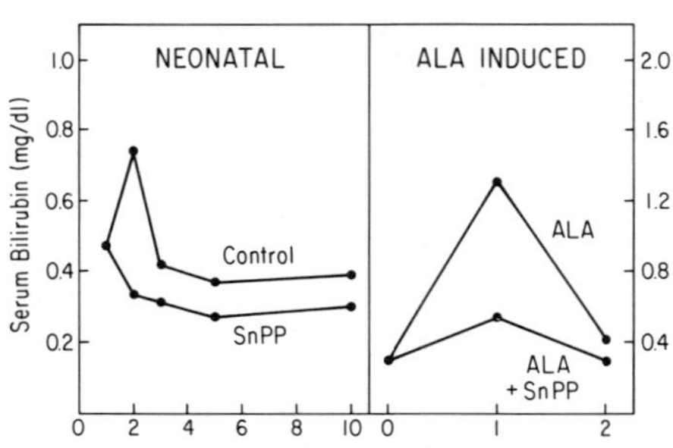

FIGURE 3. The effect of Sn-protoporphyrin on serum bilirubin levels in neonatal and ALA-induced hyperbilirubinemia in 7-day-old suckling rats. The Sn-protoporphyrin (10 μmol/kg body wt) was administered sc within 2 hr of birth. In the experimental form of hyperbilirubinemia, ALA (50 μmol/100 g body wt) was administered ip at 0, 4, and 8 hr whereas Sn-protoporphyrin (50 μmol/kg body wt) was administered sc once, at time 0.

primates[12] and humans.[10] In addition, Sn-protoporphyrin, when administered before birth, crosses the placental membranes, inhibits fetal heme oxygenase, and suppresses the transient hyperbilirubinemia that occurs in the neonate after birth.[13]

These studies clearly demonstrate the ability of Sn-protoporphyrin to significantly

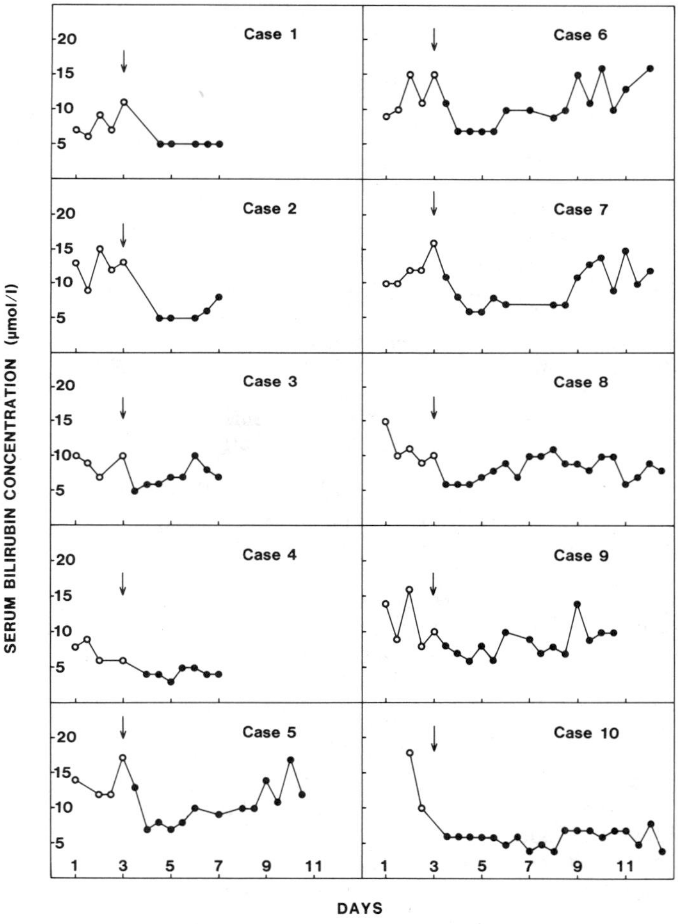

FIGURE 4. The effect of Sn-protoporphyrin on bilirubin concentration in serum of normal volunteers. The time of administration of Sn-protoporphyrin is indicated by the arrow.

decrease heme catabolism and by so doing to lower elevated levels of serum bilirubin in animals and man (FIGURE 4).

The ability of a synthetic metalloporphyrin to inhibit heme oxygenase activity *in vitro* does not, however, of itself, assure that the metalloporphyrin will suppress hyperbilirubinemia in the whole animal. Co-protoporphyrin exemplifies this point (TABLES 1 & 2). Mn-protoporphyrin also has no effect on serum bilirubin levels in the neonate[1] (this metalloporphyrin appears to facilitate the developmental increase in splenic heme oxygenase) although this compound does inhibit heme oxygenase *in vitro* (TABLE 1). Thus the demonstration of heme oxygenase inhibition *in vitro* by a synthetic metalloporphyrin does not guarantee that the compound will act in a similar fashion *in vivo*. The physicochemical characteristics of individual metalloporphyrins as well as biological variables, which may include pharmacokinetics, tissue distribution, disposition, and subcellular localization of these compounds, are all factors that appear to be important in defining the action of individual metalloporphyrins *in vivo*.

Sn-protoporphyrin suppresses the formation of the potentially toxic metabolite bilirubin *in vivo;* it concurrently enhances the disposal of heme, the untransformed substrate of the enzyme it inhibits; and its own elimination occurs without metabolic alteration—these properties suggest that it, or a similarly acting compound, can be developed into an effective therapeutic agent for use in clinical situations in which excessive hyperbilirubinemia occurs as a result of exaggerated rates of heme catabolism.

Co-protoporphyrin: Inducer of Heme Oxygenase

Of the initial series of metalloporphyrins examined in animals, three, Fe(iron)-, Cd(cadmium)-, and Co-protoporphyrin (FIGURE 5), produced a significant increase in hepatic heme oxygenase activity when this enzyme activity was measured 16 hr after metalloporphyrin administration.[1] The ability of heme (Fe-protoporphyrin) to control its own metabolism by inhibiting its own biosynthesis and stimulating its own catabolism is well documented.

A series of transition and a number of heavy metals have been shown to be inducers of hepatic heme oxygenase activity.[3] This inducing action, however, is short-lived with enzyme activity having largely returned to normal levels 48–72 hr after metal administration. In contrast, the incorporation of certain inorganic metals in the protoporphyrin macrocycle serves to markedly enhance their ability to induce hepatic heme oxygenase. Cobalt is a case in point (FIGURE 6). A single administration of inorganic cobalt produces an induction of heme oxygenase in liver that becomes maximal at 16 hr and subsides by 72 hr. This is associated with a decrease followed by a rebound increase in ALA synthase activity and a concurrent decrease in cytochrome P-450 levels.[14] In contrast, the administration of a single equimolar dose of Co-protoporphyrin produces a sevenfold increase in hepatic heme oxygenase *in vivo* within 8 hr, and this increase is sustained for prolonged periods of time, up to 10 days. A full return to normal levels of enzyme activity is not achieved until 30–35 days after the administration of a single dose of Co-protoporphyrin (FIGURE 6). The Co-protoporphyrin thus exhibits a dual action with respect to heme oxygenase activity; the metalloporphyrin is a competitive inhibitor of heme oxygenase activity *in vitro* (TABLE 1) and a

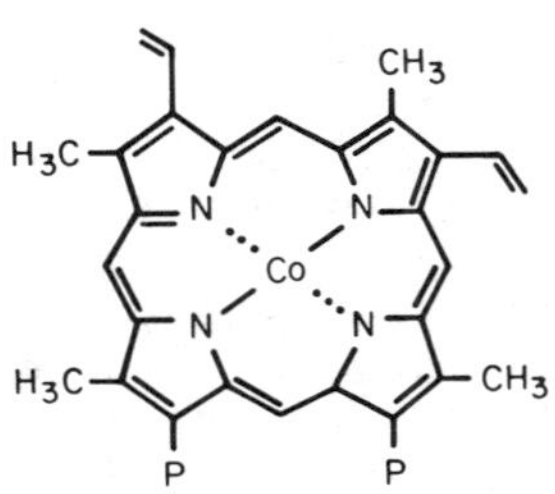

FIGURE 5. The structure of Co-protoporphyrin.

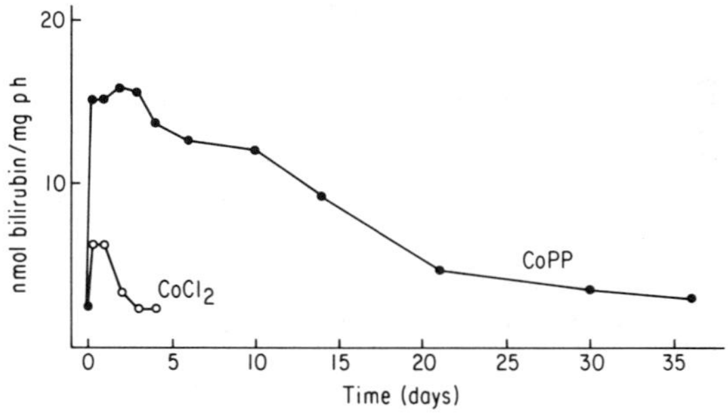

FIGURE 6. The induction of heme oxygenase in liver by administration of equimolar amounts (125 μmol/kg body wt of $CoCl_2$ and Co-protoporphyrin) to rats. Each compound was administered by sc injection at time 0.

potent and long-acting inducer of enzyme activity *in vivo* (FIGURE 6). The contrasting effects of these *in vitro* and *in vivo* effects is striking, and recent work indicates that Co-protoporphyrin induces *de novo* synthesis of substantially more enzyme protein than is measured by the spectrophotometric enzyme assay,[15] some of the heme oxygenase activity being masked by the inhibitory action of the metalloporphyrin.

A concomitant and prolonged decrease in ALA synthase activity occurs, and, as a consequence of the dual and prolonged actions on heme synthesis and catabolism, a marked and sustained decrease in the levels of cytochrome P-450 occurs. The periodic administration of Co-protoporphyrin results in a decrease of about 85% in cytochrome P-450, and this decrease can last as long as 90 days when the experiment is terminated.[16] These unique properties of Co-protoporphyrin provide an experimental model of modulating the levels of cytochrome P-450 and assessing the relative importance of cytochrome P-450-dependent metabolic activation of chemicals in determining their biological actions *in vivo.*

The intense and sustained perturbations in heme biosynthesis and degradation coupled with the pronounced and sustained decline in cytochrome P-450 content in adult rats produces significant metabolic–endocrine alterations in the host animal.[16] The relationship between the major impairment of cytochrome P-450-dependent biotransformations and the endocrine dysfunction remains to be fully elucidated. These endocrine alterations include an immediate decrease in the rate of weight gain and food intake (and, depending on the actual dose of Co-protoporphyrin administered, a loss in body weight) and a significant reduction in serum testosterone, serum thyroxine (T_4), and 3,5,3′-triiodothyronine (T_3) levels (these endocrine changes were not accompanied by reciprocal elevations in either serum luteinizing hormone or thyroid-stimulating hormone concentrations[17]); furthermore, in experiments with pair-fed control and Co-protoporphyrin-treated animals, the androgenic dysfunction produced by this metalloporphyrin is not causally related to the associated weight loss produced by the compound.[18] These effects were specific for Co-protoporphyrin and were not observed after administration in equimolar doses of inorganic cobalt or six other metalloporphyrins, including heme.[17] These endocrine actions suggest an action of Co-protoporphyrin directly, or perhaps indirectly through its derangement of cytochrome P-450 content and function, on some endocrine regulatory site(s) within the hypothalamic–pituitary axis. The aberrations described above are reversible after the administration of Co-protoporphyrin is terminated. This coincides with a return to normal of hepatic heme oxygenase activity and cytochrome P-450 content.

Thus Co-protoporphyrin through its profound and protracted effects on heme biosynthesis and catabolism and cellular cytochrome P-450 content may provide a means of elucidating possible physiological roles of cytochrome P-450, or of heme itself in the metabolism of endogenously derived chemicals. These chemicals may be involved in the maintenance of metabolic homeostasis; in the secretion, production, and disposal of hormones; and in the regulation of growth processes in the whole animal.[19]

CONCLUSION

Metalloporphyrins in which the central iron atom of heme is replaced by other elements often display markedly different actions on heme metabolism *in vivo* from those of the inorganic metals they contain. In addition, the *in vitro* and *in vivo* actions of individual synthetic metalloporphyrins may also differ markedly, as exemplified by Co-protoporphyrin, which inhibits heme oxygenase activity *in vitro* but produces a sustained induction of the enzyme *in vivo*. The presence of the metal in the protoporphyrin macrocycle also serves to increase the extent and duration of action of the metal moiety on heme metabolism, and Co-protoporphyrin is a case in point. Contrasting effects on heme metabolism exist within the metalloporphyrin species as demonstrated by the effects in the whole animal of Co-protoporphyrin and Sn-protoporphyrin on heme oxygenase activity—the former increasing enzyme activity, the latter inhibiting it. The recent demonstration that Sn-protoporphyrin, like Co-protoporphyrin, has dual actions on heme oxygenase activity *in vivo* (with the inhibitory action being expressed[15]), confirms that synthetic metalloporphyrins possess both inductive and inhibitory properties for heme oxygenase and that the biological action that is ultimately expressed *in vivo* is dependent on the nature of the central metal atom of the metalloporphyrin.

ACKNOWLEDGMENTS

The author wishes to thank his colleagues Professor Attallah Kappas, Dr. Richard A. Galbraith, Dr. M. K. Sardana, Dr. R. Blomstrand, Dr. L. Berglund, and Dr. B. Angelin for the use of their data in the preparation of this paper.

REFERENCES

1. DRUMMOND, G. S. & A. KAPPAS. 1981. Proc. Natl. Acad. Sci. USA **78:** 6466–6470.
2. SCHACTER, B. A., B. YODA & L. G. ISRAELS. 1976. Ann. Clin. Res. **8**(Suppl. 7): 28–34.
3. MAINES, M. D. & A. KAPPAS. 1977. Science **198:** 1215–1221.
4. KAPPAS, A. & M. D. MAINES. 1976. Science **192:** 60–62.
5. ROSENBERG, D. W., G. S. DRUMMOND & A. KAPPAS. 1982. Mol. Pharmacol. **21:** 150–158.
6. KAPPAS, A., G. S. DRUMMOND & M. K. SARDANA. 1985. J. Clin. Invest. **75:** 302–305.
7. KAPPAS, A., C. S. SIMIONATTO, G. S. DRUMMOND, S. SASSA & K. E. ANDERSON. 1985. Proc. Natl. Acad. Sci. USA **82:** 896–900.
8. SIMIONATTO, C. S., K. E. ANDERSON, G. S. DRUMMOND & A. KAPPAS. 1985. J. Clin. Invest. **75:** 513–521.
9. MILLEVILLE, G. S., M. D. LEVITT & R. R. ENGEL. 1985. Pediatr. Res. **19:** 94–96.
10. ANDERSON, K. E., C. S. SIMIONATTO, G. S. DRUMMOND & A. KAPPAS. 1986. Clin. Pharmacol. Ther. **39**(5): 510–520.
11. SASSA, S., G. S. DRUMMOND, S. E. BERNSTEIN & A. KAPPAS. 1985. J. Exp. Med. **162:** 864–876.
12. CORNELIUS, C. E. & P. A. RODGERS. 1984. Pediatr. Res. **18:** 728–730.
13. DRUMMOND, G. S. & A. KAPPAS. 1986. J. Clin. Invest. **77:** 971–976.

14. MAINES, M. D. & A. KAPPAS. 1975. J. Biol. Chem. **250:** 4171–4177.
15. SARDANA, M. K. & A. KAPPAS. 1987. Proc. Natl. Acad. Sci. USA **84:** 2464–2468.
16. DRUMMOND, G. S. & A. KAPPAS. 1982. Proc. Natl. Acad. Sci. USA **79:** 2384–2388.
17. SMITH, T. J., G. S. DRUMMOND & A. KAPPAS. 1987. Pharmacology **34:** 9–16.
18. GALBRAITH, R. A., G. S. DRUMMOND, L. KREY & A. KAPPAS. 1987. Pharmacology **34:** 241–249.
19. KAPPAS, A. & G. S. DRUMMOND. 1986. J. Clin. Invest. **77:** 335–339.

Role of Cytochrome P-450 in Porphyria Caused by Halogenated Aromatic Compounds[a]

HERBERT L. BONKOVSKY,[b,c] PETER R. SINCLAIR,[d,e]
W. JAY BEMENT,[d] RICHARD W. LAMBRECHT,[d]
AND JACQUELINE F. SINCLAIR[d,e]

[c]*Clinical Research Center*
and
Departments of Medicine and Biochemistry
Emory University School of Medicine
Atlanta, Georgia 30322

[d]*Veterans Administration Medical Center*
White River Junction, Vermont 05001

[e]*Department of Biochemistry*
Dartmouth Medical School
Hanover, New Hampshire 03756

INTRODUCTION

Both the human disease porphyria cutanea tarda and the hepatic porphyria caused by certain halogenated aromatic compounds[1] (HACs) are characterized by hepatic injury and hepatic accumulation and urinary overexcretion of uroporphyrin and heptacarboxyporphyrin (for reviews, see references 1–4). Iron in the liver is known to play a potentiating, but still ill-defined role in these forms of porphyria. A decrease in activity of uroporphyrinogen decarboxylase has been reported in animals with experimental porphyria caused by an HAC and in humans with porphyria cutanea tarda. However, the occurrence of humans with an inherited deficiency in uroporphyrinogen decarboxylase activity, many of whom have no symptoms or signs of porphyria, indicates that deficiency in the decarboxylase activity per se is not sufficient to cause porphyria.

A role for hepatic cytochrome P-450 in HAC-mediated porphyria was emphasized in studies using cultured chick embryo liver cells, when it was shown that inhibitors of cytochrome P-450 activity prevented cellular uroporphyrin accumulation.[5] In these studies no particular form of cytochrome P-450 was identified as being important, but it was suggested that a metabolite of an HAC, produced by the action of cytochrome P-450, caused the porphyria.

Later it was shown that at least two distinct forms of cytochrome P-450 could be induced by different classes of chemicals in these cultures: the first form (here called

[a]This work was supported by Grants RR00039, NIDDK 38825, and CA25012 from the National Institutes of Health and by research funds from the U.S. Veterans Administration.

[b]Address for correspondence: Clinical Research Center, Room GG-23, Emory University Hospital, 1364 Clifton Road, N.E., Atlanta, Georgia 30322.

cyt P-450-PB) was induced by phenobarbital (PB), propylisopropylacetamide, glutethimide, and several other compounds; the second form (here called cyt P-450-MC) was induced by 3-methylcholanthrene (MC), β-naphthoflavone, and related compounds.[6,7] Recently, these two distinct microsomal hemoproteins have been purified in our laboratory from the livers of chick embryos pretreated with glutethimide or MC and specific antibodies raised to each.[8]

Other studies with cultured chick embryo liver cells had supported a role for cytochrome P-450 by showing that treatment with MC or β-naphthoflavone[9] or with phenobarbital[10] enhanced HAC-mediated porphyria. In still other studies, however, HACs were reported to act by direct inhibition of uroporphyrinogen decarboxylase.[11,12]

In an effort to elucidate the mechanism of the porphyria caused by HACs, we reinvestigated the role of different isozymes of cytochrome P-450, using the experimental model of chick embryo liver cells in primary culture.[13,14] In this paper we summarize results that establish an obligatory role for the cyt P-450-MC rather than for the cyt P-450-PB isozyme, but that do not support the hypothesis that the porphyria is caused by irreversible inhibition of uroporphyrinogen decarboxylase (for example, by a covalently bound HAC metabolite). To integrate all our findings, we propose a model for the early steps in pathogenesis of HAC-mediated porphyria in which an "active oxidant," formed by the action of cyt P-450-MC, plays a critical role.

MATERIALS AND METHODS

Materials

Sources of the chemicals were as given previously.[13,14]

Hepatocyte Cultures

Primary cultures from livers of 16-day-old White Leghorn chick embryos were prepared and maintained in William's E medium as described previously.[15] Cells were cultured in plastic dishes with diameters of 3.5 cm unless otherwise indicated. Dishes contained 0.5–0.6 mg of protein per 3.5-cm-wide dish or 1.5–1.7 mg of protein per 6-cm-wide dish. Chemicals were added during the second day of culture after the cells had been rinsed twice with William's E medium containing dexamethasone and 3,3′,5-triiodothyronine, but not insulin. Compounds of poor water solubility were added to cultures in DMSO (not exceeding 4 μl/ml). Piperonyl butoxide was sometimes added in 50% (v/v) ethanol (6 μl/ml of medium). Controls always included appropriate amounts of solvents.

Assay of Porphyrins

Porphyrin composition was routinely determined by the spectrofluorometric procedure previously described.[13] In some studies porphyrins were separated and quantified by HPLC.[14,16] HPLC was required for experiments in which ellipticine was used because its fluorescence led to errors in the fluorometric procedure for protoporphyrin. In recent studies direct comparisons of the spectrofluorometric and HPLC methods have shown excellent agreement (coefficients of variation: <10%). Uroporphyrin and heptacarboxyporphyrin on HPLC are detected spectrofluorometrically as "uroporphyrin"; hexacarboxy-, pentacarboxy-, and coproporphyrins as "coproporphyrin"; and tri- and dicarboxyporphyrins as "protoporphyrin" (reference 14 and unpublished observations).

Enzyme Assays

δ-Aminolevulinic acid synthase (ALAS) activity was measured in intact cells.[17]

Ethoxyresorufin deethylase (EROD) activity was measured in homogenates of cells from 6-cm-wide dishes as described.[18]

Other Methods

Incorporation of ^{14}C-labeled δ-aminolevulinic acid (ALA) into heme was measured as described.[14] Proteins were measured by the method of Lowry *et al.*[19]

RESULTS

Accumulation of Uroporphyrin and Induction of Cyt P-450-MC Caused by HACs

As shown in FIGURE 1, control cells exposed to ALA accumulated mainly protoporphyrin, but cells exposed to an HAC (in this case, 2,4,5,3′,4′-pentabromobiphenyl) progressively accumulated uroporphyrin. In parallel to this shift from proto-

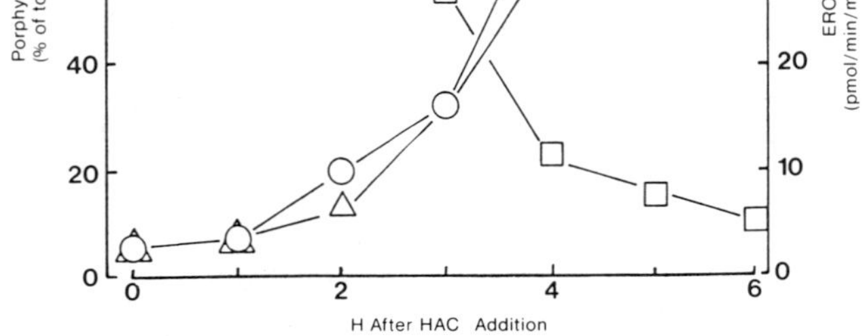

FIGURE 1. Time course of effects of an HAC on porphyrin accumulation and EROD activity in chick embryo liver cultures. Cells were prepared on 6-cm-wide dishes and harvested after the indicated time of exposure to 1.8 μM 2,4,5,3′,4′-pentabromobiphenyl. Cells were also exposed to 0.15 mM ALA for 2 hr before being harvested. Porphyrins (△: uroporphyrin; □: protoporphyrin) and EROD (○) were measured as described in the Materials and Methods section. Each value is a mean ± SD (N = 3). SD values fell within the symbols unless otherwise shown. (ALA had no effect on EROD induction by 2,4,5,3′,4′-pentabromobiphenyl.) After reference 13.

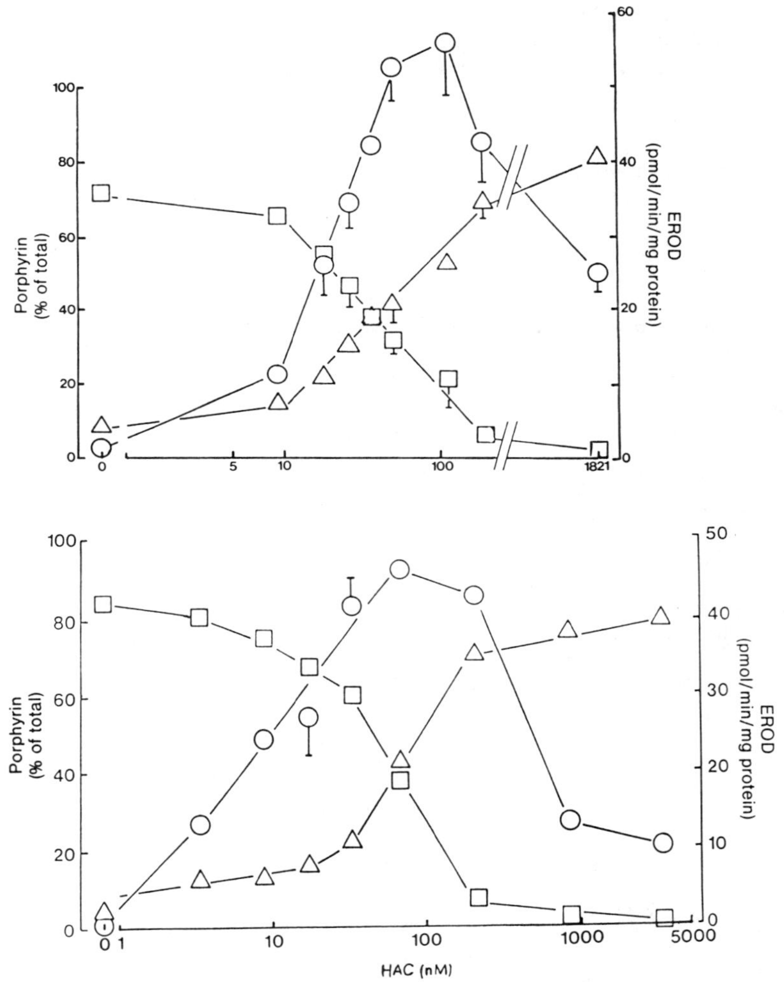

FIGURE 2. Effects of increasing the concentration of an HAC on porphyrin accumulation and EROD activity in chick embryo liver cultures. Cells were prepared on 6-cm-wide dishes. In the upper panel, cells were treated with 1.8 μM 2,4,5,3′,4,′-pentabrombiphenyl for 18 hr, then with 0.15 mM ALA for 2 hr. In the lower panel, cells were treated with 3.4 μM TCB for 22 hr, then with 0.15 mM ALA for 1 hr. Porphyrins (△: uroporphyrin; □: protoporphyrin) and EROD (○) were measured as described in the Materials and Methods section. Each value is a mean ± SD ($N = 3$). SD values fell within the symbols unless otherwise shown. After reference 13.

porphyrin to uroporphyrin accumulation was an increase in activity of EROD, an enzyme activity specific for cyt P-450-MC.[8,18]

Dose-response curves for the effects of two porphyrogenic HACs to induce EROD and to cause progressive uroporphyrin accumulation in cultured hepatocytes are shown in FIGURE 2. The decrease in EROD activities at HAC concentrations at or above 1

μM may be due partly to cell toxicity and partly to an inhibition of heme synthesis sufficient to limit cyt P-450-MC induction, as shown by the fact that EROD activities are decreased less if such cells are also exposed to exogenous heme (unpublished observations).

Effect of Preinduction of Cyt P-450-MC or Cyt P-450-PB and Inhibition of Protein Synthesis on Porphyrin Accumulation by Chick Embryo Liver Cultures

FIGURE 3 (upper portion) shows that the addition of cycloheximide, an inhibitor of protein synthesis, just before the addition of an HAC prevented the uroporphyria otherwise caused by the HAC. Thus, this effect of HACs is clearly dependent upon *de novo* protein synthesis. In contrast, when the cells had been pretreated with MC, so that cyt P-450-MC levels were already induced, addition of cycloheximide did not prevent the uroporphyria caused by HAC (FIGURE 3, lower portion). As shown in FIGURE 4, neither preinduction of cyt P-450-MC by MC nor of cyt P-450-PB by a low concentration of propylisopropylacetamide, a phenobarbital-like inducer in this system,[7,8] produced uroporphyrin accumulation. Preinduction by MC followed by an HAC, however, did cause uroporphyrin accumulation, an effect not seen after preinduction by propylisopropylacetimide followed by an HAC. These data support the conclusion that cyt P-450-MC but not cyt P-450-PB plays an obligatory role in HAC-mediated uroporphyria. (As discussed later, however, high concentrations of propylisopropylacetimide or PB do cause uroporphyrin accumulation in chick embryo liver cultures.)

In other recent studies (not presented here), we have purified a 60-kDa protein from microsomes of chick embryos pretreated with MC; raised antibodies to the protein, which inhibits EROD activity; and shown that porphyrogenic HACs induce a

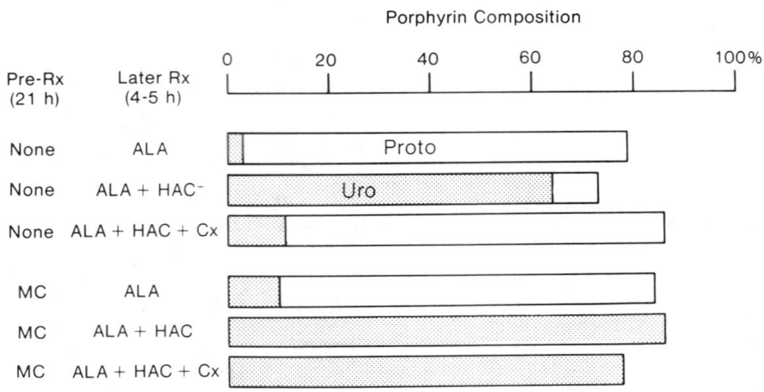

FIGURE 3. Effect of MC and cycloheximide on the pattern of porphyrins accumulated by chick embryo liver cultures exposed to ALA and an HAC. Some cells were exposed as indicated to 0.96 μM MC for 21 hr and/or 1.8 μM cycloheximide for 1 hr before addition of 3.4 μM TCB and 0.15 mM ALA. After 4 hr of further incubation, porphyrin composition was determined by spectrophotofluorometry (see Methods). In all instances, about 20% of total porphyrin was "coproporphyrin"; for clarity this fraction is not shown. Total porphyrins ranged from 0.7–1.0 nmol/dish. After reference 13.

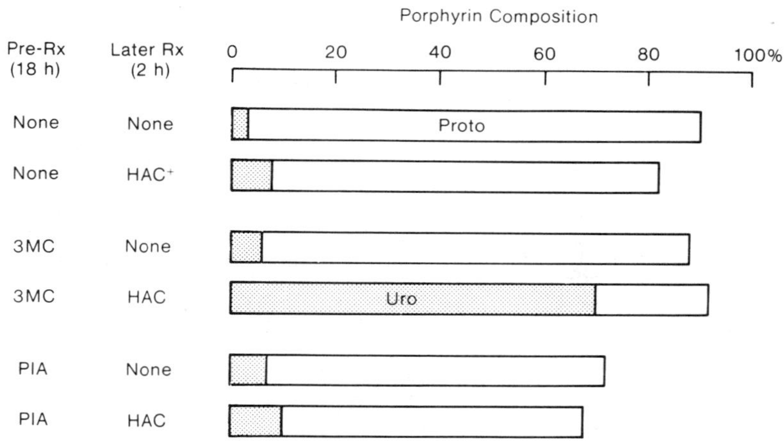

FIGURE 4. Comparison of effects of pretreatment with MC and propylisopropylacetamide on the pattern of porphyrins accumulated by chick embryo liver cultures exposed to ALA and an HAC. Some cells were pretreated for 18 hr with MC (0.96 μM) or propylisopropylacetamide (53 μM), as indicated. ALA (0.15 mM) and, to some dishes, 2,4,5,3′,4′-pentabromobiphenyl (1.8 μM) was then added for an additional 2 hr, after which porphyrin composition was determined by spectrophotofluorometry (see Methods). After reference 13.

protein of the same molecular weight that is immunologically indistinguishable from the one induced by MC.[18]

Effect of Iron Chelators and Iron Loading on Porphyrin Accumulation and Inducibility of EROD

As already mentioned, hepatic iron plays an important though still ill-defined role in the uroporphyria caused by HACs in animals. Similarly, a role for iron in the chick embryo liver cell culture model was suggested by earlier results[5] and by the porphyrin results shown in the upper portion of FIGURE 5, which shows that pretreatment with iron chelators partially prevented the HAC-mediated uroporphyrin accumulation. This effect, however, was apparently due to the chelators' preventing normal cyt P-450-MC induction as shown by the EROD activities and by the fact that the iron chelators were ineffective if cyt P-450-MC had been preinduced by MC before addition of chelator (FIGURE 5, lower portion).

In other studies, loading of cultures with iron did little to enhance the rate or extent of uroporphyrin accumulation produced by HACs (results not shown).

Effects of Selective Inhibitors of Cyt P-450-MC and Cyt P-450-PB on Uroporphyrin Accumulation by Chick Embryo Liver Cultures

Inhibitors were used to study the involvement of different isozymes of cytochrome P-450 in HAC-mediated uroporphyrin accumulation. In the first series of experiments, inhibitors were added before exposure to an HAC. FIGURE 6 shows that pretreatment with 2-dimethylaminoethyl-2,2-diphenyl valerate or cimetidine, selective inhibitors of

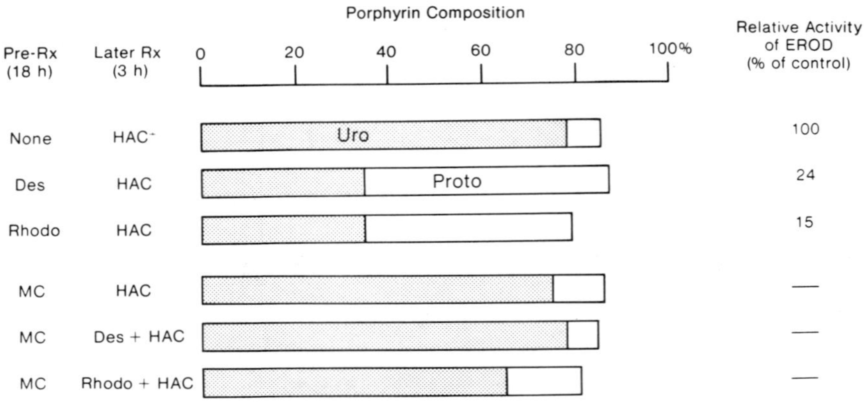

FIGURE 5. Effect of iron chelators on porphyrin accumulation and inducibility of EROD in chick embryo liver cultures exposed to ALA and an HAC. Cells were prepared on 3.5-cm-wide dishes or 6-cm-wide dishes. Some were pretreated with deferoxamine (30 μM), rhodotorulic acid (0.67 mM), or MC (0.9 μM) for 18 hr; then with TCB (3.4 μM) or TCB plus chelator for an additional 3 hr as shown. All cultures were treated with 0.15 mM ALA during the last 2 hr. Assays were performed as described in the Materials and Methods section. After reference 13.

cyt P-450-PB, did not prevent uroporphyrin accumulation; but piperonyl butoxide, an inhibitor of several isozymes, including cyt P-450-MC,[20] did prevent uroporphyrin accumulation. Ellipticine, which is selective in inhibiting cyt P-450-MC and EROD,[21] was also effective.[13,14]

As shown in FIGURE 7, inhibitors of cyt P-450-MC not only prevented, but also rapidly ended uroporphyrin accumulation already in progress if they were added to hepatocyte cultures after addition of an HAC. Virtually as soon as piperonyl butoxide was added, accumulation of uroporphyrin slowed or ceased completely, regardless of whether cells had been pretreated with 3,4,3′,4′-tetrachlorobiphenyl (TCB) (FIGURE 7B) or 2,4,5,3′,4′-pentabromobiphenyl (FIGURE 7C). In other experiments (not shown), piperonyl butoxide restored protoporphyrin synthesis nearly completely even 1) after up to 44 hr of prior exposure to TCB; 2) after 19 hr of exposure to TCB at the highest possible concentration (68 μM, the limit of solubility); and 3) in the presence of 1.8 μM cycloheximide, which markedly inhibited protein synthesis (see FIGURE 10 and reference 13).

Furthermore, we found that the ability of inhibitors to arrest uroporphyrin accumulation in TCB-pretreated cells was specific for inhibitors of cyt P-450-MC (such as piperonyl butoxide and ellipticine); as shown in FIGURE 8 (upper portion), 2-dimethylaminoethyl-2,2-diphenyl valerate, a selective inhibitor of cyt P-450-PB, was ineffective in restoring protoporphyrin synthesis in HAC-pretreated cultures. Both piperonyl butoxide and 2-dimethylaminoethyl-2,2-diphenyl valerate, however, were effective in restoring protoporphyrin synthesis by cultures pretreated with high concentrations of phenobarbital (FIGURE 8, lower portion.)

Restoration of Heme Synthesis by Piperonyl Butoxide Added after an HAC

The experiments just described established that protoporphyrin synthesis was rapidly restored by piperonyl butoxide in chick embryo liver cultures pretreated with a porphyrogenic HAC. We used two lines of evidence to show that cellular heme synthesis was also restored: increased incorporation of radioactivity from [^{14}C]ALA into heme, and decreased induction of ALAS.

Increased Incorporation of Radioactivity from [^{14}C]ALA into Heme

FIGURE 9 shows that TCB decreased heme synthesis from [^{14}C]ALA and that piperonyl butoxide reversed the decrease.

Decreased Induction of ALAS

It is now widely accepted that an important regulator of hepatic ALAS activity is the concentration of heme in a critical "regulatory heme pool."[1] Conditions such as cytochrome P-450 induction, especially with concomitant inhibition of heme synthesis, decrease feedback repression of ALAS synthase; restoration of cellular heme restores the repression. Thus, ALAS will be induced by propylisopropylacetamide (an inducer of cyt P-450-PB) plus TCB (an inhibitor of heme synthesis and inducer of cyt P-450-MC). Addition of piperonyl butoxide after the addition of propylisopropylacetamide plus TCB led to decreased induction of ALAS (FIGURE 10), again indicating that piperonyl butoxide promptly restores hepatocellular heme synthesis. The effect of piperonyl butoxide on ALAS was specific for inhibition of heme synthesis

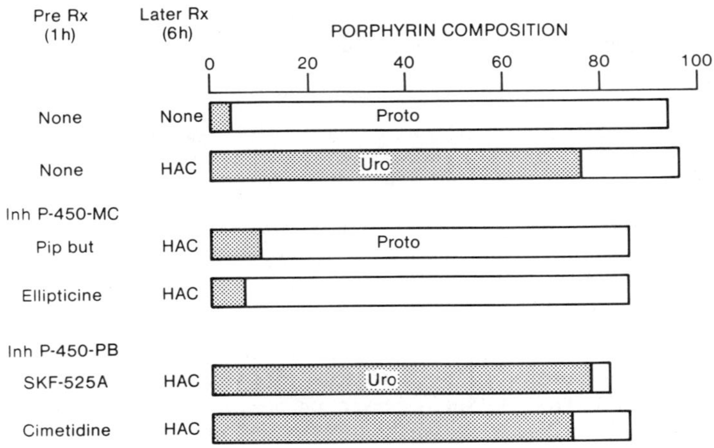

FIGURE 6. Effects of selective inhibitors of different isozymes of cytochrome P-450 to prevent accumulation of uroporphyrin produced by an HAC plus ALA. Some cells were pretreated for 1 hr with selective inhibitors of cyt P-450-PB (22 μM 2-dimethylaminoethyl-2,2-diphenyl valerate or 1 mM cimetidine) or cyt P-450-MC (30 μM piperonyl butoxide or 10 μM ellipticine). Cultures were then exposed to 0.15 mM ALA alone or to 0.15 mM ALA plus 3.4 μM TCB for an additional 6 hr before determination of porphyrin composition. Results for inhibitors except ellipticine were obtained by spectofluorometry. Results for ellipticine were obtained by HPLC of porphyrins because the fluorescence spectrum of ellipticine interfered with the rapid fluorometric assay for porphyrins.[14] After references 13 and 14.

by an HAC and not simply a manifestation of toxicity, as shown by the lack of effect on the induction of ALAS produced by propylisopropylacetamide plus deferoxamine.

DISCUSSION

The studies reviewed in this paper establish clearly that, in chick embryo liver cultures, an isozyme (or perhaps isozymes) of cytochrome P-450 (here called cyt P-450-MC) that is induced by HACs, MC, and certain other aromatic compounds plays a key role in the uroporphyria produced by HACs. This conclusion is based upon several lines of evidence: 1) similar time-course (FIGURE 1) and dose-response (FIGURE 2) curves for HAC-induced uroporphyrin accumulation and cyt P-450-MC induction, the latter conveniently measured as activity of EROD; 2) efficacy of preinduction of cyt P-450-MC, but not cyt P-450-PB, to accelerate HAC-dependent uroporphyrin accumulation (FIGURES 3 & 4); 3) prevention of uroporphyrin accumulation by pretreatment of cultured cells with inhibitors of cyt P-450-MC but not cyt P-450-PB (FIGURE 6); and 4) rapid termination of uroporphyrin accumulation and restoration of normal rates of protoporphyrin and heme synthesis by selective inhibitors of cyt P-450-MC, even after HAC-mediated uroporphyrin accumulation was already well established (FIGURES 7–10).

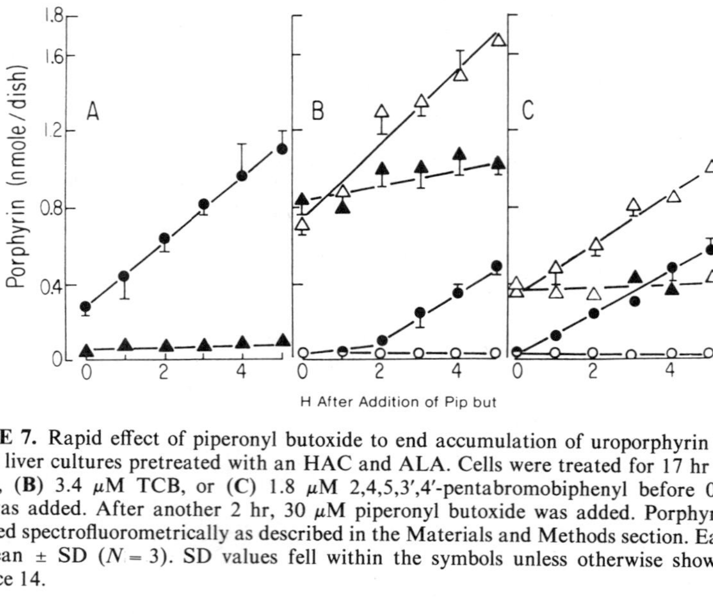

FIGURE 7. Rapid effect of piperonyl butoxide to end accumulation of uroporphyrin by chick embryo liver cultures pretreated with an HAC and ALA. Cells were treated for 17 hr with (**A**) DMSO, (**B**) 3.4 μM TCB, or (**C**) 1.8 μM 2,4,5,3′,4′-pentabromobiphenyl before 0.15 mM ALA was added. After another 2 hr, 30 μM piperonyl butoxide was added. Porphyrins were measured spectrofluorometrically as described in the Materials and Methods section. Each value is a mean ± SD ($N = 3$). SD values fell within the symbols unless otherwise shown. After reference 14.

Several other investigators have also described a role of cyt P-450-MC in the development of HAC-mediated uroporphyria, in both intact animals and cultured chick embryo liver.[4,9,22–24] However, deVerneuil *et al.* concluded that cyt P-450-PB is involved.[10] We have previously discussed[13] reasons for the apparent discrepancies between our conclusions and those of deVerneuil *et al.*

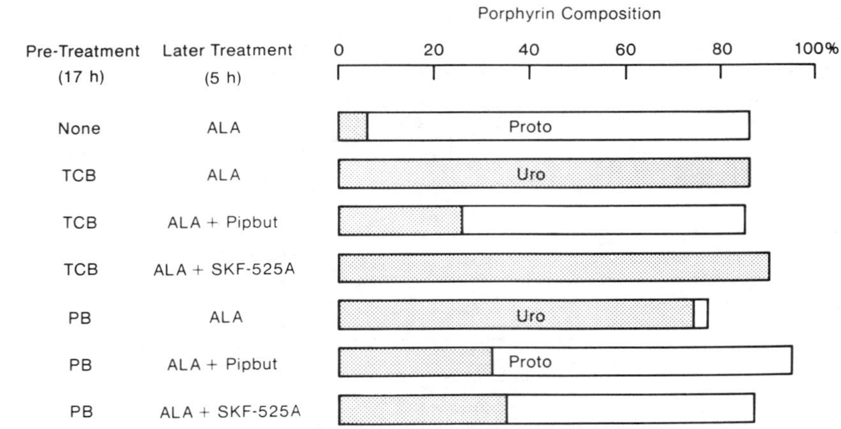

FIGURE 8. Effects of selective inhibitors of cyt P-450-MC and cyt P-450-PB on uroporphyrin accumulation caused by TCB and PB. Cells were pretreated with 3.5 μM TCB or 2 mM PB for 17 hr. Then 0.15 mM ALA, 30 μM piperonyl butoxide, and 22 μM 2-dimethylaminoethyl-2,2-diphenyl valerate were added as shown. After 5 hr further incubation, cells were harvested and porphyrins were measured spectrofluorometrically (see Methods). After reference 14.

A key unresolved question is as follows: How is induction of cyt P-450-MC linked to HAC-mediated porphyria? This in turn, raises another question: What is the biochemical mechanism underlying this form of porphyria? Two hypotheses regarding mechanism have been proposed: The first and most widely held hypothesis is that the porphyria is due to a primary decrease in activity of uroporphyrinogen decarboxylase.[1–4,11,12] The second is that the proximate cause of the porphyria is increased oxidation of uroporphyrinogen and heptacarboxyporphyrinogen to their corresponding porphyrins. The porphyrins, unlike the porphyrinogens, are not substrates for the

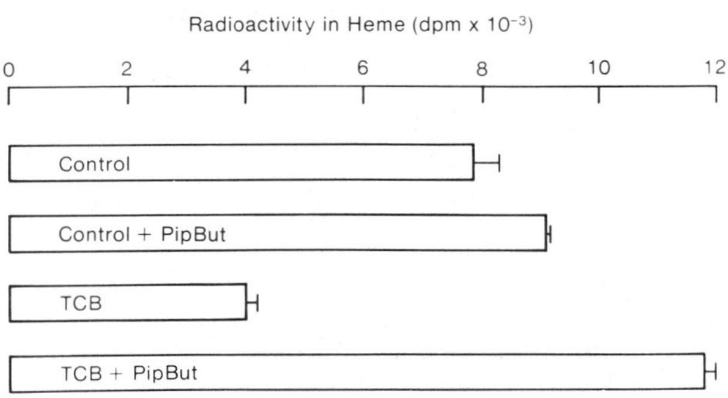

FIGURE 9. Restoration of TCB-inhibited heme synthesis by piperonyl butoxide. Cells were prepared on 6-cm-wide dishes and treated with DMSO alone (control) or 0.85 μM TCB for 4 hr. Then 30 μM piperonyl butoxide was added; after 1.7 hr, [^{14}C]ALA (1 μCi; 82 μM) was added for the final 2 hr. Cultures were harvested and heme extracted as described in reference 14. The concentration of ALA used was sufficient to swamp any endogenous pools. Average values of duplication determinations, which varied less than 10%, are presented. After reference 14.

decarboxylase and will therefore accumulate in hepatocytes. Furthermore, uroporphyrin inhibits and may inactivate uroporphyrinogen decarboxylase.[25,26] If the first hypothesis is correct, activity of the decarboxylase should be decreased substantially early in the development of the porphyria, before there is appreciable porphyrin accumulation. If the second hypothesis is correct, decarboxylase activity should be normal in the early stages of the porphyria. We now consider evidence for and against these two hypotheses.

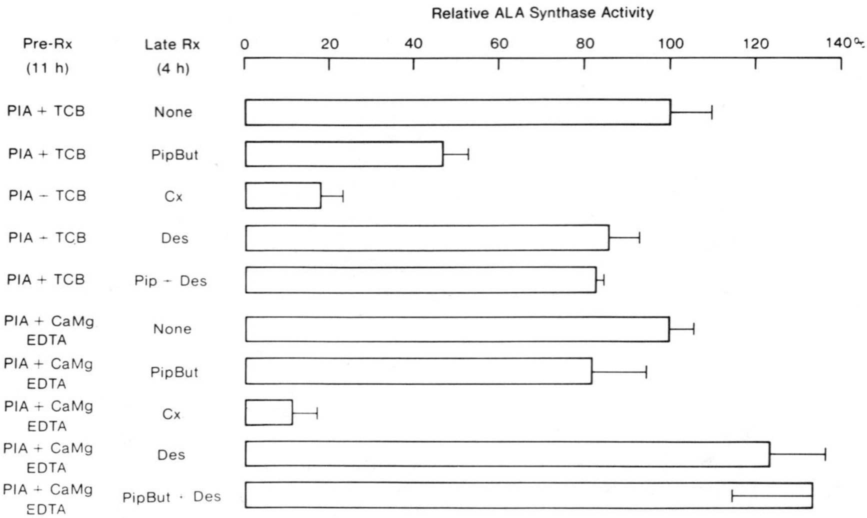

FIGURE 10. Restoration of TCB-inhibited heme synthesis by piperonyl butoxide as indicated by decreased induction of ALAS. Cells were pretreated with inducers of ALAS for 11 hr. Then 30 μM piperonyl butoxide, 1.8 μM cycloheximide, or 30 μM deferoxamine was added to some dishes for another 4 hr. ALAS was measured in intact cells as described in the Materials and Methods section. For treatment with 0.14 mM propylisopropylacetamide and 3.4 μM TCB, 100% activity was 12.1 ± 1.12 nmol ALA/hr/mg protein, and for treatment with 0.14 mM propylisopropylacetamide and EDTA, 100% activity was 8.2 ± 0.1 nmol ALA/hr/mg protein. Each value is a mean ± SD ($N = 3$). After reference 14.

Hypothesis 1: The Porphyria is Due to a Primary Decrease in the Activity of Uroporphyrinogen Decarboxylase

The question of decarboxylase activity in HAC-induced porphyria is somewhat clouded because assay of this enzyme is difficult; several methods are in current use, and there are some pitfalls in the assays, especially those in which uroporphyrinogen is used as substrate.[4,27–30] It is complicated further because decreases in enzyme activities, as measured in homogenates or extracts of cells, may not always reflect changes in activities in intact cells, and, even if they do, some changes may not be sufficient in magnitude to affect overall rates of heme formation. We have discussed these problems in some detail for other enzymes involved in hepatic heme biosynthesis.[31–33]

Previously it was found that low concentration of porphyrogenic HAC plus ALA caused uroporphyrin accumulation in chick embryo liver cultures without measurably affecting activity of the decarboxylase, measured by two methods in homogenates of the cultured cells with both uroporphyrinogen III and pentacarboxyporphyrinogen III as substrates.[34] Decreases in decarboxylase activity in homogenates were observed only at high concentrations of the HAC (>1.9 μM 2,4,5,3′,4′-pentabromobiphenyl). More recently, as described elsewhere in this volume by Lambrecht *et al.*,[35] no significant decrease in activity of uroporphyrinogen decarboxylase was found in supernatants (collected after centrifugation at 10 000 g) of the cultured liver cells that had accumulated much uroporphyrin due to exposure to high concentrations of an HAC.

The question of how decarboxylase activity is affected within intact cells remains unresolved. However, the rapid restoration of protoporphyrin and heme synthesis produced by piperonyl butoxide (FIGURES 7–10), even in the absence of new protein synthesis, clearly shows that HAC treatment cannot be causing irreversible inhibition of the decarboxylase (for example, by covalent binding of an HAC or a metabolite). Taken together, all our data suggest that activity of uroporphyrinogen decarboxylase is normal or nearly normal in the intact cultured hepatocytes treated with HACs.

Hypothesis 2: Porphyria is Primarily Due to Oxidation of Porphyrinogens to Porphyrins

This hypothesis was proposed about 30 years ago[36] and has recently been revived.[37] It is supported by the fact that porphyrins, rather than porphyrinogens, accumulate in HAC-treated cultures[14] and by the recently described lack of detectably decreased activity of uroporphyrinogen decarboxylase. Other studies, including one presented in this volume,[38] show that uroporphyrinogen can be oxidized by extracts (prepared from supernatants collected after centrifugation at 12 000 g) of livers from MC-pretreated chick embryos, but only in the presence of a porphyrogenic HAC. Extracts from control or propylisopropylacetamide-pretreated livers were ineffective. The oxidation of the porphyrinogen was inhibited by carbon monoxide, by piperonyl butoxide, and by antibodies raised against cyt P-450-MC, but not by antibodies against cyt P-450-PB. These findings thus provide a link between the evidence presented in this paper that there is a critical role for cyt P-450-MC, and the production of HAC-mediated porphyria.

Working Model for the Mechanism of HAC-Mediated Porphyria in Chick Embryo Cultures

Based upon all of our findings, we propose the model shown in FIGURE 11. Like other aromatic inducers of cyt P-450-MC, porphyrogenic HACs bind to the cytosolic *Ah* receptor, and the complex enters the nucleus where it promotes expression of the *Ah* locus.[4,39,40] As part of this pleiotropic response, a particular isozyme of cytochrome P-450 (here called cyt P-450-MC) is induced. The critical isozyme can also be induced by MC and certain other nonporphyrogenic aromatic compounds. This isozyme, acting in the presence of and probably by binding an HAC, produces an "active oxidant" derived from oxygen also bound to the hemoprotein. Formation of the active oxidant requires, in addition, electrons supplied by NADPH, presumably through NADPH–cytochrome P-450 reductase.[38] The postulated active oxidant may have several effects, including the mediation of hepatocyte toxicity and carcinogenicity, other known effects of porphyrogenic HACs.[4,39,40] Our focus, however, is on its effect to oxidize porphyrinogens to porphyrins; those with 8- and 7-carboxyl groups remain primarily within hepatocytes [13,34] where they may be hepatotoxic and may inhibit and inactivate

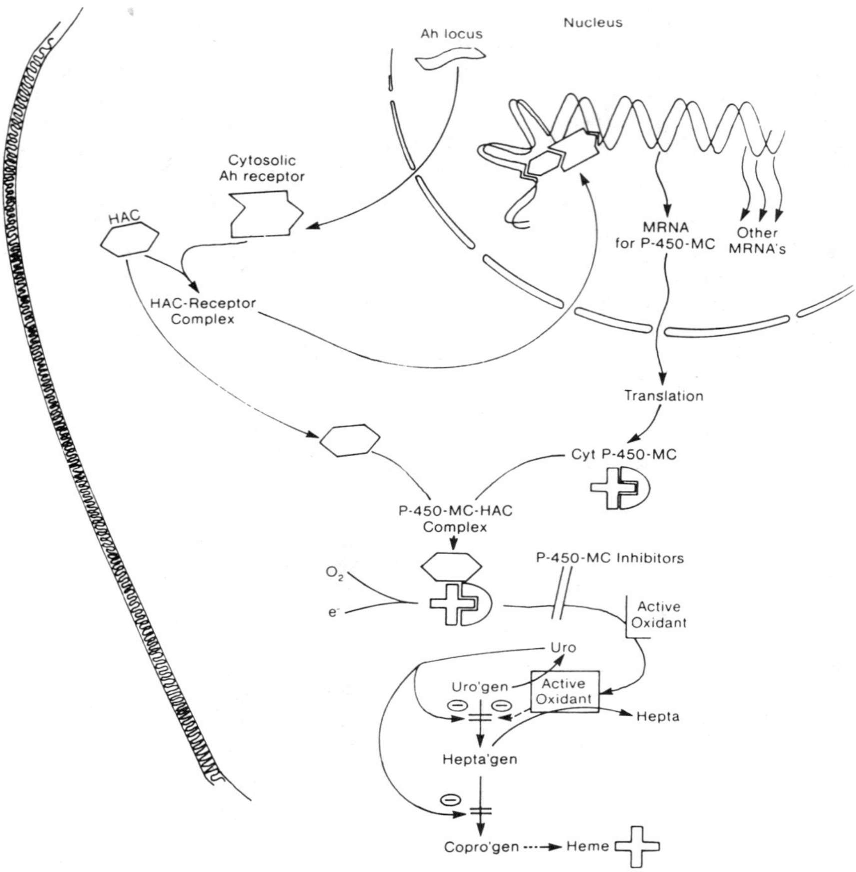

FIGURE 11. Postulated mechanism for porphyria in chick embryo liver cell cultures caused by HACs. This hypothetical scheme is most compatible with work from our laboratories presented in this paper and elsewhere in this volume.[35,38] Critical effects of HAC are 1) induction of a specific isozyme of cytochrome P-450 (here called cyt P-450-MC), which plays an essential role in forming an active oxidant, and 2) cooperation of HAC with cyt P-450-MC to increase formation of an active oxidant. Our data indicate that the active oxidant's primary role is to oxidize uroporphyrinogen and heptacarboxyporphyrinogen decarboxylase. There may also be an effect of an active oxidant, iron, an HAC, or porphyrins directly on the decarboxylase, but our data indicate that, in the intact cells, any such effect is rapidly reversible by piperonyl butoxide or other inhibitors of cyt P-450-MC for at least 2 days after exposure to a porphyrogenic HAC.

uroporphyrinogen decarboxylase, leading to more accumulation of uroporphyrinogen and uroporphyrin. The active oxidant itself may also inhibit the decarboxylase in a more direct but still reversible way.

Additional support for this model comes from our and other data showing that antioxidants are partially effective in preventing uroporphyrin accumulation by HAC-treated cultures,[13,41] although such studies must be interpreted cautiously.[13]

The ability of iron to catalyze a variety of reactions that give rise to reactive oxidants is well known,[37,42] and it is tempting to ascribe a synergistic role for hepatic iron in producing the postulated active oxidant. Chelators of iron, however, did not affect uroporphyrin accumulation in HAC-treated cultures whose cyt P-450-MC had been preinduced by MC (FIGURE 5). Thus, in our hands, the only unequivocal role of iron thus far defined has been its ability to induce cyt P-450-MC, though others have reported that iron loading enhances HAC-mediated uroporphyria in chick embryo liver cultures.[43]

A difference between HAC-induced experimental porphyria in chick embryo liver cultures and that in intact rodents is apparent: in the chick embryo system, porphyria develops within hours after HAC exposure; in contrast, in rodents and humans, a much longer period (weeks) of HAC exposure is required for massive uroporphyria.[4,29,30] Recent evidence that was obtained by using HPLC fractionation of hepatic porphyrins, however, indicates that abnormal accumulation of uroporphyrin and heptacarboxyporphyrin are detectable within a few days of feeding hexachlorobenzene to rats, whereas increases in hepatic *total* porphyrin concentrations, the measurements usually done in the past, require 40 days to be significant.[44] Thus, the difference in time course for development of abnormal HAC-mediated uroporphyrin accumulation in rodent, as against avian systems, may be more quantitative than qualitative.

Still, the degree of early uroporphyrin and heptacarboxyporphyrin accumulation in avian systems appears greater, suggesting some, perhaps secondary element, as enhancing development of HAC-mediated porphyria in chick embryo liver cultures. This could involve 1) more rapid or marked induction of ALAS, increasing flux of substrate in the pathway; 2) lower activities of uroporphyrinogen III cosynthase or decarboxylase in avian hepatocytes than in rodent hepatocytes; or 3) a greater tendency for porphyrinogen oxidation in avian embryo hepatocytes (for example, an increase in oxidation due to greater rates of formation of active oxidant or to less adequate reduction or to other protective mechanisms to keep porphyrinogens reduced in the presence of the oxidant).

Relatively high concentrations of cyt P-450-PB-inducing drugs—such as phenobarbital, propylisopropylacetamide, or glutethimide—cause considerable uroporphyrin and heptacarboxyporphyrin accumulation in culture (see FIGURE 8 and references 10, 43, and 45), but mainly cause protoporphyrin accumulation *in ovo* (from reference 43 and unpublished observations). These findings suggest that uroporphyrinogen and heptacarboxyporphyrinogen oxidation may occur more readily in chick embryo hepatocytes in culture than *in ovo* in the intact liver. Interpretation of these findings, however, is complicated by the fact that propylisopropylacetamide and similar drugs *increase* rates of heme synthesis from ALA and conversion of protoporphyrin to heme by chick embryo liver cultures.[46] Furthermore, addition of ALA to such cultures markedly increases protoporphyrin accumulation (see FIGURE 4 and reference 14). Thus, currently available evidence does not indicate that cyt P-450-PB inducers decrease conversion of uroporphyrinogen to coproporphyrinogen, protoporphyrin, and heme, despite the uroporphyrin and heptacarboxyporphyrin accumulation they produce in chick embryo liver cultures.

Recent studies by Smith *et al.* of mice with iron overload treated with a large oral dose of hexachlorobenzene have indicated that 1) induction of cytochrome P_1-450 (the isozyme analogous to what we here call cyt P-450-MC) plays a role in hepatic porphyria caused by hexachlorobenzene; 2) the development of the porphyria is diminished by treatment with piperonyl butoxide and increased by β-naphthoflavone but not PB; and 3) the porphyria does not depend critically upon metabolism of hexachlorobenzene or covalent binding of metabolites to uroporphyrinogen decarboxylase.[47] These findings are in agreement with ours, as summarized above. Furthermore, microsomes from C57B1/10 mice, a strain that is susceptible to hexachlorobenzene-mediated porphyria, showed higher rates of lipid peroxidation than microsomes from DBA/2 mice, a strain that is resistant to the porphyric effect of hexachlorobenzene. The authors conclude that an "active oxygen species," in the presence of porphyrin, leads to inactivation of uroporphyrinogen decarboxylase. Smith *et al.* observed a decrease in activity of uroporphyrinogen decarboxylase, which occurred 2 weeks after hexachlorobenzene was added, before observing an increase in *total* hepatic porphyrins, which occurred 4 or more weeks after hexachlorobenzene was added. Unfortunately, from the data presented, it is not clear whether an increase in hepatic uroporphyrin or heptacarboxyporphyrin occurred before the increase in the total porphyrin, as was recently reported by others.[44] The complexity of the problem is compounded by the fact that accumulation of porphyrin in livers of rodents treated with an HAC is not uniform; some lobes and even areas within lobes (unpublished observations) show more porphyrin accumulation than others. Thus, average porphyrin concentrations or activities of uroporphyrinogen decarboxylase in homogenates of porphyric rodent liver may not accurately reflect values of these parameters within individual hepatocytes. In contrast, in chick embryo liver cultures treated with an HAC, fluorescence microscopy shows that virtually all hepatocytes contain uroporphyrin (unpublished observations). From a consideration of all available data, we are persuaded that the question of whether a decrease in uroporphyrinogen decarboxylase is the chicken or the egg in the pathogenesis of HAC-mediated porphyria is not yet resolved.

In summary, we have demonstrated that the cyt P-450-MC, not the cyt P-450-PB, isozyme of cytochrome P-450 plays a key role in HAC-mediated uroporphyria in cultured chick embryo liver cells. The rapid termination of uroporphyrin accumulation and restoration of normal heme synthesis by addition of cyt P-450-MC inhibitors rules out irreversible inhibition of uroporphyrinogen decarboxylase as the cause of the porphyric state. Furthermore, the lack of detectably decreased activity of the decarboxylase, even in extracts of cells accumulating uroporphyrin, and studies on porphyrinogen oxidation by cell extracts support the concept that the proximate cause of the porphyria is porphyrinogen oxidation. Additional studies in chick embryo liver cell cultures and other experimental systems are needed to elucidate further the mechanism of the porphyria caused by HACs and the roles played by cytochrome P-450, porphyrinogen oxidation, and inhibition of uroporphyrinogen decarboxylase in producing this form of porphyria, and probably porphyria cutanea tarda as well.

ACKNOWLEDGMENTS

We thank G. Elder and A. Urquhart for helpful comments and discussions, J. Healey and J. Frezza for technical assistance, and M. Davis and P. Gulliford for typing the manuscript.

REFERENCES

1. ELDER, G. 1977. Semin. Hematol. **14:** 222–242.
2. BONKOVSKY, H. 1982. *In* Hepatology: A Textbook of Liver Disease. D. Zakim & T. D. Boyers, Eds.: 351–393. Saunders. Philadelphia, PA.
3. PIMSTONE, N. 1982. Semin. Liver Dis. **2:** 132–142.
4. SWEENEY, G. 1986. Clin. Biochem. **19:** 3–15.

5. SINCLAIR, P. & S. GRANICK. 1974. Biochem. Biophys. Res. Commun. **61:** 124–133.
6. ALTHAUS, F., J. SINCLAIR, P. SINCLAIR & U. MEYER. 1979. J. Biol. Chem. **254:** 2148–2153.
7. SINCLAIR, J., P. SINCLAIR & H. BONKOVSKY. 1979. Biochem. Biophys. Res. Commun. **86:** 710–717.
8. SINCLAIR, P., J. FREZZA, J. SINCLAIR, J. BEMENT, J. HEALEY & H. BONKOVSKY. 1986. Fed. Proc. **45:** 1663.
9. DEBETS, F., W. HAMERS & J. STRIK. 1980. Int. J. Biochem. **12:** 1019–1025.
10. DEVERNEUIL, H., S. SASSA & A. KAPPAS. 1983. Biochem. J. **214:** 145–151.
11. KAWANISHI, S., Y. SEKI & S. SANO. 1981. FEBS Lett. **129:** 93–96.
12. KANANISHI, S., Y. SEKI & S. SANO. 1983. J. Biol. Chem. **258:** 4285–4292.
13. SINCLAIR, P., W. BEMENT, H. BONKOVSKY & J. SINCLAIR. 1984. Biochem. J. **222:** 737–748.
14. SINCLAIR, P., W. BEMENT, H. BONKOVSKY, R. LAMBRECHT, J. FREZZA, J. SINCLAIR, A. URQUHART & G. ELDER. 1986. Biochem. J. **237:** 63–71.
15. SINCLAIR, J., P. SINCLAIR, J. HEALEY, E. SMITH & H. BONKOVSKY. 1982. Biochem. J. **204:**103–109.
16. BONKOVSKY, H., S. WOOD, S. HOWELL, P. SINCLAIR, B. LINCOLN, J. HEALEY & J. SINCLAIR. 1986. Anal. Biochem. **155:** 56–64.
17. SINCLAIR, P. & S. GRANICK. 1977. Anal. Biochem. **79:** 380–393.
18. SINCLAIR, J., P. SINCLAIR, E. SMITH, W. BEMENT, J. POMEROY & H. BONKOVSKY. 1981. Biochem. Pharmacol. **30:** 2805–2809.
19. LOWRY, O., N. ROSEBROUGH, A. FARR & R. RANDALL. 1951. J. Biol. Chem. **193:** 265–275.
20. TESTA, B. & P. JENNER. 1981. Drug Metab. Rev. **12:** 1–117.
21. LESCA, P., E. RAFIDINARIVO, P. LECOINTE & D. MANSUY. 1979. Chem. Biol. Interact. **24:** 189–198.
22. JONES, K. & G. SWEENEY. 1980. Toxicol. Appl. Pharmacol. **53:** 42–49.
23. SMITH, A., J. FRANCIS, S. KAY & J. GREIG. 1981. Biochem. Pharmacol. **30:** 2825–2830.
24. CARPENTER, H., D. WILLIAMS, M. HENDERSON, R. BENDER & D. BUHLER. 1984. Biochem. Pharmacol. **33:**3875–3881.
25. ELDER, G., J. TOVEY & D. SHEPPARD. 1983. Biochem. J. **215:** 45–55.
26. ELDER, G., A. ROBERTS & A. URQUHART. 1986. *In* Porphyrins and Porphyrias. Y. Nordmann, Ed.:147–152. J. Libbey Eurotext. London.
27. ELDER, J., J. KUSHNER & M. PRYOR. 1982. Enzyme **28:** 170–185.
28. ELDER, G. & P. WYVILL. 1982. Enzyme **28:** 186–195.
29. GARCIA, R., L. SAN MARTIN DE VIALE, J. TONIO & M. GRINSTEIN. 1973. Biochem. Biophys. Acta **309:** 203–210.
30. SMITH, A. & J. FRANCIS. 1979. Biochem. J. **183:**455–458.
31. HEALEY, J., H. BONKOVSKY, P. SINCLAIR & J. SINCLAIR. 1981. Biochem. J. **198:** 595–604.
32. BONKOVSKY, H., J. HEALEY, P. SINCLAIR, J. SINCLAIR, S. SHEDLOFSKY & G. ELDER. 1983. J. Clin. Invest. **71:** 1175–1182.
33. BONKOVSKY, H., J. HEALEY, P. SINCLAIR & J. SINCLAIR. 1985. Biochem. J. **227:** 893–901.
34. SINCLAIR, P., G. ELDER, W. BEMENT, S. SMITH, H. BONKOVSKY & J. SINCLAIR. 1983. FEBS Lett. **152:** 217–221.
35. LAMBRECHT, R., P. SINCLAIR, W. BEMENT, J. SINCLAIR & H. CARPENTER. 1987. Ann. N.Y. Acad. Sci. This volume.
36. HEIKEL, T., W. LOCKWOOD & C. REMINGTON. 1958. Nature (London) **182:** 313.
37. MUKERJI, S., N. PIMSTONE & M. BURNS. 1984. Gastroenterology **87:** 1248–1254.
38. SINCLAIR, P. & R. LAMBRECHT. 1987. Ann. N.Y Acad. Sci. This volume.
39. POLAND, A. & J. KNUTSON. 1982. Annu. Rev. Pharmacol. Toxicol. **22:** 517–554.
40. DENCKER, L. 1985. Arch. Toxicol. (Suppl.) **8:** 43–60.
41. DEBETS, F., J. REINDERS, A. DEBETS, T. LOSSBROCK, J. STRIK & G. KOSS. 1981. Toxicology **19:** 185–196.
42. HALLIWELL, B. & J. GUTTERIDGE. 1986. Arch. Biochem. Biophys. **246:** 501–514.
43. FERIOLI, A., C. HARVEY & F. DEMATTEIS. 1984. Biochem. J. **224:** 769–777.
44. KENNEDY, S., D. WIGFIELD & G. FOX. 1986. Toxicol. Lett. **31:** 235–241.

45. Marks, G., S. Follows, D. Zelt & S. Cole. 1983. Can. J. Physiol. Pharmacol. **33:** 1487–1491.
46. Shedlofsky, S., P. Sinclair, H. Bonkovsky, J. Healey, A. Swim & J. Robinson. 1987. Biochem. J. In press.
47. Smith, A., J. Francis, S. Kay, J. Greig & T. Stewart. 1986. Biochem. J. **238:** 871–878.

Patterns of Porphyrin Accumulation in Response to Xenobiotics

Parallels between Results in Chick Embryo and Rodents[a]

GERALD S. MARKS, JANE POWLES, MARTHA LYON,
STUART McCLUSKEY, ELAINE SUTHERLAND,
AND DAVID ZELT

Department of Pharmacology and Toxicology
Queen's University
Kingston, Ontario, Canada K7L 3N6

Our present understanding of heme biosynthesis makes it possible to use alterations in this pathway as an index of exposure to a variety of toxic agents.[1,2] Normally the pathway is carefully controlled,[3] and porphyrins other than protoporphyrin IX are formed in very small amounts. When the regulatory process is disturbed, however, a variety of other porphyrins and porphyrin precursors may accumulate.[1,2] In animals, these porphyrins and precursors are usually formed in the liver and excreted in the urine and feces. The usefulness of the heme biosynthetic pathway as an indicator of exposure to toxic agents has increased in recent years because of the development of sensitive techniques to separate and measure porphyrins. One of the problems associated with animal studies of porphyria is the fact that chemicals must be administered to animals for a considerable period of time.[4] Such studies are prolonged and expensive. A possible means of overcoming these problems is to use the chick embryo liver cell culture. With this system, a period of less than 24 hr is required to demonstrate porphyrin accumulation in response to chemicals.[5] Moreover, several patterns of porphyrin accumulation are observed in response to different chemicals. These patterns might be diagnostic of the mechanism of action of these chemicals.[6,7]

The objectives of this study were the following: 1) to determine whether patterns of porphyrin accumulation in the chick embryo liver cell culture were predictive of enzyme inhibition with special reference to ferrochelatase and uroporphyrinogen decarboxylase (UROD); 2) to determine whether patterns of porphyrin accumulation and enzyme inhibition in the chick embryo liver cell culture were reflective of the mechanism of action of chemicals in rodent liver.

MATERIALS AND METHODS

Source of Compounds

The analogues of 3,5-diethoxycarbonyl-1,4-dihydro-2,4,6-trimethylpyridine (DDC), that is, DDC itself (I) and the 4-ethyl (II), 4-propyl (III), 4-butyl (IV), 4-pentyl (V), 4-hexyl (VI), and 4-isobutyl (VII) analogues (FIG. 1a), were obtained as described previously.[8–10] The polychlorinated biphenyl (PCB) isomers were obtained as

[a]This work was supported by the Medical Research Council of Canada.

chlorinated dibenzo-*p*-dioxin- and dibenzofuran-free standards from RFR Corporation (Hope, RI). Aroclors (commercial mixtures of PCBs) were obtained as a gift from Dr. D. J. Ecobichon (McGill University, Montreal, Quebec).

Determination of Ferrochelatase Activity

Details of the cell culture technique have been described.[11] After an initial incubation period of 24 hr, the medium was discarded and replaced with fresh medium

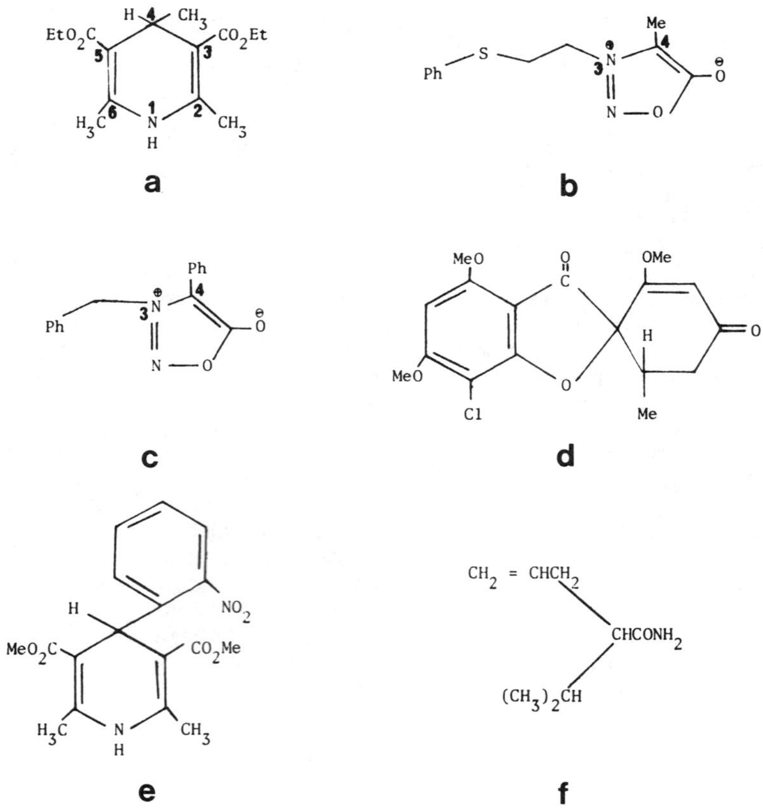

FIGURE 1. (**a**) DDC; (**b**) TTMS; (**c**) 3-benzyl-4-phenylsydnone; (**d**) griseofulvin; (**e**) nifedipine; (**f**) allylisopropylacetamide.

(15 ml). After a further 18-hr incubation period, chemicals, dissolved in 95% ethanol, were added to the dishes (maximal total volume added: 30 μl). The final concentration of drugs was 4 or 12 μM. Ferrochelatase activity was assayed 6 hr after addition of DDC or a DDC analogue as described previously[9,12] by a modification of the pyridine hemochromogen method using mesoporphyrin and iron as substrates.[13]

Determination of Porphyrin Profiles

A modification of the procedure of Zelt *et al.*[14] was used. Chick embryo liver cells were maintained in 6-cm-diameter plastic dishes containing 5 ml of the medium. Drugs were dissolved in 95% ethanol (maximal total volume: 15 μl) to give a concentration of 12 μM. Twenty-four hours later the contents of each dish were lyophilized and then the porphyrins were esterified at $-15°$ C with 5% sulfuric acid in methanol for 24 hr. The methyl esters were extracted into chloroform, washed with 5% sodium bicarbonate and with water, dried (Na_2SO_4), and then evaporated to dryness under nitrogen at 37° C. The residue was dissolved in 200 μl of a 1:1 dichloromethane:mobile phase (heptane:ethyl acetate:dichloromethane:methanol; 60:25:14:1) mixture, and 10 μl of the resulting solution was injected onto the column. In order to quantitate the individual porphyrins, several concentrations (1 to 100 pmoles) of uroporphyrin III, heptacarboxylic acid porphyrin I, hexacarboxylic acid porphyrin I, pentacarboxylic acid porphyrin I, coproporphyrin III, and protoporphyrin IX were added to cells and medium. After esterification and extraction, a series of standard curves were constructed and the results were expressed as picomoles of porphyrin per milligram of protein.

δ-Aminolevulinic Acid Synthetase Activity

Chick embryo liver cells were maintained in 10-cm-diameter disposable plastic dishes containing 15 ml of medium. After an initial incubation period of 24 hr, the medium was discarded and replaced with fresh medium. After a further 21-hr incubation period, drugs dissolved in 95% ethanol were added to the dishes (maximal total volume added: 30 μl). The final concentration of drugs was 3 μg/ml. Six hours later, δ-aminolevulinic acid synthetase (ALAS) activity was measured.[15,16] Medium from two cell culture plates was discarded, and 5 ml of an ice-cold solution containing 0.25 M sucrose, 10 mM Tris, and 1 mM EDTA at pH 7.2 was added to each plate. After centrifugation at 950 $\times$ g for 5 min at 4° C, the cell pellet was resuspended in the same buffer (0.5 ml) and homogenized (Potter–Elvehjem). The homogenate was used as the source of the enzyme.

The incubation mixture contained in a total volume of 1.0 ml the following components (final concentrations in parentheses): liver homogenate (0.5 ml), Tris-HCl (75 mM), glycine (100 mM), EDTA (7.5 mM), pyridoxal 5′-phosphate (0.4 mM), adenosine 5′-triphosphate (1.36 mM), and [2,3-^{14}C]succinate (1 μCi; 0.37 mM). The pH of the mixture was 7.5. Samples were incubated at 37° C in air, but without shaking. The ^{14}C-labeled δ-aminolevulinic acid (ALA) was eluted by ion-exchange chromatography and purified after treatment with acetylacetone to form a pyrrole derivative. Enzyme activity was expressed as pmoles of ALA formed/30 min/mg protein.

Protein Determinations

Protein was assayed by the method of Lowry *et al.*[17]

Statistical Analysis

In order to compare more than two means (such as in a study of the effects of several doses of a DDC analogue on ferrochelatase activity), a one-way analysis of variance ($p < .05$) was employed. If a significant F ratio at the .05 level was obtained, a Newman–Keuls test was used to determine which means were significantly different from each other ($p < .05$).

RESULTS AND DISCUSSION

Our initial studies were conducted with DDC and its 4-alkyl analogues (FIGURE 1a). The results in FIGURE 2 show the differences in the ability of these analogues to lower ferrochelatase activity in chick embryo liver cell culture at a concentration of 4 μM. Thus DDC (I) and its 4-ethyl (II) and 4-propyl (III) analogues were the most potent. Ferrochelatase-lowering activity decreased as the length of the 4-alkyl chain increased. Thus the 4-pentyl (V) analogue was less active than the 4-methyl (I), 4-ethyl (II), and 4-propyl (III) analogues, but more active than the 4-hexyl (VI) analogue; the 4-isobutyl (VII) analogue was inactive. Similar results were obtained using a 12 μM concentration.

The results in FIGURE 3 show the porphyrin patterns obtained with DDC and its analogues at a concentration of 12 μM in chick embryo liver cell culture. It was anticipated that protoporphyrin would be the major porphyrin to accumulate with all the DDC analogues that exhibited significant ferrochelatase-lowering activity; this was found to be the case. Thus only in the case of the 4-isobutyl (VII) analogue, which was devoid of ferrochelatase-lowering activity, was a porphyrin other than protoporphyrin, namely, coproporphyrin, the major porphyrin to accumulate. It was antic-

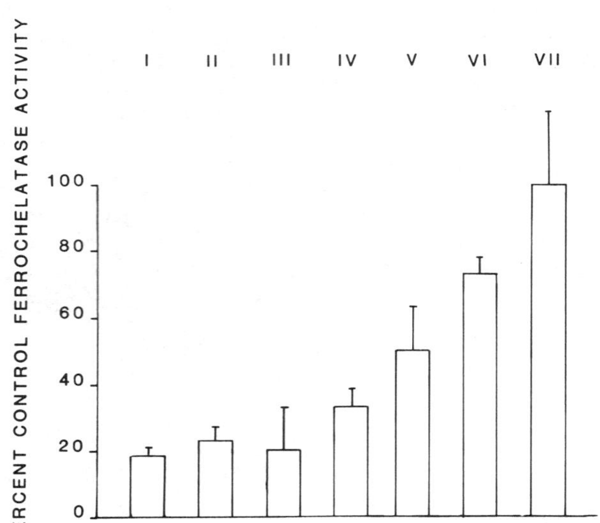

FIGURE 2. Percentage of control ferrochelatase activity in chick embryo liver cell culture 6 hr after the addition of 4 μM DDC (I), 4-ethyl DDC (II), 4-propyl DDC (III), 4-butyl DDC (IV), 4-pentyl DDC (V), 4-hexyl DDC (VI), and 4-isobutyl DDC (VII). Each bar represents a mean ± SD of four determinations. The experiment was repeated once, and similar results were obtained.

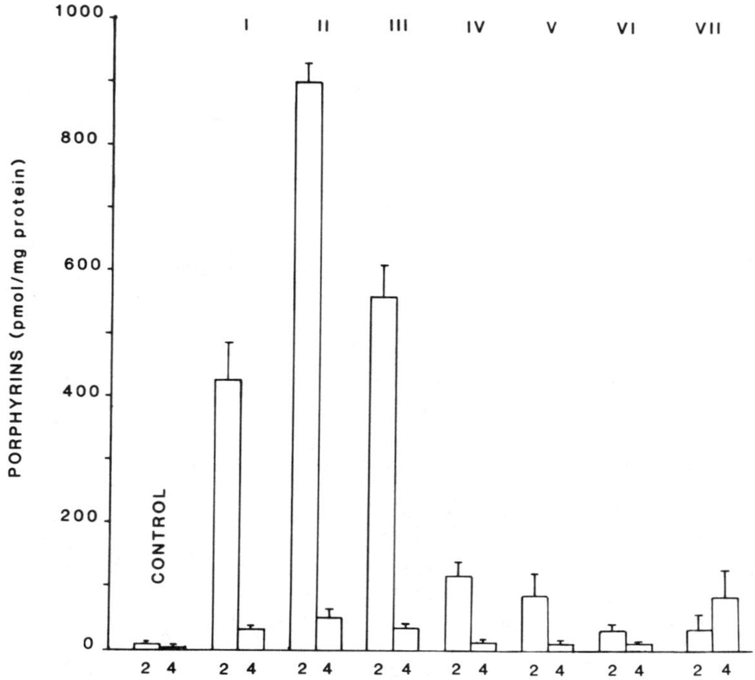

FIGURE 3. Porphyrin patterns obtained 24 hr after the addition of 12 μM DDC (I), 4-ethyl DDC (II), 4-propyl DDC (III), 4-butyl DDC (IV), 4-pentyl DDC (V), 4-hexyl DDC (VI), and 4-isobutyl DDC (VII) to chick embryo liver cell culture. Each bar represents a mean ± SD of four determinations. The numbers under the bars indicate the number of carboxyl groups (2: protoporphyrin; 4: coproporphyrin).

ipated that the amount of protoporphyrin to accumulate and the ratio of protoporphyrin to other porphyrins would be highest with the analogues that exhibited the greatest ferrochelatase-lowering activity. This was found to be the case. Thus the amount of protoporphyrin to accumulate and the protoporphyrin:coproporphyrin ratio was highest with DDC and its 4-ethyl (II) and 4-propyl (III) analogues. The amount of protoporphyrin that accumulated and the protoporphyrin:coproporphyrin ratio tended to decrease as the length of the 4-alkyl chain increased in a manner analogous to that observed with the ferrochelatase-lowering activity (FIGURE 2).

DDC increases the levels of the rate-limiting enzyme of the heme biosynthetic pathway, namely, ALAS. It has been proposed that this effect is secondary to DDC-induced lowering of ferrochelatase activity, since the decreased heme formation that results would presumably lead to diminished heme-mediated feedback repression of the synthesis of ALAS and therefore to increased ALAS activity. If this explanation is correct, one would anticipate a relationship between the degree of ferrochelatase-lowering and ALAS-inducing activity of DDC analogues. The effect of DDC analogues I, IV, V, and VI on ALAS activity is shown in FIGURE 4. The analogues of DDC studied enhance ALAS activity to a similar extent. A comparison of these results for ferrochelatase-lowering activity (FIGURE 2) shows no relationship between ferro-

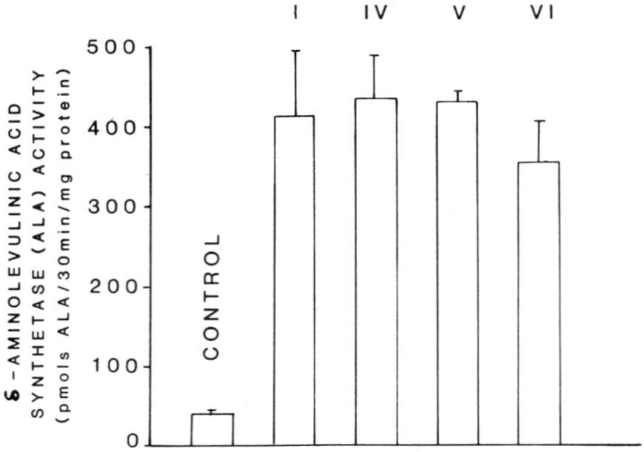

FIGURE 4. ALAS activity 6 hr after the addition of 3 μg/ml of DDC (I), 4-butyl DDC (IV), 4-pentyl DDC (V), and 4-hexyl DDC (VI) to chick embryo liver cells. Each bar represents a mean ± SD of three determinations.

chelatase-lowering (FIGURE 2) and ALAS-inducing (FIGURE 4) activity. A striking result is that obtained with 4-isobutyl DDC (VII). This analogue raises ALAS activity to approximately three times control levels at 6 hr (results not shown) although it does not exhibit ferrochelatase-lowering activity (FIGURE 2). It is therefore clear that regulatory heme levels are not lowered solely by ferrochelatase inhibition. The DDC analogues IV, V, VI, and VII are suicide substrates of cytochrome P-450 and result in heme destruction.[10] Regulatory hepatic heme levels are therefore probably lowered by DDC analogues by at least two mechanisms; namely, lowered ferrochelatase activity and heme destruction in the course of suicidal inactivation of cytochrome P-450.

3-[2-(2,4,6-Trimethylphenyl)thioethyl]-4-methylsydnone (TTMS) (FIGURE 1b) was developed as a potential antiarthritic agent. Its development, however, was halted when it was reported to produce hepatic porphyria in dogs, rats, and mice.[18] TTMS was found to be a potent inhibitor of ferrochelatase,[19] and it was anticipated that the ratio of protoporphyrin to coproporphyrin would be high; this was found to be the case (FIGURE 5). On the other hand, a related compound, 3-benzyl-4-phenylsydnone (FIGURE 1c), did not inhibit ferrochelatase, and it was anticipated that the ratio of protoporphyrin to other porphyrins would be low; this was found to be the case (FIGURE 5). Thus with the dihydropyridines and sydnones the porphyrin patterns in chick embryo liver cell culture were predictive of the inhibition of ferrochelatase. Moreover, the results with TTMS and DDC parallel those found in rodents where protoporphyrin has been identified as the major hepatic porphyrin.[18,20]

The oral administration of high doses of griseofulvin to mice resulted in liver enlargement and cellular damage.[21] This drug was subsequently shown to induce a disturbance of porphyrin metabolism in mice and rats[22,23]; protoporphyrin was shown to be the major porphyrin to accumulate in rodent liver. A positive correlation was demonstrated between porphyrin (mainly protoporphyrin) in the livers of griseofulvin-fed mice and diminished ferrochelatase activity.[24] This observation led to the suggestion that griseofulvin, like DDC, owes its porphyrinogenicity to ferrochelatase-inhibitory activity.[24] Griseofulvin is a potent porphyrinogenic agent in chick embryo liver cell culture,[5] and it was anticipated that protoporphyrin, like DDC, would be the major prophyrin to accumulate. Contrary to this expectation, however, coproporphyrin, uroporphyrin, and heptacarboxylic acid porphyrin were the major porphyrins to accumulate.[25] Moreover, unlike DDC, griseofulvin did not inhibit ferrochelatase in chick embryo liver cell culture.[25] The porphyrin pattern observed with griseofulvin suggests inhibition of UROD or coproporphyrinogen oxidase (CPPO) or both. The effect of griseofulvin on ferrochelatase activity was compared with the effect of DDC in suspensions of isolated mouse hepatocytes following a 4-hr incubation period.[26] DDC, but not griseofulvin, inhibited this enzyme. These results indicate that the difference between the effects of griseofulvin in rodents *in vivo* (inhibition) and the chick embryo liver cell culture (no effect) cannot be attributed solely to species differences.[26]

DDC, when administered to rodents[27,28] and the 17-day-old chick embryo,[29] causes the accumulation of hepatic *N*-methylprotoporphyrin IX (*N*-MePP),[27-30] a potent inhibitor of ferrochelatase. The source of the *N*-methyl substituent of *N*-MePP is the 4-methyl substituent of DDC.[31-33] DDC and several 4-alkyl analogues serve as suicide substrates for cytochrome P-450.[8] It is thought that a radical cation is produced by the one-electron oxidation of the nitrogen atom of DDC by cytochrome P-450, and this oxidation is followed by ejection of the 4-alkyl radical that alkylates the heme prosthetic group. The source of the protoporphyrin IX moiety of *N*-MePP therefore appears to be the prosthetic heme of cytochrome P-450.[8] TTMS, in an analogous manner, when administered to rodents, is catalytically activated by cytochrome P-450 leading to heme alkylation and the formation of *N*-vinylprotoporphyrin IX, which inhibits ferrochelatase.[34] It has been reported that when griseofulvin is administered to mice it causes the accumulation of a ferrochelatase-inhibitory *N*-alkyl porphyrin with chromatographic and spectral properties similar to those of *N*-MePP.[35] Although the

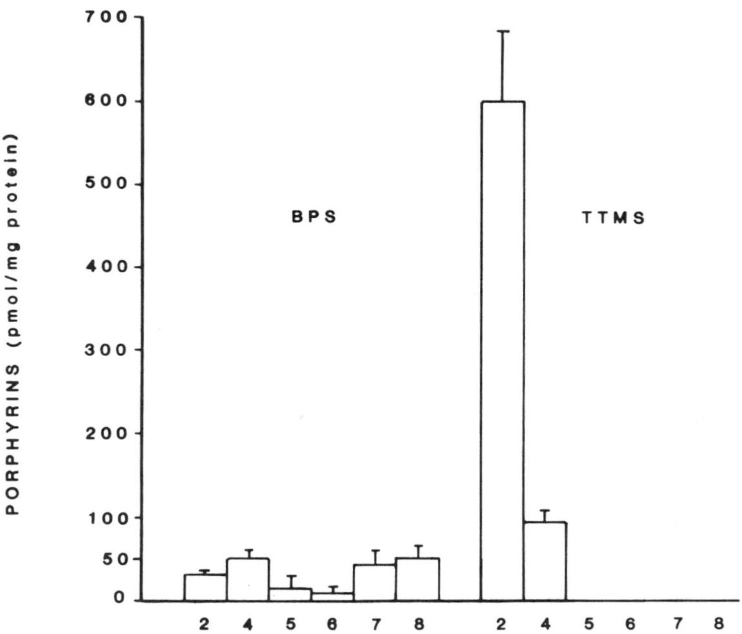

FIGURE 5. Porphyrin patterns obtained 24 hr after the addition of 3-benzyl-4-phenylsydnone (BPS) and TTMS (3 μg/ml medium) to chick embryo liver cell culture. The numbers under the bars indicate the number of carboxyl groups (2: protoporphyrin; 4: coproporphyrin; 5: pentacarboxylic porphyrin; 6: hexacarboxylic porphyrin; 7: heptacarboxylic porphyrin; 8: uroporphyrin). Each bar represents a mean ± SD of four determinations in one experiment; the results were confirmed in two additional experiments. (Adapted from reference 19, with permission.)

N-alkylporphyrins derived from DDC[30] and TTMS[34] have been thoroughly characterized by NMR and mass spectrometry, such an accumulation has yet to be demonstrated for the *N*-alkylporphyrin derived from griseofulvin. This is urgently required if progress is to be made. One possible explanation for the griseofulvin results is as follows: The pattern of hepatic cytochrome P-450 isozymes differs in different species. It is possible that the particular isozyme required for catalytic activation of griseofulvin is confined to rodents. In contrast, the particular isozyme required for catalytic activation of DDC and TTMS may be located in both rodent and chick embryo liver.

We studied the porphyrinogenic properties of a series of Aroclors in chick embryo hepatocyte culture[36] and showed that porphyrinogenicity increased with chlorine content. The porphyrin patterns obtained with the Aroclors in chick embryo liver cell culture are shown in FIGURE 6; uroporphyrin and heptacarboxylic acid porphyrin were the major porphyrins suggesting UROD inhibition. Subsequent studies indicated that this idea was correct. Thus De Verneuil *et al.*[37] showed that Aroclor 1254, a mixture of PCBs consisting predominantly of molecules with a total number of four, five, or six chlorine atoms, caused inhibition of UROD in chick embryo hepatocyte culture. Zelt[36] showed that several PCB congeners caused the accumulation of porphyrins in chick embryo hepatocyte culture; uroporphyrin and heptacarboxylic acid porphyrins were the major porphyrins to accumulate (FIGURE 7). This pattern of porphyrin accumulation suggested inhibition of UROD. This idea was shown to be correct. When UROD activity was measured 24 hr after 3,3′,4,4′-tetrachlorobiphenyl (TCBP) was added to

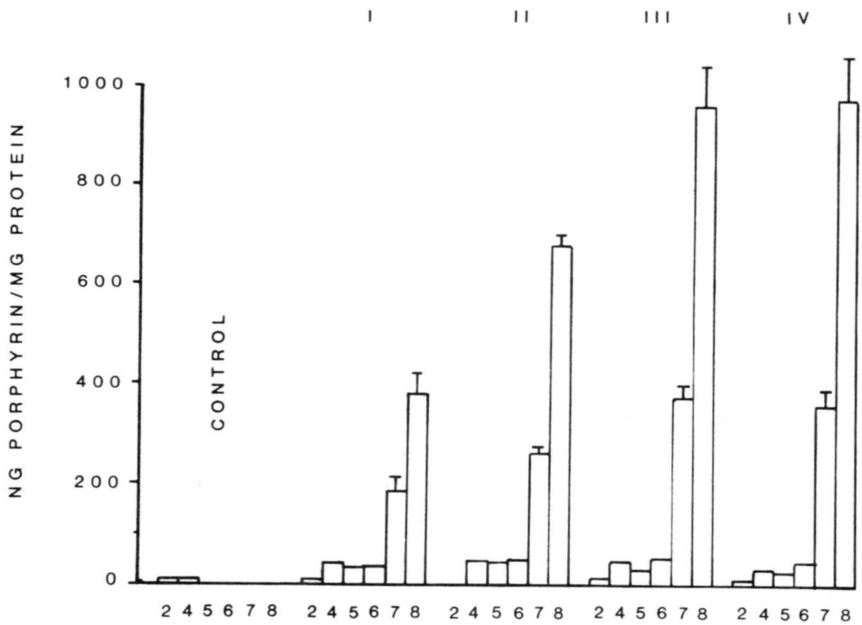

FIGURE 6. Accumulation of porphyrin intermediates in chick embryo liver cell culture in response to different Aroclors: 1221 (I), 1016 (II), 1242 (III), and 1254 (IV). The control received 2 μl acetone/ml medium. All Aroclors were administered at a dose of 7.5 μg/ml medium. The numbers under the bars indicate the number of carboxyl groups on the accumulating porphyrin. Each bar is a mean ± SEM of five determinations.

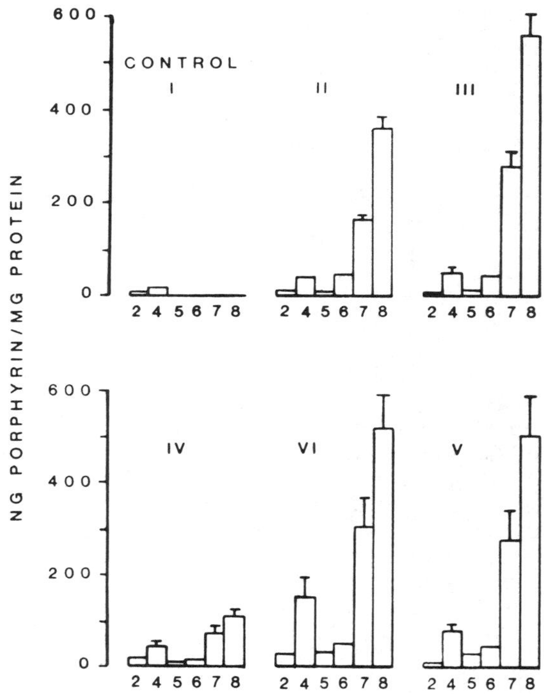

FIGURE 7. Accumulation of porphyrin intermediates in chick embryo liver cell culture in response to chlorinated dibenzo-*p*-dioxin- and dibenzofuran-free PCB congeners 3,3′,4,4′-TCBP (II), 2,3,3′,4,4′,5-HCBP (III), 2,2′3,3′4,4′-HCBP (IV), 2,2′,4,4′,5,5′-HCBP (V), and 2,2′,4,4′,6,6′-HCBP (VI). All congeners were administered at a dose of 30 nmol/ml; the controls (I) received 2 μl acetone/ml medium. Each bar is a mean ± SEM of at least nine determinations.

chick embryo liver cells, a significant inhibition was found.[38] On the other hand, when 3,3′,4,4′-TCBP was added to crude UROD, no inhibition was noted.[38] The accumulation of uroporphyrin and heptacarboxylic acid porphyrin in chick embryo liver cells after the addition of 2,3,7,8-tetrachlorodibenzo-*p*-dioxin (TCDD) is explained by the inhibition of UROD.[37] Debets *et al.*[39] have demonstrated the accumulation of uroporphyrin and heptacarboxylic acid porphyrin after the addition of hexachlorobenzene to chick embryo liver cells. The investigators attribute the porphyrin pattern to UROD inhibition, but this remains to be demonstrated.

Thus with the polyhalogenated aromatic hydrocarbons (PAHs), namely, the PCBs and TCDD, the porphyrin pattern in chick embryo liver cell culture is predictive of UROD inhibition. Moreover, the results with TCDD and hexachlorobenzene parallel those found in rodents. Thus when hexachlorobenzene was administered to rats and the levels of UROD measured as the porphyria developed (elevated levels of uroporphyrin and heptacarboxylic acid porphyrin in the urine), it was found that the levels of this enzyme fell progressively in rat liver.[4,40] Similarly, the pattern of urinary porphyrin excretion (elevated levels of uroporphyrin and heptacarboxylic acid porphyrin) induced by TCDD administration to rodents is attributable to TCDD being a potent inhibitor *in vivo* of hepatic UROD.[41,42] Because TCDD and hexachlorobenzene decrease the activity of the enzyme without altering the amount of immunoreactive enzyme protein, it is suggested that metabolites of these compounds react with a site in UROD essential for catalytic activity and form a covalently bonded enzyme–protein conjugate.[42] Several other groups of investigators have suggested that metabolites of PAHs are formed in the liver and are responsible for inactivation of UROD.[37–39] Other mechanisms have been proposed to explain UROD inhibition by PAHs, and will be discussed below.

When the activity of UROD is diminished, decarboxylation is incomplete and pentacarboxylate porphyrinogen III accumulates (FIGURE 8). The latter acts as a substrate for CPPO: the 2-propionate group is converted to a vinyl group, and the result is dehydroisocoproporphyrinogen. The vinyl group is reduced to an ethyl group and the porphyrinogen oxidized to a porphyrin to yield isocoproporphyrin (FIGURE 8). Elder[4] has suggested that detection of isocoproporphyrin may be a reliable indicator of a UROD defect. Until recently, isocoproporphyrin has only been detected in rodents with diminished UROD activity.[4] An interesting illustration of parallels between results in chick embryo and those in rodents is the detection of a porphyrin of the isocoproporphyrin series[43] in chick embryo hepatocyte culture after administration of 2,4,5,3′,5′-pentabromobiphenyl, a chemical that caused decreased UROD activity.

Following PAH administration to rodents, most investigators have either measured urinary porphyrin levels or total hepatic porphyrins and have experienced delays of as long as several months before observing significant changes.[4] Marked changes in porphyrin pattern were observed in female rat liver as early as 4 days after 5 mg/kg/day 3,3′,4,4′-TCBP administration and consisted of increases in uroporphyrin and heptacarboxylic acid porphyrin and decreases in coproporphyrin and protoporphyrin.[44] Similar changes have been reported to occur with the administration of hexachlorobenzene to female rats.[45] Moreover, Koss *et al.*[46] have reported a gradual decrease in the activity of hepatic UROD and identified uroporphyrin as the major hepatic porphyrin during the first 2 weeks of hexachlorobenzene administration to female Wistar rats. It is of interest that the elevation of uroporphyrin and heptacarboxyl porphyrin in the liver at 4 and 10 days of 3,3′,4,4′-TCBP administration is not reflected in increased levels of these porphyrins in the urine at these time periods.[44] Thus, it appears that significant inhibition of UROD may occur in the liver leading to hepatic accumulation of uroporphyrin and heptacarboxyl porphyrin without any demonstrable elevation of these porphyrins in urine. This fact sheds light on the apparent latency of porphyria development observed with PAHs in several mammalian species. An early disturbance in heme biosynthesis leading to porphyrin accumulation in the liver, that is, a disturbance induced by PAHs, will be reflected by an elevation of urinary uroporphyrin and heptacarboxyl porphyrin only at a considerably later period.

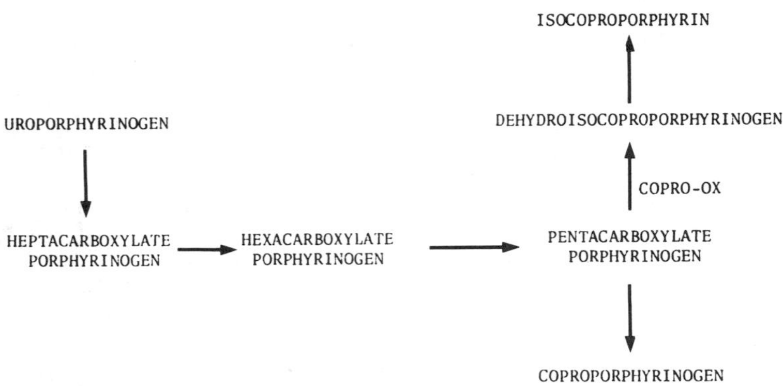

FIGURE 8. Mechanism of formation of isocoproporphyrin.

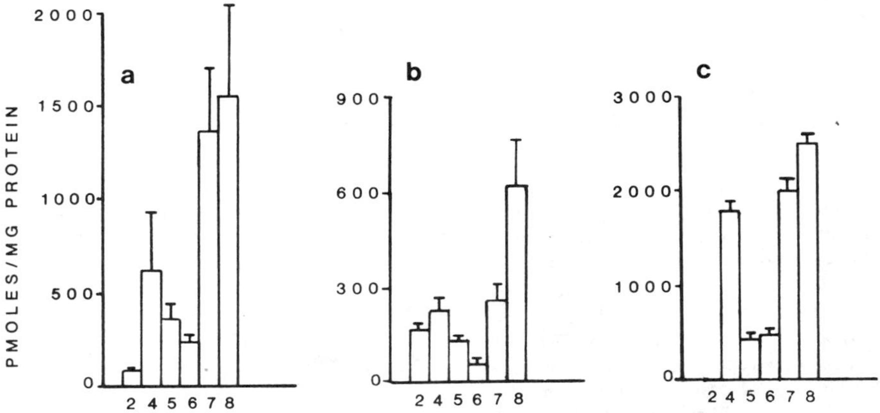

FIGURE 9. Porphyrin patterns produced by (**a**) sodium phenobarbital (232 μg/ml of medium), (**b**) nifedipine (25 μg/ml of medium), and (**c**) 3,5-diethoxycarbonyl-2,4,6-trimethylpyridine (100 μg/ml of medium) 24 hr after addition to chick embryo liver cell culture. Each bar is a mean ± SD of five determinations.

Phenobarbital, when added to chick embryo liver cell culture, causes the accumulation of uroporphyrin, heptacarboxylic acid porphyrin, and coproporphyrin (FIGURE 9a), suggesting inhibition of UROD. De Verneuil *et al.*[37] found this to be the case. Nifedipine (FIGURE 1e) and a variety of related 1,4-dihydropyridine calcium antagonists were shown to be potent porphyrinogenic agents in chick embryo liver cell culture.[47] Nifedipine, although a 1,4-dihydropyridine, did not exhibit ferrochelatase-lowering activity in chick embryo liver cell culture. Uroporphyrin, heptacarboxylic acid porphyrin, and coproporphyrin were the major porphyrins to accumulate (FIGURE 9b), suggesting inhibition of UROD; this was found by Schoenfeld *et al.*[48] to be the case. A variety of 1,4-dihydropyridine analogues of DDC, such as 4-phenyl DDC and desmethyl DDC (tested in the presence of bis[*p*-nitrophenyl]phosphate to avoid hydrolysis), resulted in the accumulation of uroporphyrin and heptacarboxylic acid porphyrin when added to chick embryo liver cells.[9] This pattern of porphyrin accumulation is similar to that produced by the pyridine analogue of DDC (FIGURE 9c) and related pyridines. This suggests that dihydropyridines that cause uroporphyrin and heptacarboxylic acid porphyrin accumulation are converted to pyridines[9] that inhibit UROD in chick embryo liver cell culture. It is of interest that an enzyme catalyzing the oxidation of 1,4-dihydropyridines to pyridines has been identified in rodent liver.[49]

It is of interest to consider the mechanism(s) by which UROD inhibition might occur in chick embryo hepatocyte culture. As discussed earlier, several investigators have attributed inhibition of this enzyme to the formation of metabolites of PAHs. The following additional hypotheses have been put forward to explain the inhibition of UROD by xenobiotics such as PAHs.[50,51] When liver microsomes are used as a source of cytochrome P-450, hydrogen peroxide is generated during the oxidation of NADPH. The hydrogen peroxide is thought to be derived[52] from the one-electron-reduced oxygen in the cytochrome P-450 cycle ($Fe^{3+}O_2^-$). The superoxide anion ferric hemoprotein complex dissociates as follows:

$$Fe^{3+}O_2 \rightarrow Fe^{3+} + O_2^{\cdot -}$$

Upon subsequent dismutation of $O_2^{\cdot-}$, hydrogen peroxide is formed as follows:

$$2O_2^{\cdot-} + 2H^+ \rightarrow H_2O_2 + O_2$$

The interaction of hydrogen peroxide and ferrous iron results in the generation of HO· radicals via the Fenton reaction as follows:

$$Fe^{2+} + H_2O_2 \rightarrow Fe^{3+} + HO\cdot + OH^-$$

Some xenobiotics enhance NADPH-dependent hydrogen peroxide formation and will therefore presumably also enhance the formation of the superoxide anion radical and the HO· radical. It has been suggested that xenobiotic-induced UROD inhibition may be caused indirectly by hydrogen peroxide or the above radicals or both by one or a combination of the following mechanisms[50,51]: 1) By interacting with endogenous hepatic substrates and producing long-lived enzyme inhibitors, which have been previously reported.[53,54] 2) By interaction of the radicals or hydrogen peroxide or both with UROD resulting in inactivation of the enzyme. 3) By interaction between the radicals or hydrogen peroxide or both with porphyrinogens resulting in their oxidation to porphyrins; porphyrins have been reported to inhibit UROD activity.[55]

The following factors are available to scavenge or degrade active oxygen species[52]: glutathione, glutathione peroxidase, superoxide dismutase, and catalase. The higher susceptibility of chick embryo liver cells than rodents to interference with heme synthesis[2] by xenobiotics and to UROD inhibition may be due to lower levels of these factors and enzymes in chick embryo liver cells than in rodent liver.

Some chemicals might induce porphyrin accumulation solely by causing an elevation in the level of the rate-limiting enzyme, ALAS.[5] When exogenous ALA was added to chick embryo liver cell culture to simulate the result of increased ALAS activity, protoporphyrin was found to predominate, being followed by coproporphyrin and uroporphyrin in abundance.[7] The large accumulation of protoporphyrin suggests that under conditions in which the rate-limiting enzyme ALAS is eliminated, ferrochelatase then becomes the rate-limiting enzyme. The conversion of ALA into heme has been studied in homogenates of livers of chick embryos, and it has been shown that at high concentrations of ALA, ferrochelatase limited the overall rate of heme synthesis.[56] Allylisopropylacetamide (FIGURE 1f) functions as a suicide substrate of cytochrome P-450. In the process, the heme moiety of cytochrome P-450 is converted to an *N*-alkylporphyrin and heme is continuously replaced with a new heme moiety. Hepatic heme levels decrease resulting in derepression of the synthesis of ALAS and in porphyrin accumulation.[57–59] One would therefore anticipate that protoporphyrin would be the major porphyrin; this was found to be the case in rodents.[20] In chick embryo liver cells, however, coproporphyrin was found in our studies to be the major porphyrin, being followed by uroporphyrin and heptacarboxylic acid porphyrin in abundance.[7]

Coproporphyrin was the major porphyrin to accumulate in chick embryo liver cells after the addition of 5α-androstan-3αol-16-one, 5β-androstan-3αol-17-one, and several aromatic amides. This pattern suggests inhibition of CPPO.[7] This enzyme remains to be carefully studied in rodents and chick embryo liver cells that have been treated with chemicals.

SUMMARY

Our objective was to determine whether patterns of porphyrin accumulation produced by chemicals in chick embryo hepatocyte culture would indicate which enzyme of heme biosynthesis was inhibited. The ferrochelatase-inhibitory potency and porphyrin patterns produced by DDC, TTMS, and their analogues were studied. The protoporphyrin:coproporphyrin ratio observed was found to correlate with ferrochelatase-inhibitory activity. The results obtained in chick embryo with TTMS and DDC parallel those found in rodents. Griseofulvin has been shown to lower ferrochelatase activity and to cause the accumulation of protoporphyrin in rodent liver. In chick embryo liver cell culture, however, coproporphyrin, uroporphyrin, and heptacarboxylic acid porphyrin accumulate and ferrochelatase activity is not lowered.

Uroporphyrin, heptacarboxylic acid porphyrin, and coproporphyrin are the major porphyrins to accumulate in response to PAHs (for example, 3,3′,4,4′-TCBP in chick embryo liver cell culture). This may be explained by inhibition of UROD, which has been observed in chick embryo and rodent liver. Some chemicals, such as phenobarbital and nifedipine, cause the accumulation of these porphyrins in chick embryo liver cell culture, and this is explained by inhibition of UROD. These chemicals have not been reported to interfere with heme biosynthesis in the intact chick embryo or rodents; possibly protective mechanisms that are not available in the cell culture system are operative in the intact animal. It was concluded that porphyrin patterns may serve as a guide to which enzyme of heme biosynthesis is inhibited in chick embryo liver cell culture. The results obtained in the culture system with certain chemicals, such as DDC and TTMS analogues and PAHs, correspond with results in rodents. In other cases, such as with griseofulvin, the results do not correspond.

REFERENCES

1. MARKS, G. S., D. T. ZELT & S. P. C. COLE. 1982. Can. J. Physiol. Pharmacol. **60:** 1017–1026.
2. MARKS, G. S. 1985. CRC Crit. Rev. Toxicol. **15:** 151–179.
3. KAPPAS, A., S. SASSA & K. E. ANDERSON. 1982. *In* The Metabolic Basis of Inherited Disease. J. B. Stanbury, J. B. Wyngaarden & D. S. Fredrickson, Eds.: 1301–1384. McGraw-Hill. New York, NY.
4. ELDER, G. H. 1979. *In* Handbook of Experimental Pharmacology. F. De Matteis & W. N. Aldridge, Eds. Vol. **44:** 157–200. Springer-Verlag, Berlin.
5. GRANICK, S. 1966. J. Biol. Chem. **241:** 1359–1375.
6. SINCLAIR, P. R. & S. GRANICK. 1974. Biochem. Biophys. Res. Commun. **61:** 124–133.
7. MARKS, G. S., S. B. FOLLOWS, D. T. ZELT & S. P. C. COLE. 1983. Can. J. Physiol. Pharmacol. **61:** 546–553.
8. AUGUSTO, O., H. S. BEILAN & P. R. ORTIZ DE MONTELLANO. 1982. J. Biol. Chem. **257:** 11288–11295.
9. COLE, S. P. C., R. A. WHITNEY & G. S. MARKS. 1981. Mol. Pharmacol. **20:** 395–403.
10. MCCLUSKEY, S. A., G. S. MARKS, E. P. SUTHERLAND, N. JACOBSEN & P. R. ORTIZ DE MONTELLANO. 1986. Mol. Pharmacol. **30:** 252–257.
11. MORGAN, R. O., P. W. F. FISCHER, J. K. STEPHENS & G. S. MARKS. 1976. Biochem. Pharmacol. **25:** 2609–2612.
12. COLE, S. P. C., E. J. VAVASOUR & G. S. MARKS. 1979. Biochem. Pharmacol. **28:** 3533–3538.
13. PORRA, R. J., K. S. VITOLS, R. F. LABBE & N. A. NEWTON. 1967. Biochem. J. **104:** 321–327.
14. ZELT, D. T., J. A. OWEN & G. S. MARKS. 1980. J. Chromatogr. **189:** 209–216.
15. DE MATTEIS, F., P. ZETTERLUND & L. WETTERBERG. 1981. Biochem. J. **196:** 811–817.
16. DE MATTEIS, F. & G. S. MARKS. 1983. FEBS Lett. **159:** 127–131.
17. LOWRY, O. H., N. J. ROSEBROUGH, A. L. FARR & R. J. RANDALL. 1951. J. Biol. Chem. **193:** 265–275.
18. STEJSKAL, R., M. ITABASHI, J. STANCK & Z. HRNBAM. 1975. Virchows Arch. B **18:** 83–100.

19. Sutherland, E. P., G. S. Marks, L. A. Grab & P. R. Ortiz de Montellano. 1986. FEBS Lett. **197:** 17–20.
20. Onisawa, J. & R. F. Labbe. 1963. J. Biol. Chem. **238:** 724–727.
21. Barich, L. L., J. Schwartz, D. J. Barick & H. G. Horowitz. 1961. Antibiot. Chemother. **11:** 566–571.
22. Lockhead, A. C., H. H. Dagg & A. Goldberg. 1967. Br. J. Dermatol. **79:** 96–102.
23. De Matteis, F. 1978. *In* Handbook of Experimental Pharmacology. F. De Matteis & W. N. Aldridge, Eds. Vol. **44:** 129–155. Springer-Verlag. Berlin.
24. De Matteis, F. & A. H. Gibbs. 1975. Biochem. J. **146:** 285–287.
25. Cole, S. P. C., D. T. Zelt & G. S. Marks. 1981. Mol. Pharmacol. **19:** 477–480.
26. Cole, S. P. C., T. E. Massey, G. S. Marks & W. J. Racz. 1981. Can. J. Physiol. Pharmacol. **59:** 1155–1158.
27. Tephly, T. R., A. H. Gibbs & F. De Matteis. 1979. Biochem. J. **180:** 241–244.
28. De Matteis, F., A. H. Gibbs & T. R. Tephly. 1980. Biochem. J. **188:** 145–152.
29. McCluskey, S. A., E. P. Sutherland, W. J. Racz & G. S. Marks. 1987. Can. J. Physiol. Pharmacol. **65:** 1500–1502.
30. Ortiz de Montellano, P. R., H. S. Beilan & K. L. Kunze. 1981. Proc. Natl. Acad. Sci. USA **78:** 1490–1494.
31. De Matteis, F., A. H. Gibbs, P. B. Farmer & J. H. Lamb. 1981. FEBS Lett. **129:** 328–331.
32. Tephly, T. R., B. L. Coffman, G. Ingall, M. S. Abou Zeit-Har, H. M. Goff, H. D. Tabba & K. M. Smith. 1981. Arch. Biochem. Biophys. **212:** 120–126.
33. Ortiz de Montellano, P. R., H. S. Beilan & K. L. Kunze. 1981. J. Biol. Chem. **256:** 6708–6713.
34. Ortiz de Montellano, P. R., A. K. Costa, A. Grab, E. P. Sutherland & G. S. Marks. 1986. *In* Porphyrins and Porphyrias. Y. Nordmann, Ed. Vol. **134:** 109–117. Colloque INSERM. John Libbey Eurotext. London.
35. De Matteis, F. & A. H. Gibbs. 1980. Biochem. J. **187:** 285–288.
36. Zelt, D. T. 1980. M. Sc. diss. Queen's University. Kingston, Ontario.
37. De Verneuil, H., S. Sassa & A. Kappas. 1983. Biochem. J. **214:** 145–151.
38. Swain, M. G., S. B. Follows & G. S. Marks. 1983. Can. J. Physiol. Pharmacol. **61:** 105–108.
39. Debets, F. M. H., J. H. Reinders, A. J. M. Debets, T. G. Lössbroeck, J. J. T. W. A. Strik & G. Koss. 1981. Toxicology **19:** 185–196.
40. San Martin de Viale, L. C., M. Del C. Rios de Molina, R. Wainstock de Calmanovici & J. M. Tomio. 1976. *In* Porphyrias in Human Diseases. M. Doss, Ed.: 445–452. S. Karger. Basel.
41. Jones, K. G. & G. D. Sweeney. 1977. Res. Commun. Chem. Pathol. Pharmacol. **17:** 631–637.
42. Elder, G. H. & D. M. Sheppard. 1982. Biochem. Biophys. Res. Commun. **109:** 113–120.
43. Sinclair, P. R., G. H. Elder, W. J. Bement, S. G. Smith, H. L. Bonkowsky & J. F. Sinclair. 1983. FEBS Lett. **152:** 217–221.
44. Taub, H., W. J. Racz, J. F. Brien, K. Nakatsu, D. W. Clarke & G. S. Marks. 1983. Proc. Can. Fed. Biol. Soc. **26:** 171.
45. Kennedy, S. W., D. C. Wigfield & G. A. Fox. 1986. Toxicol. Lett. **31:** 235–241.
46. Koss, G., S. Seubert, A. Seubert, J. Seidel, W. Koransky & H. Ippen. 1983. Arch. Toxicol. **52:** 13–22.
47. Marks, G. S., D. R. Goldman, S. A. McCluskey, E. P. Sutherland & M. E. Lyon. 1986. Can. J. Physiol. Pharmacol. **64:** 438–443.
48. Schoenfeld, N., J. Aelion, Y. Beigel, O. Epstein & A. Atsmon. 1985. Clin. Sci. **69:** 581–586.
49. Guengerich, F. P. & R. H. Böcker. 1986. Fed. Proc. **45:** 577.
50. Ferioli, A., C. Harvey & F. De Matteis. 1984. Biochem. J. **224:** 769–777.
51. Sinclair, P., W. J. Bement, H. L. Bonkowsky, R. W. Lambrecht, J. E. Frezza, J. F. Sinclair, A. J. Urquhart & G. H. Elder. 1986. Biochem. J. **237:** 63–71.
52. Bast, A. 1986. Trends Pharmacol. Sci. **7:** 266–270.
53. Rios de Molina, M. Del C., R. Wainstock de Calmanovici & L. C. San Martin de Viale. 1980. Int. J. Biochem. **12:** 1027–1032.

54. CANTONI, L., D. DAL FUIME, M. RIZZARDINI & R. RUGGIERI. 1984. Toxicol. Lett. **20:** 211–217.
55. SMITH, A. G. & J. FRANCIS. 1981. Biochem. J. **195:** 241–250.
56. HEALEY, J. F., H. L. BONKOWSKY, P. R. SINCLAIR & J. F. SINCLAIR. 1981. Biochem. J. **198:** 595–604.
57. DE MATTEIS, F. 1978. *In* Handbook of Experimental Pharmacology. F. De Matteis & W. N. Aldridge, Eds. Vol. **44:** 95–127. Springer-Verlag. Berlin.
58. ORTIZ DE MONTELLANO, P. R., B. A. MICO & G. S. YOST. 1978. Biochem. Biophys. Res. Commun. **83:** 132–137.
59. ORTIZ DE MONTELLANO, P. R. & M. A. CORREIA. 1983. Annu. Rev. Pharmacol. Toxicol. **23:** 481–503.

Effects of Chlorinated Organics on Intermediates in the Heme Pathway and on Uroporphyrinogen Decarboxylase[a]

LAVINIA CANTONI, MILENA RIZZARDINI, ANDREA GRAZIANI, CARMEN CARUGO, AND SILVIO GARATTINI

Istituto di Ricerche Farmacologiche "Mario Negri"
62-20157 Milan, Italy

INTRODUCTION

The first chlorinated organic to be shown to interfere with uroporphyrinogen decarboxylase (UROD) was hexachlorobenzene (HCB). Between 1956 and 1960, this compound, which was used as a fungicide, was the causative agent of an outbreak of porphyria cutanea tarda (PCT) in southeastern Turkey, where peasants ate bread made with contaminated flour.[1,2] PCT in humans is characterized by skin lesions due to the photosensitizing action of porphyrins and by marked uroporphyrinuria; excretion of porphobilinogen, however, remains normal. Neuropsychiatric manifestations do not occur. The disease in its sporadic form is most common in middle-aged men and is often associated with liver cell damage due to alcohol.[3]

The Turkish accident was the first clinical episode reported in the international literature of acquired porphyria due to accidental exposure to a polyhalogenated aromatic hydrocarbon (PHAH).

A porphyria with the same biochemical features, that is, excretion of mostly uroporphyrin, was later reproduced in rats fed a diet containing 0.2% wt/wt HCB,[4] and the underlying disorder was recognized as an acquired UROD deficiency.[5]

The HCB type of experimental porphyria can be induced with other chlorinated aromatic compounds too, such as hexabromobenzene, octachlorostyrene and other chlorinated benzenes, polychlorinated biphenyls (PCBs) and polybrominated biphenyls (PBBs), γ-hexachlorocyclohexane, and 2,3,7,8-tetrachlorodibenzo-*p*-dioxin (TCDD).[6,7] At present, there is still no experimental evidence of the porphyrogenicity of polychlorinated dibenzofurans, which are members of the same halogenated aromatic group as PCBs, PBBs, and TCDD, whose proposed mechanism of action proceeds via binding to a cytosolic receptor protein.[8] PHAHs are environmental contaminants,[9,10] and hence their dangers go beyond the immediate effects of industrial accidents (such as the one at Seveso, Italy in 1976, where a large populated area was contaminated with TCDD); that is, the presence of PHAHs in human and animal food and in wildlife causes concern about their long-term effects on health. Mice, rats, rabbits, chickens, and Japanese quail are known to be susceptible to the porphyrogenic effect of PHAHs.[6,7]

[a]This work was supported by Contract 85.00487.56 with the National Research Council, Rome, Italy (Target project: Preventive Medicine and Rehabilitation; Subproject: Toxicologic Risk).

UROD is the cytosolic enzyme of the heme pathway that converts uroporphyrinogen to coproporphyrinogen through the intermediate formation of heptacarboxylic, hexacarboxylic, and pentacarboxylic porphyrinogens.[11] In the normal liver, the uroporphyrinogen isomer found—isomer III—is formed through the condensation of four molecules of porphobilinogen (PBG) under the combined action of PBG deaminase and a second enzyme, uroporphyrinogen III cosynthetase.[12] Uroporphyrinogen III is also the physiological substrate for UROD. The end product of decarboxylation by UROD, coproporphyrinogen III, is the precursor of heme; no intermediate porphyrinogens accumulate. Normally heme is formed in a very slight excess, which is rapidly broken down to bile pigments.

The whole heme biosynthetic pathway is regulated by a mechanism of negative feedback depending on a small "regulatory" heme pool, and acting on the first, rate-limiting enzyme, δ-aminolevulinic acid synthase (ALAS).[13] UROD activity is not rate limiting in heme synthesis, but in PHAH porphyria it does become rate limiting because the enzyme activity is reduced,[6] and this causes a partial block in heme biosynthesis. Furthermore, PHAHs increase heme flux because they are all highly potent inducers of microsomal cytochromes.[10] At present, it still appears that all compounds that reduce UROD in the whole animal induce cytochrome P_1-450 with or without concomitant induction of other hepatic monooxygenase systems.[14]

This relationship between UROD reduction and P_1-450 induction suggests that the UROD decrease might be mediated by a P_1-450-dependent reaction. One of the possibilities that has been explored extensively is that UROD could be inactivated by some reactive metabolite formed by a P_1-450-dependent monooxygenase.[15,16] Other authors have focused on the finding that in rodents combined administration of iron and a PHAH worsens porphyrin accumulation and the degree of reduction of activity of UROD.[17] It thus appears that porphyrogenic compounds might modify UROD activity by stimulating production of reactive species, such as peroxides and free radicals, that may react with the enzyme itself.[18]

The synergism with the effect of iron appears to depend on the ability of divalent iron to catalyze peroxidative and free-radical reactions.[19] Investigations of UROD mechanism of action have shown that different isomers of the same porphyrinogen can be decarboxylated at different rates[20,21] and that porphyrins, especially oxidation products of the substrates, inhibit the enzyme.[22] Thus one can expect that changes in the isomeric type of substrate and in the extent of accumulated intermediates could affect the enzyme activity. Furthermore, UROD *in vitro* is sensitive to the action of sulfhydryl reagents, light, and reactive oxygen species.[23] This paper summarizes aspects of UROD inactivation and changes in the heme pathway that seem relevant to porphyria induced by PHAHs and presents data on the effect of combined administration of iron and HCB on the activity of xanthine oxidase (XO), which produces reactive oxygen intermediates.[24]

Reactive oxygen intermediates are known to be implicated in tissue injury, especially through the peroxidation of membrane lipids.[25] Activation of these processes by HCB could be important in establishing the hepatic damage observed in PHAH intoxication in which the reduction in UROD is a characteristic feature.

MATERIALS AND METHODS

Chemicals

TCDD and HCB (organic analytical grade) were purchased from Kor Isotopes (Boston, MA) and BDH (Poole), respectively. Imferon (50 mg Fe/ml) and its carbohydrate precursor, dextran C, were kindly provided by Dr. Piccinin (Fisons Italia, Rome). Porphyrin standards were purchased from Porphyrin Products (Logan, UT).

Animals and Treatments

Male C57Bl/6 mice, 6 weeks old, were obtained from Charles River (Calco, Como). The treated group received TCDD (25 μg/kg, ip) dissolved in a mixture of acetone and corn oil (1:6) once a week for 9 weeks; controls received the vehicle only (5 ml/kg).

Male C57Bl/10 ScSn and DBA/2 mice, 8 weeks old, were obtained from OLAC (Bicester) and Charles River (Calco, Como). The treated groups of animals were given 0.25 ml of Imferon (12.5 mg Fe/mouse) or 0.25 ml of a solution (200 mg/kg) of its carbohydrate precursor, dextran C, then fed HCB for 8 weeks as 0.02% of a diet that also contained 2% peanut oil; control animals received Imferon or dextran C alone as appropriate, with an unmedicated diet.

Mice were killed by cervical dislocation. Livers were excised, rinsed in 0.9% NaCl, and homogenized in 0.25 M sucrose. Homogenate was utilized as such or centrifuged at 40 000 × g for 30 min to obtain a supernatant fraction to be utilized as the UROD source.

Microsomes were prepared following the procedure of Kato and Takayanaghi.[26]

Biochemical Determinations

UROD activity was assayed with uroporphyrinogen III as substrate as described previously[21] utilizing a 40 000 × g supernatant from which endogenous porphyrins were eliminated through adsorption on Zerolit FF (ip) resin (BDH, Poole).

When UROD activity was measured with pentacarboxylic porphyrinogen I as substrate, the incubation conditions and the HPLC analysis of coproporphyrinogen formed were as described by Francis and Smith.[27] Activity of ALAS was measured on liver homogenates by the radiochemical method described by De Matteis *et al.*[28] using [2,3-^{14}C]succinate as substrate.

XO exists in two interconvertible forms[29]: one form is an oxidase (type O) and the other is an NADH-dependent dehydrogenase (type D). Type O activity was assayed with liver homogenate by the radiochemical method described by Ghezzi *et al.*[30] using [8-^{14}C]hypoxanthine as substrate. Total activity, that is, type O activity plus type D activity, was determined by including 1mM NAD^+ in the assay. Type O was converted to type D by adding 10 mM dithiothreitol to the incubation mixture.

Protein concentration was determined by the method of Lowry *et al.*[31] Total porphyrin content of tissue homogenate was analyzed following the method of Abbritti and De Matteis.[32] Lipid peroxidation of liver microsomes was determined *in vitro* by following the formation of malondialdehyde (MDA) by the thiobarbituric acid method using an extinction coefficient of 156 $mM^{-1} \cdot cm^{-1}$.[33]

Statistical Analysis

Statistical significance of differences between groups was assessed by Duncan's new multiple range test.

Synergism between iron and HCB on the biochemical effects measured was assessed by a two-way analysis of variance and an *F* interaction test.

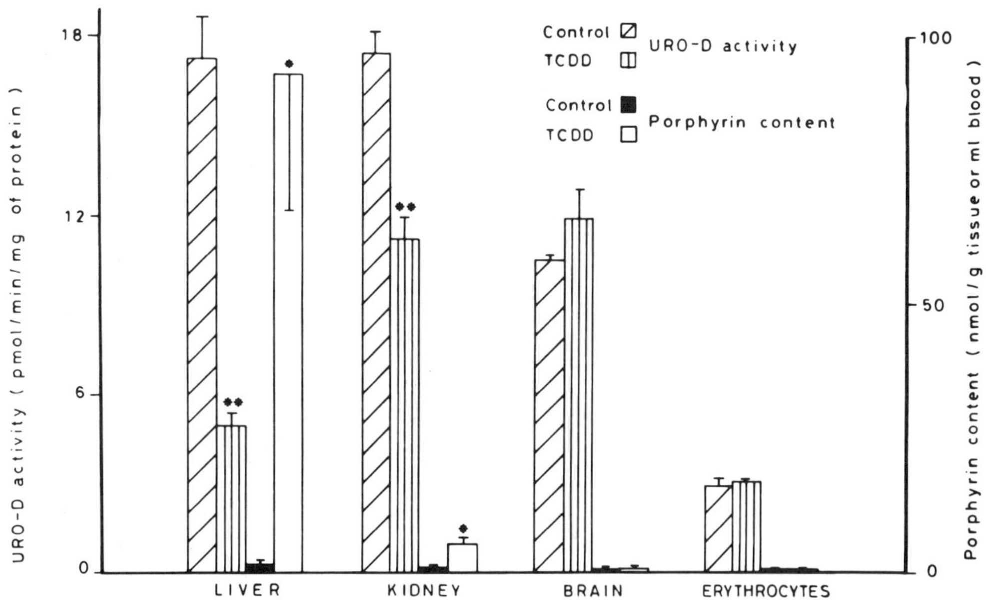

FIGURE 1. UROD activity and total porphyrin content of tissues from control and TCDD (25 μg/kg once a week for 9 weeks)-treated C57B1/6 male mice. Activity was measured as the sum of products formed (hepta-, hexa-, and pentacarboxylic porphyrinogen and coproporphyrinogen) from uroporphyrinogen III as substrate (3 μM). Each value is the mean ± SEM of four pools, each of which contained two animals. Asterisks indicate the results of Student's *t* test in comparison to the control value: *: $p < .05$; **: $p < .01$.

RESULTS

FIGURE 1 shows that treatment with TCDD markedly reduced UROD activity of liver in C57B1/6 male mice (72% reduction); in the same animals, kidney UROD was reduced 36% compared to control values, whereas brain and erythrocyte enzymes were unaffected.

The reduction of enzyme activity was reflected in the amount of porphyrins accumulated in the different tissues: the liver was the principal site of accumulation; the kidney was the site of the second greatest accumulation; the brain and erythrocytes were unaffected.

HPLC of the methyl ester derivatives of porphyrins showed that the tissues in animals with porphyria, unlike those in control animals, contained no detectable levels of coproporphyrin; massive amounts of uroporphyrin and heptacarboxylic porphyrin, however, were present (data not shown).

Hepatic activity of UROD from control and TCDD animals was measured in the presence of different concentrations of uroporphyrinogen III as substrate. The formation of total reaction products, of heptacarboxylic porphyrinogen and of coproporphyrinogen, are shown in FIGURES 2 & 3, for control and TCDD animals, respectively.

With control enzyme, the formation of total porphyrinogens plotted as a function of substrate concentration followed Michaelis–Menten kinetics. The amounts of the products of decarboxylation were determined, and those for hexacarboxylic porphyrinogen and pentacarboxylic porphyrinogen were found to be fairly constant at any substrate concentration, representing about 15% of total products (data not shown);

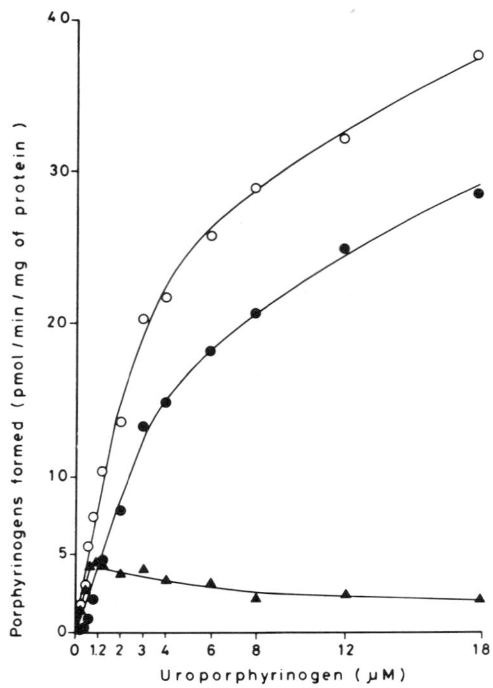

FIGURE 2. UROD activity of control C57B1/6 male mice in the presence of increasing amounts of uroporphyrinogen III as substrate. Each point represents the mean of two experimental values. O: accumulation of total products (hepta-, hexa-, and pentacarboxylic porphyrinogen and coproporphyrinogen); ●: accumulation of heptacarboxylic porphyrinogen; ▲: accumulation of coproporphyrinogen.

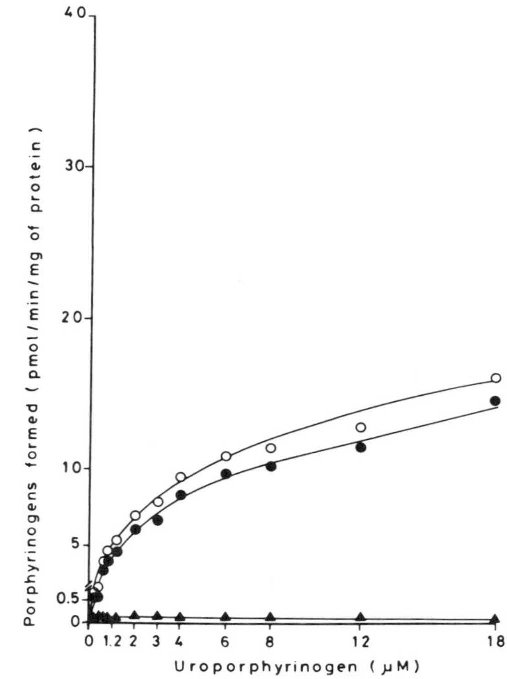

FIGURE 3. UROD activity of TCDD (25 μg/kg once a week for 9 weeks)-treated C57B1/6 male mice in the presence of increasing amounts of uroporphyrinogen III as substrate. Each point represents the mean of two experimental values; ○: accumulation of total products (hepta-, hexa-, and pentacarboxylic porphyrinogen and coproporphyrinogen); ●: accumulation of heptacarboxylic porphyrinogen; ▲: accumulation of coproporphyrinogen.

with increasing substrate concentration, the percentage of heptacarboxylic porphyrinogen rose steadily up to 80% of total products, and formation of coproporphyrinogen was progressively inhibited.

Uroporphyrinogen III decarboxylation with the enzyme from TCDD-treated mice was much slower, and heptacarboxylic porphyrinogen formation predominated at all substrate concentrations (about 90% of the total products) whereas the percentage of coproporphyrinogen formed always remained very low.

An increase in hepatic porphyrins was also noted in mice treated with HCB (TABLE 1), and this effect was much more marked in animals that received HCB after pretreatment with iron (significant interaction, $p < .01$). Hepatic UROD activity in these groups of animals was measured with pentacarboxylic porphyrinogen I as substrate (TABLE 2) at two concentrations.

UROD activity of the HCB group was significantly reduced only when the higher concentration of substrate was used whereas iron pretreatment followed by HCB resulted in a much larger reduction of the enzyme activity at any substrate concentration (significant interaction, $p < .01$). Because mouse strains with different degrees of responsiveness to induction of cytochrome P_1-450 and related monooxygenases are available, it was possible to treat a responsive strain (C57B1/10) and a nonresponsive strain (DBA/2) with iron and HCB to investigate how responsiveness correlated with susceptibility to porphyria and whether iron modulated the onset of porphyria in a nonresponsive strain. No porphyrogenic effect was observed in DBA/2 mice (TABLE 2). We next investigated whether the activity of ALAS, which is the rate-limiting enzyme in heme biosynthesis, was affected by treatment with HCB in C57B1/10 male mice because this compound could lower the level of the regulatory pool of heme as a

TABLE 1. Effect of Iron and HCB Treatment on Hepatic Porphyrin Content, ALAS Activity, and Lipid Peroxidation in C57B1/10 Male Mice

Iron	HCB	Total Porphyrin Content (nmol/g liver)	ALAS Activity[a] (pmol/min/g liver)	MDA[b] (nmol/100 mg protein)
–	–	0.68 ± 0.03	312.94 ± 24.28	119.80 ± 10.09
+	–	0.83 ± 0.17	473.88 ± 13.02	576.55 ± 52.88 NT[d]
–	+	4.22 ± 1.33	1102.16 ± 172.27 NT[c]	233.21 ± 17.10 NT[c]
+	+	424.75 ± 83.01 NT[d]IT[d]HT[d]	2729.64 ± 367.54 NT[d]IT[d]HT[d]	562.65 ± 68.56 NT[d]HT[d]

NOTE: Each value is a mean ± SEM of values obtained from four animals. All statistical comparisons have been made using Duncan's test. The abbreviations NT, IT, and HT indicate where comparisons have been made with the corresponding values obtained using no treatment, iron treatment, and HBC treatment, respectively.

[a]Activity was expressed as the amount of 5-aminolevulinate formed.

[b]Lipid peroxidation was determined *in vitro* by the formation of MDA.

[c]$p < .05$.

[d]$p < .01$.

consequence of the reduction of UROD activity. ALAS activity was induced threefold in the HCB group of animals, and this effect was potentiated by iron pretreatment (significant interaction, $p < .01$)(TABLE 1).

In the same animals, hepatic activity of both forms of XO was measured (TABLE 3). HCB treatment with or without iron treatment increased both total XO activity and the activity of type O alone in comparison to the control group. In the HCB group, however, the increase in total activity was apparently due also to an increase of type D

TABLE 2. Hepatic UROD Activity in C57B1/10 and DBA/2 Male Mice after Treatment with Iron and HCB with Two Different Substrate Concentrations

			UROD Activity[a] (pmol/min/mg protein)	
Strain	Iron	HCB	4 nmoles of Pentaporphyrinogen I	8 nmoles of Pentaporphyrinogen I
C57B1/10	–	–	25.73 ± 1.79	41.85 ± 4.29
	+	–	29.57 ± 1.27	39.70 ± 4.45
	–	+	23.31 ± 1.76 (10) IT[b]	27.36 ± 5.02 (35) NT[b]IT[b]
	+	+	7.09 ± 0.95 (73) NT[c]IT[c]HT[c]	4.18 ± 0.85 (90) NT[c]IT[c]HT[c]
DBA/2	–	–	28.99 ± 0.90	32.88 ± 3.07
	+	–	ND	36.06 ± 2.63
	–	+	ND	31.44 ± 2.00
	+	+	30.50 ± 1.20	30.46 ± 2.91

NOTE: Each value is a mean ± SEM of values obtained from four animals. See the footnote of TABLE 1 for information on statistical comparisons.

[a]Activity was expressed as the amount of coproporphyrinogen formed. The percentage of inhibition is shown in parentheses.

[b]$p < .05$.

[c]$p < .01$.

TABLE 3. Hepatic XO Activity in C57B1/10 Male Mice after Treatment with Iron and HCB

Iron	HCB	Activity[a] (nmol/min/g liver) Type O	Total	Type D	Type O Activity/ Total Activity
–	–	74.50 ± 6.22	263.80 ± 15.10	185.90 ± 14.11	0.30 ± 0.011
+	–	98.12 ± 9.54	278.39 ± 12.06	180.17 ± 11.97	0.34 ± 0.027
–	+	138.59 ± 9.42 NT[c]IT[b]	355.75 ± 11.60 NT[c]IT[c]	228.03 ± 12.63 NT[b]IT[b]	0.35 ± 0.021
+	+	173.68 ± 16.10 NT[c]IT[c]HT[b]	318.76 ± 8.68 NT[c]IT[b]HT[b]	152.17 ± 15.74 HT[c]	0.52 ± 0.049 NT[c]IT[c]HT[c]

NOTE: Each value is a mean ± SEM of values obtained from eight animals. See the footnote of TABLE 1 for information on statistical comparisons.

[a]Activity was expressed as the amount of hypoxanthine oxidized to xanthine and uric acid. The activity of type D was taken as the difference between total activity and type O activity.

[b]$p < .05$.

[c]$p < .01$.

activity, while in the group receiving iron and HCB, the increase was due only to an increase of type O. This was confirmed by the ratio between the type O activity and the total activity. This ratio was significantly higher only in the group receiving iron and HCB.

When assays were performed in the presence of dithiothreitol to reverse the nonproteolytic conversion of XO from type D to type O (TABLE 4), XO activity was still higher in treated groups than in controls. In the HCB plus iron group, the ratio between type O activity and total activity remained higher than in controls.

In vivo, another source of reactive oxygen intermediates is lipid peroxidation, and we determined extent of lipid peroxidation in the same mice by following the formation of MDA *in vitro* (TABLE 1). Iron alone and in association with HCB markedly increased MDA formation, the increase in the HCB group being smaller.

TABLE 4. Hepatic XO Activity in C57B1/10 Male Mice after Treatment with Iron and HCB in the Presence of Dithiothreitol

Iron	HCB	Activity[a] (nmol/min/g liver) Type O	Total	Type D	Type O Activity/ Total Activity
–	–	28.25 ± 3.53	221.91 ± 17.08	193.75 ± 14.25	0.13 ± 0.0085
+	–	29.94 ± 1.35	215.91 ± 7.97	190.17 ± 8.95	0.14 ± 0.006
–	+	46.23 ± 3.01 NT[c]IT[b]	272.01 ± 41.27	227.53 ± 37.86	0.16 ± 0.013
+	+	46.74 ± 6.73 NT[b]IT[b]	208.52 ± 40.12	161.78 ± 36.53	0.23 ± 0.045 NT[b]IT[b]

NOTE: Each value is a mean ± SEM of values obtained from four animals. See the footnote of TABLE 1 for information on statistical comparisons.

[a]The concentration of dithiothreitol in the incubation mixture was 10 mM. Activity was expressed as the amount of hypoxanthine oxidized to xanthine and uric acid. The activity of type D was taken as the difference between total activity and type O activity.

[b]$p < .05$.

[c]$p < .01$.

DISCUSSION

A basic pathogenetic event in PHAH-induced porphyria is a progressive decrease in UROD activity. A decrease in hepatic UROD activity may be achieved in mice with a single, low-lethality dose (75 μg/kg) of TCDD, the most potent porphyrinogenic compound yet found.[34] Hence, in this study, a TCDD-induced decrease in UROD activity was used as a model of PHAH-induced porphyria to determine which tissues, besides the liver (the organ most affected), are susceptible to the porphyrogenic affect of PHAHs. UROD activity was low in the kidney, suggesting that at least part of the porphyrin population found here was formed in kidney itself. The pattern of porphyrin accumulation in renal tissue is similar to that in urine samples taken from animals with porphyria.[35] This similarity suggests that the kidney has an active role in the overproduction of urinary porphyrins that characterizes intoxication with these compounds.

Erythrocytic UROD was not affected by TCDD treatment, thus making this experimental model of hepatic porphyria comparable to the sporadic type of human PCT syndrome and not to the familial type[36,37] (UROD deficiency in the latter is expressed in erythrocytes, and is inherited as an autosomal dominant trait). Lack of response in brain could be due to the low level of TCDD that is attained in this organ, although specific TCDD receptors have been found there.[38] It could also be due to differences—in comparison to liver and kidney—in other aspects that might be important in the porphyrogenic effect such as the induction of cytochrome P_1-450, the availability of intracellular antioxidants, and the extent of TCDD metabolism.

In PHAH-induced porphyria, reduction of the catalytic activity of UROD in the liver was reported to be unaccompanied by any reduction in the concentration of immunoreactive protein.[39] This finding is consistent with results indicating that the reduced catalytic activity of UROD, that is, the reduced rates of decarboxylation of uroporphyrinogen III by hepatic UROD (obtained from mice with porphyria), which correspond to reductions in the V_{max} value,[21] may be due to a PHAH-induced inactivation of part of the normal enzyme population in the liver.

Elucidation of possible mechanisms of this inactivation calls for extensive investigation. Jones and Sweeney have shown that in inbred strains of mice the development of porphyria appears to be inherited with the relative responsiveness to the induction of hepatic microsomal aryl hydrocarbon hydroxylase (AHH) activity, which is controlled by the *Ah* locus.[40] This locus is a set of regulatory genes that is controlled by the interactions of a cytosolic receptor protein. TCDD and related halogenated aromatic hydrocarbons are believed to exert their toxic effects, including those on the immune system and teratogenesis, through their binding to the cytosolic receptor.[10] C57B1/6 is the responsive prototype mouse strain whereas DBA/2 is the nonresponsive prototype strain. In DBA/2 mice neither HCB nor TCDD induces porphyria.[17,34] Examinations of several other strains of mice, however, showed that after administration of TCDD, porphyrogenic and necrotic effects of this compound in the liver did not completely correlate with the *Ah* phenotype.[41] Iron status of the animals appears to be important in relation to the onset of chemically induced porphyria because iron deficiency protects against porphyria induced by TCDD[42] and iron overload worsens it.[17,41] It does not appear to influence thymic involution or epidermal hyperplasia.[10] DBA/2 mice, however, are not susceptible to the porphyrogenic effects of HCB or TCDD even after an iron overload. Although it must be taken into account that TCDD predominantly accumulates in the liver[43] whereas HCB accumulates primarily in fat,[44] the difference (at least 2×10^4 times) in the amounts of these two toxins to which animals need to be exposed to develop porphyria still appears remarkable. With TCDD, we observed porphyria in rats after a total dose of 45 μg/kg or about 0.15 μmol/kg; with HCB, one of the lowest porphyrogenic doses used gave a body burden of a 732 mg/kg or 2.57 mmol/kg.[15,35]

These figures show that with TCDD, which is poorly metabolized, the amount of parent compound is very small. The hypothesis that UROD inactivation is due to a reactive metabolite thus loses credibility. However, although all PHAHs appear to induce similar pathobiochemical changes, step(s) of the porphyrogenic mechanism could differ. Another alteration of the heme pathway in this type of hepatic porphyria is induction of ALAS, which has its activity regulated through a feedback mechanism depending on the level of a pool of heme.[13] ALAS activity was high in mice with TCDD-induced porphyria.[45]

All PHAHs are potent inducers of microsomal monooxygenases. Hence, heme demand is higher in PHAH-treated animals than in controls. When heme demand is high, a slight reduction in UROD activity, such as that observed in HCB-treated mice in this study, appears to be enough to appreciably deplete the regulatory heme pool and cause an increase in ALAS activity. Such an increase is likely to contribute to the accumulation of uroporphyrinogen and its oxidation product, uroporphyrin, in the liver if at the same time UROD activity is decreased. This effect seems relevant to the question of UROD deficiency for two reasons. First, although uroporphyrin[46] cannot be decarboxylated, it can bind UROD, and has been shown to inhibit decarboxylation.[22] Thus all factors that can favor the accumulation of uroporphyrin could, theoretically, increase PHAH-induced UROD inactivation. In this study, the kinetics of accumulation of porphyrinogens from uroporphyrinogen III have been considered, and the results indicate that the last step of decarboxylation, coproporphyrinogen formation, is susceptible to inhibition when concentrations of substrate and reaction products are increased. *In vitro,* this inhibition was evident with enzyme from control animals, but with UROD from animals with porphyria too little coproporphyrinogen was formed from uroporphyrinogen for this inhibition to be detectable. In fact, coproporphyrinogen formation by UROD in mice with porphyria, measured with uroporphyrinogen III as substrate, did not appreciably change even when the assay was performed in the presence of all the porphyrinogens and porphyrins that had accumulated.[47]

Second, UROD inactivation may proceed in such a way that even a limited accumulation of uroporphyrin could be significant. A mechanism proposed by Elder *et al.*[23] to explain UROD inactivation *in vitro* suggests uroporphyrin has an active role in focusing oxidative damage on the active site of the enzyme. If such a mechanism is operative *in vivo,* even a small amount of uroporphyrin exceeding physiological demand could affect the rate of enzyme inactivation.

Uroporphyrin is the predominant porphyrin accumulated in the liver of animals with porphyria, but careful analysis of the isomer composition of uroporphyrin started only recently with the use of new HPLC techniques.[48]

In this study animals treated chronically with HCB after having been treated with iron had their livers examined for uroporphyrin content. Preliminary determinations showed that the ratio between the content of uroporphyrin I and that of uroporphyrin III was 55:45. Smith *et al.*, after administering iron and a single dose of HCB (100 mg/kg), showed that a similar ratio was reached slowly in the liver while porphyria developed, whereas in control animals uroporphyrin III predominated.[49] The rate of decarboxylation of uroporphyrinogen III as substrate is almost double that with uroporphyrinogen I.[21] Thus the changes in uroporphyrin isomer composition, which have been interpreted as an alteration of the activity of the enzyme uroporphyrinogen III cosynthetase,[49] and the concomitant changes in the rate of decarboxylation are likely to help increase porphyrin accumulation, reduce the rate of formation of coproporphyrinogen III, and hence reduce the rate of formation of heme in iron- and HCB-treated animals.

The effect of iron alone on these changes has not yet been investigated. The role of iron in exacerbating PHAH-induced porphyria as well as human PCT has long been recognized,[6] and theories involving the formation of free radicals have been formulated to explain the mechanism of UROD inactivation.[18] Iron overload alone, however, does not reduce UROD activity *in vivo*.[50] The enzyme is susceptible to inactivation *in vitro* by reactive oxygen species.[23]

Lipid peroxidation, which is one of the sources of reactive oxygen intermediates *in vivo,* has been implicated in toxic effects such as P-450 destruction[25,51] and, in some cases, has been stimulated by treatment with TCDD and HCB[52,53]; thus it was suggested as one of the processes involved in the mechanism of UROD inactivation.[53]

HCB given in association with iron did not increase the extent of lipid peroxidation observed in the group treated with iron alone despite a large increase in porphyrin accumulation. HCB alone, however, slightly stimulated lipid peroxidation. Thus it cannot be totally excluded that this process is involved, possibly very subtly such that differences could be hard to estimate by MDA determinations.

The increase in lipid peroxidation we observed in mice treated with HCB alone could be a consequence of a parallel increase in XO activity. This enzyme in its type O form is well known for its ability to generate superoxides and hydrogen peroxide, and this reduction product of molecular oxygen[24] can initiate the process of lipid peroxidation.[54] Furthermore, other reactive species can be formed by the interaction of superoxides produced by XO and iron,[19] and hence the formation of reactive species could be favored by increased iron accumulation.

In vitro, this enzyme can be converted irreversibly from type D to type O by proteolysis or reversibly by treatment with reagents for the thiol group.[29] *In vivo,* it has been suggested that after ischemia type D could be converted to type O by the action of calcium-activated proteases.[55]

In our experiments, type O was activated in the groups treated with HCB, but iron alone had no appreciable effects on the enzyme. Thus an increase in activity of type O can be dissociated from the effects of iron and does not appear to be a consequence of tissue injury due to iron-induced lipid peroxidation.

In mice treated with TCDD, the elevation of plasma alanine aminotransferase was used as an index of the hepatic damage typical of TCDD intoxication.[41,56] Such damage precedes porphyrin accumulation and is not completely prevented even by inducing a low iron state.[42] In our study, mice treated with HCB showed an increase of alanine aminotransferase in comparison to controls (8 fold), the increase being greater (80 fold) in the group treated with iron and HCB (data not reported). An increase in the type O form of XO might be a feature associated with PHAH-induced hepatic damage.

It seems worthwhile investigating further the possibility of the increase in the type O form of XO being one of the processes responsible for the formation of certain oxidant species of potential importance for UROD inactivation; if the inactivation of UROD is due to an oxidative reaction, our data suggest that XO probably achieves its effect through some iron-mediated reaction.

SUMMARY

Experimental porphyria induced by PHAHs is characterized by a progressive reduction in the activity of UROD.

After intoxication with TCDD, the most porphyrogenic compound known to date, the liver was the principal site of action, as regards both porphyrin accumulation (mostly uroporphyrin) and the degree of enzyme impairment; the kidney was the site of the second greatest accumulation; the brain and erythrocytes were unaffected.

Additional modifications of the heme pathway involved induction of the activity of ALAS and, at least in HCB-induced porphyria after iron pretreatment, may have involved reduced activity of uroporphyrinogen III cosynthetase. These changes can alter the amount and the isomeric composition of uroporphyrinogens and uroporphyrins present in the liver in a way that is likely to help reduce formation of coproporphyrinogen III in porphyric animals.

As in the human syndrome porphyria cutanea tarda, iron administration increased porphyrin accumulation and the degree of reduction of UROD activity in mice fed HCB.

Mice fed HCB also presented an activation of the type O form of XO. This activation was independent of tissue injury derived from the lipid peroxidation that was concomitant with iron administration. The increase in activity of the type O form of XO may be a characteristic feature of the liver damage found in PHAH intoxication and, in intoxicated animals, could be a source in the liver of oxidant species involved in the mechanism of UROD inactivation—if this inactivation is in fact due to an oxidative reaction.

REFERENCES

1. Schmid, R. 1960. N. Engl. J. Med. **263:** 397–398.
2. Cam, C. & G. Nigogosyan. 1963. J. Am. Med. Assoc. **183:** 88–89.
3. Elder, G. H. 1978. *In* Handbook of Experimental Pharmacology. Heme and Hemoproteins. F. De Matteis & W. N. Aldridge, Eds. Vol. **44:** 157–200. Springer-Verlag. Berlin.
4. Ockner, R. K. & R. Schmid. 1961. Nature **189:** 499.
5. De Matteis, F., B. E. Prior & C. Rimington. 1961. Nature **191:** 363–366.
6. Smith, A. G. & F. De Matteis. 1980. Clin. Hematol. **9:** 399–425.
7. Strik, J. J. T. W. A., F. M. H. Debets & G. Koss. 1980. *In* Halogenated Biphenyls, Terphenyls, Napthalenes, Dibenzodioxins and Related Products. R. D. Kimbrough, Ed.: 191–239. Elsevier. Amsterdam.
8. Safe, S. H. 1986. Annu. Rev. Pharmacol. Toxicol. **26:** 371–399.
9. Firestone, D. 1978. Ecol. Bull. **27:** 39–52.
10. Poland, A. & J. C. Knutson. 1982. Annu. Rev. Pharmacol. Toxicol. **22:** 517–554.
11. Mauzerall, D. & S. Granick. 1958. J. Biol. Chem. **232:** 1141–1162.
12. Tait, G. H. 1978. *In* Handbook of Experimental Pharmacology. Heme and Hemoproteins. F. De Matteis & W. N. Aldridge, Eds. Vol. **44:** 1–48. Springer-Verlag. Berlin.
13. Bonkovsky, H. L., P. R. Sinclair & J. F. Sinclair. 1979. Yale J. Biol. Med. **52:** 13–37.
14. Goldstein, J. A., P. Linko, J. N. Huckins & D. L. Stalling. 1982. Chem. Biol. Interact. **41:** 131–139.
15. Rizzardini, M. & A. G. Smith. 1982. Biochem. Pharmacol. **31:** 3543–3548.
16. Debets, F. M., J. H. Reinders, A. J. Debets, T. G. Lossbroek, J. J. Strik & G. Koss. 1981. Toxicology **19:** 185–196.
17. Smith, A. G. & J. E. Francis. 1983. Biochem. J. **214:** 909–913.
18. De Matteis, F. & M. Stonard. 1977. Semin. Hematol. **14:** 187–192.
19. Winterbourn, C. C. & H. C. Sutton. 1986. Arch. Biochem. Biophys. **244:** 27–34.
20. Smith, A. G. & J. E. Francis. 1979. Biochem. J. **183:** 455–458.
21. Cantoni, L., D. Dal Fiume & R. Ruggieri. 1984. Int. J. Biochem. **16:** 561–565.
22. Smith, A. G. & J. E. Francis. 1981. Biochem. J. **195:** 241–250.
23. Elder, G. H., A. G. Roberts & A. J. Urquhart. *In* Porphyrins and Porphyrias. Y. Nordmann, Ed. Colloque INSERM. Vol. **134:** 147–152. John Libbey Eurotext. London.
24. McCord, J. M. & I. Fridovich. 1968. J. Biol. Chem. **243:** 5753–5760.

25. Bus, J. S. & J. E. Gibson. 1979. *In* Reviews in Biochemical Toxicology. E. Hodgson, J. R. Bend & R. M. Philpot, Eds.: 125–149. Elsevier. Amsterdam.
26. Kato, R. & M. Takayanaghi. 1966. Jpn. J. Pharmacol. **16:** 380–390.
27. Francis, J. E. & A. G. Smith. 1984. Anal. Biochem. **138:** 404–410.
28. De Matteis, F., P. Zetterlund & L. Wetterberg. 1981. Biochem. J. **196:** 811–817.
29. Della Corte, E. & F. Stirpe. 1972. Biochem. J. **126:** 739–745.
30. Ghezzi, P., M. Bianchi, A. Mantovani, F. Spreafico & M. Salmona. 1984. Biochem. Biophys. Res. Commun. **119:** 144–149.
31. Lowry, O. H., N. J. Rosebrough, A. L. Farr & R. J. Randall. 1951. J. Biol. Chem. **193:** 265–275.
32. Abbritti, G. & F. De Matteis. 1971–72. Chem. Biol. Interact. **4:** 281–286.
33. Ohkawa, H., N. Ohishi & K. Yagi. 1979. Anal Biochem. **95:** 351–358.
34. Smith, A. G., J. E. Francis, S. J. E. Kay & J. B. Greig. 1981. Biochem. Pharmacol. **30:** 2825–2830.
35. Cantoni, L., M. Salmona & M. Rizzardini. 1981. Toxicol. Appl. Pharmacol. **57:** 156–163.
36. Elder, G. H., M. R. C. Path, G. B. Lee & J. A. Tovey. 1978. N. Engl. J. Med. **299:** 274–278.
37. de Verneuil, H., G. Aitken & Y. Nordmann. 1978. Hum. Genet. **44:** 145–151.
38. Carlstedt-Duke, J. M. B. 1979. Cancer Res. **39:** 3172–3176.
39. Elder, G. H. & D. M. Sheppard. 1982. Biochem. Biophys. Res. Commun. **109:** 113–120.
40. Jones, K. G. & G. D. Sweeney. 1980. Toxicol. Appl. Pharmacol. **53:** 42–49.
41. Greig, J. B., J. E. Francis, S. J. E. Kay, D. P. Lovell & A. G. Smith. 1984. Toxicol. Appl. Pharmacol. **74:** 17–25.
42. Jones, K. G., F. M. Cole & G. D. Sweeney. 1981. Toxicol. Appl. Pharmacol. **61:** 74–88.
43. Piper, W. N., J. Q. Rose & P. J. Gehring. 1973. Environ. Health Perspect. **5:** 241–244.
44. Courtney, K. D. 1979. Environ. Res. **20:** 225–266.
45. Goldstein, J. A., P. Hickman, H. Bergman & J. G. Vos. 1973. Res. Commun. Chem. Pathol. Pharmacol. **6:** 919–928.
46. Elder, G. H., J. A. Tovey & D. M. Sheppard. 1983. Biochem. J. **215:** 45–55.
47. Cantoni, L., D. Dal Fiume, A. Ferraroli, M. Salmona & R. Ruggieri. 1984. Toxicol. Lett. **20:** 201–210.
48. Francis, J. E. & A. G. Smith. 1985. Trends Anal. Chem. **4:** 80–86.
49. Smith, A. G. & J. E. Francis. 1986. *In* Porphyrins and Porphyrias. Y. Nordmann, Ed. Colloque INSERM. Vol. **134:** 127–136. John Libbey Eurotext. London.
50. Cantoni, L., C. Di Padova, M. Rizzardini, D. Dal Fiume, A. Graziani, P. Rovagnati & R. Tritapepe. 1986. Toxicology **38:** 187–196.
51. Levin, W., A. Y. H. Lu, M. Jacobson & R. Kuntzman. 1973. Arch. Biochem. Biophys. **158:** 842–852.
52. Stohs, S. J., M. Q. Hassan & W. J. Murray. 1983. Biochem. Biophys. Res. Commun. **111:** 854–859.
53. Alleman, M. A., J. F. Koster, J. H. P. Wilson, A. Edixhoven-Bosdijk, R. G. Slee, M. J. Kroos & H. G. V. Eijk. 1985. Biochem. Pharmacol. **34:** 161–166.
54. Thomas, C. E., L. A. Morehouse & S. D. Aust. 1985. J. Biol. Chem. **260:** 3275–3280.
55. McCord, J. M. 1985. N. Engl. J. Med. **312:** 159–163.
56. Seefeld, M. D., R. M. Albrecht, K. W. Gilchrist & R. E. Peterson. 1980. Arch. Environ. Contam. Toxicol. **9:** 317–327.

Alterations of Uroporphyrinogen Decarboxylase by Chlorinated Organics[a]

G. H. ELDER, A. G. ROBERTS, AND A. J. URQUHART

Department of Medical Biochemistry
University of Wales College of Medicine
Health Park, Cardiff CF4 4XN, Wales

INTRODUCTION

A small group of structurally diverse polyhalogenated aromatic hydrocarbons (PHAHs) cause a persistent porphyria in mammals that is characterised by accumulation of uroporphyrin and heptacarboxylic porphyrin in the liver and by increased excretion of acetate-substituted porphyrins in urine and feces. The PHAHs that are known to produce this effect are hexachlorobenzene (HCB) and hexabromobenzene, certain polychlorinated and polybrominated biphenyls, and 2,3,7,8-tetrachlorodibenzo-*p*-dioxin (TCDD).[1] Accidental ingestion of HCB or exposure to TCDD produces a cutaneous porphyria in humans that closely resembles type I porphyria cutanea tarda (PCT), a condition that occurs sporadically, often in association with alcoholism or estrogen therapy, and that cannot be attributed to exposure to toxic amounts of PHAHs.[2]

PHAH-induced porphyria and sporadic (type I) PCT both result from decreased activity of uroporphyrinogen decarboxylase EC 4.1.1.37 in the liver. Either a single dose or chronic administration of HCB or TCDD to a susceptible strain or species of mammal produces a progressive fall in the activity of this enzyme that is paralleled by hepatic accumulation and increasing excretion of porphyrins derived from its substrates.[3,4] In both conditions the development of uroporphyrinogen decarboxylase deficiency appears to be influenced by the concentration of iron in the liver. Thus iron deficiency prevents or delays the production of porphyria by PHAHs,[5,6] whereas iron overload accelerates it.[4] Similarly, removal of hepatic iron leads to clinical and biochemical remission in PCT, whereas replacement leads to relapse.[2]

The mechanism by which PHAHs produce a decrease in the activity of heptic uroporphyrinogen decarboxylase is not understood. At present there is no evidence that either the parent PHAHs or their stable metabolites directly inhibit mammalian uroporphyrinogen decarboxylase at concentrations that are likely to be present *in vivo*.[7,8] Experiments in avian hepatocyte cultures, however, suggest that short-lived, inhibitory metabolites, capable of reaction with protein sulfhydryl groups, may be formed in the liver from PHAHs and covalently bind to uroporphyrinogen decarboxylase.[8,9] Stable inhibitors that are not derived from PHAHs have also been identified in the livers of animals treated with HCB[10] or TCDD.[11] Alternatively, PHAHs may promote the formation of oxygen radicals, which cause oxidative inactivation of uroporphyrinogen decarboxylase,[12,13] or they may suppress enzyme synthesis as part of a pleiotropic response, controlled by the *Ah* locus, which follows exposure to PHAHs.[14]

[a]This work was supported by the Medical Research Council and by the U.S. Veterans Administration.

Whatever the nature of the mechanism, the end result of the toxic process that produces porphyria in mammals is decreased activity of uroporphyrinogen decarboxylase. We have therefore studied alterations of this enzyme in PHAH-induced porphyria and in model systems, in order to determine whether there are any structural changes present in the enzyme molecule that provide information about the mechanism by which PHAHs cause porphyria.

MATERIALS AND METHODS

Synthetic pentacarboxylic porphyrin III with the acetic acid substituent on ring C and uroporphyrin III were gifts from Professor A. H. Jackson (University College,

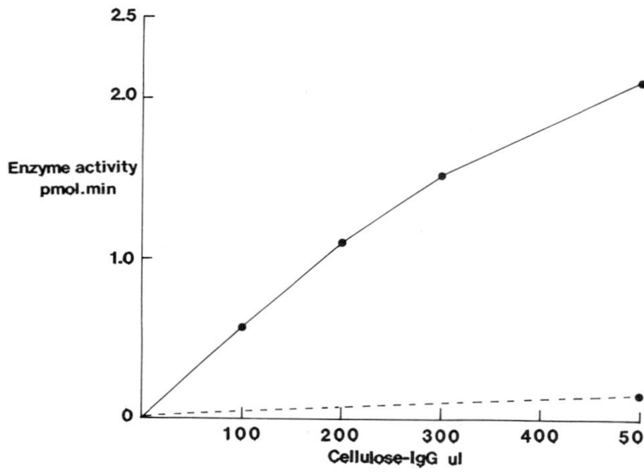

FIGURE 1. Uptake of uroporphyrinogen decarboxylase activity from rat liver supernatant fraction onto solid-phase antibody. Aliquots (0.1 ml) of liver supernatant fraction from a 20% (w/v) homogenate, diluted 10-fold, were incubated with cellulose–immunoglobulin G conjugate from 0.1–0.5 ml of suspension. Solid line shows activity bound to antiuroporphyrinogen decarboxylase immunoglobulin G conjugate; interrupted line shows binding to preimmune immunoglobulin G conjugate.

Cardiff). HCB was analytical reagent grade from BDH (Poole, Dorset). Tritium-labeled *N*-ethylmaleimide (53.3 Ci/mmol) was obtained from New England Nuclear (Boston, MA). Human uroporphyrinogen decarboxylase was purified from erythrocytes as described previously,[15] except that HPLC on an anion exchange column (TSK DEAE-5PW, Anachem, Luton, Bedfordshire) in 10 mM Tris-HCl buffer (pH 8.2) with a linear NaCl gradient (0.0–0.5 M) replaced polyacrylamide gel electrophoresis as the final step. Antisera to human erythrocyte uroporphyrinogen decarboxylase were raised in rabbits by injecting partially purified (antiserum I) or homogeneous enzyme (antiserum II).[15] Antiserum I cross-reacted with rodent liver enzyme (8–10% cross-reactivity), and an immunoglobulin G fraction from this antiserum, or from preimmune serum, was reacted with a diazonium salt of reprecipitated cellulose to prepare solid-phase reagents.[16]

Male C57/BL6 mice (20–25 g body weight) were fed either MRC diet 41B alone or the same diet containing 0.02% HCB (w/w) for 35 days. All mice were given Imferon (0.25 ml, 12.5 mg iron) intraperitoneally 3 days before starting the diet.[4] All animals were allowed free access to food and water. Liver supernatant fractions were prepared by centrifuging homogenates (20%, w/v, in 0.25 M sucrose, 10 mM potassium phosphate, pH 7.4) for 5 min at 15 000 g_{av}.

Uroporphyrinogen decarboxylase activity was measured using pentacarboxylic porphyrinogen III or uroporphyrinogen III as substrate.[17] In all experiments activities to both substrates altered in parallel. Immunoreactive human uroporphyrinogen decarboxylase was measured by electroimmunoassay.[18] To measure uptake of rodent uroporphyrinogen decarboxylase activity by solid-phase antibody, liver supernatant fraction (0.1 ml) was incubated with solid-phase reagent from 0.1 ml of suspension, and bound activity was determined as previously described.[16] In all experiments less than 15% of total enzyme activity was bound.

Reactions with sulfhydryl-group reagents (all 5 mM in 100 mM potassium phosphate buffer, pH 7.0, containing 0.5 mM dithiothreitol) were carried out at 25° C and terminated by addition of excess dithiothreitol or mercaptoethanol. Tritium-labeled enzyme was separated from unreacted [^{3}H]*N*-ethylmaleimide by gel filtration HPLC in a mixture of acetonitrile and 0.1% aqueous trifluoroacetic acid (3:2, v/v). For photoinactivation experiments, uroporphyrinogen decarboxylase (5.2 μM) in 20 mM potassium phosphate, pH 7.0, containing dithiothreitol (0.5 mM), was mixed with uroporphyrin III and irradiated with a halogen lamp at an incident light intensity of 30–35 W/m^2. Control samples contained uroporphyrin but were protected from light with aluminum foil.

RESULTS AND DISCUSSION

Immunochemical Measurement of the Specific Activity of Rodent Liver Uroporphyrinogen Decarboxylase

FIGURE 1 shows that the uptake by solid-phase antibody of uroporphyrinogen decarboxylase activity from the liver supernatant fraction from a normal rat increases with the concentration of solid-phase antibody and can be clearly distinguished from nonspecific uptake. The slope of the linear part of the curve, where the concentration of enzyme is in considerable excess over antibody, depends on the amount of enzyme bound by the solid-phase antibody and the specific activity (activity/mg enzyme protein) of the enzyme. Thus under conditions of antigen excess, the uptake of enzyme activity by the solid-phase antibody can be used as a measure of enzyme specific activity, provided there is no change in the amount of enzyme protein bound or introduction into the assay system of substances that affect the antigen–antibody reaction.[16]

Effect of HCB on the Specific Activity of Mouse Liver Uroporphyrinogen Decarboxylase

TABLE 1 shows that the uptake onto solid-phase antibody of uroporphyrinogen decarboxylase activity from liver supernatant fractions of mice receiving HCB decreases progressively as the hepatic enzyme activity falls. Possible explanations for the decrease in uptake include a decreased concentration of enzyme protein, interference by porphyrin in the antigen–antibody reaction, a change in the immunoreactivity

TABLE 1. Effect of HCB Treatment of Male C57/BL6 Mice on the Catalytic and Specific Activity of Hepatic Uroporphyrinogen Decarboxylase

Days on HCB Diet	Catalytic Activity[a] (pmol/min/g liver)	Specific Activity[b] (pmol/min/0.1 ml solid-phase antibody)
0	6.82 ± 0.46 (100)	
8	7.40 ± 0.44 (109)	1.34, 1.26–1.40 (100)
14	6.78 ± 0.28 (99)	1.41, 1.29–1.63 (105)
21	5.55 ± 0.20 (78)[c]	1.20, 1.07–1.29 (90)
28	4.83 ± 0.27 (71)[c]	0.99, 0.82–1.11 (73)[d]
35	1.52 ± 0.08 (22)[c]	0.39, 0.37–0.44 (29)[d]

[a]Each value below is a mean ± SEM for five animals. Figures in parentheses are means expressed as percentages of the catalytic activity at day 0.

[b]Each value below is a mean and range for three animals. Figures in parentheses are means expressed as percentages of the specific activity at day 8. Specific activity was measured indirectly as described in the text.

[c]Significantly different from the corresponding value for day 0 ($p < .01$).

[d]Significantly different from the corresponding value for day 8 ($p < .05$).

of uroporphyrinogen decarboxylase, or a decrease in the specific activity of the enzyme.

A decrease in the concentration of enzyme protein is unlikely to be the explanation because dilution of supernatant fractions from control mice to the same activities as those from treated mice had very little effect on uptake. A major effect of porphyrin on the antigen–antibody reaction can also be excluded because uptake was unaltered either by addition of uroporphyrin III to control samples or by removal of porphyrin from HCB-treated samples with an ion-exchange resin. A progressive decrease, however, in either the immunoreactivity or specific activity of uroporphyrinogen decarboxylase (or a combination of both these effects) as porphyria develops would explain our findings. Of these explanations, a decrease in specific activity without any major change in immunoreactivity or alteration in the concentration of enzyme protein seems the most likely because the fall in activity of the enzyme is so closely paralleled by the decrease in uptake by solid-phase antibody (TABLE 1). At the molecular level, this decrease in specific activity might reflect either an irreversible change in the kinetic properties of the enzyme or a mixed population of catalytically inactive, immunoreactive enzyme protein and normal enzyme. A similar fall in specific activity has been reported in HCB-induced porphyria in rats and in porphyria caused by TCDD in mice.[16]

These findings have a number of implications for understanding the mechanism of PHAH-induced porphyria. Because the concentration of enzyme protein apparently remains unaltered, mechanisms that involve decreased synthesis (for example, as part of a pleiotropic response determined by the *Ah* locus) are excluded. For the same reason, increased catabolism seems an unlikely explanation, although the possibility remains that the catalytically inactive immunoreactive material represents a residual fraction of altered enzyme that is only slowly degraded, perhaps because it is stabilized by binding to porphyrin. On the other hand, our results are consistent with any mechanism that produces irreversible inactivation or kinetic modification of uroporphyrinogen decarboxylase without altering its major epitopes. As a first stage in identifying the strucutral alterations of uroporphyrinogen decarboxylase in PHAH-induced porphyria, we have investigated the effects of various reagents on the catalytic and immunochemical properties of the purified human enzyme *in vitro*.

Effect of Sulfhydryl-Group Reagents on Uroporphyrinogen Decarboxylase

Purified human uroporphyrinogen decarboxylase is rapidly inactivated by *N*-ethylmaleimide but not by the more polar sulfydryl-group reagents, iodoacetic acid and iodoacetamide (FIG. 2). Inactivation is accompanied by loss of immunoreactivity. Pentacarboxylic porphyrinogen III, which is a substrate for the enzyme, protects against both inactivation and loss of immunoreactivity (data not shown), suggesting that *N*-ethylmaleimide reacts with cysteine residues that are essential for catalytic activity and that conformational changes produced by alkylation of these residues markedly impairs immunoreactivity.

There is indirect evidence that reaction with *N*-ethylmaleimide produces similar changes in mouse liver uroporphyrinogen decarboxylase. Thus supernatant fractions of mouse liver homogenates that had been inactivated by incubation with 10 mM *N*-ethylmaleimide for 30 min at 37° C did not decrease uptake of enzyme activity from untreated fractions onto solid-phase antibody.

HCB and some other porphyrogenic HCBs are converted in the liver to unstable metabolites that react with sulfhydryl groups in glutathione and proteins.[8,19] Although inactivation of uroporphyrinogen decarboxylase by such metabolites has not been directly demonstrated, reaction with catalytic sulfhydryl groups in the enzyme has been postulated as the mechanism of inactivation *in vivo*.[8,9] Our results, by showing that covalent bonding of hydrophobic groups to essential cysteine residues in the enzyme leads to loss of immunoreactivity, contradict results showing effects *in vivo,* and therefore provide evidence against this hypothesis.

Oxidative Inactivation of Uroporphyrinogen Decarboxylase

A number of porphyrins catalyze the photoinactivation of uroporphyrinogen decarboxylase, presumably by a singlet oxygen-mediated reaction.[13,20] FIGURE 3 shows that photoinactivation proceeds in the presence of an equimolar concentration of uroporphyrin III without loss of immunoreactivity. At much higher molar ratios (uroporphyrin III:enzyme; 60:1), there is more rapid and complete destruction of both catalytic activity and immunoreactivity. Inactivation by xanthine:xanthine oxidase, a system that generates superoxide anion radicals, similarly destroys both functions.[13] These results suggest that oxidative inactivation of uroporphyrinogen decarboxylase is accompanied by loss of immunoreactivity, unless the destructive oxygen radicals are produced predominantly in the vicinity of the active center. Uroporphyrin III is a

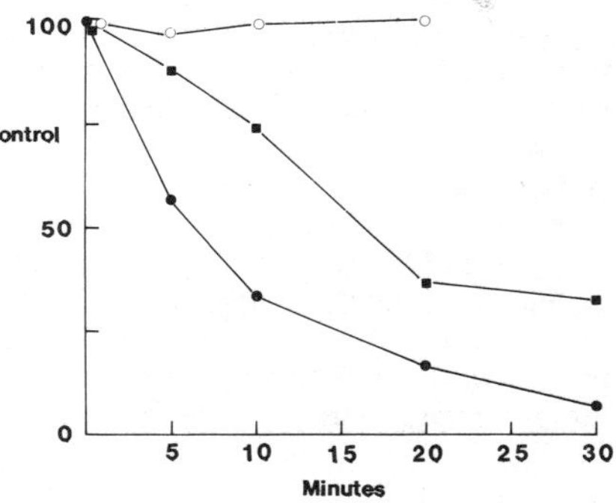

FIGURE 2. The effect of sulfhydryl-group reagents on human uroporphyrinogen decarboxylase. Catalytic activity was measured after incubation of purified enzyme with 5 mM iodoacetic acid (○) or 5 mM *N*-ethylmaleimide (●). Immunoreactive enzyme concentration (■) was measured after incubation with 5 mM *N*-ethylmaleimide. Enzyme activities and concentrations are means of duplicates expressed as percentages of mean control values.

competitive inhibitor of uroporphyrinogen decarboxylase[21] and may bind to the active center.[15] At low molar ratios, uroporphyrin may therefore have two functions in the photoinactivation reaction: it may act as a catalyst for the production of singlet oxygen by light and as a means of focusing oxidative damage around the uroporphyrin-binding site, at the active center and away from major epitopes. The structural basis of this specific photoinactivation reaction has not been identified. A number of protein amino acids are sensitive to photooxidation by porphyrins: cysteine, histidine, methionine, tryptophan, and tyrosine.[22] Reaction of photooxidized enzyme with [^{3}H]*N*-ethylmaleimide led to loss of immunoreactivity and an incorporation of tritium (1635 cpm/μg enzyme) similar to that obtained for unoxidized enzyme (1766 cpm/μg enzyme), indicating that cysteine residues are unlikely to be involved in oxidation of the active center.

These experiments extend previous reports that uroporphyrinogen decarboxylase, like many other enzymes,[22] is inactivated by oxygen radicals. If this type of process is important in the pathogenesis of PHAH-induced porphyria in mammals,[6,12] some mechanism for focusing oxidative damage at the active center is likely to be required in order to produce catalytically inactive, immunoreactive enzyme.

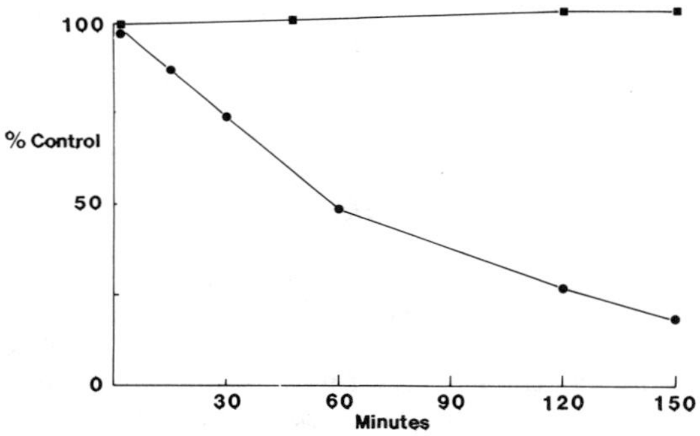

FIGURE 3. Photooxidation of human uroporphyrinogen decarboxylase. Purified enzyme (5.2 μM) was irradiated in the presence of uroporphyrin III (5.0 μM) before measurement of enzyme activity (●) or concentration of immunoreactive enzyme (■). Enzyme activities and concentrations are means of duplicates expressed as percentages of mean values for control samples containing uroporphyrin but not exposed to light.

CONCLUSION

Any explanation of the mechanism of the porphyrogenic action of PHAHs needs to explain the part that iron plays in this process. Although iron-dependent metabolism of PHAHs has been suggested,[23] it seems more likely that iron, in view of its importance in the pathogenesis of other uroporphyrinogen decarboxylase defects (such as PCT, in which PHAHs do not appear to be involved), has a more general function, perhaps acting as a catalyst for the formation of short-lived, highly reactive oxygen radicals.[2,4,6,12,13] Our results indicate that the oxidative inactivation of uroporphyrinogen decarboxylase, if promoted by PHAH, is likely to be selective for the catalytic region of the molecule. Planar iron–porphyrin chelates are effective catalysts for the formation of hydroxyl radicals from superoxide anion radicals or hydrogen peroxide in biological systems.[24] Experiments with avian hepatocytes[12,25] and in mice[4,13] suggest that PHAHs may promote the formation of oxygen radicals and the oxidation of porphyrinogens to porphyrins in liver. Oxidation of porphyrinogens at neutral pH in the presence of metal ions leads to the formation of metal chelates.[26] The possibility that such compounds are formed in the liver in PHAH-induced porphyria requires investigation. If they are, an iron–porphyrin chelate capable of binding close to the active center of uroporphyrinogen decarboxylase might then inactivate the enzyme by catalyzing the local formation of hydroxyl radicals, which, due to their short half-life, might not diffuse far enough to oxidize epitopes in other regions of the molecule.

Catalytically inactive, immunoreactive uroporphyrinogen decarboxylase has also been identified in the liver in PCT.[18] During the clinically active phase of this condition, there is a marked fall in the specific activity of hepatic uroporphyrinogen decarboxylase, and, as with PHAH-induced porphyria, there is some evidence that selective inactivation of the catalytic site of the enzyme is iron-dependent.[18] The alteration in uroporphyrinogen decarboxylase that we have detected in PHAH-induced porphyria may therefore be common to both conditions and represent the end result of a process that is initiated, perhaps in different ways, by a variety of hepatotoxic agents. Elucidation of the nature of the structural alterations of uroporphyrinogen decarboxylase in these conditions should help to explain their pathogenesis.

REFERENCES

1. MARKS, G. S. 1984. Crit. Rev. Toxicol. **154:** 151–179.
2. SWEENEY, G. D. 1986. Clin. Biochem. **19:** 3–15.
3. ELDER, G. H . 1978. *In* Handbook of Experimental Pharmacology. F. De Matteis & W. N. Aldridge, Eds. Vol. **44:** 157–200. Springer-Verlag. Berlin.
4. SMITH, A. G., J. E. FRANCIS, S. J. E. KAY, J. B. GREIG & F. P. STEWART. 1986. Biochem. J. **238:** 871–878.
5. JONES, K. G., F. M. COLE & G. D. SWEENEY. 1981. Toxicol. Appl. Pharmacol. **61:** 74–88.
6. WAINSTOK DE CALMANOVICI, R., S. B. BILLI, C. A. ALDONATTI & L. C. SAN MARTIN DE VIALE. 1986. Biochem. Pharmacol. **35:** 2399–2405.
7. DE VERNEUIL, H., S. SASSA & A. KAPPAS. 1983. J. Biol. Chem **258:** 2454–2460.
8. DEBETS, F. M. H., J. H. REINDERS, A. J. M. DEBETS, T. C. LOSSBROEK, J. J. T. W. A. STRIK & G. KOSS. 1981. Toxicology **19:** 185–196.
9. DE VERNEUIL, H., S. SASSA & A. KAPPAS. 1983. Biochem. J. **214:** 145–151.
10. RIOS DE MOLINA, M. C., R. WAINSTOK DE CALMANOVICI & L. C. SAN MARTIN DE VIALE. 1980. Int. J. Biochem. **12:** 1027–1032.
11. CANTONI, L., D. DAL FIUME, M. RIZZARDINI & R. RUGGIERI. 1984. Toxicol. Lett. **20:** 211–217.
12. FERIOLI, A., C. A. HARVEY & F. DE MATTEIS. 1984. Biochem. J. **224:** 769–777.
13. ELDER, G. H., A. G. ROBERTS & A. J. URQUHART. 1986. *In* Porphyrins and Porphyrias. Y. Nordmann, Ed.: 147–152. Colloque INSERM/John Libbey Eurotext. Paris.
14. JONES, K. G. & G. D. SWEENEY. 1980. Toxicol. Appl. Pharmacol. **53:** 42–49.
15. ELDER, G. H., J. A. TOVEY & D. M. SHEPPARD. 1983. Biochem. J. **215:** 45–55.
16. ELDER, G. H., J. A. TOVEY & D. M. SHEPPARD. 1982. Biochem. Biophys. Res. Commun. **109:** 113–120.
17. ELDER, G. H. & P. C. WYVILL. 1982. Enzyme **28:** 186–195.
18. ELDER, G. H., A. J. URQUHART, R. E. DE SALAMANCA, J. J. MUNOZ & H. L. BONKOVSKY. 1985. Lancet **ii:** 229–232.
19. SHIMADA, T. & Y. SAWABE. 1983. Toxicol. Appl. Pharmacol. **70:** 486–493.
20. BATTLLE, A. M., C. DEL, R. E. DE SALAMANCA, S. CHINARRO, S. G. AFONSO & A. M. STELLA. 1985. Int. J. Biochem. **18:** 143–147.
21. SMITH, A. G. & J. E. FRANCIS. 1981. Biochem. J. **195:** 241–250.
22. SPIKES, J. D. 1984. *In* Porphyrins in Tumor Phototherapy. A. Andreoni & R. Cubeddu, Eds.: 51–60. Plenum. New York, NY.
23. SINCLAIR, P. & S. GRANICK. 1974. Biochem. Biophys. Res. Commun. **61:** 124–133.
24. SADRZADEH, S. M. H., E. GRAF, S. S. PANTER, P. E. HALLAWAY & J. W. EATON. 1984. J. Biol. Chem. **259:** 14354–14358.
25. SINCLAIR, P. R., W. J. BEMENT, H. L. BONKOVSKY, R. W. LAMBRECHT, J. E. FREZZA, J. F. SINCLAIR, A. J. URQUHART & G. H. ELDER. 1986. Biochem. J. **237:** 63–71.
26. FALK, J. E. 1964. Porphyrins and Metalloporphyrins. BBA Library. Vol. **2:** 35–36. Elsevier. Amsterdam.

Inhibitory Effect of Tetrachloro-*p*-Hydroquinone and Other Metabolites of Hexachlorobenzene on Hepatic Uroporphyrinogen Decarboxylase Activity with Reference to the Role of Glutathione[a]

GÜNTER KOSS,[b] MARGA LOSEKAM,[b] JUTTA SEIDEL,[b]
KLAUS STEINBACH,[c] AND WOLFGANG KORANSKY[b]

[b]*Institute of Toxicology and Pharmacology*
[c]*Department of Chemistry*
Philipps University
D-3550 Marburg, Federal Republic of Germany

INTRODUCTION

Studies of hexachlorobenzene (HCB)-induced disturbances of the hepatic porphyrin pathway have long assumed inhibition of hepatic uroporphyrinogen decarboxylase (UROD) activity as the primary metabolic defect.[1,2] Increasing evidence has accumulated that the impairment of the enzyme activity is closely related to the biotransformation of HCB.[3-5] Surprisingly though, *in vivo* experiments failed to demonstrate any porphyrinogenic action of phenolic and sulfur-containing metabolites of HCB.[6,7] Unfortunately, no attempts were made in these experiments to assay enzyme activity in the exposed experimental animal. To address the question of enzyme activity, several *in vitro* experiments were performed that suggested metabolites of hexachlorobenzene did indeed exert an inhibitory effect on UROD activity.[8,9]

Nevertheless, these *in vitro* experiments were flawed, we believe, by a comparatively high ratio between the concentration of tested metabolite and that of cytosolic protein, as well as by the use of disproportionately high concentrations of glutathione. For this reason, it was thought advantageous to reconsider the question of the inhibitory effect of HCB metabolites on the enzyme activity in light of these previous studies. The aim of this research then was to study the response of liver UROD to metabolites of HCB in the absence of glutathione.

Particular interest was focused on tetrachloro-*p*-hydroquinone (TCH), a substance with chemical properties characteristic of the hydroquinones. In addition, its oxidation product, chloranil,[10] was included in the experiments, and the impairment of enzyme activity, the behavior of the compounds in the incubation mixture, and the possible interactions between these compounds and the constituents of the test samples were studied.

[a]This study was supported by the Deutsche Forschungsgemeinschaft (Bonn) and P. G. Unna-Stiftung (Göttingen).

To contribute to the understanding of the role of glutathione in HCB-induced porphyria,[3,11–13] the concentrations of glutathione in the livers of HCB- and TCH-treated experimental animals were determined. Additionally, the behavior of glutathione was studied *in vitro*.

MATERIALS AND METHODS

Animals and Treatment

Female Wistar rats weighing 150 g and male B6C3F1 mice weighing 25 g were used (Zentralinstitut für Versuchstierzucht, Hannover). The animals were given tapwater and an Altromin diet (Lage) *ad libitum*. Rats were treated every other day for 3 months with different doses of HCB (TABLE 1). Mice were treated daily for 3 days with 75.8 μmol (20 mg) of TCH/kg body wt. The substances were dissolved in olive oil (DAB 8). Controls received olive oil only. The substances were given by way of stomach tube between 8:00 and 9:00 A.M.; 24 hr (rats) and 6 hr (mice) after the last administration, the animals were sacrificed by decapitation and were exsanguinated. The liver was perfused *in situ* with ice-cold physiological saline to minimize the content of erythrocytes. Liver tissue of female Wistar rats (150 g) without treatment was used for *in vitro* studies.

TABLE 1. Hepatic Content of HCB and GSH

HCB dosage (μmol/kg)	Content in Fresh Liver Tissue: HCB (μmol/g)	Content in Fresh Liver Tissue: GSH (μmol/g)
—	—	7.5
8.9	0.04	6.7
17.7	0.14	6.6
35.5	0.34	6.5
70.9	0.66	4.7
141.8	1.10	2.9

NOTE: Female Wistar rats were treated orally with doses of 8.9–141.8 μmol of HCB/kg body wt every other day for 3 months. HCB and GSH were determined 24 hr after the last treatment with HCB. For determinations, pooled fresh liver tissue of three or four animals was used. Determinations were performed in duplicate.

Liver Cytosol Preparation

Fresh liver samples were homogenized in five volumes (v/w) of 0.154 M potassium chloride (pH 6.8). Homogenates were first centrifuged at 12 000 g for 10 min and then at 100 000 g for 60 min at 4°C. Supernatants (cytosol) were used as enzyme preparations. To obtain cytosol free of reduced glutathione (GSH), cytosolic protein was separated by ammonium sulfate precipitation.[14] Although the protein still contained about 2 nmol of GSH/mg protein, it was referred to as GSH-free cytosolic protein.

UROD Assay

Incubation mixtures contained 0.067 M potassium phosphate buffer (pH 6.8), 4 μmol of added GSH, 4 μmol of EDTA, about 2 nmol of uroporphyrinogen I, and 0.5 ml of liver cytosol (about 9 mg of protein) in a final volume of 4 ml. When GSH-free cytosol (see above) was used, no GSH was added. Assays were performed under nitrogen at 37°C for 30 min in the dark with mechanical shaking. They were quenched with concentrated HCl.[14] Porphyrins were analyzed by means of HPLC.[15]

Chemicals

HCB (MW: 282, based on ^{35}Cl), pentachlorophenol (MW: 264), and pentachlorothiophenol (MW: 280) were purchased from Aldrich (Milwaukee, WI). TCH (MW: 246) was obtained from Kodak (Rochester, NY), and pentachlorothioanisole (MW: 294) was purchased from Lancaster Synthesis. Chloranil (MW: 244) was obtained from Merck-Schuchardt (Munich), and 1-methyl(2,3,4,5,6-pentachlorophenyl)sulfoxide (MW: 310) and 1-methyl(2,3,4,5,6-pentachlorophenyl)sulfone (MW:326) were synthesized.[16] The substances were purified by recrystallization from organic solvents. Tests by means of gas chromatography and of photometry (chloranil[17]) showed the substances to be of greater than 99.5% chemical purity. Diazomethane in diethyl ether was used to convert OH and SH groups into corresponding OCH_3 and SCH_3 groups. Porphyrins, NADPH, the reductase for oxidized glutathione (GSSG), L-cysteine, and L-methionine were purchased from Sigma (Munich). GSH and 1-chloro-2,4-dinitrobenzene (CDNB) were obtained from Merck-Schuchardt (Munich).

Analytical Determinations

GSH was determined after its nonenzymatic conjugation with CDNB.[18] Analysis was performed by means of HPLC (Millipore, Eschborn). Separation of the conjugate was obtained on a 10 cm × 8 mm (10 μm) Radial Pak 8 chromatographic cartridge (Millipore, Eschborn). Elution was performed with an 80:20 mixture (by volume) of 23 mM ammonium formate buffer (pH 5.0) and acetonitrile at a flow rate of 1 ml/min. Detection was carried out at 340 nm. GSSG was determined by recording the oxidation of NADPH at 340–400 nm in a Model 34 spectrophotometer (Beckman, Fullerton, CA) after addition of GSSG reductase.[19] Chlorinated compounds were determined by means of a Model 5730A gas chromatograph equipped with an electron capture detector (Hewlett Packard, Avondale, CA).[20,21] Chloranil, which was in chloroform, was determined using a Model DU-40 spectrophotometer (Beckman, Fullerton, CA). The absorbance was measured at 292 nm, and the log of the extinction coefficient was 4.4.[17] Combined gas chromatography and mass spectrometry (GC-MS) was performed with an LKB 2091 instrument (Bromma, Sweden) equipped with a PDP11 computer. GC separation was obtained with a 250-cm-long glass column (inner diameter: 2.2 mm) with 3% QF–1 on 100–120 mesh GC Q. Oven temperature was 200°C. Helium, the carrier gas, was used at a flow rate of 30 ml/min.

Electron spin resonance spectra were recorded on an E9 X-band spectrometer (Varian, Palo Alto, CA) at a frequency of 9.13 GHz. Samples were examined in a 1-mm-wide quartz cell.

Protein was determined using bovine plasma gamma globulin (Bio-Rad, Richmond, CA) as standard.[22]

RESULTS

HCB-Induced GSH Depletion

To examine the influence of HCB on the hepatic GSH content and its dose dependency, female Wistar rats were treated with the drug for 3 months. Subsequently, the concentrations of HCB and GSH in the liver tissue were determined. As expected, elevated HCB doses accounted for an increase in the concentration of HCB in liver tissue. In all treated animals, the cytosolic GSH in the liver was depleted. It is noteworthy that even animals that received a dose of only 8.9 μmol (2.5 mg) of HCB/kg body wt showed a 12% depletion of GSH content (TABLE 1). In the same animals the UROD activity was reduced to 80% of the control level.[23]

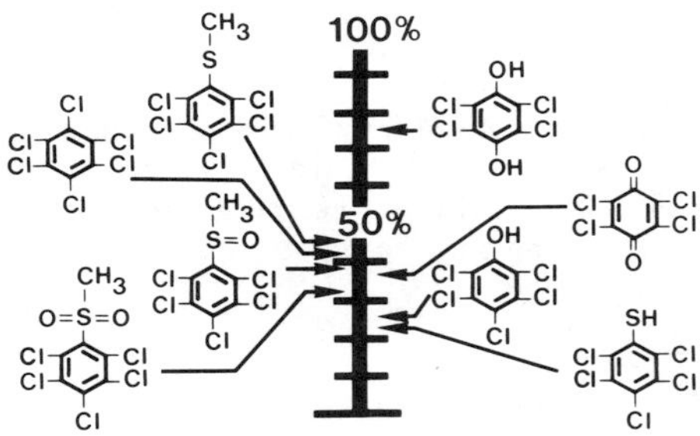

FIGURE 1. The inhibitory effect *in vitro* of HCB and its metabolites on the UROD activity in rat liver. Incubation mixtures (4 ml) contained 1.04 mg of GSH-free rat liver cytosolic protein, 1.9 nmol of uroporphyrinogen I, and 4 nmol of each chlorinated compound. The inhibitory effect of each chlorinated compound is expressed relative to the amount of coproporphyrinogen (measured as coproporphyrinogen tetramethylester) formed in the control incubation mixture: 100% inhibition: no coproporphyrinogen formed; 0% inhibiton: the amount of coproporphyrinogen formed in the control incubation mixture (0.27 nmol/30 min/mg of protein), which did not contain chlorinated compounds. Each value represents the mean of results from four incubations.

Influence of TCH on the Liver GSH Content

Four male B6C3F1 mice were treated with 78.5 μmol (20 mg) of TCH/kg body wt for 3 days. Six hours after the last treatment, 1 g of liver tissue was removed from each rat; on average, 0.48 nmol (0.12 μg) of TCH and 9.1 μmol of GSH were isolated. Each of the three nontreated mice had, on average, 8.9 μmol of GSH/g of liver tissue. Obviously, TCH exerted no measurable influence on the GSH content in mouse liver.

Inhibition of the UROD Activity in GSH-Free Rat Liver Cytosol by Metabolites of HCB

Rat liver cytosol protein without GSH was incubated with TCH, chloranil, pentachlorophenol, several sulfur-containing metabolites, and uroporphyrinogen I. For comparative reasons, HCB was also tested. The criterion of UROD activity was the amount of coproporphyrin detected after the incubation. As shown in FIGURE 1, all metabolites impaired the conversion of uroporphyrinogen to coproporphyrinogen. It was of interest that even the unchanged parent compound (HCB) revealed a significant inhibitory effect on the UROD activity. The most pronounced effect was exhibited by TCH and its oxidation product, chloranil. For this reason, the response of the enzyme toward comparatively low concentrations of TCH and chloranil was tested. The results showed that only 0.3 nmol (80 ng) of TCH and 3 nmol (800 ng) of chloranil in an incubation mixture containing 0.9 mg of protein was needed to inhibit the coproporphyrinogen formation (FIG. 2).

UROD Activity Inhibition in Vivo *by TCH*

Pooled native GSH-containing liver cytosol (0.5 ml with 11 mg of protein) from mice treated with TCH (see above) was incubated with 0.2 nmol of uroporphyrinogen I. After the incubation, 0.12 pmol coproporphyrin/mg protein/30 min were detected. Liver cytosol of untreated mice was found to produce 0.32 pmol of coproporphyrin/mg of protein/30 min. The comparison of the results showed that the hepatic UROD activity was diminished by 62.5%.

When using GSH-free liver cytosol from the same animals, a practically identical inhibitory effect (59%) was observed in the treated mice compared with the controls. The conclusion was drawn that GSH did not play a significant role when added to incubation mixtures for detecting enzyme activity inhibition induced *in vivo*.

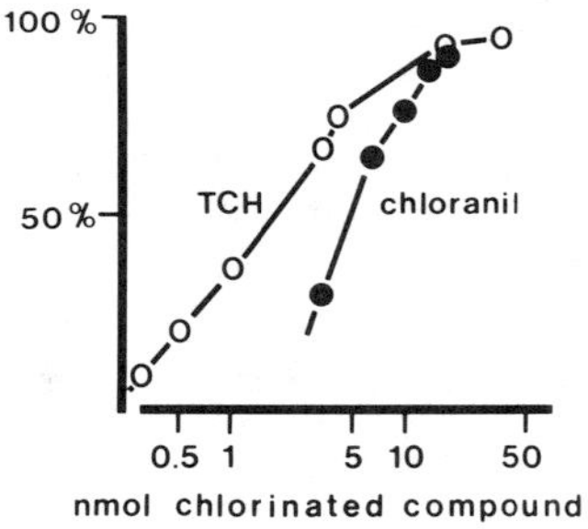

FIGURE 2. The inhibitory effect *in vitro* of TCH and chloranil on the UROD activity in rat liver. Incubation mixtures (4 ml) contained 0.93 mg of GSH-free rat liver cytosolic protein, 1.4 nmol of uroporphyrinogen I, and the indicated amounts of TCH and chloranil. The inhibitory effect of each compound is expressed, as in FIGURE 1, relative to the amount of coproporphyrinogen formed in the control incubation mixture (0.24 nmol/30 min/mg of protein), which did not contain chlorinated compounds.

Recovery of the Chlorinated Compounds and GSH from Aqueous and Protein-Containing Mixtures

To obtain more insight as to the interaction between the chlorinated compounds and constituents of incubation mixtures used to assay UROD activity, both the recovery of the tested substances and the recovery of GSH were determined after a 30-min incubation. The recoveries were at least 75% for pentachlorophenol, HCB, and sulfur-containing compounds from buffer and from buffer containing GSH or protein. The corresponding recoveries were less than 75% when the incubation mixture contained GSH and protein. It was remarkable that the recoveries of TCH and chloranil were small or nonexistent when the incubation mixture contained GSH and protein (TABLE 2).

GSH itself was completely recovered from buffer (100%). Only about half the amount was isolated when the buffer contained cytosolic protein (55.4%). When the buffer contained chlorinated compounds, the recovery was reduced to 57.1%. It was less than that when cytosolic protein was added. The recovery of GSH from the complete incubation mixture was extremely poor when it contained TCH or chloranil (TABLE 2). In some incubation mixtures the relative amount of GSSG formed from GSH during the incubation was determined. From the complete incubation mixture

TABLE 2. Recovery of Chlorinated Compounds and GSH from Buffer and Protein-Containing Mixtures

		Buffer & GSH			Protein-Containing Mixture & GSH	
	Recovery of CCs from Buffer	Recovery of CCs	Recovery of GSH	Recovery of CCs from Protein-Containing Mixture	Recovery of CCs	Recovery of GSH
TCH	8.7	2.2	57.1	5.8	ND	6.3
Chloranil	79.5	ND	60.4	84.3	ND	9.7
Pentachlorophenol	97.5	99.6	92.6	97.5	89.5	43.5
Pentachlorothiophenol	98.5	85.0	86.9	99.0	77.0	56.0
Pentachlorothioanisol	86.5	81.3	76.0	75.0	60.0	57.5
Methylpentachloro-phenylsulfoxide	99.0	87.0	84.6	98.5	62.8	47.9
Methylpentachlorophenylsulfone	100.0	98.2	89.6	100.0	7.0	46.4
HCB	99.5	79.5	86.0	97.5	83.5	59.0
Control[a]	—	—	100.0	—	—	55.4

NOTE: The chlorinated compounds (CCs) were added to the 4-ml incubation mixtures in 100-μl volumes of methanol. Each 100-μl volume of methanol contained 0.15 μmol of each CC. There were four incubation mixtures: buffer alone (0.067 M potassium phosphate buffer, pH 6.8, containing 4 μmol of EDTA); buffer and 0.5 μmol of GSH; buffer and 1 mg of cytosolic protein; and buffer, 1 mg of cytosolic protein, and 4 μmol of GSH. The incubations were carried out under nitrogen at 37°C for 30 min in the dark with mechanical shaking. Subsequently, the recoveries of CCs and GSH were determined. Each result represents the mean of two incubations.

[a]No CCs were added.

without chlorinated compounds, 24% of the added GSH was recovered in its oxidized form. This relative portion increased to 38% when HCB was added, and there was a further increase to 53% when the incubation mixture contained TCH. The results suggest that GSH is consumed and that there are interactions between the chlorinated compounds and the constituents of the incubation mixtures.

Chemical Alterations in Vitro *of the Chlorinated Compounds*

The poor recoveries for the chlorinated compounds from the incubation mixture led to the question of whether the interaction suggested above resulted in metabolically altered substances. For this reason, the extracts of the incubation mixtures were examined for additional chlorinated compounds. Several compounds were studied for their inhibitory effect on UROD, and GC revealed that all had undergone metabolic conversion. So far, conversion products of HCB, TCH, and chloranil have been identified. From the incubation mixture to which HCB was added, pentachlorothiophenol and pentachlorothioanisol were isolated, accounting for about 3% of the incubated HCB. The extracts from the incubation mixtures with TCH and chloranil also contained additional compounds. After the extracts were mixed with etheral diazomethane, a substance was identified as methylated trichloromercapto-*p*-hydroquinone (FIG. 3). Additional photometric analysis showed considerable amounts of chloranil in samples incubated with TCH and of TCH in samples incubated with chloranil. Obviously, one compound was formed from the other.

To elucidate the origin of the sulfur in the compound containing the SH group, two samples of TCH and chloranil were incubated—the first with GSH (without protein), the second with protein (without GSH). Examination of the extracts revealed that the sulfur of the metabolite was derived not only from GSH but also from protein. It was not fully excluded, however, whether the trace amounts of GSH in the precipitated protein contributed to the formation of the sulfur-containing metabolite. Apart from protein and GSH, L-cysteine and L-methionine (0.5 μmol/4 ml) were also incubated with TCH and chloranil (0.15 μmol/4ml). Astonishingly, after adding diazomethane the extracts of all incubation mixtures contained trichloromercapto-*p*-hydroquinone in its methylated form. Its amount accounted for 3–5% of the TCH and chloranil added to the incubation mixture.

To determine whether there were sulfur-sharing conjugates of TCH or chloranil, the incubation mixtures were treated at an alkaline pH value to achieve cleavage of conjugates. This treatment, however, did not result in considerably elevated concentrations of the sulfur-containing metabolite in the samples. Obviously the conditions during the incubation and the chemical properties of TCH and chloranil favored the

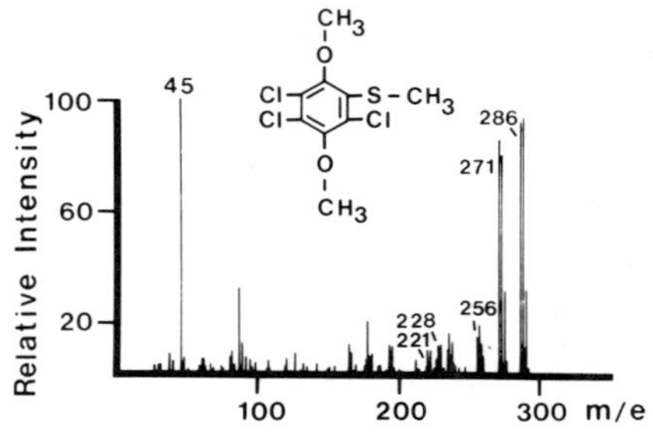

FIGURE 3. Mass spectrum analysis of trichloromercapto-*p*-hydroquinone formed during an incubation of TCH and GSH in an aqueous solution (pH 6.8) at 37°C. After its isolation, the compound was methylated in an ethereal solution of diazomethane. The figures indicate signals at *m/e* 286 (molecular ion, M^+), *m/e* 271 (M^+-CH_3), *m/e* 256 ($M^+-CH_3-CH_3$), *m/e* 228 (M^+-2CH_3-CO), and *m/e* 221 ($M^+-CH_3-Cl-CH_3$).

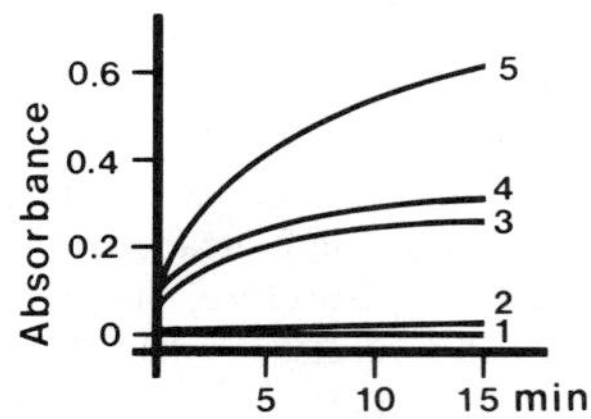

FIGURE 4. Spectrophotometric determination of the uroporphyrin I formation from uroporphyrinogen I enhanced by chloranil and delayed by GSH. To the cuvette with 2 ml deaerated 0.067 M potassium phosphate buffer (pH 7.4), 1) 40 nmol chloranil in 20 μl of methanol, 2) 3.4 nmol of uroporphyrinogen I in 120 μl of buffer and 200 nmol of GSH in 60 μl of buffer, 3) 3.4 nmol of uroporphyrinogen I, 4) 3.4 nmol of uroporphyrinogen I, 40 nmol of chloranil, and 200 nmol of GSH, or 5) 3.4 nmol of uroporphyrinogen I and 40 nmol of chloranil were added. The absorbance was determined at 405 nm. The temperature was 22°C.

formation of sulfur-sharing conjugates and the cleavage of these conjugates to form the sulfur-containing aromatic compounds.

As a further step to elucidate the chemical behavior of TCH and chloranil, potassium phosphate buffer containing these compounds was examined by means of electron spin resonance spectroscopy. The signals recorded suggest the occurrence of the tetrachloro-*p*-benzosemiquinone radical. The intensity of the signal of the TCH-derived radical was comparatively higher, indicating the higher concentration of this radical in the sample. The spontaneously occurring radical confirmed the high reactivity of TCH and chloranil.

Photometric Detection of the Oxidation of Uroporphyrinogen I by Chloranil

Immediately after addition of an uroporphyrinogen I solution to phosphate buffer, the sample displayed a light absorbance at 405 nm that progressively increased in intensity over time. This indicated the well-known formation of uroporphyrin from its corresponding reduced form. This process was drastically enhanced when chloranil was added to the uroporphyrinogen-containing buffer. When GSH was added, the uroporphyrinogen oxidation, which either occurred spontaneously or when mediated by chloranil, was clearly delayed (FIG. 4). This experiment confirmed the oxidative property of chloranil and the protective function of GSH.

To obtain insight as to the fate of chloranil, the ultraviolet spectra of the incubation mixtures were examined. The spectra revealed a minor absorbance at 292 nm, indicating a drastic decrease in the chloranil concentration. In contrast, there was a comparatively distinct absorbance at 323 nm, indicating the formation of TCH from chloranil during the incubation with uroporphyrinogen I. It was estimated that about 20% of the chloranil was converted to TCH during the 15-min incubation. When chloranil alone was incubated in buffer (see also TABLE 2), only about 5% was recovered in the form of TCH after 15 min.

DISCUSSION

The distinct inhibitory effect of the HCB metabolites on the UROD activity in the GSH-free cytosolic protein emphasized the relevance of the GSH depletion in the

HCB-exposed experimental animal. One of its major causes can be explained by the formation of GSH adducts with HCB and its metabolites, which is catalyzed by GSH sulfotransferases[24,25] or which occurs nonenzymatically as shown for TCH. The concentration of HCB and its metabolites, however, appeared to be much lower than that of GSH in liver tissue.[3] Moreover, the biotransformation rate of HCB was shown to be relatively slow.[20] It was assumed, therefore, that various other processes in which GSH is involved contributed to the observed GSH depletion. An increased hepatic malondialdehyde content in the HCB-treated rats (results not shown) may indicate the occurrence of oxygen radicals and their reaction with lipids (lipid peroxidation).[26] These radicals—generated by the microsomal NADPH-cytochrome *c* reductase, cytochrome P-450, or other cell constituents—are detoxified at the expense of GSH, which is oxidized to GSSG. The disulfide is reduced back to GSH by the NADPH-dependent GSH reductase.[27] If sufficient reducing equivalents are not available because of their consumption by processes that are stimulated or induced by HCB,[28] the GSSG reduction may be impaired. GSSG would then accumulate in the cell, to be subsequently released into bile, resulting in a further depletion of intracellular GSH.[19]

It was shown that TCH and chloranil changed their structure to resemble radical compounds that can easily be attacked by nucleophilic substances like GSH to form trichloromercapto-*p*-hydroquinone, already identified as an urinary metabolite in HCB-treated rats.[29] It is also possible that TCH oxidized to form the tetrachloro-*p*-benzosemiquinone radical, which in turn generated the superoxide anion radical, or other reactive oxygen species,[30,31] which, also at the expense of GSH, may be reduced to water. The further oxidation of the benzosemiquinone to tetrachloro-*p*-benzoquinone would likewise be accompanied by the generation of oxygen radicals. This process, possibly supported by recycling of the benzoquinone (chloranil), could also contribute to the GSH depletion.

Moreover, TCH, its conversion products, or any other reactive metabolite of HCB could impair the biosynthesis of GSH. We provided evidence of the reaction of TCH and chloranil with L-cysteine and L-methionine—a reaction possibly resulting in a decrease in the bioavailability of these GSH precursors.[32]

The utilization of GSH-free cytosolic protein of liver tissue to study the inhibitory effect of HCB metabolites on the UROD activity was justified for three reasons: 1) HCB was found to induce a dose-dependent GSH depletion in liver tissue. 2) GSH is not evenly distributed within the liver lobule, which accounts for higher content in the periportal and for lower contents in the centrilobular region.[33] It cannot be regarded as established, but it is assumed that HCB-induced GSH depletion affects hepatocytes of the centrilobular region comparatively drastically. 3) We, as well as others, have observed that enzyme preparations used to assay UROD activity, which contain GSH, or are free of GSH, decarboxylated uroporphyrinogen through coproporphyrinogen with about the same velocity.[34]

The utilization of GSH-free cytosolic protein permitted a clearer picture of the behavior, strength, and mode of action of the compounds tested. TCH appeared to be the most effective metabolite of HCB—a result that confirms experiments in which the incubation mixtures contained relative large amounts of GSH.[9] It may be deduced from both studies that TCH finds its way to the UROD irrespective of the cellular GSH content. Among other factors, this property of TCH certainly played a role in experiments in which the compound was given to the experimental animal in combination with HCB, or in which a combination of HCB and pentachlorophenol (the latter is also metabolized to TCH[35]) was administered. Both studies revealed a comparatively early onset of the disturbances of the hepatic porphyrin pathway.[36,37]

Similarly distinct was the inhibitory effect of chloranil. Its cause may depend on the structural properties that were already discussed for TCH. Additionally, a further aspect may be considered. Chloranil is a known oxidant.[38] One may expect, therefore, that chloranil will oxidize porphyrinogens to porphyrins. In fact, the oxidation of uroporphyrinogen was supported by evidence confirming an earlier observation on the oxidation of urinary porphyrinogens.[39] In the presence of sufficient intracellular chloranil, porphyrinogens may undergo oxidation, and uroporphyrinogen may cease to be a substrate of UROD. As a result, the porphyrins formed may accumulate in the cell. The oxidation of porphyrins by chloranil, therefore, could contribute to the porphyrin pathway disturbance.

Like TCH, the other conversion products of HCB may also undergo metabolic alterations. Pentachlorothiophenol and pentachlorophenol could be converted into radicals (thiyl[40] and phenoxyl,[41] respectively) and, as such, interact with the UROD protein. The thiophenol may form disulfide bridges with SH groups of the enzyme.[42–44] To obtain evidence on a proposed binding of the metabolites of HCB and of the parent compound itself to the enzyme protein, further analytical investigations are being performed. These investigations are based on observations made in the present study, as well as on those made in an earlier experiment in which a covalent binding of still unknown labeled substances to liver cytosolic protein of [^{14}C]HCB-treated rats was detected.[45]

It was surprising that even HCB inhibited the UROD activity. Apart from the aspects mentioned above, another point of view is worth considering: HCB , as well as a number of its metabolites, appears to exhibit planar structure, a property that is also shown by porphyrinogens.[46] It may be assumed, therefore, that the planar halogenated aromatic competes with the substrate for the catalytic site(s) of the UROD and thus inhibits the decarboxylation of uroporphyrinogen.[47] Likewise, this phenomenon could be of relevance to experiments concerning the polychlorinated biphenyl-impaired enzyme activity. These experiments indicate that the probability of coplanarity is higher for porphyria-inducing congeners than for inactive ones.[48]

From the presented results, it may be deduced that the inhibition of the hepatic UROD activity in the HCB-exposed organism is primarily induced by conversion products of HCB and that this functional impairment occurs first in the centrilobular region of the liver. This result may support the hypothesis that considers the biotransformation of HCB as a prerequisite for the onset of porphyria.[28] When, after high dosing with HCB or after chronic HCB exposure, a sufficient amount of metabolites is available, the enzyme activity of the periportal liver lobule region could decrease. If there is an additional GSH depletion, even HCB could diminish the UROD activity. Apart from this defect, HCB and its metabolites may induce several other functional and morphological alterations,[49–52] which, although not involved in the initial impairment of the UROD activity, after a long period of time could certainly enhance the persistence of the disturbance of the porphyrin pathway.

SUMMARY

Exposure of rats to HCB caused a dose-dependent depletion of GSH. Chlorophenolic and sulfur-containing metabolites of HCB incubated with GSH-free rat liver cytosolic protein drastically diminished the UROD activity. In addition, HCB also exhibited inhibitory potency. The most effective compounds studied were TCH and its oxidation product, chloranil. Incubation of liver cytosolic protein and of GSH with HCB and its metabolites yielded results that suggested interaction between the compounds and cell constituents—an interaction that may cause inhibition of the hepatic UROD activity in the HCB-exposed organism.

REFERENCES

1. Elder, G. H., J. O. Evans & S. A. Matlin. 1976. Clin. Sci. Mol. Med. **51:** 71–80.
2. von Tiepermann, R., G. Koss & M. Doss. 1980. Hoppe-Seyler's Z. Physiol. Chem. **361:** 1217–1222.
3. Debets, F., J.-H. Reinders, G. Koss, J. Seidel & A. Strik. 1981. Chem.-Biol. Interact. **37:** 77–94.
4. Carpenter, H. M., D. E. Williams, M. C. Henderson, R. C. Bender & D. R. Buhler. 1984. Biochem. Pharmacol. **33:** 3875–3881.
5. Wainstok de Calmanovici, R., M. D. C. Rios de Molina, M. C. Taira de Yamasato, J. M. Tomio & L. C. San Martin de Viale. 1984. Biochem. J. **218:** 753–763.
6. Goerz, G., W. Vizethum, K. Bolsen, T. Krieg & R. Lissner. 1978. Arch. Dermatol. Res. **263:** 189–196.
7. Koss, G., S. Seubert, A. Seubert, W. Koransky, H. Ippen & M. Straub. 1979. *In* Chemical Porphyria in Man. J. J. T. W. A. Strik & J. H. Koeman, Eds.: 175–180. Elsevier Biomedical Press. Amsterdam.
8. del Carmen, M., R. de Molina, R. Wainstok de Calmanovici & L. C. San Martin de Viale. 1980. Int. J. Biochem. **12:** 1027–1032.
9. Billi, S. C., G. Koss & L. C. San Martin de Viale. 1986. Res. Commun. Chem. Pathol. Pharmacol. **51:** 325–336.
10. Staub, H. 1926. Biochem. Z. **178:** 167–180.
11. Wolf, M. A., R. Lester & R. Schmid. 1962. Biochem. Biophys. Res. Commun. **8:** 278–279.
12. Lissner, R., W. Brunner, K. Bolsen, H. Merk & G. Goerz. 1985. Arzneim.-Forsch. Drug Res. **35:** 713–715.
13. Ingebrigtsen, K., J. K. Skaare, I. Nafstad, K. Grave & M. Forde. 1986. Gen. Pharmacol. **17:** 19–24.
14. Tomio, J. M., R. C. Garcia, L. C. San Martin de Viale & M. Grinstein. 1970. Biochim. Biophys. Acta **198:** 353–363.
15. Seubert, A. & S. Seubert. 1982. Anal. Biochem. **124:** 303–307.
16. Hinsberg, G. 1908. Ber. Dtsch. Chem. Ges. **41:** 2836–2839.
17. Braude, E. A. 1945. J. Chem. Soc. (London) Part II: 490–497.
18. Baars, A. J. & D. D. Breimer. 1983. Pharm. Weekbl. **5:** 145–148.
19. Akerboom, T. P. M., M. Bilzer & H. Sies. 1982. J. Biol. Chem. **257:** 4248–4252.
20. Koss, G., W. Koransky & K. Steinbach. 1976. Arch. Toxicol. **35:** 107–114.
21. Koss, G., W. Koransky & K. Steinbach. 1979. Arch. Toxicol. **42:** 19–31.
22. Bradford, M. M. 1976. Anal. Biochem. **72:** 248–254.
23. Koss, G., B. Stoffelns, S. Seubert, A. Seubert, W. Koransky & H. Ippen. 1986. *In* Hexachlorobenzene. IARC Scientific Publications. No. 77. C. R. Morris & J. R. P. Cabral, Eds.: 493–495. Oxford University Press. Oxford.
24. Koss, G., W. Koransky & K. Steinbach. 1979. Arch. Toxicol. **42:** 19–31.
25. Renner, G. 1981. Xenobiotica **11:** 435–446.
26. Comporti, M. 1985. Lab. Invest. **53:** 599–623.
27. Reed, D. J. 1986. Biochem. Pharmacol. **35:** 7–13.
28. Strik, J. J. T. W. A., F. M. H. Debets & G. Koss. 1980. *In* Halogenated Biphenyls, Terphenyls, Naphthalenes, Dibenzodioxins and Related Products. R. D. Kimbrough, Ed.: 191–239. Elsevier Biomedical Press. Amsterdam.
29. Koss, G., S. Seubert, A. Seubert, M. Herbert, W. Koransky & H. Ippen. 1980. *In* Mechanisms of Toxicity and Hazard Evaluation. B. Holmstedt, R. Lauwerys, M. Mercier & M. Roberfroid, Eds.: 517–520. Elsevier Biomedical Press. Amsterdam.
30. Witte, I., U. Juhl & W. Butte. 1985. Mutat. Res. **145:** 71–75.
31. Beckman, J. & J. N. Siedow. 1985. J. Biol. Chem. **260:** 14604–14609.
32. Reed, D. J. & S. Orrenius. 1977. Biochem. Biophys. Res. Commun. **77:** 1257–1264.
33. Smith, M. T., N. Loveridge, E. D. Wills & J. Chayen. 1979. Biochem. J. **182:** 103–108.
34. Garcia, R. C., L. C. San Martin de Viale, J. M. Tomio & M. Grinstein. 1973. Biochim. Biophys. Acta **309:** 203–210.
35. Ahlborg, U. G. 1980. CRC Crit. Rev. Toxicol. **7:** 1–35.

36. CARPENTER, H. M., M. J. HARVEY & D. R. BUHLER. 1985. J. Toxicol. Environ. Health **15:** 81–92.
37. DEBETS, F. M. H., J. J. T. W. A. STRIK & K. OLIE. 1980. Toxicology **15:** 181–195.
38. HUCK, H. & H.-L. SCHMIDT. 1981. Angew. Chem. **93:** 421–422.
39. SCHWARTZ, S., P. EDMONDSON, B. STEPHENSON, D. SARKAR & H. FREYHOLTZ. 1976. Ann. Clin. Res. **8:** 156–161.
40. HARMAN, L. S., C. MOTTLEY & R. P. MASON. 1984. J. Biol. Chem. **259:** 5606–5611.
41. FISCHER, V., P. R. WEST, L. S. HARMAN & R. P. MASON. 1985. Environ. Health Perspect. **64:** 127–137.
42. MAUZERALL, D. & S. GRANICK. 1958. J. Biol. Chem. **232:** 1141–1162.
43. DE VERNEUIL, H., S. SASSA & A. KAPPAS. 1983. J. Biol. Chem. **258:** 2454–2460.
44. STRAKA, J. G. & J. P. KUSHNER. 1983. Biochemistry **22:** 4664–4672.
45. KOSS, G., S. SEUBERT, A. SEUBERT, W. KORANSKY, P. KRAUS & H. IPPEN. 1980. Int. J. Biochem. **12:** 1003–1006.
46. FUHRKOP, J. H. & J. SUBRAMANIAN. 1976. Philos. Trans. Roy. Soc. London Ser. B **273:** 335–352.
47. SMITH, A. G. & J. E. FRANCIS. 1981. Biochem. J. **195:** 241–250.
48. SASSA, S., O. SUGITA, N. OHNUMA, S. IMAJO, T. OKUMURA, T. NOGUCHI & A. KAPPAS. 1986. Biochem. J. **235:** 291–296.
49. COURTNEY, K. D. 1979. Environ. Res. **20:** 225–266.
50. BÖGER, A., G. KOSS, W. KORANSKY, R. NAUMANN & H. FRENZEL. 1979. Virchows Arch. Pathol. Anat. Histol. **382:** 127–137.
51. HANSTEIN, W. G., T. D. HEITMANN, A. SANDY, H. L. BIESTERFELDT, H. H. LIEM & U. MÜLLER-EBERHARD. 1981. Biochim. Biophys. Acta **678:** 293–299.
52. SMITH, A. G., J. E. FRANCIS, D. DINSDALE, M. M. MANSON & J. R. P. CABRAL. 1985. Carcinogenesis **6:** 631–636.

Biochemical Studies on Griseofulvin-Induced Protoporphyria

TOKIO SHIMOYAMA AND SHIGEO NONAKA

Department of Dermatology
Nagasaki University School of Medicine
Nagasaki, Japan

INTRODUCTION

Griseofulvin (GF), an antifungal drug that has been distributed worldwide, has been used mainly in the treatment of dermatophytoses. Since the discovery of GF-induced protoporphyria in 1963, GF has also been utilized in investigations of porphyric photosensitivity, in therapeutic treatments for human erythropoietic protoporphyria, and in studies of synergistic effects on chemically induced porhyrinopathies. In this paper, the history of studies on GF-induced protoporphyria is reviewed, and the biochemistry of and the direction of future studies on this phenomenon are discussed.

HISTORY

Weston-Hurst and Paget[1] reported that pigments with red fluorescence were observed in the liver cells when a feed containing 2.5% GF was given to mice. They speculated that these pigments consisted of protoporphyrin and pointed out that GF induced abnormalities of porphyrin metabolism. DeMatteis and Rimington[2] observed a marked elevation of fecal protoporphyrin and urinary coproporphyrin when a feed containing GF was given to mice. They confirmed that a large amount of GF induced an acceleration of porphyrin metabolism. Furthermore, Lochhead *et al.*[3] reported an elevation of δ-aminolevulinic acid dehydrase activity in the liver when mice were given a feed containing from 1.25 to 2.5% GF. From their data, it was obvious that GF caused an increase of protoporphyrin levels in the liver and erythrocytes of mice. Nakao *et al.*[4] studied the relationship between erythrocytic and hepatic protoporphyrin levels in GF-induced protoporphyria. They concluded that the source of the porphyrins responsible for the increased erythrocytic protoporphyrin levels in mice with experimentally induced porphyria might not be the erythrocytes themselves, but the liver, where overproduction of porphyrin actually occurred. Poh-Fitzpatrick *et al.*[5] studied variations in erythrocytic protoporphyrin levels in patients with erythropoietic protoporphyria and in mice with GF-induced protoporphyria. They observed that the porphyrin content diminished rapidly with the age of erythrocytes in erythropoietic protoporphyria. On the other hand, the aging erythrocyte population became progressively more porphyrin rich in GF-induced protoporphyria. They agreed with the opinion of Nakao *et al.* that source of the erythrocytic protoporphyrin in GF-induced protoporphyria was the liver.

BIOCHEMISTRY

The mechanism of GF-induced protoporphyria is still not completely clear. In an early study, when feeds containing GF at a concentration of 2.5% were given to mice, it

TABLE 1. Hepatic and Erythrocytic Porphyrin Contents in Mice Fed GF

GF Concentration in Feed (%)	Duration of Feeding (days)	Number of Animals	Number of Animals with Abnormal Porphyrin Metabolism	Liver Weight: Body Weight (%)	Hepatic Porphyrin ($\mu g/g$ wet weight)			Erythrocytic Porphyrin ($\mu g/dl$)	
					UP	CP	PP	CP	PP
0	0	27	0	5.9	ND	0.11	0.38	ND	19.6
0.1	38–450	34	3	5.8 (31)	ND	0.17	0.35 (31)	3.1	33.2
				6.0 (3)			21.78 (3)		
0.5	27–240	13	13	12.8	0.8	9.3	292.8	44.4	684.4
1.0	17–60	16	16	14.8	0.8	4.6	250.1	22.2	706.7

NOTE: Each figure in parentheses indicates the number of animals to which the entry applies. UP: uroporphyrin; CP: coproporphyrin; PP: protoporphyrin; ND: not detected.

was observed that a disturbance of porphyrin metabolism appeared as an elevation of fecal protoporphyrin, coproporphyrin, and hepatic protoporphyrin.[2] Furthermore, it was very interesting that the protoporphyrin level in the erythrocytes was markedly elevated, although there was no remarkable abnormality in urinary porphyrins and porphyrin precursors. Enzymatic defects in GF-induced protoporphyria include a significant decrease in hepatic heme synthetase and an increase in hepatic δ-aminolevulinic acid dehydrase (ALAD).[3] The activities of δ-aminolevulinic acid synthetase (ALAS) and heme synthetase in the spleen, however, were unchanged.[4] Nakao *et al.*[4] concluded that the overproduction of protoporphyrin in the liver induced by the porphyria-inducing chemicals caused a rise in the concentration of plasma protoporphyrin, which was apparently taken up by the red blood cells of mice, resulting in an increased erythrocyte protoporphyrin concentration. Poh-Fitzpatrick *et al.*[5] carried out very interesting studies of the erythrocytic protoporphyrin levels in humans with erythropoietic protoporphyria and in mice with GF-induced protoporphyria. In erythropoietic protoporphyria, porphyrin content diminished rapidly with age. On the other hand, in GF-induced protoporphyria, the aging erythrocyte population became progressively more porphyrin rich. The conclusion reached from these studies was that GF-induced protoporphyria is hepatic in origin.

There have been no published studies on the relationship between concentration of GF and porphyrinopathy. In order to determine the lowest concentration of GF needed to induce abnormal porphyrin metabolism, we gave mice feeds containing GF in concentrations of 0.1%, 0.5%, and 1.0%.[6] The results are shown in TABLE 1. The findings obtained with low concentrations of GF differed from those obtained with high ones. Liver swelling was more marked in high concentrations; that is, the liver: body weight ratio was elevated with increased concentration of GF. The liver:body weight ratio in the 1.0% GF concentration group showed the highest rise of the three groups, even when the feeds were given for a short period. There were no differences between the ratios for the 0.1% GF group and those for the control group, nor were there any differences between the ratios for the group fed over a short period and those for the group fed over a long period. The values for coproporphyrin and protoporphyrin in the liver were markedly elevated in both the 0.5% and 1.0% GF groups. The elevation of protoporphyrin was more prominent than that of coproporphyrin. Protoporphyrin in the erythrocytes was also elevated in the 0.5% and 1.0% GF groups. When the feed was given for long periods in the 0.5% GF group, the erythrocytic and hepatic porphyrin levels became higher than those for the mice in the 1.0% GF group given feed for a short period.

From these results, we concluded that the changes in the 0.5% GF group were basically similar to those of the 1.0% GF group; that is, an increase in protoporphyrin and abnormalities in porphyrin metabolism were seen in 100% of the mice given feeds containing 0.5% GF. On the other hand, the bulk of the mice given feeds containing 0.1% GF were not affected; only a few cases showed any abnormality of porphyrin metabolism in the liver when compared to the control group. Furthermore, there was no tendency toward an increase in the incidence of abnormal porphyrin metabolism with such a low concentration of GF as 0.1%, even when the feeds were given to the mice over a long period of time. It is unknown why only a few mice developed an abnormal porphyrin metabolism with the low concentration of GF. Many factors may be involved, such as the absorption of GF into the intestinal tract, the concentration of GF in the serum, and the reactivity of the liver cells to GF. Finally, it appears that a concentration of 0.1% GF might be suitable for experiments in which low concentrations of GF would be an advantage. Such an experiment on GF-induced protoporphyria would involve studying the influence of other chemicals in combination with GF.

To investigate the potentiation of chemicals in porphyrin metabolism, iron, ethyl

TABLE 2. Hepatic and Erythrocytic Porphyrin Contents in Groups of Mice Fed Various Chemicals

Group	Duration of Feeding (days)	Number of Animals	Number of Animals with Abnormal Porphyrin Metabolism	Liver Weight: Body Weight[a] (%)	Hepatic Porphyrin (μg/g wet weight)		Erythrocytic Porphyrin (μg/dl)	
					CP	PP	CP	PP
Control		27	0	5.9 ± 1.1	0.11	0.38	ND	19.6
0.1% GF-alone	38–450	34	3	5.8 (31) 6.0 (3)	0.70	0.35 (31) 21.78 (3)	3.1	33.2
Ethanol–iron	20–400	31	1	5.5 ± 1.7	0.29 ± 0.09	0.46 ± 0.41 2.25 (1)	31.9	49.2
Ethanol–iron–0.1% GF	50–330	33	7	5.8 ± 1.9	0.27 (24) 1.47–4.17 (4)	0.53 (29) 2.12–230.1 (4)	30.3	64.1
Estrogen-alone	15–85	11	1	8.0 ± 3.1		0.21	ND	12.7
Estrogen–0.1% GF	30–250	24	5	8.4 ± 1.6	0.22 (17) 0.50–2.28 (4)	0.36 (19) 1.92–56.08 (4)	25.0	64.8 637.7–771.2 (2)
PCB-alone	140–150	20	0	5.0 ± 1.2	0.13	0.19		
PCB–0.1% GF	130–150	12	8	10.1 ± 2.0	0.74	0.90	28.4	61.0

[a]Each value below is a mean ± SEM.

alcohol, estrogen, and polychlorinated biphenyls (PCBs), all of which are known to influence porphyrin metabolism, were given to the mice with or without 0.1% GF.[7] The results are shown in TABLE 2. In this study we gave ethanol and iron simultaneously. Both chemicals seem to have a slight potential for inducing porphyria, and so we attempted to investigate their interactive effect. The action of the ethanol–iron combination was relatively weak, as we expected. In the ethanol–iron-alone group, no swelling of the liver was seen. There were no differences between the ethanol–iron-alone and ethanol–iron–0.1% GF groups. The incidence of abnormal liver porphyrin levels, however, was elevated in the ethanol–iron–0.1% GF group. This profile of abnormalities in liver porphyrin is similar to that in mice treated with higher concentrations of GF. These data, then, seem to prove that the ethanol–iron combination enhances the action of GF. To investigate the potentiation of estrogen on porphyrinopathy, we administered both estrogen and 0.1% GF. The incidence of abnormal porphyrin metabolism in the estrogen–0.1% GF group was both greater than that in the estrogen-alone group and greater than that in the 0.1% GF-alone group. The abnormalities in porphyrin metabolism in mice of the estrogen–0.1% GF group were similar to those in mice fed high concentrations of GF; that is, both these groups showed a predominance of protoporphyrin in the liver and the erythrocytes. Therefore, estrogen enhances the action of low concentrations of GF. The difference between the ethanol–iron combination and estrogen, however, was in liver swelling. There was remarkable liver swelling in the estrogen-alone group. Because this swelling was recognized in both the estrogen–0.1% GF and estrogen-alone groups, there seems to be no relationship between porphyrin metabolism and liver swelling in our study.

The interaction of the PCBs and 0.1% GF was remarkable. In the 0.1% GF-alone group, the incidence of abnormal porphyrin metabolism was relatively low. There were no changes in the PCB-alone group, but an elevation in liver coproporphyrin and protoporphyrin was seen in 8 of 12 mice in the PCB–0.1% GF group. Unlike the situation with high concentrations of GF, the major abnormality in porphyrin metabolism due to the PCBs and 0.1% GF was an elevation of coproporphyrin in the liver. Some mice showed an inverse coproporphyrin:protoporphyrin ratio in the liver. Normally, the liver and erythrocytes in mice treated with high concentrations of GF always show a predominance of protoporphyrin over coproporphyrin. From these data, we concluded that 0.1% GF enhances the action of PCBs rather than the opposite: that PCBs enhance the action of 0.1% GF. Furthermore, these porphyrinopathic substances may be potentiated by combining with other chemicals. Animals treated with low concentrations of GF may be useful for investigating such a possibility.

Previously, we gave an account of 26 patients who had porphyria cutanea tarda,[8] 4 of whom had a history of chemotherapy for tuberculosis. The most common chemotherapy was a combined therapy of *p*-aminosalicylic acid, streptomycin, and isonicotinic hydrazine (INH). Therefore, one of our collegues investigated the porphyrinopathy of INH.[9] The results are shown in TABLE 3. INH was mixed into drinking water at a concentration of 0.05%. The INH–GF group comprised 8 mice that were given drinking water containing INH and feed containing 0.1% GF; the INH-alone group, 12 mice that were given drinking water containing INH and normal feed; and the GF-alone group, 12 mice that were given ordinary drinking water and feed containing 0.1% GF. Hepatic uroporphyrin, coproporphyrin, and protoporphyrin levels in the mice of the INH–GF group were both higher than those in the mice of the INH-alone group and higher than those in the mice of the GF-alone group. Notably, the hepatic protoporphyrin level was relatively high; that is, a protoporphyria was induced. However, erythrocyte protoporphyrin levels in the INH–GF group were elevated to only twice the level in the GF-alone or the INH-alone group. The hepatic protoporphyrin level in INH-alone group was slightly elevated. Hepatic red fluorescence was seen in the INH-GF group and the INH-alone group. The severity was more marked in the

TABLE 3. Hepatic and Erythrocytic Porphyrin Contents in Mice Fed INH and GF

Treatment	Number of Mice	Liver Weight: Weight[a] (%)	Hepatic Porphyrin[a] (μg/g wet weight)			Erythrocytic Porphyrin[a] (μg/dl)		Positive Rate of Hepatic Fluorescence (%)
			UP	CP	PP	CP	PP	
0.05% INH and 0.1% GF	8	5.10 ± 0.34	0.25 ± 0.09[b]	0.30 ± 0.06[b]	9.75 ± 2.63[c]	6.28 ± 1.14	93.42 ± 14.92[c]	87.5
0.05% INH	12	4.54 ± 0.13	0.04 ± 0.01	0.04 ± 0.01	0.38 ± 0.07[c]	5.03 ± 0.90	48.78 ± 8.13	58.3
0.1% GF	12	4.85 ± 0.27	0.02 ± 0.01	0.08 ± 0.01	0.08 ± 0.01	6.49 ± 1.34	49.15 ± 5.56	0

NOTE: Adapted from Watanabe *et al.*[9]
[a]Each value below is a mean ± SEM.
[b]$p < .01$.
[c]$p < .001$.

INH–GF group than in the INH-alone group. From these results, INH seems capable of inducing a slight porphyrinopathy, and this porphyrinopathy seems to be boosted by GF. There is a tremendous variety of chemicals in our environment, and some of these chemicals may be able to potentiate porphyrinopathies. It is necessary to investigate the potentiation in combination with many chemicals.

The susceptibility to porphyrinopathies varies greatly among animals. The most sensitive animal to GF is the mouse. We tried to induce protoporphyria by GF in rats and hamsters, but the protoporphyria in these animals had a lower degree of severity. We also noticed that the severity of a porphyrinopathy varied greatly among some strains of mice.[10] The results are shown in TABLE 4. The HR/De and HRS/J strains of hairless mice were used in this study. Two concentrations of GF, that is, 0.1% and 0.5%, were administered to these mice. There was no difference between the hepatic and erythrocyte porphyrin levels in the control HR/De mice and those of the control HRS/J mice. When the feed containing 0.1% GF was given to the two strains, a slight elevation in the hepatic and fecal porphyrin levels was seen in the HR/De mice. Furthermore, the levels of hepatic, erythrocytic, and fecal porphyrins increased markedly in HRS/J mice. When the feed containing 0.5% GF was given to the mice, a great difference of porphyrin levels existed between the HR/De and HRS/J mice. From these results, it appears that HRS/J mice are more sensitive to GF than HR/De mice. The HRS/J mice may be utilized more conveniently for the investigation of porphyric photosensitivity. Further studies should be carried out concerning the influence of strain on porphyrinopathy.

With regard to hepatic protoporphyrin secretion, an extremely interesting paper was published in 1983.[11] Poh-Fitzpatrick *et al.* reported that an apparent reduction of hepatic protoporphyrin content was significant when 0.5% cholic acid was given to mice treated with 2.0% GF. They suggested that in selected circumstances, hepatic protoporphyrin secretion might be enhanced in protoporphyria disease states by bile salt supplementation.

There has been no report as to how hepatic protoporphyrin content decreases in GF-induced protoporphyria when the GF is discontinued. The reversibility of GF-induced protoporphyria is not yet well understood. One of our colleagues, however, has performed a comparative study of the hepatic and erythrocytic protoporphyrin levels in GF-induced protoporphyria after a discontinuation of the treatment with GF.[12] The results are shown in FIGURE 1. Feed containing 0.5% GF was given to the mice for 36 days, and it was confirmed that protoporphyria was induced in all the mice. Erythrocytic protoporphyrin levels were elevated 60 times and hepatic protoporphyrin levels 300 times the control levels. After discontinuation, the erythrocytic protoporphyrin level decreased to within twice the control level in 2 weeks, whereas the hepatic protoporphyrin level retained levels of over 30 times the control level 90 days after the discontinuation. From these results, it became apparent that there is a different trend in both decreasing curves of hepatic and erythrocytic protoporphyrins, although GF-induced protoporphyria by the concentration of 0.5% and the term of 1 month were normalized by the discontinuation of GF. First, we expected a similar decrease of hepatic and erythrocytic protoporphyrin. There was, however, a discrepancy between the two curves. Considering that the decline of hepatic protoporphyrin was relatively gradual, the possibility exists that hepatic protoporphyrin is not easily excreted in the feces. This possibility was indicated by the report that protoporphyrin deposited in the liver was crystallized.[13] On the other hand, it is possible that there are two mechanisms for the normalization of the erythrocytic protoporphyrin level. One is a mechanism by the destruction of erythrocytes. The declining curve of the erythrocytic protoporphyrin level will coincide with a curve of erythrocyte life span if the decline of the erythrocytic protoporphyrin level occurs by this mechanism. It has been suggested that the life span

TABLE 4. Hepatic, Erythrocytic, Cutaneous, and Fecal Porphyrin Contents in Two Strains of Hairless Mice

	Control		0.1% GF		0.5% GF	
	HR/De	HRS/J	HR/De	HRS/J	HR/De	HRS/J
Number of mice	5	3	5	8	8	8
Intake of GF (days)	0	0	86	50	38–82	19–90
Liver weight:						
body weight (%)	5.64 ± 0.14	6.11 ± 0.37	5.90 ± 0.12	6.80 ± 0.06[a]	13.29 ± 1.79	17.68 ± 1.38[a]
Hepatic porphyrin (μg/g wet weight)						
UP	0.05 ± 0.02	0.06 ± 0.02	0.05 ± 0.05	0.04 ± 0.02	1.00 ± 0.16	2.58 ± 0.51[b]
CP	0.08 ± 0.02	0.10 ± 0.02	0.16 ± 0.02	0.39 ± 0.08[a]	4.87 ± 0.68	9.66 ± 1.54[b]
PP	0.13 ± 0.04	0.09 ± 0.03	0.61 ± 0.28	5.29 ± 1.36[a]	181.99 ± 30.59	535.13 ± 109.37[b]
Erythrocytic porphyrin (μg/dl)						
CP	9.57 ± 2.75	3.65 ± 1.21	6.46 ± 3.72	4.53 ± 1.68	46.99 ± 11.59	331.02 ± 74.51[b]
PP	13.62 ± 5.88	27.12 ± 1.93	17.43 ± 6.82	42.18 ± 5.24[b]	558.15 ± 136.54	4113.74 ± 1190.73[b]
Cutaneous porphyrin (μg/100 g wet weight)						
CP	4.09 ± 1.67	3.85 ± 0.77	1.30 ± 1.30	17.98 ± 4.95[a]	4.83 ± 4.30	82.6 ± 27.68[a]
PP	24.40 ± 18.42	9.86 ± 5.61	40.31 ± 18.36	135.61 ± 72.35	110.54 ± 95.53	370.14 ± 117.49
Fecal porphyrin (μg/g et weight)						
UP	0.06 ± 0.01	0.05 ± 0.03	Trace	0.013 ± 0.01	0.04 ± 0.04	0.04 ± 0.04
CP	0.24 ± 0.04	0.34 ± 0.03	0.36 ± 0.08	0.91 ± 0.27	1.45 ± 0.22	1.17 ± 0.10
PP	0.14 ± 0.03	0.27 ± 0.06	0.98 ± 0.28	2.72 ± 0.92	9.66 ± 0.54	30.97 ± 21.91

NOTE: Each value, with the exception of values for number of mice and GF intake, is a mean ± SEM. Adapted from Yamashita *et al.*[10]

[a] $p < .05$.

[b] $p < .01$.

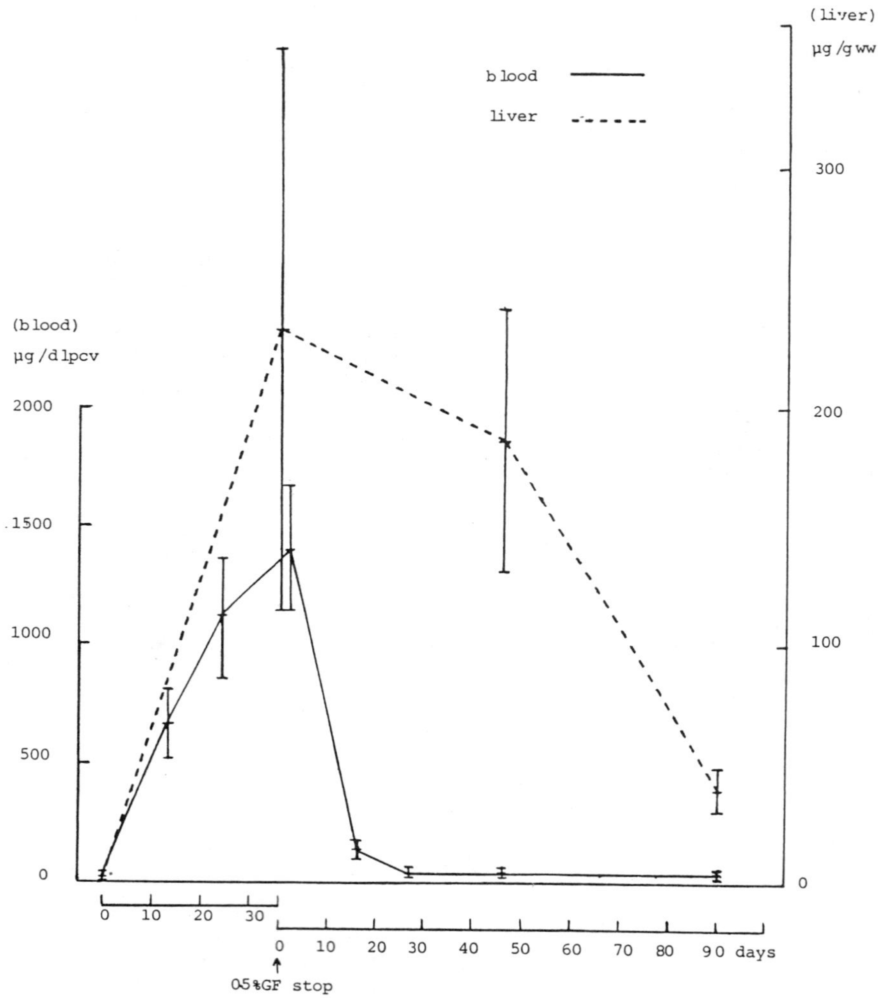

FIGURE 1. Hepatic and erythrocytic protoporphyrin levels in GF-induced protoporphyria. Adapted from Irifune.[12]

of erythrocytes in mice is 30 to 40 days. The decrease of erythrocytic protoporphyrin seems to occur too quickly to be attributed to the destruction of erythrocytes with age. The other possibility is a decrease caused by an active mechanism in which the porphyrin is excreted or transferred from the erythrocytes. More investigation is necessary to determine how the porphyrins could pass through the erythrocytic membrane.

Biochemical studies on GF-induced protoporphyria are still incomplete. We should carry out biochemical investigations on mice in which protoporphyria may be induced by GF. This experimental animal could be utilized in many studies, including studies not described in the introduction.

REFERENCES

1. WESTON-HURST, E. & G. E. PAGET. 1963. Protoporphyrin, cirrhosis and hepatomata in the livers of mice given griseofulvin. Br. J. Dermatol. **75:** 105–112.
2. DE MATTEIS, F. & C. RIMINGTON. 1963. Disturbance of porphyrin metabolism caused by griseofulvin in mice. Br. J. Dermatol. **75:** 91–104.
3. LOCHHEAD, A. C., J. H. DAGG & A. GOLDBERG. 1967. Experimental griseofulvin porphyria in adult and foetal mice. Br. J. Dermatol. **79:** 96–102.
4. NAKAO, K., O. WADA, F. TAKAKU, S. SASSA, Y. YANO & G. URATA. 1967. The origin of the increased protoporphyrin in erythrocytes of mice with experimentally induced porphyria. J. Lab. Clin. Med. **70:** 923–932.
5. POH-FITZPATRICK, M. B., A. A. LAMOLA, G. L. ZALAR, M. WEISTEIN, F. DOLEIDEN & M. FREEMAN. 1977. Comparative study of protoporphyrins in erythropoietic protoporphyria and griseofulvin-induced murine protoporphyria. J. Clin. Invest. **60:** 380–389.
6. SHIMOYAMA, T., S. NONAKA, T. HONDA, F. MURAYAMA, T. OHGAMI & H. YOSHIDA. 1984. A biochemical study of experimental porphyria. I. The influence of griseofulvin at various concentrations on porphyrin metabolism. J. Dermatol. (Tokyo) **11:** 49–55.
7. SHIMOYAMA, T., S. NONAKA, T. HONDA, T. OHGAMI, F. MURAYAMA & H. YOSHIDA. 1985. Biochemical studies of experimental porphyria. II. The influence of porphyrinogenic substances in mice treated with low concentrations of griseofulvin. J. Dermatol. (Tokyo) **12:** 416–420.
8. NONAKA, S., T. OHGAMI, N. NAGATO, M. WATANABE, H. YOSHIDA, F. MURAYAMA, H. IRIFUNE & K. YAMASHITA. 1987. Fecal porphyrin analyses in patients with porphyria cutanea tarda. J. Dermatol. (Tokyo) **14:** 291–296.
9. WATANABE, M., S. NONAKA, T. OHGAMI, N. NAGATO, H. YOSHIDA & K. YAMASHITA. 1987. Investigation of the porphyrinogenicity on isonicotine hydrazide (INH) combined a low concentration of griseofulvin. Nippon Hifuka Gakkai Zasshi **97:** 340.
10. YAMASHITA, K., F. MURAYAMA, T. OHGAMI, H. IRIFUNE, M. WATANABE & S. NONAKA. 1986. Induction of experimental porphyrin by griseofulvin: A difference of porphyrinogenecity on the strains of mice. Nippon Hifuka Gakkai Zasshi **96:** 436.
11. POH-FITZPATRICK, M. B., J. A. SKLAR, C. GOLDMAN & J. H. LEFKOWITCH. 1983. Protoporphyrin hepatopathy: Effects of cholic acid ingestion in murine griseofulvin-induced protoporphyria. J. Clin. Invest. **72:** 1449–1458.
12. IRIFUNE, H. 1987. Experimental protoporphyria induced by griseofulvin. I. Reversibility of protoporphyria induced by griseofulvin at the concentration of 0.5% for one month. Nippon Hifuka Gakkai Zasshi **97:** 801–804.
13. GSHNAIT, F., K. KONRAD, H. HONIGSMANN, H. DENK & K. WOLFF. 1975. Mouse model for protoporphyria. I. The liver and hepatic protoporphyrin crystals. J. Invest. Dermatol. **65:** 290–299.

Porphyrins as Indicators of Chemical Injury and Exposure

ELLEN K. SILBERGELD

Toxic Chemicals Program
Environmental Defense Fund
Washington, DC 20036

The central role of the heme biosynthetic pathway in cellular metabolism establishes its importance in the normal processes of physiology and in responses to toxic exposure. The interplay between these heme-related mechanisms of cellular toxicity and the background provided by genetics, nutrition, sexual differentiation, and other factors may constitute components of susceptibility and individual response. The critical differences in the toxic response to polyhalogenated aromatic hydrocarbons conferred by the *Ah* locus demonstrate this at the genetic level. The existence of the inherited diseases involving metabolic alterations in heme synthesis suggests that porphyrinopathy may be mechanistically involved in toxicity observed after exposure to a range of agents, including metals, polyhalogenated hydrocarbons, and pharmacologic enzyme inducers.

There is another major role for heme synthesis in studies of chemical intoxication. This is as a source of biological markers of exposure and cellular response. Recent developments in biochemical toxicology have focused on the identification and exploitation of molecular interactions between toxicants and biological processes as markers, or signals, of encounters between humans and their chemical environment. In chemical carcinogenesis, for instance, much research is now directed toward the study of DNA adducts, or molecules that signal the absorption, metabolism, and penetration of electrophiles to cellular DNA. The demonstration that a toxin has reached its molecular target not only greatly refines our approximations of dosimetry for purposes of epidemiologic research but also serves to improve estimates of risk. Biochemical markers are useful particularly if they are measurable in compartments that can be readily sampled from human subjects. Several large-scale studies are underway to qualify levels of DNA adducts with exposure to specific carcinogens (such as cisplatin, in cancer patients, or benzo[*a*]pyrene, in smokers and coke-oven workers). Coupled to longitudinal studies of outcome, chemical markers may provide a major bridge between the inferences of exposure assessments and the extrapolation of risk assessments.

So far, the new technology of molecular epidemiology has been applied only to the study of chemical-induced carcinogenesis. Yet the porphyrin pathway has a longer history in medicine and epidemiology as a source of biochemical markers of internal dosimetry and biological response. In the screening, diagnosis, and management of lead intoxication in adults and children, effective use has been made of measuring the activity of δ-aminolevulinate dehydratase in red cells and urine and the concentration of protoporphyrin IX in erythrocytes. These markers correlate well with external and internal exposures to lead. Moreover, this pathway may be related to possible factors governing susceptibility as well as cellular response to a range of drugs and toxins. The heme pathway is also highly accessible, in that it exists in cells that can be sampled and cultured, and its products are excreted.

Apart from the large clinical literature on lead, however, relatively little use has been made of porphyrins and related substances as chemical markers. Some attention has been given to measuring the activity of aryl hydrocarbon hydroxylase in the lymphocytes and tissues of persons exposed to cigarette smoke or to 2,3,7,8-tetrachlorodibenzo-*p*-dioxin. Porphyrins have been measured in other persons with potential exposure to porphyrinogenic toxins, but no clear use as markers has been made of these clinical data.

There are problems in the simple use of heme-related parameters as markers. Heme synthesis also responds to genetic, nutritional, and other factors in addition to possible chemical exposure. Moreover, the many feedback mechanisms in this pathway tend to produce adjustments in enzyme activity and precursor utilization that result in apparently normal levels of products measured under steady-state conditions or after chronic exposure. In order to exploit the synthesis of knowledge exemplified in this symposium, it is therefore important to determine to what extent responses of the pathway are *specific* as well as *sensitive* to specific chemical exposures, and to develop methods for evaluating the *dynamic* status of heme syntheses that more adequately reflect the level of biological function.

Critical to the interpretation of markers is an understanding of the sequence of alterations in biochemistry. This volume approaches the fundamental question: What is the relevance of this marker or indicator system to predicting adverse health effects? A major section of this volume is devoted to discussing the role of heme synthesis in the function of organ systems in order to direct consideration of the issue of biological and clinical significance. To the extent that this volume succeeds, we hope that there will be useful and innovative consideration of this pathway and its response as a source of information bridging the critical areas of exposure and outcome in clinical medicine and toxicology.

Metal- and Metalloid-Induced Porphyrinurias

Relationships to Cell Injury

BRUCE A. FOWLER,[a,b] AGNETA OSKARSSON,[c] AND JAMES S. WOODS[d]

[b]*National Institute of Environmental Health Sciences Research Triangle Park, North Carolina 27709*

[c]*National Food Administration Uppsala, Sweden*

[d]*Battelle Seattle Research Center Seattle, Washington 98105*

INTRODUCTION

The value of correlating specific porphyrin excretion patterns with other data on cell and organelle functionality rests with providing better insight into the mechanism(s) of the porphyrinuria and delineating the biological significance of this phenomenon with regard to cell injury. The heme biosynthetic pathway enzymes δ-aminolevulinic acid synthetase (ALAS), coproporphyrinogen oxygenase, and ferrochelatase are localized in the mitochondrion,[1–3] which also performs a number of other functions for the cell.[4] Furthermore, it is becoming increasingly clear that there is a close relationship between normal mitochondrial structure and biochemical functionality[5] and that any physical perturbation of the mitochondria (such as osmotic swelling) will also disrupt essential processes (such as NAD-linked substrate oxidation).[6] An example of the potential significance of this with respect to the heme biosynthetic pathway and the development of porphyrinurias concerns the dependence of ferrochelatase activity upon the reduction of ferric iron to ferrous iron by the mitochondrial electron transport chain during transport of this element across the membrane.[7] Hence any disruption of mitochondrial respiration and the attendant flow of electrons will also influence the activity of this enzyme and perhaps the other mitochondrial enzymes with subsequent development of coproporphyrinuria. Similar arguments concerning the importance of organelle structure to biochemical function have also recently been made[8,9] with regard to putative relationships between the metal-induced disruption of the structure of endoplasmic reticulum and the induction of microsomal heme oxygenase, the rate-limiting enzyme in the heme degradation pathway. The main point here is that there are abundant basic data from a number of laboratories that confirm a close set of interdependent relationships between the structure of organelles and the processes localized in organelles (such processes include mitochondrial respiration, microsomal drug metabolism, and the biosynthesis of heme). The following review will attempt to examine the effects of specific metals and metalloids on enzymes in the heme biosynthetic pathway in relation to perturbations in

[a]Present address: Toxicology Program, University of Maryland, 660 West Redwood Street, Baltimore, Maryland 21201.

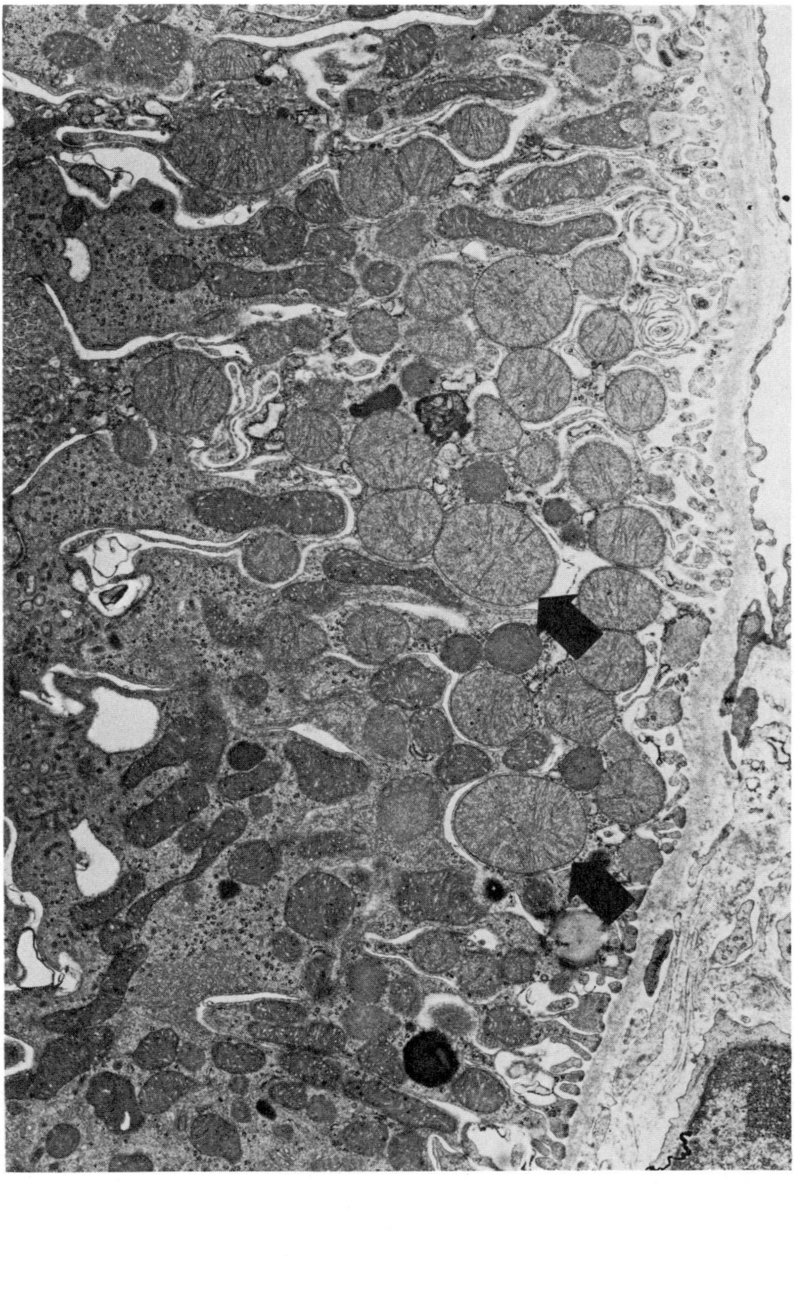

FIGURE 1. Electron micrograph of a kidney proximal tubule cell from a rat exposed to lead-contaminated drinking water for 6 months. Note the swollen mitochondria (arrows). Original magnification: ×11 334; reduced to 77% of original size.

other ultrastructural–biochemical indices of organelle dysfunction and the development of element- or mixture-specific porphyrinurias.

LEAD

Lead is the most well studied of all the toxic metals with regard to effects on the heme biosynthetic pathway, porphyrinuria patterns, and target organ toxicity. Studies from a number of laboratories[10–13] have shown that lead inhibits the mitochondrial enzymes ALAS and ferrochelatase as well as the soluble enzyme δ-aminolevulinic acid dehydratase (ALAD) and that such inhibition causes increased urinary excretion of δ-aminolevulinic acid (ALA) and coproporphyrin and a rise in erythrocytic zinc protoporphyrin (ZPP). Other studies have demonstrated that the effects of lead on these enzymes are associated with perturbation in mitochondrial structure–function relationships in target cell populations both *in vivo*[13–15] and *in vitro*.[16] FIGURE 1 shows swollen mitochondria in a renal proximal tubule cell after prolonged dietary exposure to lead. This characteristic morphological response was associated with statistical alterations in the activities of ALAS and ferrochelatase,[12,13] increased urinary excretion of ALA and coproporphyrin, and an elevation in ZPP. *In vitro* studies (FIGS. 2A & 2B) have shown that such perturbations in mitochondrial structure are also directly associated with marked perturbations in a number of other mitochondrial biochemical functions such as energy-linked structural transformations (TABLE 1), reductions in membrane potential, and changes in both heme pathway and non–heme pathway membrane marker enzyme activities (FIG. 3). The point of these observations is that there is both ultrastructural and biochemical evidence, both *in vivo* and *in vitro*, that measureable derangements in the heme biosynthetic pathway associated with a metal-specific porphyrinuria pattern are linked with deleterious changes in the organelle containing the heme pathway enzymes responsible for the porphyrinuria. This relationship between derangements in the pathway and deleterious changes in the organelle indicates that a porphyrin excretion pattern resulting from exposure to a metal may be linked to a host of other subcellular derangements.

MERCURY

Prolonged exposure to both organic and inorganic mercurials has been shown to produce porphyrinuria in human populations showing clinical evidence of mercury toxicity.[17–20] Studies conducted on the Minamata patients[17] demonstrated a marked increase in urinary urobilinogen whereas studies conducted on workers exposed to mercury vapor showed increased coproporphyrin excretion.[18–20] Experimental studies[21] involving prolonged methyl mercury exposure in rodents showing no evidence of overt neurological dysfunction demonstrated a characteristic porphyrinuria pattern: increased excretion of coproporphyrin and uroporphyrin (the excretion of the former exceeding that of the latter) but no significant change in the excretion of ALA or porphobilinogen (PBG). Analysis of heme biosynthetic pathway enzymes in various tissues showed that the kidney was the most likely source for this pattern. Correlative ultrastructural–biochemical studies[22,23] confirmed a dose-related increase renal mitochondrial volume density that was also associated with a reduction of mitochondrial respiratory function. These studies, which were in agreement with *in vitro* results,[21,24,25] demonstrated that mercurial disruption of renal heme biosynthesis and associated porphyrinuria are concomitant with other quantitative structural–functional perturba-

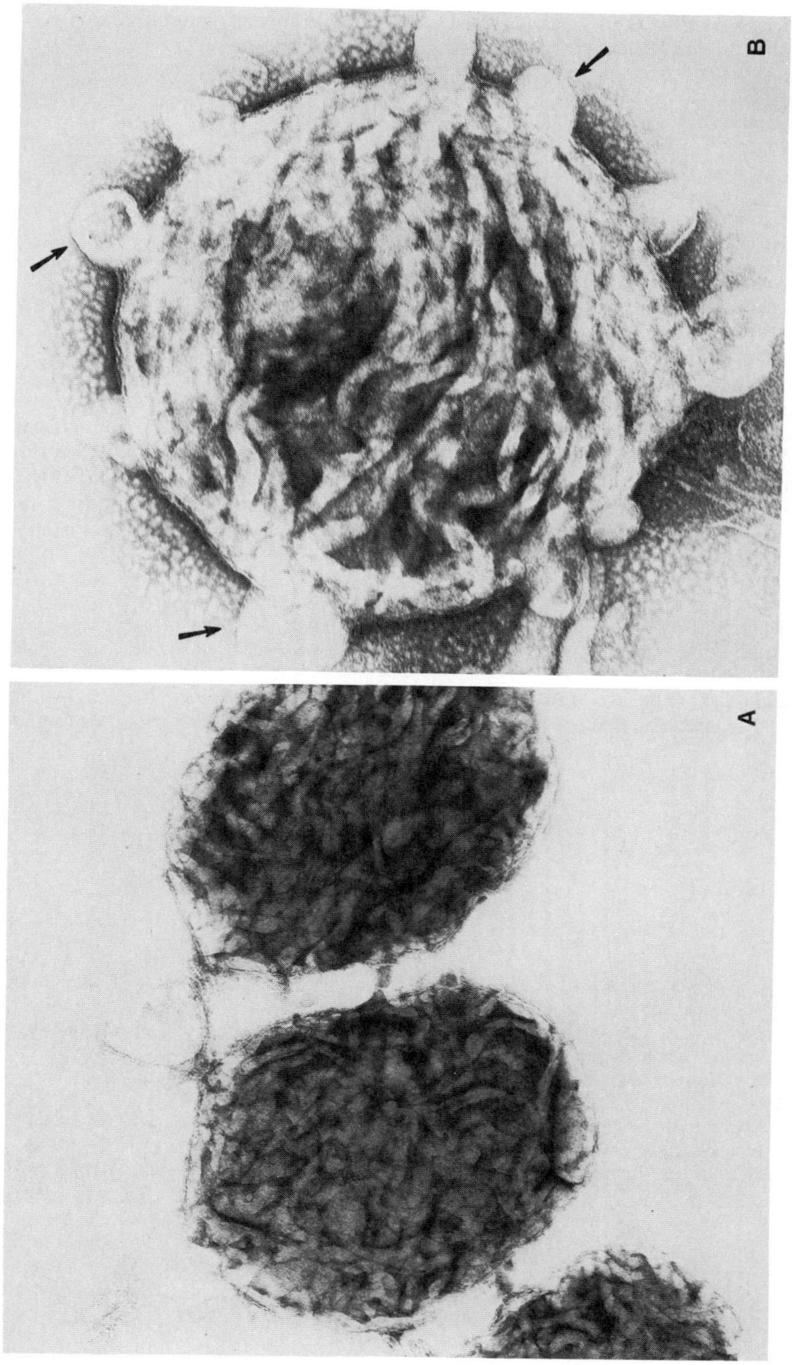

FIGURE 2. **(A)** Electron micrograph of negatively stained renal mitochondria from a control rat showing normal cristae configuration. **(B)** Electron micrograph of a mitochondrion from the same mitochondrial incubation following incubation with 10^{-4} M lead. Note the blebbing of the outer membrane and the disruption of the normal cristae pattern. Original magnification: ×136 050; reduced to 47% of original size.

TABLE 1. *In Vitro* Effects of Lead on ANS and EtBr Fluorescence in Renal Mitochondria under Nonrespiring and Respiring Conditions

In Vitro Lead Concentration (M)	Change in ANS Fluorescence ($\triangle F$/mg protein)		Change in EtBr Fluorescence ($\triangle F \times 10^{-2}$/mg protein)	
	Buffer[a]	Succinate[b]	Buffer[a]	Succinate[b]
Control	0.66	0.11	5.6	2.0
10^{-5}	0.64	0.08	5.3	1.9
10^{-4}	0.61	0.05	5.3	1.0
10^{-3}	0.54	−0.10	4.8	0.0

NOTE: ANS: 1-anilino-8-naphthalene sulfonic acid; EtBr: ethidium bromide.

[a]Net increase in fluorescence upon addition of mitochondrial suspension to the reaction cuvette.

[b]Net increase in fluorescence upon addition of succinate to the reaction cuvette.

tions of essential cellular organelles (mitochondria) in a target cell population. Furthermore, in the case of the rodent, these toxic manifestations occur before the development of overt neurological or renal dysfunction. The main point here is that the early detection of large increases in a mercury-specific pattern of urinary porphyrins may provide a novel noninvasive approach to specific assessment of cellular toxicity before the onset of overt clinical disease.

ARSENIC

Prolonged oral exposure in rats and mice to arsenate has previously been shown to produce a characteristic hepatic porphyrinuria pattern: increased excretion of uropor-

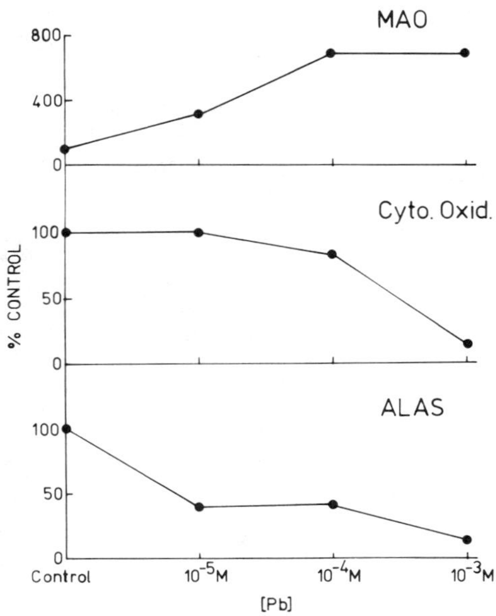

FIGURE 3. *In vitro* effects of lead on the outer membrane marker enzyme monoamine oxidase (MAO) and on the inner membrane marker enzymes cytochrome oxidase (Cyto. Oxid.) and δ-aminolevulinic acid synthetase (ALAS). These effects occurred in concert with the morphological changes in mitochondrial membrane structure described in FIGURE 2.

phyrin and coproporphyrin (the excretion of the former exceeding that of the latter) but no change in the excretion of ALA or PBG.[26] This unique porphyrin pattern was correlated with ultrastructural evidence of high-amplitude mitochondrial swelling *in situ* (FIG. 4), decreased mitochondrial NAD-linked substrate oxidation, and reduced respiratory control ratios.[27,28] Extensive studies[27] conducted on rat liver mitochondria showed increased mitochondrial membrane protein synthesis, which was correlated with a proliferation of the mitochondrial inner membrane, as measured by ultrastruc-

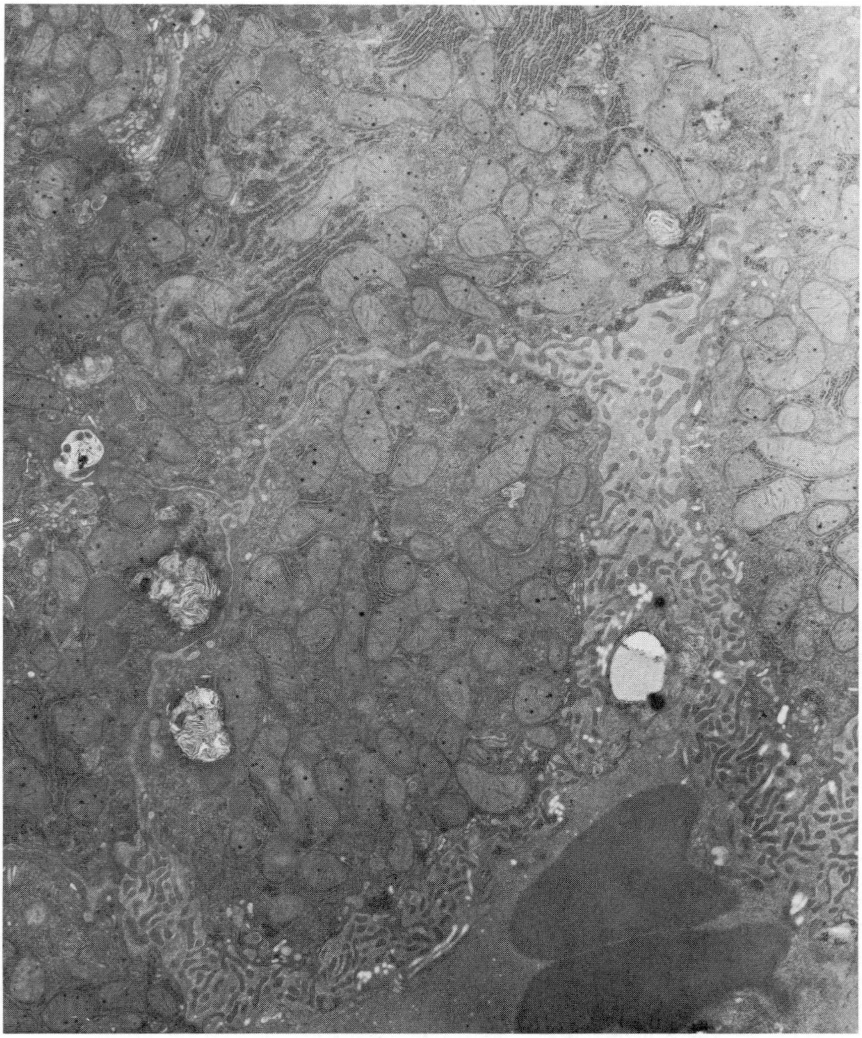

FIGURE 4. Electron micrograph of a hepatocyte from a rat that received 40 ppm arsenate in its drinking water. Note the swelling *in situ*. Original magnification: ×16 758; reduced to 58% of original size.

tural morphometry (FIG. 5); decreased ability to undergo energy-linked transformations; and increased activities for the marker enzymes of the inner and outer membranes but not for those of the matrix. These data again demonstrate the close correlation between the specific porphyrinuria pattern produced by this common environmental agent and other ultrastructural–biochemical indices of cellular injury in the target cell population that occur before the development of overt clinical organ toxicity. The main point is again that the observed agent-specific porphyrinuria represents an excellent noninvasive biological marker for an ongoing toxic process at the hepatocellular level and that it was correlated with a number of structural–functional perturbations in an essential organelle system such as the mitochondria.

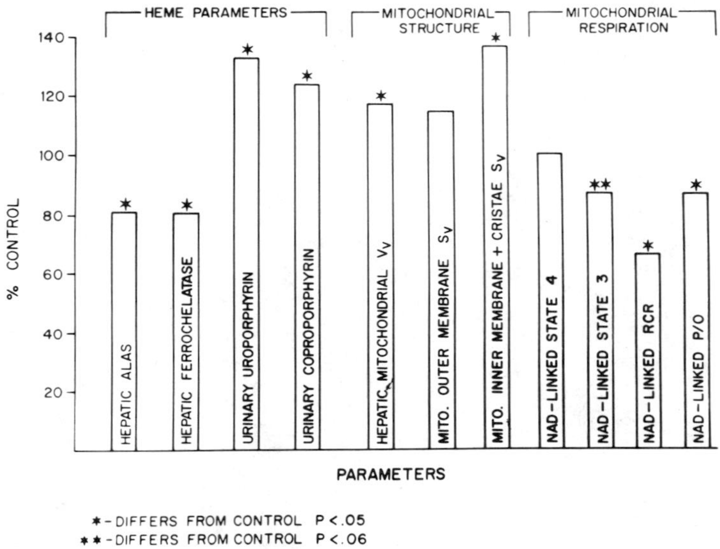

FIGURE 5. Effects of prolonged oral exposure to 40 ppm arsenate in drinking water on the activities of rat liver mitochondrial heme biosynthetic pathway enzymes, on the excretion patterns of urinary porphyrin, and on other ultrastructural morphometric–biochemical parameters. The mitochondrial preparations in which these parameters were examined also showed perturbations in a number of parameters involving the inner membrane (see references 26 and 28 for more details).

MULTIELEMENT EXPOSURES

A desirable feature of any biological indicator is the ability to detect and fingerprint an ongoing toxic process in a situation where more than one agent is present. Relatively few studies[29–34] have used porphyrins to assess toxicity from mixtures of metals. Data from these investigations are extremely promising, however, both for the more common toxic metallic agents (such as lead, cadmium, and arsenic[29–31]) and for some "new" III-V binary compounds (such as gallium arsenide), which are undergoing expanded use in the semiconductor industry.[32,33] In a lead, cadmium, and arsenic factorial design study,[29–31] lead and arsenic were found to

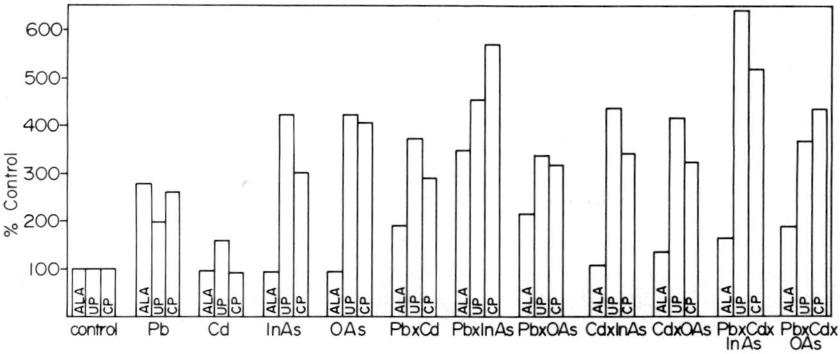

FIGURE 6. Porphyrin excretion patterns from a lead–cadmium–arsenic interaction study. The effects of different combinations of these metals on the excretion levels of δ-aminolevulinic acid (ALA), uroporphyrin (UP), and coproporphyrin (CP) are shown; these excretion levels are expressed as percentages of the control levels. Note that no two of these patterns are identical and that the relative magnitude of porphyrin excretion is greatest in those treatment groups that received all three elements in the diet. Reproduced from Fowler and Mahaffey.[30] See also reference 31 for more details.

produce the distinctive porphyrinuria patterns described above whereas cadmium did not alter porphyrin excretion, presumably because of its extensive binding to metallothionein. Combinations of these three elements, however, produced highly specific porphyrin excretion patterns; that is, each mixture produced a unique pattern (FIG. 6). The mixtures, after the overall magnitude of porphyrin excretion produced by each one was determined, could be ranked in decreasing order—the mixtures producing the highest excretion levels first, the mixtures producing the lowest excretion levels last—as follows: three metals, two metals, one metal, control.

The expanded use of binary metal compounds from groups III and V of the periodic table for the production of high-speed semiconductor circuits is of growing toxicological concern because little is known about the health risks associated with exposure to compounds such as gallium arsenide, indium arsenisde, or indium antimonide. Recent studies,[32] which used monitoring of porphyrins as an index of arsenic exposure after

TABLE 2. Summary of Porphyrinuria Patterns Produced by Various Metals and Metalloids

Heme–Porphyrin Parameter	Metal or Metalloid				
	Pb	Hg	As	Pb×Cd×As	GaAs
Urine					
ALA	↑	NC	NC	↑	↑
Uroporphyrin	±↑	↑	↑↑	↑[a]	↑
Coproporphyrin	↑↑	↑↑	↑	↑[a]	↑
Blood					
ALAD Activity	↓	NC	NC	NM	↓
ZPP	↑	NM	NM	NM	NM

NOTE: NC: no change with respect to the control; NM: not measured.
[a]See FIGURE 6.

acute intratracheal administration of gallium arsenide, reported the same excretion pattern described above for arsenic, thus providing a means for monitoring the biological activity of the arsenic moiety of this compound. More recent studies,[33-35] which monitored changes in the excretion of ALA and measured ALAD activity in blood, showed a concomitant increase in ALA primarily associated with inhibition of red cell ALAD activity, which was shown to result from the action of gallium on the enzyme. Overall, the data shown indicate a urinary excretion pattern in which the excretion of ALA and uroporphyrin exceeds that of coproporphyrin. This pattern was also associated with ultrastructural manifestations of renal proximal tubule cell injury.

A general summary of some commonly observed metal- or mixture-specific disturbances in heme metabolism, as made evident by different porphyrinuria patterns, is given in TABLE 2.

SUMMARY AND PROSPECTS FOR THE FUTURE

It should be clear from the above discussion that various metals and metalloids alone or in combination produce specific "fingerprint" perturbations in the heme biosynthetic pathway; moreover, correlative ultrastructural pathobiology studies have demonstrated that these disturbances are reflective of a number of other subcellular derangements in essential metabolic systems. These data also suggest that the observed porphyrinurias are hence more than simply a reflection of chemical exposure but rather represent ongoing deleterious effects at the biochemical and organelle levels of biological organization within target cell populations. There is at present no reported case where disturbances of heme metabolism and attendant porphyrinurias were not associated with changes in other subcellular processes when such organelle or biochemical processes were evaluated. The possibility of detecting such "silent damage" below an overt clinical threshold or of observing tissue damage at the light microscope level, however, presents those responsible for making risk assessment and risk management decisions with the dilemma of determining at what point one can conclude such changes represent adverse health effects. Clearly, the improved capability to detect ongoing toxic processes via biological indicators such as porphyrins generates the necessity for a value judgement as to whether mortality, overt clinical disease, histopathological damage in a given target organ, or production of deleterious metabolic disturbances at subcellular levels will be the threshold for regulatory action. More research is needed to better delineate the relationships that must exist between subcellular dysfunctions such as those described above and irreversible damage at the cell, tissue, or organ levels in order to clarify these questions.

In terms of the future, it is also clear toxicologists will be under increasing pressure to make assessments concerning noncarcinogenic effects of chemicals and mixtures of both inorganic and organic compounds. Some of these mixtures will involve well-known agents such as lead, mercury, cadmium, and arsenic whereas others will involve agents such as gallium arsenide for which there is relatively little toxicological information but whose use in industry is rapidly expanding. Based on the studies conducted to date, it appears that disturbances in the heme–porphyrin pathway represent an extremely sensitive and specific class of biological indicator for achieving both goals. In combination with other correlative ultrastructural–biochemical studies, these metabolic products meet the three main criteria for biological indicators of toxicity: sensitivity, specificity, and relationship to toxic response.[36] The importance of porphyrins as a major class of biological indicators for prospective evaluation of ongoing toxicity below the clinical level cannot be overstated because these molecules provide an important tool for assessing chemical effects long *before* the onset of clinical disease.

REFERENCES

1. WOODS, J. S. & V. V. MURTHY. 1975. δ-Aminolevulinic acid synthetase from fetal rat liver: Studies on the partially purified enzyme. Mol. Pharmacol. **11:** 70–78.
2. ELDER, G. E. & J. O. EVANS. 1978. A radiochemical method for the measurement of coproporphyrinogen oxidase and the utilization of substrates other than coproporphyrinogen III by the enzyme from rat liver. Biochem. J. **169:** 205–214.
3. JONES, M. S. & O. T. G. JONES. 1968. Evidence for the localization of ferrochelatase on the inner membrane of rat liver mitochondria. Biochem. Biophys. Res. Commun. **31:** 977–982.
4. FOWLER, B. A., G. W. LUCIER & A. W. HAYES. 1982. Organelles as tools in toxicology. *In* Methods in Toxicology. A. W. Hayes, Ed.: 635–658. Raven Press. New York, NY.
5. HACKENBROCK, C. R. 1972. States of activity and structure in mitochondrial membranes. Ann. N. Y. Acad. Sci. **195:** 492–505.
6. MATLIB, M. A. & P. A. SRERE. 1976. Oxidative properties of swollen rat liver mitochondria. Arch. Biochem. Biophys. **174:** 705–712.
7. FLATMARK, T. & I. ROMSLO. 1975. Energy-dependent accumulation of iron by isolated rat liver mitochondria. J. Biol. Chem. **250:** 6433–6438.
8. FOWLER, B. A., R. KARDISH & J. S. WOODS. 1983. Alteration of hepatic microsomal structure and function by acute indium administration: Ultrastructural morphometric and biochemical studies. Lab. Invest. **48:** 471–478.
9. WOODS, J. S. & B. A. FOWLER. 1986. Alterations in hepatocellular structure and function by thallium chloride: Ultrastructural morphometric and biochemical studies. Toxicol. Appl. Pharmacol. **83:** 218–229.
10. HAEGER-ARONSEN, B. 1960. Studies on the excretion of δ-aminolevulinic acid and other heme precursors in lead workers and lead-intoxicated rabbits. Scand. J. Clin. Lab. Invest. **12** (Suppl. 47): 1–128.
11. MOORE, M. R., P. A. MEREDITH & A. GOLDBERG. 1980. Lead and heme biosynthesis. *In* Lead Toxicity. R. L. Singhal & J. A. Thomas, Eds.: 79–118. Urban & Schwarzenberg. Baltimore, MD.
12. FOWLER, B. A., C. A. KIMMEL, J. S. WOODS, E. E. MCCONNELL & L. D. GRANT. 1980. Chronic low-level lead toxicity in the rat. III. An integrated toxicological assessment with special reference to the kidney. Toxicol. Appl. Pharmacol. **56:** 59–77.
13. OSKARSSON, A. & B. A. FOWLER. 1986. Effects of lead on the heme biosynthetic pathway in rat kidney. Exp. Mol. Pathol. **43:** 397–408.
14. BULL, R. J. 1980. Lead and energy metabolism. *In* Lead Toxicity. R. L. Singhal & J. A. Thomas, Eds.: 119–168. Urban & Schwarzenberg. Baltimore, MD.
15. PIOMELLI, S. & J. GRAZIANO. 1980. Laboratory diagnosis lead poisoning. Pediatr. Clin. North Am. **27:** 843–853.
16. FOWLER, B. A., K. S. SQUIBB & A. OSKARSSON. 1981. Mitochondrial membrane potential and energy-linked membrane transformation: Inhibition by lead binding *in vitro*. J. Cell Biol. **91:** 287A.
17. TOKOUMI, H. 1968. Minamata disease in human adults. *In* Minamata Disease: 37–72. Kumamoto University Press. Kumamoto.
18. GOLDWATER, L. J. & M. M. JOSELOW. 1967. Absorption and excretion of mercury in man. XIII. Effects of mercury exposure on urinary excretion of coproporphyrin and δ-aminolevulinic acid. Arch. Environ. Health **15:** 327–331.
19. SUZUKI, T. 1962. Exposure to inorganic mercury and urinary excretion of coproporphyrin. Jpn. J. Exp. Med. **32:** 45–53.
20. TSUCHIYA, K. 1964. Coproporphyrins in lead and mercury workers. Ind. Health **2:** 162–171.
21. WOODS, J. S. & B. A. FOWLER. 1977. Renal porphyrinuria during chronic methyl mercury exposure. J. Lab. Clin. Med. **90:** 266–272.

22. Fowler, B. A., H. W. Brown, G. W. Lucier & M. R. Krigman. 1975. The effects of chronic oral methyl mercury exposure on the lysosome system of rat kidney: Morphometric and biochemical studies. Lab. Invest. **32;** 313–322.
23. Fowler, B. A. & J. S. Woods. 1977. Ultrastructural and biochemical changes in renal mitochondria during chronic oral methyl mercury: The relationship to renal function. Exp. Mol. Pathol. **27:** 403–412.
24. Israels, L. G. 1972. Inhibition of heme synthesis in the kidney by organomercurials. Biochem. Pharmacol. **21:** 434–435.
25. Woods, J. S., D. L. Eaton & C. B. Lukens. 1984. Studies on porphyrin metabolism in the kidney: Effects of glutathione on renal uroporphyrinogen decarboxylase. Mol. Pharmacol. **26:** 336–341.
26. Woods, J. S. & B. A. Fowler. 1978. Altered regulation of mammalian hepatic heme biosynthesis and urinary porphyrin excretion during prolonged exposure to sodium arsenate. Toxicol. Appl. Pharmacol. **43:** 361–371.
27. Fowler, B. A. & J. S. Woods. 1979. The effects of prolonged oral arsenate exposure on liver mitochondria of mice: Morphometric and biochemical studies. Toxicol. Appl. Pharmacol. **50:** 177–187.
28. Fowler, B. A., J. S. Woods & C. M. Schiller. 1979. Studies of hepatic mitochondrial structure and function: Morphometric and biochemical evaluation of *in vivo* perturbation by arsenate. Lab. Invest. **41:** 313–320.
29. Mahaffey, K. R. & B. A. Fowler. 1977. Effects of concurrent administration of dietary lead, cadmium and arsenic in the rat. Environ. Health Perspect. **19:** 165–171.
30. Fowler, B. A. & K. R. Mahaffey. 1978. Interactions between lead, cadmium and arsenic in relation to porphyrin excretion patterns. Environ. Health Perspect. **25:** 87–90.
31. Mahaffey, K. R., S. G. Capar, B. C. Gladen & B. A. Fowler. 1981. Concurrent exposure to lead, cadmium and arsenic: Effects on toxicity and tissue metal concentrations in the rat. J. Lab. Clin. Med. **98:** 463–481.
32. Webb, D. R., I. G. Sipes & D. E. Carter. 1984. *In vitro* solubility and *in vivo* toxicity of gallium arsenide. Toxicol. Appl. Pharmacol. **76:** 96–104.
33. Goering, P. L. & B. A. Fowler. 1986. Effect of intratracheal gallium arsenide (GaAs) administration on δ-aminolevulinic acid dehydratase (ALAD) in rats: Relationship to urinary excretion of δ-aminolevulinic acid (ALA). Pharmacologist **28:** 229.
34. Fowler, B. A. & P. L. Goering. 1986. Mechanism of gallium arsenide (GaAs) inhibition of δ-aminolevulinic acid dehydratase (ALAD). Pharmacologist **28:** 229.
35. Goering, P. L. & B. A. Fowler. 1987. Mechanism of urinary excretion of δ-aminolevulinic acid after intratracheal instillation of gallium arsenide. Ann. N. Y. Acad. Sci. This volume.
36. Fowler, B. A., J. Abel, C.-G. Elinder, H.-J. Hapke, J. H. R. Kagi, J. Kleiminger, Y. Kojima, A. J. M. Schoot Uiterkamp, E. K. Silbergeld, S. Silver, K. H. Summer & R. J. P. Williams. 1984. Structure, mechanism and toxicity. *In* Changing Metal Cycles and Human Health. J. Nriagu, Ed.: 391–404. Dahlem Workshop Reports. Springer-Verlag. Heidelberg.

Turkish Epidemic Hexachlorobenzene Porphyria

A 30-Year Study

HENRY PETERS,[a] DEREK CRIPPS,[a] AYHAN GÖCMEN,[b]
GEORGE BRYAN,[a] ERDOGAN ERTÜRK,[a]
AND CARL MORRIS[c]

[a]*Department of Neurology*
University of Wisconsin Center of Health Sciences
Madison, Wisconsin

[b]*Department of Pediatrics*
Hacettepe University Medical School
Ankara, Turkey

[c]*U.S. Environmental Protection Agency*
Washington, DC

INTRODUCTION

Hexachlorobenzene (HCB)-treated seed grain diverted for bread production between 1955 and 1959 during periods of relative famine in southeastern Turkey produced an epidemic of porphyria cutanea tarda (PCT) and mixed porphyria, which, according to Cam,[1-4] affected about 3000 to 4000 persons and had a mortality rate of 10%.

Cam first described the porphyria as resembling the congenital erythropoietic type, but subsequently realized that it was acquired from ingesting HCB contained in fungicides imported from western Europe in 1954 to combat *Tilletia tritici,* a wheat fungus. The fungicides were named Cholorable and Surmesan, and contained 10% HCB. Two kilograms of fungicide was mixed with 1000 kg of seed, and because the treated seed wheat had arrived past planting time, it was diverted to replace edible stores of wheat.

Cam suspected that porphyria turcica (PT), the Turkish porphyria, was the result of HCB intoxication, and this was demonstrated experimentally by Kantemir *et al.,*[5] Ockner and Schmid,[6] and Wray *et al.,*[7] who produced porphyria in experimental animals. Dogramaci[8] reported that 5% of the patients were less than 5 years of age, that 83% of the patients were between the ages of 6 and 16, and that the incidence of adults over the age of 20 in the acute phase was less than 6%.

Even though use of the fungicide was terminated in 1959, Peters *et al.*[9] reported continued activity of the porphyric process in patients seen between 1961 and 1963, and noted clinical improvement during and after therapy with intravenous and oral edetic acid (EDTA) in seven unselected cases of HCB-induced porphyria. In addition, children born to mothers with HCB-induced porphyria and those born to mothers known to have ingested HCB died within a year of a condition called pembe yara (pink sore), and this eliminated children between the ages of 2 and 5 years in many villages between 1955 and 1960. Cam[1-4] believed that pembe yara was another manifestation of HCB intoxication because these children did not demonstrate porphyrinuria—rather, they all died of weakness, convulsions, and localized cutaneous annular erythema. Peters[10] demonstrated, by gas chromatography, a marked increase in HCB in the maternal milk of four affected mothers of children with pembe yara.

The interval between HCB ingestion and the development of symptoms appeared to be approximately 6 months. Initial symptoms included weakness and the inability to handle eating utensils, rise from a squat, or climb stairs. The patients had loss of appetite, photosensitivity with bullous and milia formations on the sun-exposed areas of the skin, hyperpigmentation, and porphyrinuria with red or brown urine.

Hyperpigmentation was most pronounced in exposed areas of skin, but was also present in unexposed areas, whereas hypertricosis—principally of the forehead, the cheeks, the arms and legs, and occasionally the entire body—resulted in the children being labeled as "monkey children." Fragility of the skin with bullae up to 5 cm in size (kara yara or black sore) caused mutilating scars on healing. There was a positive Nikolsky sign in which the skin would layer and easily separate with pressure or abrasion. Arthritis was prominent in younger children, with swelling and spindling of the fingers, but little pain.

In 1964, Dogramaci[8] described 547 patients who showed bullae in 99.6%, porphyrinuria in 95.4%, weakness in 94.7%, hypertrichosis in 82.3%, hyperpigmentation in 64.2%, and arthritic changes in 54.8% of their number. The hepatomegaly was the rule early in the disease, but decreased in time, and the incidence of goiter seemed greater than expected.

In the same year, Bleiberg *et al.*[11] described a Turkish-style porphyria in New Jersey factory workers. These workers had been exposed to 2,4-dichlorophenoxyacetic (2,4-D) and 2,4,5-trichlorophenoxyacetic (2,4,5-T), both of which structurally resemble HCB. Their symptoms disappeared on cessation of exposure.

Between 1977 and 1986, we studied a total of 225 patients in eastern Turkey who had a history of PCT between 1958 and 1963. This included four out of seven patients previously treated in the acute phase with intravenous and oral EDTA. In addition, we have acquired historical and, in many instances, personal observations on over 1000 family members, including progeny, spouses, and siblings. We studied patients from the endemic area as well as patients who had emigrated from the area after the acute epidemic. Fetal wastage, child death rate, and health status of children born to porphyric mothers and fathers were evaluated. Urine and stool specimens for porphyrin determinations including porphyrin precursors were obtained from patients, some family members, and controls. Maternal milk samples and occasional fat biopsies were analyzed by gas chromatography for HCB concentration. Urine and fecal porphyrin determinations were carried out by the method of Rimington[12] and as described by Cripps and Peters.[13] Porphyrin precursors, including porphobilinogen (PBB) and δ-aminolevulinic acid (ALA), were determined by the method of Mauzerall and Granick.[14] Blood was obtained for determinations of iron, lead, copper, zinc, uroporphyrinogen synthetase, and creatine phosphokinase levels, and for thyroid function studies. Several skin biopsy specimens were obtained from the dorsum of a finger, fixed with 10% formaldehyde solution, and stained with hematoxylin–eosin and periodic acid–Schiff stains. Clinically, symptoms and historical data were rated independently, and a consensus was reached.

CLINICAL OBSERVATIONS

The majority of patients were located in 12 villages surrounding the city of Diyarbakir. Out of 225 patients, there were 144 males and 81 females. The average age was 32.8 years. The average age at onset was 10.3 years, and the average duration of most symptoms was 2.3 years. Dermatological signs and symptoms included the following: hyperpigmentation (in 67.5% of the patients); severe scarring (84.0%); hirsutism (46.2%); pinching of the facies (40.0%); and fragile skin (35.5%). Striking was the shortened stature, as compared to uninvolved siblings, in 42.6% of those who developed PT before puberty. Arthritic changes were seen in 68.4%, and small hands were seen in 66.2% of the patients. Neurologic symptoms and signs in the form of weakness were present in 63.0%, and paresthesias were acknowledged in 53.3% of the patients; one could detect cogwheeling of the arms in 26.0%, sensory shading in 51.5%, and mechanical myotonia on forearm percussion in 29.7% of the patients. Striking was the incidence of enlarged thyroid glands in 36.0% of the patients, this involving 23.5% of the males and 58.7% of the females. An enlarged liver was noted in 4.0% of the patients. There were 17 patients who described recent occasional red urine, though objective confirmation was not available.

During the period of study, two patients died from liver failure, one being a 27-year-old male with cirrhosis, and the other a 54-year-old female who succumbed after treatment with rifamycin and isoniazide for pulmonary tuberculosis.

Out of 188 pregnancies in 42 porphyric mothers, there were 15 fetal deaths, and 31 children died in the first several years after delivery.

The cogwheeling of elbows and wrists was associated with some reduction of associated movements at the wrist, but not by bradykinesia, loss of static reflexes, or reduction in vocal power as seen in parkinsonism. The myotonia could not be explained on the basis of the slight reduction of thyroid function demonstrated in some of the patients.

LABORATORY FINDINGS

Urine and stool porphyrins were determined in most patients. Nine patients had clearly abnormal increases of porphyrins, and eight patients had a significant increase of just one of the abnormal porphyrins. The remainder of the patients were within the limits of the controls. The mean level of ALA for the 112 patients studied was 4.76 mg/liter, and that for the controls was 4.35 mg/liter. On the other hand, almost 50% of the PT patients showed elevated ALA levels. Burns *et al.*[15] reported on the induction of Turkish porphyria in vegetable sprayers who got an inordinate amount of HCB as a contaminant. Although the vegetable sprayers developed changes in porphyrin metabolism analogous to those reported by Bleiberg[11] in factory workers making 2,4-D and 2,4,5-T, they did not develop clinical symptoms of PT. On the other hand, the vegetable sprayers did show a correlation between the excretion of ALA and the blood levels of HCB. Burns *et al.*[15] suspected this finding was artifactual, and hence rejected it, but our studies suggest that their observations were valid. Testing for ALA levels in PT patients may actually allow us to estimate the amount of HCB circulating through and the total amount of HCB within the body. We will, therefore, continue to monitor these precursors. PBG levels were not elevated.

HCB levels were determined in 56 samples of human milk obtained from porphyric mothers; the average value for the HCB level was .51 ppm (the standard deviation was 0.75 ppm, the highest value being 2.8 ppm). Although values decreased during lactation, they again reached high levels in successive pregnancies. Control maternal milk samples from unaffected family members averaged .07 ppm. This was in sharp contrast to low or absent HCB levels in human milk obtained from other parts of Turkey where the epidemic had not occurred. A biopsy specimen of fat tissue from a male porphyric patient who died of cirrhosis contained 0.7 ppm. We should keep in mind that the allowable concentration of HCB in cow's milk set by the World Health Organization is 0.020 ppm.[16] X-rays of 10 patients with unusual arthritis showed osteoporosis of carpal, metacarpal, and phalangeal bones, particularly at the ends, with narrowing of joint spaces.

DISCUSSION

Unlike genetic hepatic porphyria, which becomes clinically active in humans during maturity or the early teens, HCB-induced hepatic porphyria becomes clinically active very early in life, with porphyrinuria being present in children only several years old. In addition, the ingestion of HCB by mothers resulted in transmission of the chemical in large amounts in the maternal milk, killing the youngster suckled within 2 years. Although clinical porphyrinuria was seldom reported in these human offspring, animals such as the rat that had been fed milk from HCB-dosed dams by Cripps *et al.*[17] not only had a fatal outcome, but developed porphyrinuria (including the precursors ALA and PBG) and died in convulsions. In addition, we will measure the level of kynurinic acid in PT to see whether the porphyria due to HCB influences tryptophan metabolism, and hence has a property in common with acute intermittent porphyria (AIP).[18]

Although nerve conduction and other studies have not been possible on most PT patients in the field, Sufit *et al.*[19] have done nerve conduction studies that indicate rats fed HCB develop sensory motor axonal neuropathy. Berkeley and MacGee[20] describe a severe neurosensory neuropathy in a farmer who used 2,4-D as a spray, and Eberstein and Goodgold[21] describe three patients who developed painful numbness, aching, and weakness in the legs after exposure through the skin to and possible inhalation of 2,4-D. Electromyograms showed there was minimal deinnervation. Had the exposures in these cases been more chronic, it is possible that more severe deinnervation might have resulted. Albers *et al.*[22] have demonstrated a high incidence of axonal neuropathy in AIP, a result that parallels the one reported by Sufit *et al.*[19]

The second neurological finding—of myotonia on forearm percussion—is mirrored in the studies done by Eberstein and Goodgold[21] in which 2,4-D, a relative of HCB, was able to induce myotonia in muscle preparations. Hildebrand,[23] Bucher,[24] and Sollmann[25] all described the production of myotonia by 2,4-D in experimental animals such as mice and dogs. In a human patient described by Park *et al.*[26] and Prescott *et al.*,[27] intoxication with 2,4-D and mecoprop resulted in clincial myotonia that lasted until detoxification by alkaline diuresis 3 months later. In our PT patients, the relationship between myotonia and hypothyroidism was difficult to evaluate because some of the patients had a minor reduction of thyroid activity, and many had enlarged thyroid glands. At the same time, the patients were not myxedematous—a condition that is usually associated with the presence of myotonia. The myotonia was not clinically debilitating in PT patients, and after-contraction of grip was not demonstrable.

The myotonic involvement was not generalized as in Thompson's disease, nor were there associated symptoms of gonadal atrophy, muscle wasting of dystrophic type, cataracts of the eyes, or other phenomena such as those seen in myotonia dystrophica or Steinert's disease. It remains to be seen to what extent this symptom will explain itself with the passage of time. As also demonstrated by Sufit *et al.*,[19] marked production of positive waves analogous to a myotonic discharge (pseudomyotonia) was noted on electromyography of rats fed 800 ppm of HCB for a period of 20 weeks.

We noted the presence of cogwheeling of the elbows in 20% of the PT patients. This cogwheeling was not accompanied by static or resting tremor, bradykinesia, or loss of associated movements. The effects that might be seen with the passage of time (say, 30 years) in patients with such a condition, assuming that the HCB would continue to exert a deleterious effect on the central nervous system because of its fat solubility, might include more overt extrapyramidal symptoms such as parkinsonism or dystonia.

We noted that the incidence of seizure activity in 1 out of 225 patients would indicate that seizure frequency activity in the PT patients is not elevated above that of the general population. According to Göcmen,[28] however, infants who developed pembe yara died within 1–2 years after birth. Pembe yara was caused by HCB dosage both transplacentally and to an even greater extent by means of mother's milk[10] that contained large amounts of HCB. These pembe yara patients died of cardiorespiratory failure, which was often with associated with seizure activity. In this regard, the pembe yara sequence described by Cripps *et al.*[17] in rats resulted in the pups dying with convulsive activity. Had pembe yara patients survived, would they have shown later development of epileptic activity? Will the levels of HCB in maternal milk now have a long-term epileptic effect on the current progeny of PT patients?

In terms of the general mentation in the PT patients, the neuropsychiatric finding of schizophrenia in 1 out of 225 patients is to be noted together with a certain elevation in the hypochondriasis score. Such a parallel would seem natural in view of the marked psychological stress that many of the PT patients experienced because of their disfiguring neurocutaneous and arthritic symptoms. Despite the disfigurement, the great majority of the patients married and were absorbed into community activities. Although some of the jobs acquired by these patients may well have been below what they ordinarily might have achieved, some PT patients went on to become, for example, pharmacists, bank tellers, and teachers (such progress, however, was not made by the patients most severely involved). The level of intelligence, we felt, was average to above average in some instances.

We would also comment on the psychoneurotic preocccupation with disease that seems to characterize some of the patients studied. On the other hand, we were also impressed with the community of feeling that seemed to characterize those patients suffering from PT and by the support that they found so often in their families and in the villages in which they lived. One patient, for example, who had horrible disfiguring lesions on his face that required plastic surgery, was deliberately put in charge of a tea house so that he would have to stay in the social mainstream and interact with his friends every day.

Finally, we wish to review the clinical pattern of HCB-induced porphyria as first encountered in 1961, so that we can detail the improvement in seven patients with PT that followed a chelation regimen with EDTA during the acute phase. It was our desire to try this mode of treatment, which had seemed successful in the treatment of AIP and chronic intermittent porphyria,[10] that gave rise to our initial study of this epidemic in 1961. Between 1961 and 1963, over 30 patients were studied, but only 7 patients were available for a chelation regimen with EDTA. This regimen was administered to four of these seven patients for 1 week via the intravenous route at a dosage of 1.5 g/day. This was followed by 1.5 g orally for a period of 10 months. The other three patients were given only oral EDTA. The patients were not restricted in food intake or exposure to sunlight, and yet EDTA administration was followed by complete reversal of symptoms and signs that had, until the reversal, persisted for 14 months. One patient began to relapse 15 months after oral therapy was stopped, but after a brief resumption of oral EDTA, symptoms again disappeared. Improvement in all of the treated PT patients included rapid loss of photosensitivity, clearing of cutaneous symptoms with healing of bullae, cessation of the Nikolsky sign, loss of hyperpigmentation, and reversal of the hypertrichosis. Of this group, four were available for follow-up during our studies from 1977 to the present. Two additional patients from the early period who had not received EDTA were also reassessed as part of the follow-up on the 225 patients studied. Although all the treated patients appeared better than many untreated patients, the variability of the clinical pattern in PT and the inability to estimate the body burden of HCB in all but lactating mothers made firm conclusions impossible. Moreover, in AIP and also in PT, we do not see a clear correlation between symptoms and signs and altered porphyrin excretion, with the PT patients showing the most overt porphyrinuria frequently being the least systematic. Some treatment attempts with oral EDTA in PT in the last several years have suggested that the patients felt better. Further studies of this type of therapy as well as further evaluation of neurological, oncological, teratological, dermatological, and endocrine changes are contemplated.

It should be noted that Klüver[29] and Solomon and Figge[30] demonstrated that there is deposition of porphyrins within the white matter of both the central nervous system and the peripheral nerves, the deposition paralleling the formation of myelin in the nervous system. We reported in the past on the favorable influence of chelating therapies in the treatment of AIP[9,31,32]as well as in the treatment of porphyria cutanea tarda,[33,34] PT,[10] and of course lead poisoning. From the success of these therapies evolved the theory that disturbance of porphyrin metabolism in AIP is due to withdrawal of porphyrins for purposes of chelation and a block in porphyrin synthesis meaning that there is no longer adequate replenishment of porphyrins within the nervous system. In this respect, it has been shown that AIP patients, when they are acutely symptomatic with involvement of the central nervous system and peripheral nerves, may show an excess urinary excretion of zinc, which will normalize during recovery or following EDTA therapy. The possibility that an excess of zinc or some other heavy metal could block the pathways of porphyrin synthesis seemed reasonable, and could explain the unblocking effect of chelation often resulting in immediate clinical response.[31,32] Indeed, in AIP, an enzymatic block has been identified in the form of a reduction in uroporphyrinogen synthetase, although, in our experience, this is not a completely reliable test to rule out AIP if the levels of uroporphyrinogen synthetase are within the normal range. In a small group of patients with PT that we treated with EDTA between 1961 and 1963,[9] we felt that we could demonstrate a reversal of clinical symptomatology. Excretion of heavy metals at that time was studied on a very limited basis, and failed to show either a zinc diuresis before treatment or a convincing cupruria prior to EDTA administration. During EDTA therapy, marked diuresis of zinc and copper was noted. Twenty to 25 years later, we were still unable to demonstrate clear evidence of increased zinc, copper, or lead excretion despite the ALA levels having become elevated above the norm in 50% of the patients studied.[22] According to Lambrecht *et al.*,[33] the feeding of EDTA plus HCB to rats reduces coproporphyrin, uroporphyrin, zinc, and copper levels in liver slices and is accompanied by a reduction of porphyrin fluorescence. For further information on PT, see the papers by Peters and co-workers.[34–36]

CONCLUSIONS

Twenty-five to 30 years after the patients being acutely intoxicated with HCB, we continue to see, in some patients, evidence of disturbed porphyrin metabolism in urine and feces, as well as elevated ALA levels in over 50% of the patients in the absence of heavy metal intoxication. In addition, there is significant evidence of sensory motor neuropathy and myotonia that is replicated in animals fed HCB; this evidence parallels findings in animals and humans intoxicated with 2,4-D. There is evidence of cogwheeling, suggesting early extrapyramidal involvement, the significance of which remains to be seen. What long-term effects of HCB on the newest generation will result from both transplacental and maternal milk transmission of HCB must be studied because levels of HCB transported in maternal milk are not sufficiently high at this stage to induce pembe yara. Could seizures develop in the progeny from low maternal milk levels of HCB at a later date?

Cabral *et al.*[37,38] reported carcinogenesis in hamsters, mice, and rats fed a HCB diet. Tissues susceptible to carcinogenic activity include the thyroid gland, liver, lymphoreticular system, and kidney.[37–40] In addition, studies by Smith and Cabral[39] on rats fed HCB demonstrated an increased incidence of urinary tumors. HCB toxicity and carcinogenicity in mice, rats, and hamsters was dose-, species-, and sex-related.[40] Although we have not encountered definite malignancies in the PT patients, two porphyric patients, who had undergone thyroidectomy (and showed no malignant changes), experienced a recurrence of the thyroid enlargement within 1 year. These patients are being studied further. EDTA administration has been therapeutic in PT during the acute phase and has protected rats that were fed HCB.

REFERENCES

1. CAM, S. 1960. Une nouvelle dermatose epidemique des enfants. Ann. Dermatol. Syphiligr. **87:** 393–397.
2. CAM, S. 1958. Cases of skin porphyria related to hexachlorobenzene intoxication. Saglik Derg. **32:** 215–216.
3. CAM, S. 1959. Cutaneous porphyria related to intoxication. Dirim **34:** 11–15.
4. CAM, S. & G. NIGOGOSYAN. 1963. Acquired toxic porphyria cutanea tarda due to hexachlorobenzene: Report of 348 cases caused by this fungicide. J. Am. Med. Assoc. **183:** 88–91.
5. KANTEMIR, I., S. GINER & O. KAYAALP. 1960. Investigations with hexachlorobenzene and organic mercury compounds. Turk Hij. Tecr. Biyol. Derg. **20:** 19.
6. OCKNER, R. K. & R. SCHMID. 1961. Acquired porphyria in man and rat due to hexachlorobenzene intoxication. Nature **189:** 499.
7. WRAY, J. D., Y. MUFTU & I. DOGRAMACI. 1962. Hexachlorobenzene as a cause of porphyria turcica. Turk. J. Pediatr. **40:** 132.
8. DOGRAMACI, I. 1964. Porphyrias and porphyrin metabolism with special reference to porphyria in childhood. Adv. Pediatr. **13:** 11.
9. PETERS, H. A., S. A. M. JOHNSON & S. CAM. 1966. Hexachlorobenzene-induced porphyria: Effect of chelation on the disease, porphyrin and metal metabolism. Am. J. Med. Sci. **251:** 314–322.
10. PETERS, H. A. 1954. BAL therapy of acute porphyrinuria (role of zinc in pathogenesis). Neurology **4:** 477–479.
11. BLEIBERG, J., M. WALLEN & R. BRODKIN. 1964. Industrially acquired porphyria. Arch. Dermatol. **98:** 793–797.
12. RIMINGTON, C. 1961. Quantitative determinations of porphobilinogen and porphyrins in the urine and feces. Assoc. Clin. Pathol. Broadsheet: 36.
13. CRIPPS, D. J. & H. A. PETERS. 1967. Fluorescing erythrocytes and porphyrin screening test on urine, stool, and blood. Arch. Dermatol. **96:** 712.
14. MAUZERALL, D. & S. GRANICK. 1956. The occurrences and determination of δ-aminolevulinic acid and porphobilinogen in the urine. J. Biol. Chem. **219:** 435.
15. BURNS, J. E., F. M. MILLER & E. D. GOMES. 1974. Hexachlorobenzene exposure from contaminated DCPA in vegetable spraymen. Arch. Environ. Health **24:** 1210–1214.
16. COURTNEY, K. D. 1979. Hexachlorobenzene (HCB): A review. Environ. Res. **20:** 225.
17. CRIPPS, D. J., H. A. PETERS & A. GÖCMEN. 1981. Experimental pembe yara. J. Invest. Dermatol. **76:** 323.
18. PRICE, J. M., R. R. BROWN & H. A. PETERS. 1959. Tryptophan metabolism in porphyria, schizophrenia, and variety of neurologic and psychiatric diseases. Neurology **9:** 456–468.
19. SUFIT, R. L., R. HODACH, R. ARENDS, *et al.* 1986. Decreased conduction velocity and

pseudomyotonia in HCB-fed rats. *In* Hexachlorobenzene: Proceedings of an International Symposium. C. R. Morris & J. R. P. Cabral, Eds.: 361–363. IARC. Lyons.
20. BERKELEY, M. C. & K. R. MACGEE. 1963. Neuropathy following exposure to a dimethylamine salt of 2,4-D. Arch. Intern. Med. **111:** 351–352.
21. EBERSTEIN, A. & J. GOODGOLD. 1979. Experimental myotonia induced in denervated muscles by 2,4-D. Muscle Nerve **2:** 364–368.
22. ALBERS, J. W., W. C. ROBERTSON & J. R. DAUBE. 1978. Electrodiagnostic findings in acute porphyric neuropathy. Muscle Nerve **1:** 292–296.
23. HILDEBRAND, E. M. 1946. War on weeds. Science **103:** 465–468.
24. BUCHER, N. L. R. 1946. Effects of 2,4-dichlorophenoxyacetic acid on experimental animals. Proc. Soc. Exp. Biol. Med. **63:** 204–205.
25. SOLLMANN, T. 1957. Manual of Pharmacology and Its Applications to Therapeutics and Toxicology. 8th edit. W. B. Saunders. Philadelphia, PA.
26. PARK, J. J., I. DARRIEN & L. F. PRESCOTT. 1977. Pharmacokinetic studies in severe intoxication with 2,4-D and mecprop. Clin. Toxicol. **18:** 154–155.
27. PRESCOTT, L. F., J. PARK & I. DARRIEN. 1979. Treatment of severe 2,4,-D and mecprop intoxication with alkaline diuresis. Br. J. Clin. Pharmacol. **7:** 111–116.
28. GÖCMEN, A., H. A. PETERS, D. J. CRIPPS, C. R. MORRIS & I. DOGRAMACI. 1986. Porphyria turcica. I. Hexachlorobenzene-induced prophyria. *In* Hexachlorobenzene: Proceedings of an International Symposium. C. R. Morris & J. R. P. Cabral, Eds.: 567-573. IARC. Lyons.
29. KLÜVER, H. 1944. On naturally occurring porphyrins in the central nervous system. Science **99:** 482.
30. SOLOMAN, H. M. & F. H. FIGGE. 1958. Occurrence of porphyrins in peripheral nerves. Proc. Soc. Exp. Biol. Med. **97:** 329.
31. PETERS, H. A., P. L. EICHMAN & H. H. REESE. 1958. Therapy of acute, chronic, and hepatic porphyria in patients with chelating agents. Neurology **8:** 621.
32. PETERS, H. A., S. M. WOODS, P. L. EICHMAN & H. H. REESE. 1957. The treatment of acute porphyria with chelating agents: A report of 21 cases. Ann. Intern. Med. **47:** 889–899.
33. LAMBRECHT, R. W., E. ERTÜRK, H. A. PETERS, *et al.* 1986. Effects of ethylenediaminetetraacetic acid on hexachlorobenzene-induced changes in rats. *In* Hexachlorobenzene: Proceedings of an International Symposium. C. R. Morris & J. R. P. Cabral, Eds.: 505–507. IARC. Lyons.
34. PETERS, H. A., P. J. CRIPPS, A. GÖCMAN, E. ERTÜRK, G. T. BRYAN & C. R. MORRIS. 1986. Neurotoxicity of hexachlorobenzene-induced porphyria turcica. *In* Hexachlorobenzene: Proceedings of an International Symposium. C. R. Morris & J. R. P. Cabral, Eds.: 575–579. IARC. Lyons.
35. PETERS, H. A., A. GÖCMAN, P. J. CRIPPS & C. R. MORRIS. 1986. Porphyria turcica: Hexachlorobenzene-induced porphyria (neurologic manifestations and therapeutic trials of ethylenediaminetetraacetic acid in the acute syndrome). *In* Hexachlorobenzene: Proceedings of an International Symposium. C. R. Morris & J. R. P. Cabral, Eds.: 581–583. IARC. Lyons.
36. GÖCMAN, A., H. A. PETERS, P. J. CRIPPS, C. R. MORRIS & I. DOGRAMACI. 1986. Porphyria turcica: Hexachlorobenzene-induced porphyria. *In* Hexachlorobenzene: Proceedings of an International Symposium. C. R. Morris & J. R. P. Cabral, Eds.: 585–587. IARC. Lyons.
37. CABRAL, J. R. P., P. SHUBIK & T. MOLLNER. 1977. Carcinogenic activity of hexachlorobenzene in hamsters. Nature **269:** 510–511.
38. CABRAL, J. R. P., T. MOLLNER & F. RAITANO. 1979. Carcinogenicity of hexachlorobenzene in mice. Int. J. Cancer **23:** 47–51.
39. SMITH, A. G. & J. R. P. CABRAL. 1980. Liver cell tumors in rats fed hexachlorobenzene. Cancer Lett. **11:** 169–172.
40. ERTÜRK, E., R. W. LAMBRECHT, H. A. PETERS, *et al.* 1986. Oncogenicity of hexachlorobenzene. *In* Hexachlorobenzene: Proceedings of an International Symposium. C. R. Morris & J. R. P. Cabral, Eds.: 417–423. IARC. Lyons.

Lead Effects on the Heme Biosynthetic Pathway

Relationship to Toxicity

MICHAEL R. MOORE AND ABRAHAM GOLDBERG

University Department of Medicine
Gardiner Institute
Western Infirmary
Glasgow, Scotland

ALBERT A. C. YEUNG-LAIWAH

Department of Pharmacology and Internal Medicine
University of Alberta
Edmonton, Canada

INTRODUCTION

The anemia of lead poisoning was first noted over 100 years ago, and since that time lead has been shown to have many dramatic effects on heme biosynthesis. In 1895, Stokvis found increased urinary porphyrins in lead-poisoned rabbits.[1] Since then, human studies have shown similarly that lead poisoning results in increased urinary excretion of coproporphyrin[2,3] and δ-aminolevulinic acid (ALA).[4] There is also accumulation of protoporphyrin in erythrocytes.[5] These abnormalities are a result of lead inhibiting the activity of certain enzymes of heme biosynthesis. In particular, there is marked depression of ALA dehydratase, ferrochelatase, and coproporphyrinogen oxidase. The activity of the rate-controlling enzyme of the pathway, ALA synthase, is consequently increased. The alterations in production of porphyrins and precursors are therefore explained by these enzymic changes (FIG. 1).[5]

The precise mechanisms by which lead acts on these enzymes of heme biosynthesis is not known. ALA dehydratase is the enzyme in the heme pathway that is most susceptible to the inhibitory effects of lead. It is likely that the lead binds to essential sulfhydryl groups, blocking the active site on the enzyme.[5] Zinc is also necessary for full activity of ALA dehydratase.[6,7] The addition of zinc to erythrocytes previously exposed to lead has been shown to result in reactivation of the enzyme. It is, therefore, tempting to speculate that zinc is involved at the active site, probably in close relationship to the sulfhydryl group, and that lead and zinc may compete for the same binding site. Boudène and co-workers[8] studied the activity of immunologically reacting ALA dehydratase protein in erythrocytes of lead-exposed subjects. They found that the enzyme protein was increased in moderately lead-poisoned subjects, suggesting increased synthesis as a means of compensating for the inhibited activity. The mechanism by which lead inhibits the activities of coproporphyrinogen oxidase and ferrochelatase is unknown, though sulfhydryl groups are also necessary for ferrochelatase activity.[9] The ALA synthase activity rises as a compensatory response to the diminished synthesis of free heme, a response that is enhanced by increases in the activity of heme oxygenase.[10]

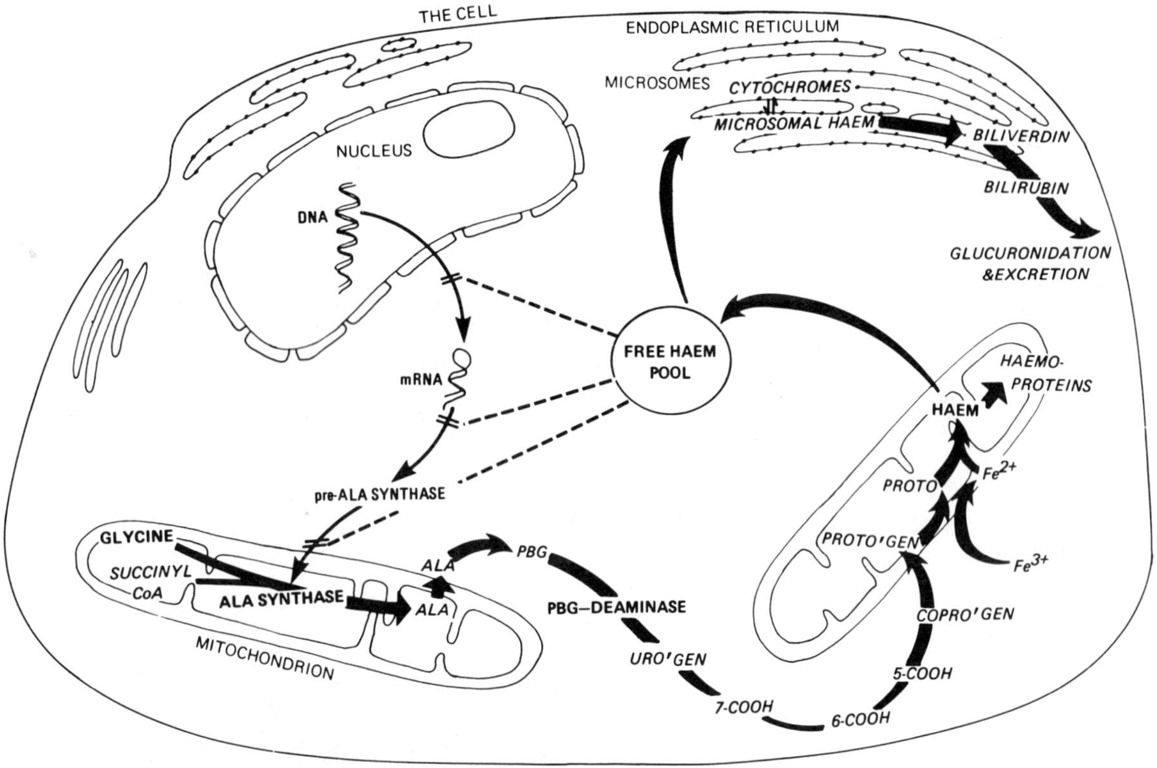

FIGURE 1. Control of heme synthesis in the cell. This shows the role of the free heme pool in controlling the rate-limiting enzyme, ALA synthase. In lead exposure, a decrease of this pool (associated with a decrease in the ability of ALA dehydratase to synthesize PBG; of coproporphyrinogen oxidase to synthesize coprophyrinogen; and of ferrochelatase to synthesize heme) results in increase of ALA synthase activity. The size of the pool is further lowered by the enhancement of heme oxygenase, which breaks down heme to biliverdin.

Many of the clinical features of lead poisoning are a consequence of effect of lead on heme biosynthesis. Depression of heme production is probably the main cause of the anemia, although hemolysis and depression of globin synthesis may also be important.[5] Heme is also essential for the formation of cytochrome P-450 for the mixed function oxidase system. Patients with lead poisoning have been shown to have prolonged phenazone (antipyrine) half-lives, which are consistent with decreased hepatic cytochrome P-450 and impaired drug metabolism in lead-treated rats (TABLE 1).[11] Many of the clinical features of lead poisoning (such as abdominal pain, constipation, and neuropathy) are similar to those of the acute porphyrias in which there are hereditary abnormalities of heme biosynthesis with overproduction of porphyrins and precursors. These common features of the two conditions can all be explained by neurodysfunction. It is this last condition that poses the most intriguing questions about the nature of lead effects on heme biosynthesis.

NEUROTOXICITY

A number of different concepts may explain the neurotoxicity of lead. The direct effects or biochemical effects, such as those on neurotransmitters, are outside the scope

TABLE 1. Antipyrine Metabolism in Lead Poisoning and in Acute Porphyria

	Blood Lead (μM)		Leukocyte ALA Synthase (μmol/mg protein/hr)		Antipyrine Half-Life (hr)	
	Mean	Range	Mean	Range	Mean	Range
Lead poisoning ($N = 10$)						
Before chelation	5.3	3.7–9.4	530	312–1040	13.8	8.0–22.1
After chelation	2.8	2.1–3.8			9.8	7.8–13.6
Acute porphyria ($N = 8$)						
Attack	0.6	0.3–1.4	3327	746–7812	15.1	7.7–34.9
Remission			747	419–2542	8.1	4.6–15.5
Normal		0.1–1.2		50–300		8.5–15.1

of this article. Within heme biosynthesis, there are three possibilities that are not mutually exclusive. The likely answers to the pathogenesis of the neurotoxicology of lead may thus be sought in one or more of the following concepts: 1. That the excessive production of the precursor, ALA, is neurotoxic. 2. That neurotoxic products are synthesized from porphyrins and precursors. 3. That deficiency of heme influences nerve cell function.

ALA

A feature of lead poisoning is the excessive production and urinary excretion of ALA. The liver is considered to be the main site of ALA production. If this hypothesis is true, then this precursor must cross the blood–brain barrier to exert neurotoxic effects. Experimental studies in normal rats have shown that the blood–brain barrier is weakly permeable to ALA. For instance, after intraperitoneal administration, the maximal ALA concentration attained in brain ranged from 4% to 12% of maximal blood levels.[11–13] McGillion *et al.*[12] have also shown that brain concentrations of ALA remained elevated after the blood levels had returned to normal. More significantly, during the acute attack of porphyria in man, ALA and porphobilinogen (PBG) can be detected in cerebrospinal fluid (CSF), where they are not normally detectable. CSF levels of ALA in acute porphyria (10^{-5} to 10^{-7} M) have been reported to be considerably lower than in blood (10^{-3} to 10^{-5} M).[14] But, interestingly, *in vitro* experiments have shown that brain slices and cultured neurons can actively concentrate ALA from the surrounding medium, suggesting that neuronal concentration may be much higher than in CSF.[15]

Acute administration of ALA either intraperitoneally to mice or into the cerebral ventricles of rats has been shown to produce transient excitatory effects.[11,12,16] When mice were treated with ALA, no morphological changes were found in the nerves, but electrical nerve conduction was transiently impaired, and there were alterations in behavior.[17–19] In the rat, ALA has also been linked to a hypotensive response, thought to be due to capillary dilatation.[11] A number of studies involving oral and parenteral administration of ALA and PBG, however, have failed to produce any significant pharmacological effects except for some skin sensitivity.[20,21] In allylisopropylacetamide-induced chemical porphyrias, in which there are high levels of endogenously produced ALA, no neurological changes have been observed. In contrast, behavioral changes have been reported in experimental animals after ALA administration.[17,18,22,23] The concentrations of ALA required to produce these "*in vivo*" effects, however, are many times higher than those occurring in cerebrospinal fluid of subjects with lead poisoning. These experimental situations constitute brief exposures to ALA and are clearly not comparable to the more prolonged exposure prevailing in lead poisoning.

Conditions other than Lead Poisoning

There are two other diseases in which large amounts of ALA are produced and excreted in the urine: the acute porphyrias and hereditary tyrosinemia.[9,24,25] In these disorders, accumulation of ALA occurs as a consequence of inhibition of PBG deaminase and ALA dehydratase, respectively.[24,26,27] In both conditions, patients can develop clinical features resembling those of lead poisoning, namely abdominal pain, motor weakness, tachycardia, and hypertension. In hereditary tyrosinemia, both plasma concentration and urinary excretion of ALA are increased whereas PBG levels remain unchanged. The occurrence of a syndrome like acute porphyria in certain patients with hereditary tyrosinemia suggests, as in plumboporphyria (homozygous ALA dehydratase deficiency), that PBG or its derivative, porphobilin, are unlikely to be the cause of the neurotoxicity.

Pharmacological Effects of ALA

Whereas clinical and *in vivo* experimental evidence for porphyrin precursor toxicity has been unimpressive, much information is now available to show that ALA can exert certain pharmacological effects *in vitro*. These experiments have employed tissues from different animal species, including invertebrates and amphibians.[28,29] High concentrations of ALA have had to be used, often exceeding 100 times that observed in patients' serum. The clinical significance of these studies remains questionable.

Direct Effects of ALA

Even after nerve transmission is blocked with tetrodotoxin, ALA can inhibit rat duodenal muscle tone and both spontaneous and induced contractions.[30] This has been considered to be partly due to the effects of ALA on the α-adrenergic receptors.[30] On the other hand, no such effect was observed in guinea-pig ileum,[12] thus highlighting the interspecies variation. ALA can also inhibit frog gastrocnemius muscle contraction to direct electrical stimulation.[31]

Neurotransmitter Release and ALA

Experiments using rat phrenic nerve hemidiaphragm preparations have shown that ALA inhibits Na^+,K^+-augmented spontaneous release of acetylcholine and the release of this evoked by nerve impulses.[32,33] The responses of pithed and anesthetized rats are qualitatively similar to histamine response, but there is no evidence that ALA has histamine-releasing capability.[34] ALA and PBG also depolarize frog muscle fibers, which suggests that ALA might have a clinically significant effect at the synaptic level.[35]

Neurotoxicity of ALA

ALA (10 μM) was found to be neurotoxic to both neuronal and glial components of brain cell cultures from the chick embryo.[15] Such a concentration could possibly occur in humans because a porphyric patient with a CSF level of 21 μM has been recorded.[36] Other workers, however, have generally measured much lower levels.[37] ALA neurotoxicity can certainly offer an explanation for the histopathological changes seen in acute porphyria, but results obtained with avian cells *in vitro* do not necessarily reflect the *in vivo* human situation, where cells may be more resistant to toxic agents.

ALA and γ-Aminobutyric Acid (GABA)

The pharmacological relationship between the structurally similar ALA, GABA, and glutamic acid is complex and somewhat confusing. GABA is an inhibitory neurotransmitter in both spinal cord and cerebral cortex, whereas glutamic acid exerts an excitatory effect.[38] ALA selectively competes with GABA for GABA receptor sites on rat brain synaptosomes[39] and can mimic GABA action by increasing chloride ion conductance in the crayfish stretch receptors.[40] At relatively high concentrations (100 μM), ALA can release both GABA and glutamic acid from unstimulated brain cortex synaptosomes, but at lower concentrations (1–5 μM), it can only liberate GABA.[41] ALA has also been shown to impede Na^+,K^+-ATPase activity in rabbit brain preparations.[42] Since inhibition of Na^+,K^+-ATPase can influence neuromuscular transmission by increasing neurotransmitter release, ALA seems to act as a GABA agonist. But at higher concentrations, ALA can paradoxically inhibit K^+-stimulated release of GABA from preloaded synaptosomes.[43] GABA activity is terminated through a reuptake mechanism using a high-affinity transport system in both nerve and glial cells. ALA has been shown to inhibit GABA uptake by brain synaptosomes at high concentrations (1 mM),[43,44] but not at lower concentrations.[15] In rabbit jejunum, however, the action of both acetylcholine and GABA has been shown to differ from that of ALA; this difference suggests no such interaction with GABA receptors.[30]

PORPHYRINS

It has been argued that excess of protoporphyrin is a health effect, but one can equally argue that excess protoporphyrin is only an indication that changes have occurred in the activity of ferrochelatase. Unlike erythropoietic protoporphyria, dermatological features are not found in lead poisoning, and so there are no obvious health effects that may be clearly associated with excessive protoporphyrin concentrations.

Indeed in experimental animals or "*in vitro,*" porphyrin administation has had no demonstrable pharmacological, electroencephalographic, or behavioral effect, or alteration of blood pressure.[45,46] However, some reported effects of porphyrins include dose-dependent neurotoxicity from high (mM) concentrations of uroporphyrin I, coproporphyrin I, and protoporphyrin IX on sensory neurons in tissue culture[47]; toxic changes in muscle end plates of mice after treatment with a synthetic porphyrin; inhibition of cell division in specific cell lines by hematoporphyrin derivative (HPD); stimulation of collagen biosynthesis by uroporphyrin I; and inhibition of K^+-augmented acetylcholine release by uroporphyrin I.[46] Selman *et al.*[48] presented evidence that HPD in the presence of light may decrease blood flow in smooth muscle. This might be ascribed to phototoxic effects of porphyrins on the complement system.[49] Protoporphyrin and photooxidized porphyrins have also been found to inhibit membrane-bound Na^+,K^+- and Mg^{2+}-ATPase[50] and to damage membrane proteins[51]; this, if occurring in the small intestine, would cause depolarization of the smooth muscle. The highest concentration of protoporphyrin employed in more recent studies (2.3 mM), after initially reducing tone, produced an effect closely resembling the response to sodium pump inhibitors.[52]

Activation of guanylate cyclase may also be expected to influence the tone of intestinal smooth muscle because the guanylate cyclase cyclic GMP system is coupled to muscarinic receptors and affects the activity of transport and binding sites for calcium ions. Protoporphyrin has been shown to activate soluble and plasma membrane guanylate cyclase.[53,54]

Riopelle and Kennedy[47] were protagonists of the theory that the effects of ALA were consequent upon its *in vivo* conversion to porphyrins. They found that there was a dose-dependent neurotoxicity of protoporphyrin IX, uroporphyrin I, and coproporphyrin I in primary cultures of sensory neurons. They were unable to detect any effect from ALA at concentrations up to 1.5 mM. They therefore suggested that ALA may manifest its effects only by *in vitro* conversion to porphyrin, perhaps by the synthetic pathway within Schwann cells described by Whetsell *et al.*[55] in lead-treated cultures. Sima *et al.*[19] came to a similar conclusion after studying effects of synthetic porphyrins and ALA upon motor nerve conduction velocity in dark-maintained mice. The effects produced by ALA were associated with a transient, generalized fluorescence, suggestive of extrahepatic porphyrin synthesis.

The mechanism by which porphyrins might produce pharmacological changes cannot at present be fully interpreted because of the small number of pharmacological investigations on these compounds. The principal evidence against them rests in the lack of neurological change found in the nonacute porphyrias in which great excesses of both circulating and excreted porphyrin are found, but in which there are no neurological features.[24]

HEME DEFICIENCY

Although brain and nervous tissue have heme biosynthetic capability, the activities of these enzymes in the rat brain are, quantitatively, much lower than those in the liver.[56–58] Brain tissue contains little ALA, PBG, total porphyrins, heme, and cytochrome P-450, but it does perform certain hemoprotein-dependent functions. In consideration of this, two key questions need to be answered. First, does heme deficiency occur in neural tissues of patients during lead exposure? Second, does it fluctuate with the severity of exposure? If the neuropathy at lead poisoning is due to impaired intraneuronal heme biosynthesis, then the precipitating factors must be capable of modifying neuronal heme biosynthesis. In this respect, Paterniti *et al.*[56] were unable to demonstrate induction of rat brain ALA synthase after fasting or intraperitoneal injections of ethanol, allylisopropylacetamide, or 3,5-dicarbethoxy-1,4-dihydrocollidine, although subsequent work using hematin has shown the brain enzyme to be inducible.[59]

The "Heme Pool"

Until now, direct measurement of the intracellular "heme pool" has not been possible, its existence being based on indirect evidence. The "free heme pool" is currently believed to be the regulatory component of the total heme in the liver. Deficiency of hepatic heme, however, does not necessarily entail a similar state of depletion in neural tissues unless both organs share a common control mechanism for heme biosynthesis. Earlier studies have reported that brain ALA synthase could not be induced in rats after either dietary starvation or treatment with known potent inducers of hepatic ALA synthase such as phenobarbital, ethanol, and allylisopropylacetamide.[58] These findings suggested that the brain is perhaps endowed with a different heme biosynthetic regulatory mechanism. Paradoxically, De Matteis *et al.*[59] have provided evidence to indicate that regulation of rat brain ALA synthase is controlled by negative feedback by heme. They showed that ALA synthase activity altered during the neonatal period in parallel with the physiological heme requirements of brain tissue. Inhibition of cerebellar heme biosynthesis by succinylacetone, a specific inhibitor of ALA dehydratase, also led to ALA synthase induction, whereas intraventricular injection of hematin caused a depression of its activity. Indeed, the addition of hematin to rat brain preparations results in a marked increase in its monooxygenase function. The contraction of the heme pool, which occurs in lead poisoning, induces ALA synthase activity, in addition to which there is an increase in heme oxygenase.[60] Studies in which the fate of [^{14}C]ALA was followed after intraventricular injection in the rat brain have supported the existence of a heme pool in that organ.[62]

Tryptophan Pyrrolase

An alternative aspect of heme deficiency relates to its requirement for tryptophan pyrrolase activity (tryptophan 2,3-dioxygenase, EC 1.13.11.11).[61] Activity is reduced when free heme availability falls and has been linked to both the regulatory free heme pool and to the effects of porphyrinogenic drugs.[62,63] This enzyme is rate-limiting in tryptophan degradation. Any diminution in activity, therefore, may be associated with increases in the concentration of tryptophan and neuroactive tryptophan metabolites such as 5-hydroxytryptamine (5-HT). It is of interest that one such metabolite, 5-hydroxyindole acetic acid, was shown to be increased in the urine of patients with acute porphyria,[64] and that 5-HT administered to humans results in severe abdominal pain, psychomotor disturbance, nausea, and dysuria. This lends some support to this

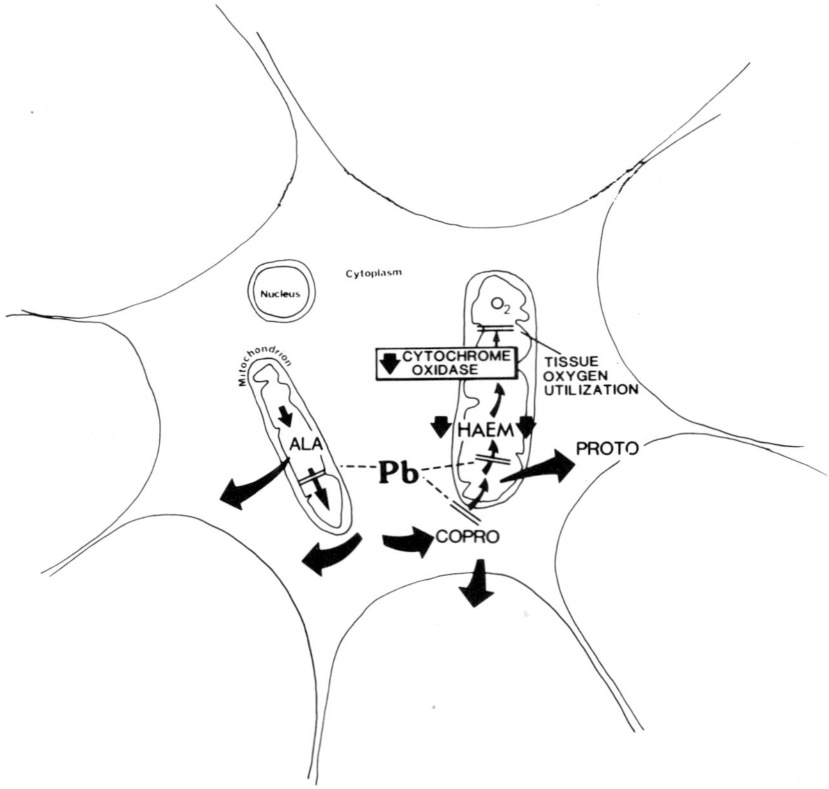

FIGURE 2. The influence of lead on neuronal heme synthesis. Lessened synthesis of heme together with contraction of the free heme pool could result in diminished synthesis of the cytochromes used in terminal oxidative pathways. The proper functioning of nerve cells would be compromised by this diminution, and would be further compromised by the pharmacological actions of ALA.

being the mechanism whereby the endogenous alteration in metabolism in lead poisoning and acute porphyrias results in the neurological features of this disease.[65]

The Cytochromes

In the human liver, at least, there is some indirect evidence in favor of the present hypothesis; this evidence reveals the neurological features of the dyssynthesis of the microsomal cytochrome P-450-dependent mixed function oxidase system. As assessed by antipyrine clearance, the system was significantly depressed in patients with lead poisoning and acute porphyria (TABLE 1).[5,11,66]

Cytochrome *c* oxidase (cytochrome aa_3) activity was reported to be depressed in muscle biopsies obtained from patients with lead exposure and acute porphyria.[67] Yeung-Laiwah and co-workers hypothesized that a deficiency of cytochrome oxidase would lend support to the concept of a functional neuronal block operating in the genesis of the clinical manifestations of acute porphyria,[68] and the same would apply to lead poisoning. These findings are supported by the work of Bull and co-workers,[69] who showed alterations in oxygen utilization in lead-treated mitochondria, and also by the work of Bonkowsky and co-workers,[70] who found that NADH oxidation was impaired in skin fibroblasts in culture from patients with acute intermittent porphyria. As a consequence of intracellular heme deficiency, other metabolic disturbances can be expected to arise in those pathways dependent on hemoproteins. For instance, lead and other porphyrinogenic chemicals have been shown to impair mitochondrial NADH oxidation[69,71,72] and to deplete hepatic ATP.[73] It is of note that a yeast mutant of *Saccharomyces cerevisiae* deficient in the cytochrome oxidase accumulates porphyrins, presumably because of a deficiency in uroporphyrinogen decarboxylase.[74] The consequent heme deficiency, which may be the root cause of the cytochrome oxidase deficiency, can be corrected by hemin administration to such yeasts, which restores mitchondrial membrane synthesis.[75] No direct evidence is yet available, however, to show that lead either inhibits NADH oxidation or ATP production in human nervous tissue.

CONCLUSIONS

In conclusion, the toxic effects of lead related to changes in the nervous system are of great interest. Many if not most of these effects are unrelated to alterations in heme synthesis, and those that have been shown to be related are only linked secondarily to such alterations. Current evidence supports the hypothesis that some of the neurological manifestations of lead exposure arise as a result of heme deficiency within the nerve cell. An intracellular deficiency of heme-containing compounds, such as the cytochromes, could link some of the biochemical, clinical, and neuropathological findings of lead poisoning. Such a conclusion does not exclude the possibility of ALA acting concurrently as a pharmacological agent to compound these effects of heme deficiency (FIG. 2).

REFERENCES

1. STOKVIS, B. J. 1985. Zur pathogenese der hämatoporphyrinurie. Z. Klin. Med. **28:** 1–21.
2. DUESBERG, R. 1931. Über die Anämien. I. Porphyrie und Erythropoese. Arch. Exp. Pathol. **162:** 249.
3. GROTEPASS, W. 1932. Zur Kenntnis des im Harn auftretenden Porphyrins bei Bleivergiftung. Hoppe–Seyler's Z. Physiol. Chem. **205:** 193–197.
4. HAEGER-ARONSEN, B. 1960. Studies in urinary excretion of δ-aminolaevulinic acid and other haem precursors in lead workers and lead-intoxicated rabbits. J. Clin. Lab. Invest. **12**(Suppl. 47): 1–28.
5. MOORE, M. R. & A. GOLDBERG. 1985. Health implications of the hematopoietic effects of lead. *In* Dietary and Environmental Lead: Human Health Effects. K. R. Mahaffey, Ed.: 261–299. Elsevier. Amsterdam.
6. CHEH, A. & J. B. NEILANDS. 1973. Zinc: An essential metal ion for beef liver δ-aminolevulinate dehydratase. Biochem. Biophys. Res. Commun. **55:** 1060–1063.

7. ABDULLA, M. & B. HAEGER-ARONSEN. 1971. ALA dehydratase activation by zinc. Enzyme **12:** 708–710.
8. BOUDÈNE, C. I., N. DESPAUX-PAGES, E. COMOY & C. I. BOHUON. 1984. Immunological and enzymatic studies of erythrocytic 5-aminolaevulinate dehydratase: Comparison of results obtained in normal and lead-exposured subjects. Int. Arch. Occup. Environ. Health **55:** 87–96.
9. GOLDBERG, A. & M. R. MOORE. 1980. The porphyrias. *In* Clinics in Haematology 9. Saunders. London.
10. MEREDITH, P. A. & M. R. MOORE. 1979. The influence of lead on haem biosynthesis and biodegradation in the rat. Biochem. Soc. Trans. **7:** 637–639.
11. MOORE, M. R. & A. GOLDBERG. 1980. Lead and heme biosynthesis. *In* Lead Toxicity. R. L. Singhal & J. A. Thomas, Eds.: 79–117. Urban and Schwartzenberg. Baltimore, MD.
12. MCGILLION, F. B., M. R. MOORE & A. GOLDBERG. 1975. Some pharmacological effects of δ-aminolaevulinic acid on blood pressure in the rat and on rabbit isolated ear arteries. Clin. Exp. Pharmacol. Physiol. **2:** 365–371.
13. SHANLEY, B. C., A. C. NEETHLING, V. A. PERCY & H. CARSTENS. 1975. Neurochemical aspects of porphyria. I. Studies on the possible neurotoxicity of δ-aminolaevulinic acid. S. Afr. Med. J. **49:** 576–680.
14. MEREDITH, P. A., M. R. MOORE, B. C. CAMPBELL, G. G. THOMPSON & A. GOLDBERG. 1978. δ-Aminolaevulinic acid metabolism in normal and lead-exposed humans. Toxicology **9:** 1–9.
15. PERCY, V. A., M. C. L. LAMM & J. J. F. TALJAARD. 1981. δ-Aminolaevulinic acid uptake, toxicity and effect on [^{14}C]γ-aminobutyric acid uptake into neurons and glia in culture. J. Neurochem. **36:** 69–76.
16. JARRETT, A., C. RIMINGTON & D. A. WILLOUGHBY. 1956. 5-Aminolaevulinic acid and porphyria. Lancet **i:** 125–127.
17. MCGILLION, F. B., M. R. MOORE & A. GOLDBERG. 1972. The effects of δ-aminolaevulinic acid on the spontaneous activity of mice. Scott. Med. J. **18:** 133–134.
18. CUTLER, M. G., M. R. MOORE & F. G. EWART. 1979. Effects of δ-aminolaevulinic acid administration on social behavior in the laboratory mouse. Psychopharmacology **61:** 131–135.
19. SIMA, A. A. F., J. C. KENNEDY, D. BLAKESLEE & D. M. ROBERTSON. 1981. Experimental porphyric neuropathy: A preliminary report. Can. J. Neurol. Sci. **8:** 105–114.
20. BERLIN, N. I., A. NEUBERGER & J. J. SCOTT. 1956. The metabolism of 5-aminolaevulinic acid. II. Normal pathways studied with the aid of ^{14}C. J. Biochem. **64:** 90–100.
21. MEYER, U. S., L. J. STRAND, M. DOSS, A. C. REES & H. S. MARVER. 1972. Intermittent acute porphyria: Demonstration of a genetic defect in porphobilinogen metabolism. N. Engl. J. Med. **286:** 1277–1282.
22. MOORE, M. R. & P. A. MEREDITH. 1976. The association of δ-aminolaevulinic acid with the neurological and behavioural effects of lead exposure. *In* Trace Substances in Environmental Health X. D. D. Hemphill, Ed.: 363–371. University of Missouri Press. Columbia, MO.
23. PIERACH, C. A. & P. S. EDWARDS. 1978. Neurotoxicity of δ-aminolaevulinic acid and porphobilinogen. Exp. Neurol. **62:** 810–814.
24. DISLER, P. B. & M. R. MOORE. 1985. Porphyria. *In* Clinics in Dermatology. Vol. 3: 1–163. Lippincott. Philadelphia, PA.
25. GENTZ, J., S. JOHANSSON, S. LINDBLAD, S. LINDSTEDT & R. ZETTERSTROM 1969. Excretion of δ-aminolaevulinic acid in hereditary tyrosinaemia. Clin. Chim. Acta **23:** 257–263.
26. SILBERGELD, E. K., R. E. HRUSKA, D. BRADLEY, J. M. LAMON & B. C. FRYKHOLM. 1982. Neurotoxic aspects of porphyrinopathies: Lead and succinylacetone. Environ. Res. **29:** 459–471.
27. SASSA, S. & A. KAPPAS. 1983. Hereditary tyrosinemia and the heme biosynthetic pathway: Activity by succinyl acetone. J. Clin. Invest. **71:** 625–634.
28. LOOTS, J. M., D. M. BECKER, B. J. MEYER, N. GOLDSTUCK & S. KRAMER. 1975. The effect of porphyrin precursors on monosynaptic reflex activity in the isolated hemisecreted frog spinal cord. J. Neurol. Trans. **36:** 71–81.

29. NICHOLL, R. A. 1976. The interaction of porphyrin precursors with GABA receptors in the isolated frog spinal cord. Life Sci. **19:** 521–526.
30. CUTLER, M. G., M. MCLAUGHLIN, E. MCNEIL & M. R. MOORE. 1985. Effects of δ-aminolaevulinic acid on contractile activity in the isolated small intestine of the rabbit. Neuropharmacology **24:** 1005–1009.
31. CUTLER, M. G., J. M. DICK & M. R. MOORE. 1978. Effect of δ-aminolaevulinic acid on frog nerve–muscle function. Life Sci. **23:** 2233–2238.
32. FELDMAN, D. S., R. D. LEVERE & J. S. LIEBERMAN. 1968. Presynaptic neuromuscular inhibition by δ-aminolaevulinic acid, a porphyrin precursor. Trans. Am. Neurol. Assoc. **93:** 206–208.
33. BORNSTEIN, J. C., J. B. PICKETT & I. DIAMOND. 1979. Inhibition of the evoked release of acetylcholine by the porphyrin precursor, δ-aminolaevulinic acid. Ann. of Neurol. **5:** 94–96.
34. MOORE, M. R., F. B. MCGILLION & A. GOLDBERG. 1976. Some pharmacological and behavioural effects of δ-aminolaevulinic acid. *In* Porphyrins in Human Diseases. M. Doss, Ed.: 148–154. Karger. Basel.
35. BECKER, D. M., N. GOLDSTUCK & S. KRAMER. 1975. Effect of δ-aminolaevulinic acid on the resting membrane potential of frog sartorius muscle. S. Afr. Med. J. **49:** 1790–1792.
36. SWEENEY, V. P., M. A. PATHAK & A. K. ASBURY. 1970. Acute intermittent porphyria: Increased ALA synthetase activity during an attack. Brain **93:** 369–380.
37. BONKOWSKY, H. L., D. P. TSCHUDY, A. COLLINS, J. DOHERTY, I. BOSSENMAIER, R. CARDINAL & C. J. WATSON. 1971. Repression of the overproduction of porphyrin precursors in acute intermittent porphyria by intravenous infusions of hematin. Proc. Natl. Acad. Sci. USA **68:** 2725–2729.
38. WALKER, J. E. 1983. Glutamate, GABA and CNS disease: A review. Neurochem. Res. **8:** 521–550.
39. MULLER, W. E. & S. H. SNYDER. 1977. δ-Aminolaevulinic acid influences on synaptic GABA receptor-binding may explain CNS symptoms of porphyria. Ann. Neurol. **2:** 340–342.
40. DICHTER, H. N., L. TADDEINI, S. LIN & G. F. AYALA. 1977. δ-Aminolaevulinic acid: Effects of porphyrin precursor on an isolated neuronal preparation. Brain Res. **126:** 189–195.
41. BECKER, D. M., D. VILJOEN & S. KRAMER. 1976. Porphyrin precursors and their effect *in vitro* in some aspects of nerve function. *In* Porphyrins in Human Disease. M. Doss, Ed.: 163–172. Karger. Basel.
42. RUSSELL, V. A., M. C. L. LAMM & J. J. F. TALJAARD. 1983. Inhibition of Na^+,K^+ATPase activity by 5-aminolaevulinic acid. Neurochem. Res. **8:** 1407–1415.
43. BRENNAN, M. J. W., R. C. CANTRILL & S. KRAMER. 1980. Effect of δ-aminolaevulinic acid on GABA receptor-binding in synaptic plasma membranes. Int. J. Biochem. **12:** 833–835.
44. BRENNAN, M. J. W. & R. C. CANTRILL. 1979. The effect of δ-aminolaevulinic acid on the uptake and efflux of [^{3}H]GABA in rat brain synaptosomes. J. Neurochem. **32:** 1781–1786.
45. GOLDBERG, A., W. D. H. PATON & S. W. THOMPSON. 1954. Pharmacology of porphyrins and porphobilinogen. Br. J. Pharmacol. **9:** 91–99.
46. MOORE, M. R., K. E. L. MCCOLL, C. RIMINGTON & A. GOLDBERG. 1987. Disorders of Porphyrin Metabolism. Plenum. New York, NY.
47. RIOPELLE, R. J. & J. C. KENNEDY. 1982. Some aspects of porphyrin neurotoxocity *in vitro*. Can. J. Physiol. Pharmacol. **60:** 707–714.
48. SELMAN, S. H., M. KREIMER-BIRNBAUM, J. E. KLAUNIG, P. J. GOLDBLATT, R. W. KECK & S. L. BRITTON. 1984. Blood flow in transplantable bladder tumors treated with hematoporphyrin derivative and light. Cancer. Res. **44:** 1924–1927.
49. LIM, H. W., L. YOUNG, M. HAGAN & I. GIGLI. 1985. Delayed phase of hematoporphyrin-induced phototoxicity: Modulation by complement, leuocyctes and antihistamines. J. Invest. Dermatol. **84:** 114–117.
50. AVNER, D. L., R. LARSEN & M. M. BERENSON. 1983. Inhibition of liver surface membrane Na^+,K^+-adenosine triphosphatase, Mg^{2+}-adenosine triphosphatase and 5-nucleotidase activities by protoporphyrin. Gastroenterology **85:** 700–706.

51. DUBBLEMAN, T. M., A. F. DE GOEIJ & J. VAN STEVENINCK. 1980. Protoporphyrin-induced photodynamic effects on transport processes across the membrane of human erythrocytes. Biophys. Acta **595:** 133–139.
52. CULTER, M. G. & S. ARROL. 1987. Pharmacological effects of haem biosynthetic intermediates in isolated intestinal preparations. Eur. J. Pharmacol. **134:** 249–256.
53. IGNARRO, L. J., K. S. WOOD & M. S. WOLIN. 1982. Activation of purified soluble guanylate cyclase by protoporphyrin IX. Proc. Natl. Acad. Sci. USA **79:** 2870–2873.
54. LACOMBE, M. L. & C. EBERENTZ-LHOMME. 1983. Protoporphyrin IX activates the Mg^{2+}-dependent guanylate cyclase from rat liver plasma membranes. Biochem. Biophys. Res. Commun. **116:** 47–53.
55. WHETSELL, W. O., S. SASSA, D. BICKERS & A. KAPPAS. 1978. Studies on porphyrin–heme biosynthesis in organotypic cultures of chick dorsal root ganglion. I. Observations on neuronal and nonneuronal elements. J. Neuropathol. Exp. Neurol. **37:** 497–507.
56. PATERNITI, J. R., J. J. SIMONE & D. S. BEATTIE. 1978. Detection and regulation of δ-aminolaevulinic acid synthetase activity in the rat brain. Arch. Biochem. Biophys. **189:** 96–101.
57. SASSA, S., G. L. ZALAR & A. KAPPAS. 1978. Studies in porphyria. VII. Induction of uroporphyrinogen I synthase and expression of the gene defect of acute intermittent porphyria in mitogen-stimulated human lymphocytes. J. Clin. Invest. **61:** 499.
58. PERCY, V. A. & B. C. SHANLEY. 1979. Studies on haem biosynthesis in rat brain. J. Neurochem. **33:** 1267–1274.
59. DE MATTEIS, F., P. ZETTERLUND & L. WETTERBERG. 1981. Brain 5-aminolaevulinate synthase: Developmental aspects and evidence for a regulatory role. Biochem. J. **196:** 811–817.
60. BISSELL, D. M. & L. E. HAMMAKER. 1977. Effect of endotoxin on tryptophan pyrrolase and δ-aminolaevulinate synthase: Evidence for an endogenous regulatory heme fraction in rat liver. J. Biochem. **166:** 301–304.
61. DE MATTEIS, F. & R. E. RAY. 1982. Studies on cerebellar haem metabolism in the rat *in vivo*. J. Neurochem. **39:** 551–556.
62. BADAWY, A. A. B. & M. EVANS. 1975. Regulation of rat liver tryptophan pyrrolase by its cofactor haem: Experiments with haematin and 5-aminolaevulinate and comparison with the substrate and hormonal mechanisms. Biochem. J. **150:** 511–520.
63. BADAWY, A. A. B., A. N. WELCH & C. J. MORGAN. 1981. Tryptophan pyrrolase in haem regulation. Biochem. J. **198:** 309–314.
64. LUDWIG, G. D. & I. S. EPSTEIN. 1961. A genetic study of two families having the acute intermittent type of porphyria. Ann. Intern. Med. **55:** 81–90.
65. LITMAN, D. A. & M. A. CORREIA. 1985. Elevated brain tryptophan and enhanced 5-hydroxytryptamine turnover in acute hepatic heme deficiency: Clinical implications. J. Pharmacol. Exp. Ther. **232:** 337–345.
66. BIRNIE, G. G., K. E. L. MCCOLL, G. G. THOMPSON, M. R. MOORE, A. GOLDBERG & M. J. BRODIE. 1987. Antipyrine metabolism in acute hepatic porphyria in relapse and remission. Br. J. Clin. Pharmacol. **23:** 358–361.
67. GOLDBERG, A., D. DOYLE, A. C. YEUNG-LAIWAH, M. R. MOORE & K. E. L. MCCOLL. 1985. A study of cytochrome oxidase activity in muscle biopsies from patients with acute intermittent porphyria. Q. J. Med. **57:** 799.
68. YEUNG-LAIWAH, A. A. C., M. R. MOORE & A. GOLDBERG. 1987. Pathogenesis of acute porphyria. Q. J. Med. **63:** 377–392.
69. BULL, R. J. 1980. Lead and energy metabolism. *In* Lead Toxicity. R. L. Singhal & J. A. Thomas, Eds.: 119–167. Urban and Schwartzenberg. Baltimore, MD.
70. BONKOWSKY, H. L., D. P. TSCHUDY, E. C. WEINBACH, P. S. EBERT & J. M. DOHERTY. 1975. Porphyrin synthesis and mitochondrial respiration in acute intermittent porphyria: Studies using cultured human fibroblasts. J. Lab. Clin. Med. **85:** 93–102.
71. DA PALMA-CARLOS, A. G. 1968. A regulasao da biossintesa das porfirinas e do heme. Ph.D. diss. Lisbon.
72. LABBE, R. F., J. NUTTER, M. L. COWGER & L. D. NIELSEN. 1970. Control of heme biosynthesis: Effect of nicotinamide adenine dinucleotide on the induction of δ-aminolaevulinate synthetase. Biochem. Med. **3:** 465–474.

73. Gajdos, A. & M. Gajdos-Torok. 1969. The quantitative regulation of the biosynthesis of porphyrins by intracellular ATP concentration. Biochem. Med. **2:** 372–388.
74. Arrese, M. R., D. Vojensky & J. Mattoon. 1982. Characterisation of a cytochrome oxidase-deficient yeast mutant which accumulates porphyrins. Biochem. Biophys. Res. Commun. **107:** 848–855.
75. Gopalan, G., J. Jayaraman & C. Rajamanickan. 1984. Effect of exogenous addition of hemin on the biogenesis of mitochondrial membranes during glucose repression of *Saccharomyces cerveisiae*. Arch. Biochem. Biophys. **235:** 159–166.

Porphyrinurias and Occupational Disease[a]

MANFRED O. DOSS

Department of Clinical Biochemistry
Faculty of Medicine
University of Marburg
D-3550 Marburg, Federal Republic of Germany

INTRODUCTION

The disturbances of porphyrin metabolism are principally differentiated into porphyrias and secondary porphyrinopathies, that is, porphyrinurias and porphyrinemias. Porphyrinuria is a symptom in erythropoietic and erythrohepatic porphyrias, in acute hepatic porphyrias and lead poisoning, and in chronic hepatic porphyrias leading to porphyria cutanea tarda. Acquired, secondary coproporphyrinurias, however, are nonspecific disorders without immediate clinical significance, in contrast to porphyrinurias related to a porphyria disease. Such "secondary porphyrinurias" are generally confined to excessive urinary coproporphyrin excretion, and secondary porphyrinemias to excessive zinc-protoporphyrin concentration in erythrocytes.

PORPHYRIAS VERSUS SECONDARY PORPHYRINURIAS

In the hepatic and erythropoietic porphyrias, characterized by specific inherited or acquired, especially toxic, enzyme deficiency (TABLE 1), the symptoms and clinical manifestations are related to the disturbance of the heme biosynthetic pathway.[1] Pathophysiologic signs in acute hepatic porphyrias, as molecular regulatory diseases, and cutaneous symptoms in porphyria cutanea tarda (PCT) and erythropoietic porphyrias, as porphyrin-accumulation diseases, originate from metabolic consequences of the deranged porphyrin pathway (TABLE 2). In this context, lead poisoning is regarded as toxic or toxogenic acute hepatic porphyria.[2]

PCT is also called chronic hepatic porphyria (CHP) because 1) the main site of the PCT disease process concentrates on the liver, 2) a functional and morphologic liver injury is almost regularly associated with this chronic porphyria ("membrane disease"), 3) the term CHP characterizes the clinical latent phase with the absence of skin lesions more precisely than the dermatological term PCT, 4) the toxic form due to occupational and environmental exposure takes mainly a subclinical course—with the occasional exception of a severe epidemic outbreak,[3] and 5) human CHP imitates experimental CHP in animals.[4]

Individuals with autosomal dominant acute and chronic porphyrias produce and excrete heme precursors at high rates during the clinical phase of the disease.[2] In contrast, porphyrin overproduction and excretion usually does not increase to such an extent in secondary coproporphyrinurias as observed in porphyrias in phases of the clinical disease. If this is accepted, disturbance of hepatic heme biosynthesis in secondary coproporphyrinurias seems to be modulated in minor amplitudes, leading to

[a]This work was supported by research grants from the Deutsche Forschungsgemeinschaft (Do 134), Bonn, Federal Republic of Germany.

TABLE 1. Porphyrinopathies Related to Enzyme Deficiencies and Disturbances According to the Heme Biosynthetic Chain

Enzyme Deficiency/Inhibition	Disease/Disorder	Genetic Trait
Porphobilinogen synthase (ALA dehydratase)	Hereditary porphobilinogen synthase defect porphyria	Recessive
	Lead poisoning	(Toxogenetic in heterozygotes)
Porphobilinogen deaminase	Acute intermittent porphyria	Dominant
Uroporphyrinogen III cosynthase (isomerase)	Congenital erythropoietic porphyria	Recessive
Uroporphyrinogen decarboxylase	CHT (PCT); hepatoerythropoietic porphyria	Dominant; recessive
Coproporphyrinogen oxidase	Hereditary coproporphyria	Dominant
	Secondary coproporphyrinuria	(Acquired)
Protoporphyrinogen oxidase	Variegate porphyria	Dominant
Ferrochelatase	Erythropoietic (erythrohepatic) protoporphyria	Dominant

low-grade or moderate porphyrinuria over a wide spectrum of conditions and causes (TABLE 3). Symptomatic porphyrinurias are particularly connected with a metabolic response to toxins, to hepatic dysfunction, or to increased erythropoiesis, in addition to certain individual factors. These features may explain why secondary coproporphyrinurias in themselves do not develop and produce clinical symptoms and manifestations.

HISTORICAL BACKGROUND

The statement by Franke and Fikentscher in 1935 that even minor fluctuations in porphyrin excretion may be important in basic and clinical research is certainly still

TABLE 2. Different Clinical Manifestations and Enzyme Deficiencies in the Porphyrias

Acute Symptoms	Cutaneous Symptoms
Abdominal colics, panneuropathy	Photosensitivity
Acute intermittent porphyria (porphobilinogen deaminase ↓)	Congenital erythropoietic porphyria (uroporphyrinogen III cosynthase ↓)
Lead poisoning (porphobilinogen synthase ↓, coproporphyrinogen oxidase ↓, ferrochelatase ↓)	Erythropoietic(-hepatic) protoporphyria[a] (ferrochelatase ↓)
Porphobilinogen synthase defect acute porphyria (porphobilinogen synthase ↓↓)	PCT (CHP)[a] and hepatoerythropoietic porphyria (uroporphyrinogen decarboxylase ↓↓)
Hereditary coproporphyria (coproporphyrinogen oxidase ↓)	Hereditary coproporphyria (coproporphyrinogen oxidase ↓)
Variegate porphyria (protoporphyrinogen oxidase ↓)	Variegate porphyria (protoporphyrinogen oxidase ↓)
Strong induction of hepatic ALA synthase	Little or no induction of hepatic ALA synthase

[a]Abdominal symptoms associated with liver and bile duct disease, not primarily caused by porphyria.

true in 1986.[5] Thus, 50 years ago, quantitative porphyrin assessment was established as an extraordinarily subtle test of liver function based on observations of alcohol-stimulated porphyrinuria as well as changes in porphyrin excretion owing to nutritional factors. A reasonable definition and description of porphyrinuria was proposed by Goldberg and Rimington in 1962:

> In the past the term 'porphyria' was reserved for those diseases of porphyrin metabolism caused by an "inborn error of metabolism," that is, a genetic defect. "Porphyrinuria" was the term used to describe minor disorders of porphyrin metabolism induced by other diseases, or certain drugs or chemicals. . . . The term "porphyrinuria" should be confined to those minor disorders of porphyrin metabolism caused by another disease, or certain drugs or chemicals, in which the clinical features are not directly attributable to the porphyrin abnormality.[6]

TABLE 3. Various Causes and Conditions Related to Secondary Coproporphyrinurias

Intoxications	Hematologic diseases
Alcoholism (acute and/or chronic); foreign and environmental chemicals such as hexachlorobenzene, polyhalogenated biphenyls, dioxins (TCDD), vinyl chloride, carbon tetrachloride, benzene, chloroform; heavy metals such as lead, arsenic, mercury; drugs	Hemolytic, sideroachrestic, sideroblastic, aplastic anemias; ineffective erythropoiesis (intramedullary hemolysis); pernicious anemia; thalassemia; leukemia; erythroblastosis; iron deficiency anemia
Liver diseases	Diabetes mellitus
Cirrhosis, active chronic hepatitis, toxic and infectious hepatitis, fatty liver, alcoholic liver syndromes, drug injury, cholestasis, cholangitis, biliary cirrhosis	Malignancies
Adverse effect of drugs	Hepatocellular tumors, hepatic metastases, pancreatic carcinoma, lymphomatosis, other systemic diseases
Analgesics, sedatives, hypnotics, anesthetics, sex hormones	Disturbance of iron metabolism
Infectious diseases	Hemosiderosis, idiopathic and secondary hemochromatosis
Mononucleosis, acute poliomyelitis	Hereditary hyperbilirubinemias
	Dubin–Johnson syndrome, Rotor's syndrome
	Pregnancy
	Fasting
	Bronze baby syndrome
	Myocardial infarction
	Erythrohepatic protoporphyria

PORPHYRINURIAS RELATED TO VARIOUS TOXIC, DRUG-MEDIATED, CLINICAL, AND METABOLIC CONDITIONS

A large number of factors are capable of increasing porphyrin excretion, owing to different and multiple causes and etiologies (TABLE 3). Generally, a differentiation should be made between secondary coproporphyrinuria with normal uroporphyrin excretion, and secondary copro-uroporphyrinuria as a transitional state to CHP or subclinical PCT. Hepatic factors include the following: 1) disturbance in coproporphyrinogen metabolism, which results from inhibition of coproporphyrinogen oxidase as well as from the rapid loss from, and diminished utilization of coproporphyrinogen in, the hepatocytes, which may also explain why coproporphyrin, its autooxidation product, predominates physiologically in the urine; 2) decreased biliary excretion of coproporphyrin leading to a compensatory urinary excretion, so that the coproporphyrin isomer ratio (I:III) becomes a sensitive index for impaired liver function and intrahepatic cholestasis[7]; 3) disturbed activity of hepatic uroporphyrinogen decarboxylase (UD), leading to a secondary copro-uroporphyrinuria with transition to latent CHP; 4) increased availability of δ-aminolevulinic acid (ALA) after enhancement of ALA synthase activity. In itself, secondary coproporphyrinuria is not associated with porphyria symptoms of a hepatologic–gastroenterologic, neurologic, or dermatologic order, even though coproporphyrinuria can occur with such symptoms. In this case, these are not due to acute or chronic hepatic porphyria, but still, from the clinical point of view, they may require differential diagnosis. Urinary porphyrin precursors are generally not elevated in secondary coproporphyrinuria. Symptomatic coproporphyrinurias develop secondarily to a basic disease, for example, in chronic liver and blood diseases, disturbances in iron metabolism, intoxication (alcohol, foreign chemicals, lead), drug side effects, and tumors (TABLE 3). Secondary coproporphyrinuria is predominantly of hepatic origin. Urinary ALA excretion is usually not increased in secondary coproporphyrinuria and latent CHP conditions, with the exception of the hepatic reaction to lead and some cases of severe acute alcoholism.

HUMAN PORPHYRINURIA AFTER ORAL ALA LOADING

Physiologically, about 70% of urinary coproporphyrin consists of isomer III. Oral intake of ALA maximizes hepatic porphyrinogen synthesis and expands the excretion pattern quantitatively, making it clearer in quality, but producing no unexpected changes in composition. Maximal excretion is observed 24 hr after ALA ingestion (12 mg/kg of body weight). Total urinary porphyrin excretion in healthy persons increases about 30-fold, from 0.1 up to 3.3 μmol/l, with the constellation of the individual porphyrins being similar to normal: uroporphyrin: 15% of total porphyrins; hepta-, hexa-, and pentacarboxylic porphyrin: 2, 3, and 18%; and coproporphyrin: 62%. The total isomer III content was 90%; the uroporphyrin isomer III content, 64%; and the coproporphyrin isomer III content, 94%.[8] These values show that porphyrinogen III synthesis from exogenous ALA is stimulated by an elevated porphyrin-precursor supply as has also been found in acute hepatic porphyria of the three autosomal types and of the rare recessive type (TABLE 1) and in lead poisoning as well. In all these acute porphyria conditions the relative coproporphyrinogen isomer III content increases endogenously by 85 to 95%, reflecting the compensatory dynamic changes in porphyrinogen biosynthesis, beginning with the induction of hepatic ALA synthase.[2] This counterregulatory substrate overproduction can be imitated by oral precursor loading. The dominance of coproporphyrin among urinary porphyrins even under a forced stimulation of porphyrinogen biosynthesis after ALA intake shows a metabolic expansion of the physiologic excretion pattern. The results reflect the function of the various enzymatic sequences in the liver as well as the easy diversion of coproporphyrinogen from hepatocytes and its autooxidation to excreted coproporphyrin. From these studies it may be concluded that a secondary coproporphyrinuria with physiologic isomer composition can be produced by increasing the availability of substrate at the beginning of the heme pathway, certainly by an induction of hepatic ALA synthase, or by an inhibition of coproporphyrinogen oxidase. The increase in coproporphyrinuria after exogenous ALA is dose-dependent. Thus, mild to moderate stages of secondary coproporphyrinuria can be imitated in humans. These modulations in porphyrinogen biosynthesis by an exogenous precursor do not evoke clinical symptoms, and represent one possible pathogenetic model for the metabolic disorder defined as secondary coproporphyrinuria.

DISTURBANCE IN COPROPORPHYRINOGEN METABOLISM

The mitochondrial enzyme coproporphyrinogen oxidase is inhibited by alcohol, as tested in leukocytes from patients with acute alcohol ingestion and also in chronic alcoholics.[9,10] It seems probable that the molecular injury to mitochondria caused by ethanol and acetaldehyde changes the activity of the hepatic enzyme, resulting in a pathologic coproporphyrinuria; alcohol interferes with functions of the inner mitochondrial membrane.[11] It follows that the change in the redox state of the cell with an increase in the hepatic NADH:NAD ratio cannot serve as an explanation of coproporphyrinuria because NADH is not a cofactor of mammalian coproporphyrinogen oxidase. On the other hand, coproporphyrinogen oxidase is highly sensitive to changes in oxygen concentrations. Factors affecting the metabolism of coproporphyrinogen III are intriguing[4]: A decrease in the amount of the active enzyme and competitive inhibition have both been discussed in connection with increased oxidation and decreased uptake of coproporphyrinogen by the inner membranes of the mitochondrion. Hepatic coproporphyrinogen oxidase reacts not only to alcohol and lead, but also to drugs, other heavy metals, foreign chemicals, starvation, and fat in the diet.

Certain drugs and environmental chemicals such as lead can induce secondary coproporphyrinuria, depending on dose and duration of exposure or treatment, on hepatotoxicity, on the hepatic microsomal system, on inhibition of coproporphyrinogen oxidase followed by substrate loss with autooxidation and compensatory increase in substrate availability after induction of hepatic ALA synthase, and on varying individual susceptibility. These proposed mechanisms underlying human coproporphyrinuria have been only partly elucidated in experimental studies, and are difficult to investigate in man. Lead inhibits hepatic coproporphyrinogen oxidase.[12]

In the hexachlorobenzene-induced CHP process, coproporphyrinuria develops first and precedes the initial stage of CHP,[13] corresponding to alcohol-induced coproporphyrinuria in rats; it is followed by an increase in uroporphyrin, indicating a transition to CHP.[14] Both experiments are intended as models for chemically induced coproporphyrinuria in man. Independent of coproporphyrinogen metabolism, drug- and alcohol-mediated coproporphyrinuria may result partly from impaired excretory liver function with increased excretion of urinary coproporphyrin isomer I.

COPROPORPHYRINURIA WITH TRANSITION TO CHP

The absolute and especially the relative increase in heptacarboxyporphyrin within the entire porphyrin pathway shows a transition to CHP caused by an incipient lack of UD, that is, a disturbance in hepatic UD activity leading first to clinically occult CHP, type A (TABLE 4). The following urinary and hepatic porphyrin ratios are biochemical indicators for the transition of secondary coproporphyrinuria to CHP: The urinary uroporphyrin:coproporphyrin ratio as well as the hepatic uroporphyrin: protoporphyrin ratio is less than 1 in secondary coproporphyrinuria; a progression into CHP can be documented by the inversion of these ratios. A secondary coproporphyrinuria is found in one-third of all patients with chronic liver damage, although CHP is seen in less than 15% of all patients with chronic hepatic diseases.

Two requirements are necessary for the transition of a secondary coproporphyrinuria to CHP: 1) genetic disposition of a dominantly inherited autosomal defect in UD, which becomes subclinically or clinically relevant only through alcohol or estrogens (including oral contraceptives) in connection with liver injury[2,15]; 2) toxic inhibition or inactivation of UD in the liver caused by polyhalogenated hydrocarbons, such as

TABLE 4. Transition of Secondary Coproporphyrinuria to CHP Reflected by Urinary Porphyrin Excretion: Progression (↓) Induced by Alcohol, Polyhalogenated Biphenyls, Dioxin, and Vinyl Chloride

Condition	Constellation of Urinary Porphyrins	Total Porphyrins (μmol/24 hr)	Uro (%)	Hepta (%)	Copro (%)	Porphyrin Accumulation in the Liver (Uro, Hepta)	Liver Injury	Cutaneous Symptoms
Normal	Copro ≫ Uro ≫ Hepta	<0.2	<30	<5	>60	–	–	–
Secondary copro-porphyrinuria	Copro ≫ Uro > Penta	<1.0	<20	<3	↓>80	–	+	–
			Transition					
CHP A	Copro > Uro > Hepta	<1.0	<40	↓>5	↓>40	(+)	+	–
CHP B	Uro > Copro > Hepta	<1.5	↓>40	↓<20	↓<40	+	+	–
CHP C	Uro > Hepta > Copro	<2.0	>50	↓>20	↓>10	++	+	–/(+)
CHP D (PCT)	Uro > Hepta ≫ Copro	>2.5	>50	>20	↓<10	+++	+	+/++

TABLE 5. Liver and Red Blood Cell Enzymes in Hexachlorobenzene-Induced CHP in Rats

	Enzyme Activity (mU/g protein)		
	Porphobilinogen Synthase	Porphobilinogen Deaminase	Uroporphyrinogen Decarboxylase
Liver	90.9[a]	1.93[a]	0.03[a]
Controls	47.5	1.16	0.66
Hemolysate	9.22	3.93	0.77
Controls	7.63	3.43	0.75

[a]Significantly different from the controls ($p < .01$).

hexachlorobenzene, polychlorinated and polybrominated biphenyls, dioxins, and probably also methyl and vinyl chloride in combination with liver damage, already present, or caused by one of these agents. This has been demonstrated by experimental studies in hexachlorobenzene- and alcohol-treated rats. The decrease in hepatic UD is dose-dependent, and is restricted to the hepatic enzyme in both conditions (TABLES 5 & 6).[13,14] The progressive inhibition of UD by hexachlorobenzene[13,16] corresponds to the progressive increase in urinary porphyrins. These results represent models for chemically induced porphyrinuria in man.

In contrast to the inherited UD deficiency, which disposes toward the genetic or familial type of PCT,[17,18] the enzyme disturbance should be restricted to the liver in the so-called sporadic, toxic, or acquired type (FIG. 1).[2] Recent studies on immunoreactive hepatic UD in man are consistent with the view that the enzyme abnormality appears to be acquired through the operation of an inactivation process that is restricted to the liver.[19]

TOXIC AND OCCUPATIONAL CONDITIONS

In addition to alcohol,[20] various environmental contaminants show porphyrinogenic activity; these contaminants include hexachlorobenzene, polyhalogenated (polychlorinated and -brominated) biphenyls, chlorodibenzodioxins (especially 2,3,7,8-tetrachlorodibenzo-*p*-dioxin (TCDD)), and vinyl and methyl chloride.[21–23]

Alcohol

Coproporphyrinuria in acute alcohol ingestion is due mainly to an inhibition of coproporphyrinogen oxidase[9] in the liver, counterregulated by an increase in ALA synthase, whereas a stereospecific carrier disturbance in the renal and biliary excretion

TABLE 6. UD Activity in Rat Liver and Spleen after Chronic Alcohol Ingestion

	UD Activity (nmol coproporphyrin/g protein · hr)		
	Alcohol	Control	Significance
Liver	11	20	$p < .01$
Spleen	9	11	$p < .05$

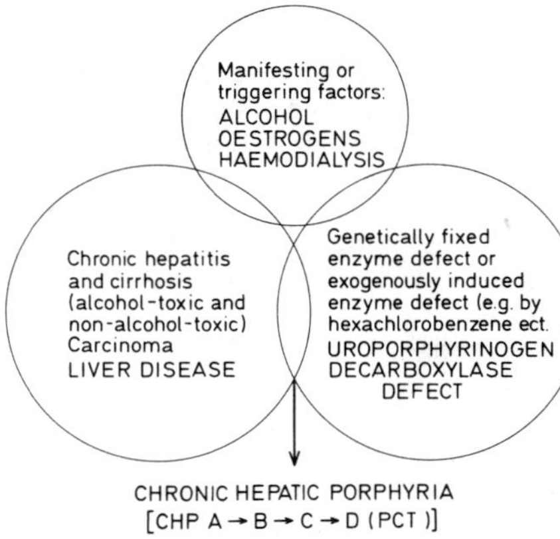

FIGURE 1. Pathobiochemical development of CHPs, including the coincidence of endogenous, especially genetic, and exogenous, especially toxic, factors. Exclusively exogenous factors may be responsible in CHP cases induced by environmental chemicals. Hepatic UD activity is decreased in all cases; however, the activity of this enzyme in blood cells and in other tissues is decreased only in the genetically determined type of CHP or PCT.

of the symmetrical coproporphyrinogen I isomer leads to coproporphyrinuria in chronic alcohol liver disease.[24] Coproporphyrinuria after acute alcohol ingestion is comparable with the hepatic reaction to lead.[2] An interference with lead is possible; this causes coproporphyrinuria as well as ALA and zinc protoporphyrinemia, owing to triple enzyme inhibition (TABLE 7). Thus, the metabolic effects of lead and alcohol on the heme biosynthetic sequences catalyzed by ALA synthase, porphobilinogen synthase, coproporphyrinogen oxidase, and ferrochelatase are similar in principle. The magnitude of alcohol-induced coproporphyrinuria in humans can be compared with that in subclinical lead intoxication. In patients with uncharacteristic abdominal complaints and pathologic coproporphyrinuria, an isomer assay can rule out acute hepatic porphyria and lead intoxication because the percentage of isomer III is increased in both of these porphyrinopathies. Marked symptomatic porphyrinuria with an increase in isomer I is found in primary biliary cirrhosis, and also in erythrohepatic protoporphyria with liver cirrhosis. Excessive coproporphyrin isomer I in the urine is a useful sign of acquired intrahepatic cholestasis, especially in alcoholic liver cirrhosis, and also indicates an inherited derangement of bilirubin transport mechanisms.[1]

Regarding its toxic potential in the pathobiochemical development and clinical manifestation of CHP, alcohol is less porphyrinogenic than hexachlorobenzene, if the findings in metabolite and enzyme changes after chronic alcohol ingestion are compared with those obtained in hexachlorobenzene porphyria in rats.[20] After 25 days of alcohol feeding under the conditions described, one cannot expect the development

TABLE 7. Interference between Alcohol- and Lead-Induced Enzyme Inhibition (↓) and Corresponding Hereditary Enzyme Deficiencies in the Heme Pathway: Toxic Porphyrinopathy versus Genetic Porphyria

Enzyme	Ethanol	Lead	Genetic Deficiency/Disorder
Porphobilinogen synthase	↓	↓↓	Heterozygotes without clinical symptoms but with increased sensitivity to lead Homozygotes with acute intermittent porphyria syndrome
Coproporphyrinogen oxidase	↓	↓↓	Hereditary coproporphyria
Ferrochelatase	↓	↓↓	Erythrohepatic (erythropoietic) protoporphyria

of experimental CHP, considering the fact that it is necessary to feed hexachlorobenzene for more than 30 days in order to simulate a state of latent CHP.[4,13,14] The coproporphyrinuria following chronic alcohol consumption in rats with a coproporphyrin:uroporphyrin ratio of 4:1 associated with a decreased activity of hepatic UD[14] most likely imitates the transition phase between secondary coproporphyrinuria and CHP found in human alcohol liver syndromes.[7] The action of increasing the fluidity of hepatocyte membranes by ethanol or hexachlorobenzene[25] appears to be an important model of the pathogenesis of PCT as a membrane disease[2] in connection with either or both the alcohol-induced reduction or genetically predisposed reduction of UD (FIG. 1).

Secondary coproporphyrinuria is most frequent in acute and chronic alcohol consumption, in alcoholic liver disease, and after occasional intense transient alcohol consumption (from experience in West Germany). Alcohol-induced porphyrinuria syndrome in liver, that is, secondary coproporphyrinuria (copro-uroporphyrinuria with transition to CHP), is the main symptomatic disturbance in porphyrin metabolism. Alcohol inhibits hepatic UD,[14] this inhibition being a prerequisite for the development of CHP and PCT; another prerequisite is probably liver injury (FIG. 1). Potentially, every secondary coproporphyrinuria can turn into CHP via an inactivation of hepatic UD.

Polyhalogenated Biphenyls

Polyhalogenated aromatic hydrocarbons are capable of inducing secondary coproporphyrinuria and subclinical CHP in humans.[26,27] Coproporphyrinuria has been interpreted as being a consequence of the specific inductive properties of polychlorinated biphenyls.[27]

In our contribution to the study of coproporphyrinuria and CHP type A observed in Michigan farm families exposed to polybrominated biphenyls,[26] we found secondary coproporphyrinuria in 47% and initial CHP in 16% of the affected individuals. The transition of secondary coproporphyrinuria into CHP is explained by an inhibition of hepatic UD because it has been shown in hepatocytes that polychlorinated biphenyl compounds induce ALA synthase and inhibit UD, resulting in an accumulation of uro- and heptacarboxylic porphyrin.[28] Pretreatment of liver cultures with phenobarbital potentiated the inhibitory effects of polychlorinated biphenyls, and also of TCDD, on UD activity. The authors suggest that an inhibitor is generated from the parent polyhalogenated aromatic hydrocarbons by metabolic activation of certain cytochrome P-450 isoenzymes.[29]

Vinyl Chloride

A chronic hepatic disorder of porphyrin metabolism was found in 36 workers with vinyl chloride-induced hepatic injury following long-term industrial exposure.[23] Pathologic porphyrinuria, especially secondary coproporphyrinuria with transition to subclinical CHP, was a consistent pathobiochemical parameter for the recognition of vinyl chloride-induced hepatic lesions.[30] The porphyrinuria is of diagnostic value for the incipient toxic phase. Erythrocyte UD activity studied in six cases with initial CHP was normal, suggesting that vinyl chloride inactivates this enzyme only in the liver (TABLE 8).

The pathogenesis of vinyl chloride-induced coproporphyrinuria with a tendency toward CHP may include several factors: 1) inhibition of coproporphyrinogen oxidase

TABLE 8. Dioxin-, Vinyl Chloride-, and Polybrominated Biphenyl-Induced Porphyrinuria

	Dioxin			Vinyl Chloride		After Polybrominated Biphenyl Exposure
	After Exposure (1976)	Follow-up (1980)	Follow-up[a] (UD)	After Exposure (~10 years)	Follow-up[a] UD	
Number of subjects with normal porphyrin excretion	48	7		10		10
Number of subjects with secondary coproporphyrinuria	8	4[b]	8	32	32	6
Number of subjects with CHP (A)	4	1[b]	4	4	4	3
Total	60	12	12	46	36	19

[a] UD activity in erythrocytes not decreased.
[b] Among those with the same condition in 1976.

TABLE 9. Course of CHP in Dioxin-Exposed Family Members from 1977 to 1984

Subject	Family Member	1977	1978	1982	1984	UD in Erythrocytes
1	Father	N	D	C	D	Decreased
2	Mother	N	N	N	N	Normal
3	Son	D	C	A	C	Decreased
4	Daughter	D	C	B	A	Decreased
5	Daughter	N	N	N	N	Decreased

NOTE: N: normal; CHP type D: PCT.

associated with impaired liver function; 2) inhibition of UD associated with pathomorphologically established liver disease[30]; 3) loss of cytochrome P-450 owing to heme destruction by vinyl chloride, which may also influence porphyrin synthesis by induction of ALA synthase. Molecular oxygen is needed for the metabolism of vinyl chloride: A reduced NADPH-dependent, vinyl chloride-mediated destruction of heme bound to cytochrome P-450 and "free" heme[31] will not remain without negative influence on the oxygen-sensitive function of coproporphyrinogen oxidase.

Dioxin

TCDD, a contaminant of various environmental chemicals, is the most potent porphyrinogenic agent known.[32] It induces hepatic ALA synthase[33] and inhibits hepatic UD.[28,34,35]

After an accident in a chemical plant near Seveso, Italy, two persons exposed to low levels of TCDD developed CHP type D and PCT, respectively. They belong to a family of 66 members living partly in the TCDD-contaminated area.[36] Evidence of genetic predisposition was obtained by determining erythrocyte UD[14] in affected and unaffected family members.[37] Erythrocyte activity was diminished in 1978, 1982, and 1984 to the same level (~50%), reflecting a genetic abnormality (TABLE 9). The father and one sister also showed a UD deficiency. In 1977, the proband (son, born 1953, TABLES 9 & 10) developed PCT. One year later, after remission of clinical symptoms under chloroquine therapy, the porphyria process was in the CHP type C phase, declining to a subclinical type A and B without clinical relevance in 1982 (TABLE 9). Porphyrin precursor excretion (ALA and porphobilinogen) was normal in all cases.

The manifestation of PCT in subjects 3 and 4 was probably triggered by the exposure to dioxin on the basis of an inherited UD deficiency. The father did not develop PCT after the Seveso accident; his PCT was triggered by alcohol in 1978 and 1984 (TABLE 10). Furthermore, the renewed progression of the CHP disease process in

TABLE 10. Causes of PCT Manifestations in Members with Inherited Disposition for CHP

Subject[a]		1977		1978	1982		1984
1		N	Alcohol →	D	C	Alcohol →	PCT
3	Dioxin →	D		C	A	Alcohol →	C
4	Dioxin →	D		C	B		A

[a]See TABLE 9.

the proband from type A to C between 1982 and 1984 was probably aggravated by moderate alcohol intake against the background of hereditary enzyme deficiency.

Twelve (20%) of 60 persons from Seveso (not belonging to the above-mentioned family) who were exposed to TCDD had a secondary coproporphyrinuria, and 4 of these 12 showed a transition constellation to CHP type A with an increase of uro- and heptacarboxyporphyrin and coproporphyrin dominance (TABLE 8). Follow-up studies in 1980 revealed a retrogression of CHP and secondary coproporphyrinuria to normal in seven cases. Erythrocyte UD activity was not found to be diminished in these subjects with pathologic porphyrinuria, suggesting that exposure to TCDD could induce PCT only in individuals with an inherited UD deficiency, but not in persons with the normal enzyme. Taking into consideration the porphyrinuria investigations after polybrominated biphenyls and vinyl chloride, one can conclude that progression to the clinically overt phase is observed only in genetically susceptible persons. The purely toxic response to foreign and environmental chemicals seems to be limited in most cases to an increase in urinary porphyrin excretion without clinical relevance (FIG. 2).

The observation that a single acute exposure to TCDD will not evoke PCT in persons without genetic susceptibility has an experimental explanation: TCDD porphyria studies in the rat, that is, studies comparing acute and chronic exposure to TCDD, demonstrated quite clearly that CHP is induced only by chronic, and not by a single, administration of TCDD.[38] Urinary porphyrins and their porphyrin pattern

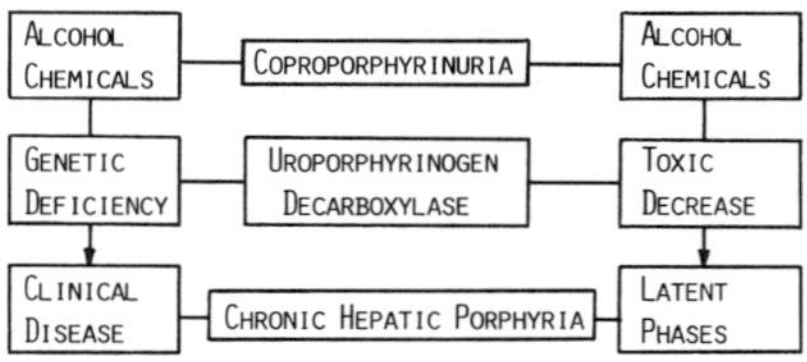

FIGURE 2. Exposure and gene–environment interaction.

represent a highly sensitive indicator for the different degrees of intoxication with TCDD[39]: Chronic administration of varying amounts of TCDD to rats proved that the development of coproporphyrinuria depends on the amount and duration of the intoxication, which will result in CHP after several months, with an increase in heptacarboxyporphyrin and uroporphyrin excretion. The continuous transition to CHP, as reproduced in animal experiments with hexachlorobenzene[13] and TCDD,[39] represents a valid model for the pathogenesis of organochemically induced chronic disturbance in human hepatic porphyrin metabolism.

Further discussion concerning PCT and exposure to TCDD, based on a reevaluation of the literature, shows that the relation between acquired PCT and TCDD has to be questioned.[40] Critical comments support our conclusion that individuals exposed to TCDD can develop a secondary coproporphyrinuria with transition to latent CHP as a toxic hepatic response, and probably do not suffer from the clinical stage of PCT unless a hereditary disposition exists.

SUMMARY

Pathologic porphyrinuria in man is based on a complex etiology and pathogenesis. In hepatic porphyrias, coproporphyrinuria is usually only one of the pathognomostic porphyrin parameters in the urine. Secondary coproporphyrinuria means that an increased excretion of coproporphyrin occurs as the biochemically dominant symptom of a disturbance in porphyrin and heme metabolism during an intoxication, individual condition, or basic disease.

Certain foreign and environmental chemicals, such as hexachlorobenzene, polyhalogenated aromatic hydrocarbons, vinyl chloride, and dioxin, alter the heme pathway functionally. Increased porphyrinuria can follow as a toxic response that is differentiated into secondary coproporphyrinuria and chronic hepatic porphyria. This is characterized by a simultaneous increase in hepatic and urinary uroporphyrin and heptacarboxylic porphyrins, owing to inhibition of hepatic uroporphyrinogen decarboxylase. Most of the coproporphyrinurias observed in man are caused by alcohol ingestion. Dioxin, vinyl chloride, and polyhalogenated biphenyls induce an incipient subclinical stage of chronic hepatic porphyria in persons with normal red cell uroporphyrinogen decarboxylase. In contrast, exposure to dioxin on the part of persons with inherited uroporphyrinogen decarboxylase deficiency can cause latent chronic hepatic porphyria to develop into PCT. Coproporphyrinuria and latent chronic hepatic porphyria do not produce clinical symptoms. Secondary porphyrinuria with transition to chronic hepatic porphyria is a metabolic response following various toxic and pathologic conditions; it serves as a sensitive index for chemical exposure and occupational disease.

ACKNOWLEDGMENT

The author wishes to thank Miss Angelika Nitz for her excellent technical assistance.

REFERENCES

1. Kappas, A., S. Sassa & E. E. Anderson. 1983. The porphyrias. *In* The Metabolic Basis of Inherited Disease. J. B. Stanbury, J. B. Wyngaarden, D. S. Frederickson, J. L. Goldstein & M. S. Brown, Eds.: 1299–1384. McGraw-Hill. New York, NY.
2. Doss, M. 1982. Hepatic porphyrias: Pathobiochemical, diagnostic, and therapeutic implications. *In* Progress in Liver Diseases. H. Popper & F. Schaffner, Eds.: 573–597. Grune & Stratton. New York, NY.
3. Bickers, D. R. 1982. Environmental and drug factors in hepatic porphyria. Acta Dermato.–Venerol. Suppl. (Stockholm) **100:** 29–41.
4. Elder, G. H. 1976. Acquired disorders of haem synthesis. Essays Med. Biochem. **2:** 75–114.
5. Franke, K. & R. Fikentscher. 1935. Die Bedeutung der quantitativen Porphyrinbestimmung mit der Lumineszenzmessung für die Prüfung der Leberfunktion für Ernährungsfragen. Münch. Med. Wochenschr. **82:** 171–172.
6. Goldberg, A. & C. Rimington. 1962. Diseases of Porphyrin Metabolism. C. C. Thomas. Springfield, IL.
7. Doss, M. 1980. Pathobiochemical transition of secondary coproporphyrinuria to chronic hepatic porphyria in humans. Klin. Wochenschr. **58:** 141–148.
8. Doss, M. & E. Schermuly. 1976. Urinary porphyrin excretion pattern and isomer distribution of I and III in human porphyrin disorders. *In* Porphyrins in Human Diseases. M. Doss, Ed.: 189–204. Karger. Basel.
9. McColl, K. E. L., G. G. Thompson, M. R. Moore & A. Goldberg. 1980. Acute ethanol ingestion and heme biosynthesis in healthy subjects. Eur. J. Clin. Invest. **10:** 107–112.
10. McColl, K. E. L., M. R. Moore, G. G. Thompson & A. Goldberg. 1981. Abnormal haem biosynthesis in chronic alcoholics. Eur. J. Clin. Invest. **11:** 461–468.

11. CEDERBAUM, A. I. & E. RUBIN. 1975. Molecular injury to mitochondria produced by ethanol and acetaldehyde. Fed. Proc. **34:** 2045–2051.
12. GOLDBERG, A., P. A. MEREDITH, S. MILLER, M. R. MOORE & G. G. THOMPSON. 1978. Hepatic drug metabolism and haem biosynthesis in lead-poisoned rats. Br. J. Pharmacol. **62:** 529–536.
13. VON TIEPERMANN, R., G. KOSS & M. DOSS. 1980. Uroporphyrinogen decarboxylase deficiency in experimental chronic hepatic porphyria. Hoppe–Seyler's Z. Physiol. Chem. **361:** 1217–1222.
14. DOSS, M., R. VON TIEPERMANN, G. STUTZ & R. TESCHKE. 1981. Alcohol-induced decrease in uroporphyrinogen decarboxylase activity in rat liver and spleen. Enzyme **26:** 24–31.
15. SIXEL-DIETRICH, F. & M. DOSS. 1985. Hereditary uroporphyrinogen decarboxylase deficiency predisposing porphyria cutanea tarda (chronic hepatic porphyria) in females after oral contraceptive medication. Arch. Dermatol. Res. **278:** 13–16.
16. SMITH, A. G. & J. E. FRANCIS. 1983. Synergism of iron and hexachlorobenzene inhibits hepatic uroporphyrinogen decarboxylase in inbred mice. Biochem. J. **214:** 909–913.
17. KUSHNER, J. P., A. J. BARBUTO & G. R. LEE. 1976. An inherited enzyme defect in porphyria cutanea tarda: Decreased uroporphyrinogen decarboxylase activity. J. Clin. Invest. **58:** 1089–1097.
18. DE VERNEUIL, H., G. AITKEN & Y. NORDMANN. 1978. Familial and sporadic porphyria cutanea: Two different diseases. Hum. Genet. **42:** 145–151.
19. ELDER, G. H., A. J. URQUHART, R. E. DE SALAMANCA, J. J. MUNOZ & H. L. BONKOVSKY. 1985. Immunoreactive uroporphyrinogen decarboxylase in the liver in porphyria cutanea tarda. Lancet **ii:** 229–232.
20. DOSS, M. O. 1985. Alcohol and porphyrin metabolism. *In* Alcohol-Related Diseases in Gastroenterology. H. K. Seitz & B. Kommerell, Eds.: 232–252. Springer-Verlag. Berlin.
21. STRIK, J. J. T. W. A. 1978. Porphyrinogenic action of polyhalogenated aromatic compounds, with special reference to porphyria and environmental impact. *In* Diagnosis and Therapy of Porphyrias and Lead Intoxication. M. Doss, Ed.: 151–166. Springer-Verlag. Berlin.
22. MARKS, G. S. 1985. Exposure to toxic agents: The heme biosynthetic pathway and hemoproteins as indicators. CRC Crit. Rev. Toxicol. **15**(2): 151–179.
23. DOSS, M., C.-E. LANGE & G. VELTMAN. 1984. Vinyl chloride-induced hepatic coproporphyrinuria with transition to chronic hepatic porphyria. Klin. Wochenschr. **62:** 175–178.
24. KAPLOWITZ, N., N. JAVITT & A. KAPPAS. 1972. Coproporphyrin I and III excretion in bile and urine. J. Clin. Invest. **51:** 2895–2899.
25. KOSZO, F., L. I. HORVATH, N. SIMON, C. S. SIKLOSI & M. KISS. 1982. The role of possible membrane damage in porphyria cutanea tarda: A spin-label study of rat liver cell membranes. Biochem. Pharmacol. **31:** 11–17.
26. STRIK, J. J. T. W. A., M. DOSS, G. SCHRAA, L. W. ROBERTSON, R. VON TIEPERMANN & E. G. HARMSEN. 1979. Coproporphyrinuria and chronic hepatic porphyria type A found in farm families from Michigan (USA) exposed to polybrominated biphenyls (PBB). *In* Chemical Porphyria in Man. J. J. T. W. A. Strik & J. H. Koeman, Eds.: 29–53. Elsevier/North-Holland. Amsterdam.
27. COLOMBI, A., M. MARONI, A. FERIOLI, M. CASTOLDI, LIU KE JUN, C. VALLA & V. FOA. 1982. Increase in urinary porphyrin excretion in workers exposed to polychlorinated biphenyls. J. Appl. Toxicol. **2:** 117–121.
28. DE VERNEUIL, H., S. SASSA & A. KAPPAS. 1983. Effects of polychlorinated biphenyl compounds, 2,3,7,8-tetrachlorodibenzo-*p*-dioxin, phenobarbital and iron on hepatic uroporphyrinogen decarboxylase: Implications for the pathogenesis of porphyria. Biochem. J. **214:** 145–151.
29. SASSA, S., H. DE VERNEUIL & A. KAPPAS. 1984. Inhibition of uroporphyrinogen decarboxylase activity in polyhalogenated aromatic hydrocarbon poisoning. *In* Biological Mechanisms of Dioxin Action. Banbury Report 18. A. Poland & R. D. Kimbrough, Eds.: 215–224. Cold Spring Harbor Laboratory. Cold Spring Harbor, NY.
30. POPPER, H. & L. B. THOMAS. 1975. Alterations of liver and spleen among workers exposed to vinyl chloride. Ann. N. Y. Acad. Sci. **246:** 172–194.
31. GUENGERICH, F. P. & T. W. STRICKLAND. 1977. Metabolism of vinyl chloride: Destruction

of the heme of highly purified liver microsomal cytochrome P-450 by a metabolite. Mol. Pharmacol. **13:** 993–1004.
32. Goldstein, J. A., P. Hickman, H. Bergman & J. G. Vos. 1973. Hepatic porphyria induced by 2,3,7,8-tetrachlorodibenzo-*p*-dioxin in the mouse. Res. Commun. Chem. Pathol. Pharmacol. **6:** 919–928.
33. Poland, A. & E. Glover. 1973. 2,3,7,8-Tetrachlorodibenzo-*p*-dioxin: A potent inducer of δ-aminolevulinic acid synthetase. Science **179:** 476–477.
34. Sinclair, P. R. & S. Granick. 1974. Uroporphyrin formation induced by chlorinated hydrocarbons (lindane, polychlorinated biphenyls, tetrachlorodibenzo-*p*-dioxin): Requirements for endogenous iron, protein synthesis and drug-metabolizing activity. Biochem. Biophys. Res. Commun. **61:** 124–133.
35. Smith, A. G., J. E. Francis, S. J. E. Kay & J. B. Greig. 1981. Hepatic toxicity and uroporphyrinogen decarboxylase activity following a single dose of 2,3,7,8-tetrachlorodibenzo-*p*-dioxin to mice. Biochem. Pharmacol. **30:** 2825–2830.
36. Strik, J. J. T. W. A., M. M. T. Janssen & A. M. Colombi. 1980. The incidence of chronic hepatic porphyria in an Italian family. Int. J. Biochem. **12:** 879–881.
37. Doss, M., H. Sauer, R. von Tiepermann & A. M. Colombi. 1984. Development of chronic hepatic porphyria (porphyria cutanea tarda) with inherited uroporphyrinogen decarboxylase deficiency under exposure to dioxin. Int. J. Biochem. **16:** 369–373.
38. Goldstein, J. A., P. Linko & H. Bergman. 1982. Induction of porphyria in the rat by chronic versus acute exposure to 2,3,7,8-tetrachlorodibenzo-*p*-dioxin. Biochem. Pharmacol. **31:** 1607–1613.
39. Cantoni, L., M. Salmona & M. Rizzardini. 1981. Porphyrogenic effect of chronic treatment with 2,3,7,8-tetrachlorodibenzo-*p*-dioxin in female rats: Dose–effect relationship following urinary excretion of porphyrins. Toxicol. Appl. Pharmacol. **57:** 156–163.
40. Jones, R. E. & M. Chelsky. 1986. Further discussion concerning porphyria cutanea tarda and TCDD exposure. Arch. Environ. Health **41:** 100–103.

Porphyrins in Urine as an Indication of Exposure to Chlorinated Hydrocarbons

J. J. T. W. A. STRIK

*National Institute of Public Health
and Environmental Hygiene
Bilthoven, the Netherlands*

INTRODUCTION

In the past much progress has been made in the ability to detect smaller and smaller amounts of chemicals in our environment. In addition, the safety testing of chemicals as a part of toxicology has made enormous progress. Years ago, only the acute toxicity was determined; since then, chronic toxicity testing and reproduction studies have gradually come to be required for many compounds. Mutagenic effects and effects on the immune system and the nervous system, including effects on behavior, have become a major concern.

CHLORINATED HYDROCARBONS

Chlorinated hydrocarbons have a widespread and essential role in the chemical industry and in a variety of manufacturing operations, such as those involving plastics, pesticides, or solvents. Toxicological effects of exposure consist in general of damage to the skin, central nervous system, and liver.[1,2]

CHEMICAL PORPHYRIA

Chronic hepatic porphyria in man is a disorder of hepatic porphyrin metabolism that can either be inherited as a congenital anomaly or be caused by exposure to certain chemical compounds. Among the so-called porphyrinogenic chemicals, which mostly belong to the group of halogenated hydrocarbons, one finds vinyl chloride, hexachlorobenzene, polybrominated biphenyls (PBBs), polychlorinated biphenyls (PCBs), tetrachlorodibenzodioxin (TCDD), chlorinated naphthalenes, and some organophosphorus and organochlorine pesticides. The mechanism of chemical hepatic porphyria is based on inhibition of the enzyme uroporphyrinogen decarboxylase, which belongs to the heme-synthesizing pathway. This results in an increased accumulation and excretion of uroporphyrin and heptacarboxylic porphyrin.

Populations and workers exposed to different halogenated hydrocarbons were monitored for urinary porphyrins in order to detect any induction of hepatic porphyria or to find out if this parameter could be relevant in assessing possible exposure. Measured were total porphyrins and the porphyrin composition. This consists of uroporphyrin; heptacarboxylic, hexacarboxylic, and pentacarboxylic porphyrin; and coproporphyrin. Compounds not known for but suspected of having porphyrinogenic potential were screened by use of chick embryo liver cell culture. These studies are presented in detail elsewhere.[3,4]

TOTAL URINARY PORPHYRINS

Patients exposed to porphyrinogenic chemicals have been compared to appropriate controls in several studies. For example, a design shared by several studies involves comparing brominated and chlorinated hydrocarbon exposed workers to unexposed workers, both groups being from the same plant. Furthermore, PBB-exposed inhabitants of Michigan were compared to controls from Wisconsin; Yusho patients (patients exposed to PCB-contaminated rice oil) in Japan, to controls in Japan; workers at the Icmesa plant and inhabitants of Seveso that were exposed to TCDD, to controls from the Seveso region; TCDD-exposed State Forestry Management workers, to unexposed State Forestry Management workers; and TCDD-exposed workers from the Hembrug affair, to different control groups. From these studies, it appears that a total porphyrin excretion of up to 200 μg/l is normal. Higher levels indicative for disease status[5] have not been found in the above-mentioned studies.

PORPHYRIN PATTERN

In the control groups mentioned above, porphyrin patterns in urine were analyzed. In all these different control groups, "increased" uroporphyrin and heptacarboxylic porphyrin levels were found in a few to up to 19% of the subjects. The term increased is based on the interpretation by Doss[5] of the urinary porphyrin excretion by patients in different stages of hepatic porphyria. In young people (students) and children, the incidence of increased uroporphyrin and heptacarboxylic porphyrin was much lower or absent. A higher than normal incidence of the above-mentioned porphyrins is considered to presage chemically induced porphyria, but does not appear to presage symptoms or illnesses attributable to hepatic porphyria as described in detail by Doss.[5] Furthermore, the increase in uroporphyrin and heptacarboxylic porphyrin found in the control groups is believed to be attributable, at least in part, to factors other than chemically induced porphyria (for example, age, sex, alcohol, and estrogens).[5] A higher incidence of these porphyrins was found in the Michigan and Seveso studies. No differences in incidence of these porphyrins were found in the studies with the brominated and chlorinated hydrocarbon exposed workers, the study with the Yusho patients, the study with the State Forestry Management workers, or the study with the Hembrug workers.

CONCLUSION

Qualitative changes in urinary porphyrin patterns are an early indication for a possible chemically induced disturbance in heme synthesis and may be useful in assessing exposure to porphyrinogenic chemicals as demonstrated in the Michigan and Seveso studies. Still, more research is needed to validate the diagnostic significance of the first deviations from normal porphyrin excretion and to understand how various factors affect the range of a normal porphyrin pattern and effect changes in porphyrin patterns within "normal" populations.

REFERENCES

1. HAMILTON, A. & H. L. HARDY. 1974. Industrial Toxicology. 3rd edit. Publishing Sciences Group. New York, NY.

2. NICHOLSON, W. J. & J. A. MOORE, EDS. 1979. Health Effects of Halogenated Aromatic Hydrocarbons. Vol. 320. New York Academy of Sciences. New York, NY.
3. STRIK, J. J. T. W. A. 1981. Health status of factory workers with long-term exposure to chlorinated hydrocarbons. *In* Toxicology of Halogenated Hydrocarbons: Health and Ecological Effects. M. A. Q. Khan & R. H. Stanton, Eds. Pergamon. New York, NY.
4. STRIK, J. J. T. W. A. & J. H. KOEMAN, EDS. 1979. Chemical Porphyria in Man. Elsevier/North-Holland Biomedical Press. Amsterdam.
5. DOSS, M., ED. 1978. Diagnosis and Therapy of Porphyrins and Lead Intoxication. Springer. Berlin.

Mechanism of PCB-Induced Porphyria and Yusho Disease

YOSHIHIKO SEKI, SHOSUKE KAWANISHI,
AND SEIYO SANO

Department of Public Health
Faculty of Medicine
Kyoto University
Kyoto 606, Japan

INTRODUCTION

In 1968, Yusho disease occurred in northern Kyushu, the western part of Japan. Victims of the disease had ingested a particular brand of rice bran oil that had been accidentally contaminated with Kanechlor-400, a mixture of polychlorinated biphenyls (PCBs). About 1800 people were affected.[1] Unfortunately, PCB-induced porphyrias were not well understood at the time of the accident. Urinary determinations of porphyrins in such disorders, for example, had not yet been reported. In 1970, however, Vos and Koeman[2] first reported that oral administration of commercial PCBs to chickens caused hepatic-type porphyria, which was characterized by excretion and hepatic accumulation of uroporphyrin.

In this paper, we describe our studies of the structural requirement of synthetic PCBs for inducing porphyria in cultured chick embryo liver cells. In addition, we describe studies in which the inhibitory effect of PCB isomers on uroporphyrinogen decarboxylase (UROD) was investigated by using cultured liver cells supplemented with exogenous δ-aminolevulinic acid (ALA) and by the purified enzyme from chicken erythrocytes. To clarify the role of inhibition of UROD, we have investigated 1) the time course of induction of ALA synthetase (ALAS), cytochrome P-450, and mixed function oxidases, and 2) the inhibition of UROD in the liver of C57BL/6 mice and ddY mice fed PCBs.

METHODS

Several symmetrical PCB isomers were synthesized from chloroaniline, which was treated with sodium nitrate. The sodium nitrate was then replaced by iodine, and Ullmann condensation followed. 3,4,5,3′,4′-Pentachlorobiphenyl, 5-hydroxy-2,3,4,3′,4′-pentachlorobiphenyl, and 2-hydroxy-3,4,5,3′,4′,5′-hexachlorobiphenyl were generously provided by Dr. T. Mizutani, Kyoto Prefectural University, Kyoto. Commercial PCB mixtures such as Kanechlor-400 and -500 were obtained from Kanegafuchi Chemical Industries. 2,3,7,8-Tetrachlorodibenzo-*p*-dioxin (TCDD) was purchased from Cambridge Isotope Laboratories. 2,7-Dichlorodibenzo-*p*-dioxin (DCDD) and 1,2,3,4-TCDD were obtained from Nishio Industries. Uroporphyrin III octamethyl ester was kindly supplied by Dr. T. K. With, Central County Hospital, Svendborg. Uroporphyrinogen III was prepared by reduction of uroporphyrin III with freshly prepared 3% sodium amalgam according to the method of Sano.[3]

Chick embryo liver cell cultures were prepared according to a method described by

Granick.[4] Purification of UROD from chicken erythrocytes was accomplished by chromatography on DEAE-cellulose, ammonium sulfate fractionation, chromatography on Sephadex G-100, and chromatofocusing.[5] UROD activity was measured as previously reported and expressed as the amount of coproporphyrinogen III formed.[5] The amounts of uroporphyrin III; hepta-, hexa-, and pentacarboxylic porphyrin III; and coproporphyrin III were measured by using high-pressure liquid chromatography (HPLC).[5] Mitochondrial ALAS activity was measured colorimetrically according to the method of Sassa *et al.*[6] employing a modified Ehrlich reagent.[7] Activities of 7-ethoxyresorufin deethylase and 7-ethoxycoumarin deethylase were assayed by a direct fluorometric method using a fluorescence spectrophotometer as described by Prough *et al.*[8] The microsomal cytochrome P-450 content was determined by the method of Omura and Sato[9] in 0.1 M phosphate buffer, pH 7.4. Kanechlor-500 content in liver cytosol of mice was measured by the method of Veierov and Aharonson[10] using a gas chromatograph coupled to an electron capture detector. Protein concentration of various tissue fractions was determined by the method of Lowry *et al.*[11] using bovine serum albumin as a standard.

The urine samples of 20 Taiwanese Yusho patients (Yu-Cheng patients) were kindly supplied by Professor Ta-Cheng Tung, and porphyrins were analyzed in Kyoto using the HPLC technique.

RESULTS

Porphyrin-Inducing Activity of Synthetic PCBs in the Chick Embryo Liver System

TABLE 1 presents the porphyrin-inducing activities of synthetic PCBs in chick embryo liver cell cultures. It was found that 3,4,3′,4′-tetrachlorobiphenyl, 3,4,5,3′,4′-pentachlorobiphenyl, and 3,4,5,3′,4′,5′-hexachlorobiphenyl were the most active inducers. 3,5,3′,5′-Tetrachlorobiphenyl was a moderate inducer. 4,4′-Dichlorobiphenyl; 2,3,2′,3′-, 2,4,2′,4′-, 2,5,2′,5′-, and 2,6,2′,6′-tetrachlorobiphenyl; and 2,3,4,2′,3′,4′- and 2,4,6,2′,4′,6′-hexachlorobiphenyl isomers were weak or inactive inducers.

TABLE 1 also presents the porphyrin-inducing activity of hydroxyderivatives of PCB isomers as the possible metabolites. 2-Hydroxy-3,4,5,3′,4′,5′-hexachlorobiphenyl was a moderate inducer, one that was not so active as 3,4,5,3′,4′,5′-hexachlorobiphenyl. 5-Hydroxy-2,3,4,3′,4′-pentachlorobiphenyl was a poor inducer. It was far less active than we expected 2,3,4,3′,4′-pentachlorobiphenyl would be.

When cultured chick embryo liver cells were supplemented with ALA, a large amount of protoporphyrin accumulated. Addition of 3,4,3′,4′-tetrachlorobiphenyl in the medium resulted in the accumulation of uroporphyrinogen III (see FIGURE 1 of reference 12), indicating the possibility that the active PCBs may block the heme biosynthetic pathway at the level of UROD.

Inhibitory Effect of PCB and TCDD Isomers on Purified UROD

The strongest porphyrin inducers, 3,4,3′,4′-tetrachloro- and 3,4,5,3′,4′,5′-hexachlorobiphenyl, strongly inhibited UROD purified from chicken erythrocytes. The moderate inducer, 3,5,3′,5′-tetrachlorobiphenyl, moderately inhibited the enzyme. Addition of weak inducers resulted in a partial inhibition of the enzyme (see TABLE 3 of reference 5). These results indicate that there is a relationship between the uroporphyrin induction in cultured liver cells and the inhibition of purified UROD by PCB isomers.

The inhibitory effect of 3,4,5,3′,4′,5′-hexachlorobiphenyl on UROD was compared with the inhibitory effects of dioxin isomers (TABLE 2). 2,7-DCDD and 1,2,3,4-TCDD inhibited the enzyme to the same extent as 3,4,5,3′,4′,5′-hexachlorobiphenyl, whereas 2,3,7,8-TCDD, which is known to be the strongest porphyrin inducer among polychlorinated aromatic compounds,[13] markedly inhibited the enzyme activity (80% inhibition by 0.3 μM of 2,3,7,8-TCDD). It was calculated, on a molar basis, that one molecule of the enzyme was inhibited by about two molecules of 2,3,7,8-TCDD (molar inhibition: 35%), suggesting the possibility of special interaction of the TCDD with the enzyme.

TABLE 1. Effect of PCB Isomers on Porphyrin Formation in Chick Embryo Liver Cells

Compound Added	Porphyrin Content (pmol/mg protein)
After Incubation for 24 hr	
None	13 ± 1.5
Biphenyl	15 ± 2.4
4,4′-Dichlorobiphenyl	56 ± 20
2,3,2′,3′-Tetrachlorobiphenyl	80 ± 20
2,4,2′,4′-Tetrachlorobiphenyl	81 ± 29
2,5,2′,5′-Tetrachlorobiphenyl	15 ± 3.0
2,6,2′,6′-Tetrachlorobiphenyl	62 ± 16
3,4,3′,4′-Tetrachlorobiphenyl	3400 ± 480
3,5,3′,5′-Tetrachlorobiphenyl	420 ± 170
2,3,4,2′,3′,4′-Hexachlorobiphenyl	200 ± 65
2,4,6,2′,4′,6′-Hexachlorobiphenyl	140 ± 35
3,4,5,3′,4′,5′-Hexachlorobiphenyl	2600 ± 410
After Incubation for 16 hr	
None	11 ± 1.4
3,4,3′,4′-Tetrachlorobiphenyl	1300 ± 140
3,4,5,3′,4′-Pentachlorobiphenyl	1300 ± 100
2,3,4,2′,3′,4′-Hexachlorobiphenyl	180 ± 34
3,4,5,3′,4′,5′-Hexachlorobiphenyl	1100 ± 71
2-Hydroxy-3,4,5,3′,4′,5′-hexachlorobiphenyl	750 ± 160
5-Hydroxy-2,3,4,3′,4′-pentachlorobiphenyl	34 ± 29

NOTE: Liver cells were prepared and incubated as described in the text. After the medium was changed, a series of PCB isomers dissolved in 25 μl of dimethyl sulfoxide were added. The concentration of the PCB isomer was 2 μg/ml in every case. Each value represents the mean ± SEM for five experiments.

Accumulation of Porphyrins in the Liver of Mice Treated with Kanechor-500 and Other Synthetic PCBs

Male C57BL/6, C57BL/10, and ddY mice, all of which were 7 weeks old, were fed a diet containing 300 ppm of 3,4,5,3′,4′,5′-hexachlorobiphenyl for 3 weeks. Four animals in each group were killed after 1 day of starvation. TABLE 3 summarizes the porphyrin content in the liver. The oral administration of 3,4,5,3′,4′,5′-hexachlorobiphenyl caused a remarkable accumulation of uroporphyrin in the livers of C57BL/6 and C57BL/10 mice but not in the liver of ddY mice (C57BL/6 mice were more sensitive than C57BL/10 mice to PCBs).

TABLE 2. The Inhibitory Effects of 3,4,5,3′,4′,5′-Hexachlorobiphenyl and Dioxins on Purified UROD

Chemical	Amount Added μg	Amount Added μM	Inhibition (%)
3,4,5,3′,4′,5′-Hexachlorobiphenyl	1	1.7	0
	10	17	83
2,3,7,8-Tetrachlorodibenzo-*p*-dioxin	0.01	0.03	35
	0.1	0.3	80
	1	3	100
2,7-Dichlorodibenzo-*p*-dioxin	0.1	0.4	0
	1	4	14
	10	40	57
1,2,3,4-Tetrachlorodibenzo-*p*-dioxin	0.1	0.3	3
	1	3	17
	10	30	66

NOTE: The assay system contained 0.3 nmol of uroporphyrinogen III, 1 mM glutathione, 0.1 mM EDTA, 0.1 M potassium phosphate (pH 6.8), 0.6 μg of UROD (15 nM), and the PCB or one of the dioxins dissolved in 20 μl of dioxane in a total volume of 1 ml. After preincubation for 5 min, the substrate was added and the assay was allowed to proceed at 37°C for 30 min in the dark.

Then, C57BL/6 and ddY mice were fed a diet containing 500 ppm of Kanechlor-500 in order to prepare them for an investigation of porphyrin metabolism. In C57BL/6 mice, the hepatic uroporphyrin level became elevated within 1 week, and an extremely large amount of uroporphyrin accumulated after 3 weeks of feeding (FIGURE 1). By contrast, porphyrin had not accumulated in ddY mice after 3 weeks, was only slightly increased after 6 weeks (FIGURE 1), and remained constant after 10 weeks (data not shown). The amounts of PCB in the liver cytosol for C57BL/6 and ddY mice after 6 weeks of feeding were 28 ± 1.8 and 33 ± 1.9 μg/g of liver, respectively, suggesting that the marked strain difference in uroporphyrin accumulation could not be explained by the amount of PCB retained in the liver cytosol.

TABLE 3. Porphyrin Content in the Liver of Mice Treated with 3,4,5,3′,4′,5′-Hexachlorobiphenyl

Strain	Treatment	Porphyrin (μg/g of liver) Uroporphyrin	Coproporphyrin	Protoporphyrin
C57BL/6	Control	0.11 ± 0.02	0.08 ± 0.02	0.26 ± 0.02
	HCB	12.52 ± 1.69	0.78 ± 0.13	0.99 ± 0.09
C57BL/10	Control	0.11 ± 0.02	0.17 ± 0.02	0.10 ± 0.03
	HCB	2.09 ± 0.56	0.84 ± 0.18	0.47 ± 0.12
ddY	Control	0.10 ± 0.03	0.06 ± 0.02	0.17 ± 0.04
	HCB	0.26 ± 0.03	0.06 ± 0.02	0.24 ± 0.07

NOTE: Seven-week-old male mice were treated with a diet containing 300 ppm of 3,4,5,3′,4′,5′-hexachlorobiphenyl (HCB). After 3 weeks, the animals were killed and porphyrin levels in the liver were determined. Each result is a mean ± SEM of values from four mice.

Cytochrome P-450 Levels and ALAS and UROD Activities in the Liver of Mice Treated with Kanechlor-500

Administration of 500 ppm of Kanechlor-500 to C57BL/6 and ddY mice for 1 week led to increases in the microsomal cytochrome P-450 levels to 4.2 and 2.3 times the control level, respectively (FIGURE 2). A downfield shift of the reduced cytochrome P-450:CO difference spectrum absorption maximum from 450 to 448–449 nm was observed in both strains. In C57BL/6 mice, 7-ethoxyresorufin deethylase and 7-ethoxycoumarin deethylase activities strongly increased and maximized within the first week (135 and 35 times the control values, respectively). These levels were maintained for 3 weeks and then gradually declined (FIGURE 2A). These changes in the activities of mixed function oxidases were in good agreement with changes in the cytochrome P-450 content. In ddY mice, however, the activities of 7-ethoxyresorufin deethylase and 7-ethoxycoumarin deethylase and the cytochrome P-450 content were not increased as much as those of C57BL/6 mice from 1 to 3 weeks, but the activity of 7-ethoxycoumarin deethylase became elevated and exceeded the level in C57BL/6 mice after 6 weeks (FIGURE 2B).

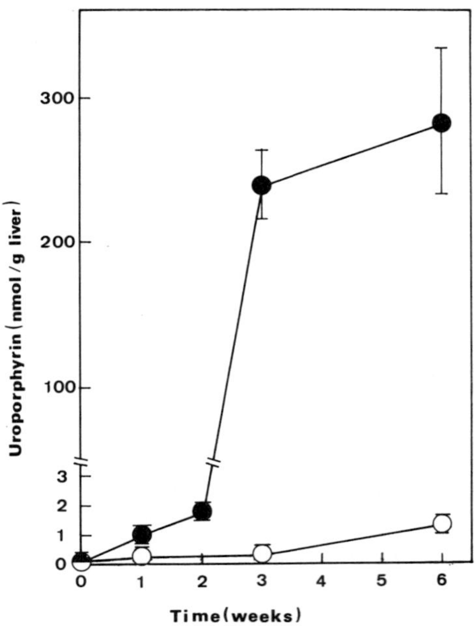

FIGURE 1. Accumulation of uroporphyrin in the liver of male C57BL/6 (●) and ddY (○) mice fed a diet containing 500 ppm of Kanechlor-500. Each point represents a mean ± SEM of values from four mice. There were no increases in uroporphyrin in mice not fed PCBs.

In C57BL/6 mice fed 500 ppm of Kanechlor-500, the activity of mitochondrial ALAS activity gradually increased for 2 weeks (5.7 times the control) and then rapidly rose for 3 weeks (20 times the control). These increases were accompanied by the rapid development of porphyria (FIGURE 3A). The activity of UROD was depressed approximately 40% within 1 week and 80% within 3 weeks. The results suggest that the remarkable induction of ALAS observed from week 2 to week 3 (FIGURE 3A) was due to a decrease in the regulatory heme pool. This decrease seems to have been caused by a greater inhibition of UROD activity because no more induction of cytochrome P-450

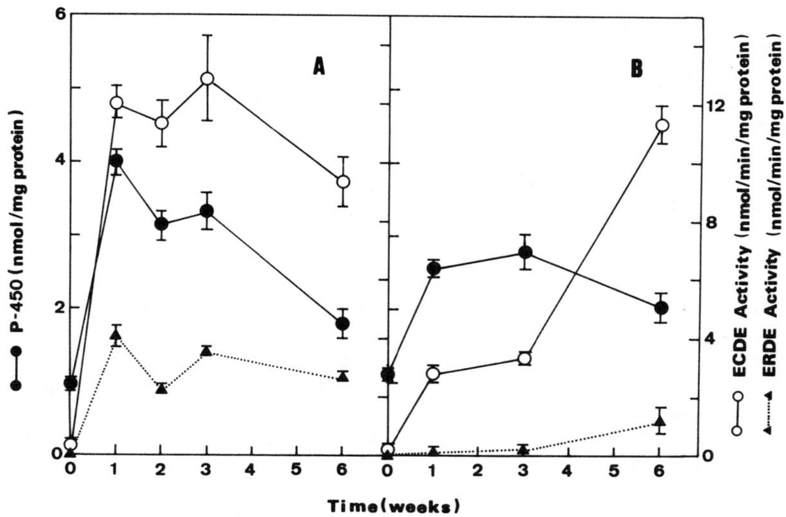

FIGURE 2. Time course of microsomal cytochrome P-450 content (●) and activities of 7-ethoxycoumarin deethylase (○) and 7-ethoxyresorufin deethylase (▲) in the liver of male C57BL/6 (**A**) and ddY (**B**) mice fed a diet containing 500 ppm of Kanechlor-500. Each point represents a mean ± SEM of values from four mice.

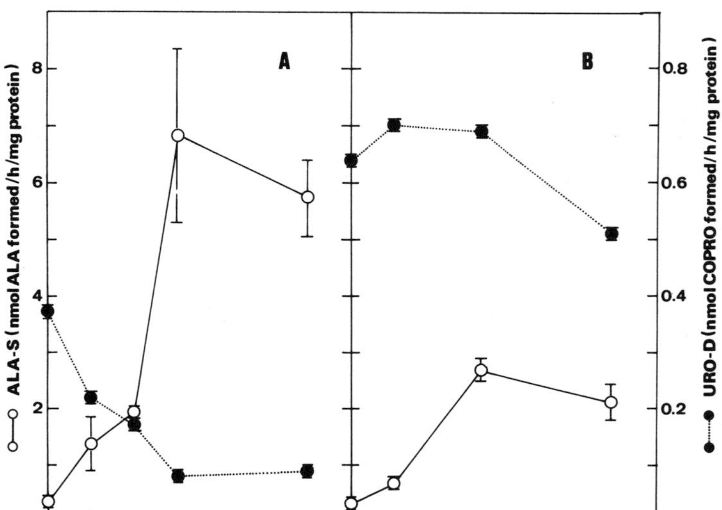

FIGURE 3. Time course of activities of ALAS (○) and UROD (●) in the liver of male C57BL/6 (**A**) and ddY (**B**) mice fed a diet containing 500 ppm of Kanechlor-500. Each point represents a mean ± SEM of values from four mice.

was observed during this period (FIGURE 2A). In ddY mice, UROD activity was unaffected for 3 weeks, and a moderate decrease was observed after 6 weeks (20% inhibition/mg of protein and 35% inhibition/g of liver) (FIGURE 3A). The ALAS activity increased but remained within moderate limits (it was eight times the control activity after 3 weeks), and its changes with time resembled those of cytochrome P-450 (FIGURE 2B), suggesting that the induction of ALAS in ddY mice is mainly caused by heme depletion through (apo)cytochrome P-450 induction, and that this moderate inhibition of UROD (after 6 weeks) does not influence the concentration of heme in the regulatory heme pool.

Is Accumulated Porphyrin in the Liver Uroporphyrinogen or Uroporphyrin?

The amounts of uroporphyrinogen and uroporphyrin in the liver were measured using a specially designed Thunberg-type vessel equipped with a Potter–Elvehjem-type homogeneizer. Laparotomy of the C57BL/6 mice fed PCB for 3 weeks revealed a liver showing strong immediate red fluorescence not only in the external surface but also in the cut surface of the liver. The liver was homogenated, and porphyrins were extracted

TABLE 4. Amounts of Porphyrins in Urine of Taiwanese Patients with PCB Poisoning After 2 Years of PCB Intoxication

	Sex	Age	Uroporphyrin (μg/l)	Coproporphyrin (μg/l)
Patients				
1	M	12	64	38
2	M	17	Trace	2
3	M	17	50	62
4	M	18	15	16
5	M	26	16	61
6	M	30	4	48
7	M	54	32	255
8	M	65	2	6
9	F	14	6	3
10	F	14	4	17
11	F	16	139	30
12	F	16	6	7
13	F	19	14	5
14	F	26	5	34
15	F	26	7	127
16	F	26	14	80
17	F	30	14	5
18	F	31	18	91
19	F	50	24	2
20	F	68	27	14
Controls				
1	M	33	7	25
2	M	34	22	48
3	M	39	10	39
4	M	46	7	20
5	M	47	2	2
6	M	55	15	11

NOTE: Urinary free porpyhrins were separated by HPLC and quantified fluorometrically.

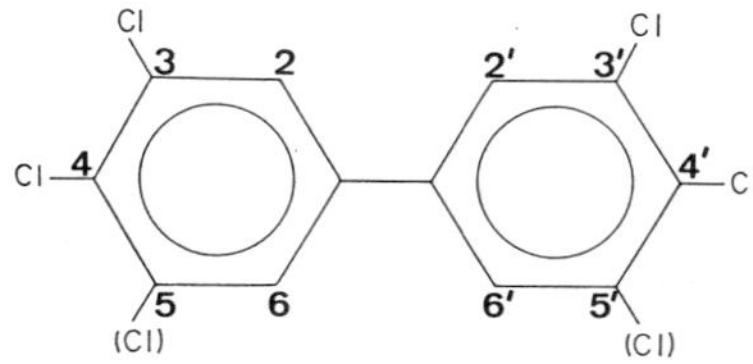

FIGURE 4. Structural requirement of PCBs for porphyrin-inducing activity.

under completely anaerobic conditions. The results show that more than 90% of the uroporphyrinogen formed in C57BL/6 mice fed a diet containing 500 ppm of Kanechlor-500 (3 weeks treatment) was already oxidized to uroporphyrin *in vivo*.

Yusho Disease with a Reference of Porphyria and Clinical Response

The outbreak of Yusho disease in 1968 was the first large-scale outbreak of PCB poisoning in the general population. The rice oil that caused the outbreak was contaminated with Kanechlor-400 that had leaked through pinholes in a duct that had contained PCBs. These PCBs were constituents of a heat-transfer medium that was used during the refining process. It was found that the rice oil contained about 150–1000 ppm of PCBs, 2–7 ppm of polychlorinated dibenzofurans (PCDFs), and 500–800 ppm of polychlorinated quaterphenyls (PCQs).[14]

Yusho disease is clinically characterized by acneiform eruptions; pigmentary deposits on nails, gingiva, and lips; edema of eyelids; and enlargement of follicular openings. Disorders of gastrointestinal, respiratory, and nervous systems are also seen.[15]

Unfortunately, an episode of food poisoning similar to the Yusho outbreak occurred in the central part of Taiwan in 1979 and affected about 1700 people.[16]

TABLE 4 shows the uroporphyrin and coproporphyrin levels in the urine of Taiwanese patients after 2 years of PCB intoxication. Three of the 20 patients examined showed a slightly elevated uroporphyrin level in urine (over 50 μg/l).

DISCUSSION

We studied the structural requirement of synthetic PCBs for inducing porphyria in cultured liver cells and found that 3,4,3′,4′-tetrachlorobiphenyl, 3,4,5,3′,4′-pentachlorobiphenyl, and 3,4,5,3′,4′,5′-hexachlorobiphenyl were the most active (TABLE 1). On the basis of these data, we have established the kind of structure a PCB with porphyria-inducing properties is likely to have (FIGURE 4). First, a coplanar structure of the chlorobiphenyl is essential for porphyrinogenic activity. The most potent inducers, which include 3,4,3′,4′-tetrachlorobiphenyl, 3,4,5,3′4′-pentachlorobiphenyl, and 3,4,5,3′,4′,5′-hexachlorobiphenyl, have such a structure. Second, occupation of both *para* and *meta* positions by chlorine atoms is required for porphyrin induction. 4,4′-Dichlorobiphenyl, for example, is inactive whereas 3,4,3′,4′-tetrachlorobiphenyl, 3,4,5,3′4′-pentachlorobiphenyl, and 3,4,5,3′,4′,5′-hexachlorobiphenyl are active.

The question arises whether porphyrin-inducing activity is due to the direct action of added chemicals or to some active metabolites formed *in vivo*. We previously studied what kind of PCB isomers would accumulate in the mouse liver by feeding mice a diet containing PCB isomers.[17] We classified PCB isomers into three types based on PCB accumulation as shown in FIGURE 5. Isomers of the first type included the 3,4,3′,4′-,

2,3,2′,3′-, 2,5,2′,5′-, and 2,6,2′,6′-tetrachlorobiphenyls. These isomers did not accumulate appreciably in the liver. These four compounds have unsubstituted pairs of carbon atoms in the molecule and hence are likely to be metabolized to the hydroxy derivatives. Isomers of the second type included 2,4,2′,4′-tetrachlorobiphenyl and 2,3,4,2′,3′,4′-hexachlorobiphenyl. The accumulation of these isomers was intermediate between the accumulation of the first type and that of the third. Although these isomers possess unsubstituted pairs of carbon atoms, they are not so easily metabolized. They have chlorine atoms at the *ortho-ortho′* positions and hence possess sterically hindered structures. Isomers of the third type include 3,5,3′,5′-tetrachlorobiphenyl and the 2,4,6,2′,4′,6′- and 3,4,5,3′,4′,5′-hexachlorobiphenyls. The accumulation of these isomers in the liver was marked. These compounds contain only isolated unsubstituted carbon atoms, which may explain why these compounds are only metabolized with difficulty.

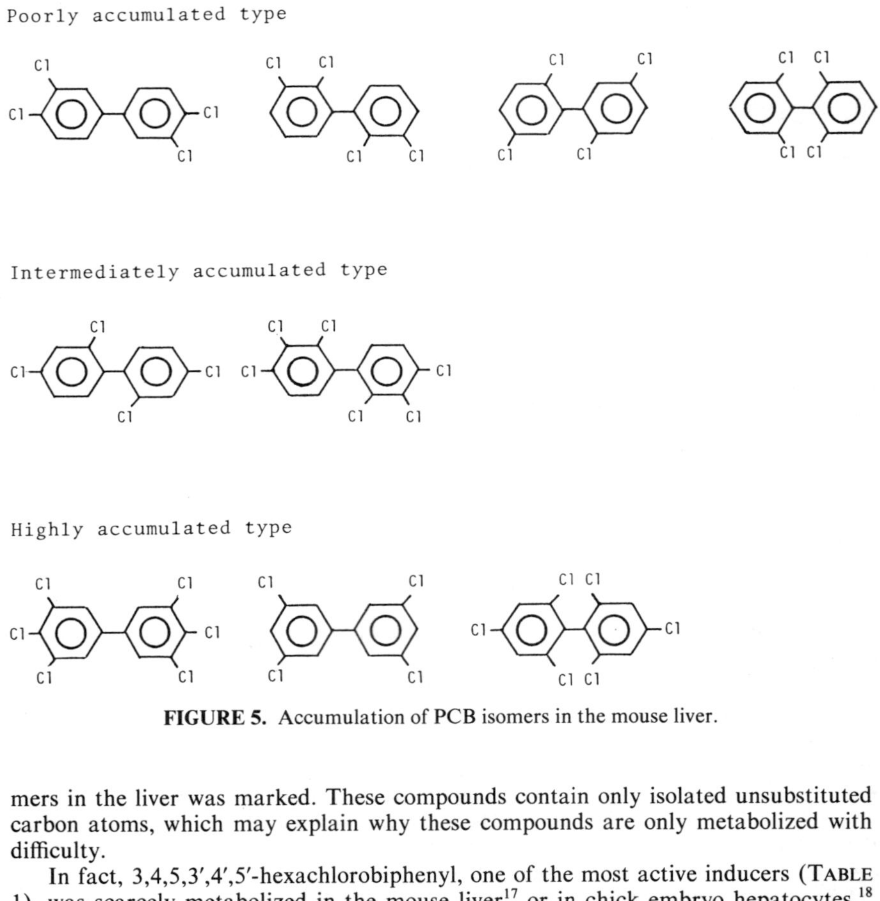

FIGURE 5. Accumulation of PCB isomers in the mouse liver.

In fact, 3,4,5,3′,4′,5′-hexachlorobiphenyl, one of the most active inducers (TABLE 1), was scarcely metabolized in the mouse liver[17] or in chick embryo hepatocytes.[18] 2-Hydroxy-3,4,5,3′,4′,5′-hexachlorobiphenyl, a possible hydroxy derivative of 3,4,5,3′,4′,5′-hexachlorobiphenyl, was a moderate porphyrin inducer, and its inducing activity was not much stronger than that of the original hexachlorobiphenyl. Therefore, it seems likely that porphyrin induction by 3,4,5,3′,4′,5′-hexachlorobiphenyl is due to the original chemical and not to its metabolites. It is important to note that

3,4,5,3′,4′,5′-hexachlorobiphenyl does directly inhibit UROD *in vitro* (TABLE 2). The same results were obtained with 2,3,7,8-TCDD.

There is, however, another hypothesis: the active metabolite of PCBs may inhibit the enzyme. This hypothesis was based on observations that PCB added either to homogenously purified UROD from human erythrocytes or to homogenates from chick embryo liver cell cultures did not inhibit the enzyme activity.[19,20] The details of the experiments in which these observations were made were not given, so we cannot be sure why these observations are inconsistent with our own results. It may be that we used a different enzyme concentration. We could observe the inhibitory effects of PCB or TCDD only when both a low concentration of enzyme and a low concentration of PCB were used.

According to an alternative hypothesis proposed by Ferioli *et al.*,[21] polychlorinated aromatic compounds may cause the metabolic block indirectly, by stimulating the production of reactive species such as peroxides and free radicals. Our present results show that the autoxidation of uroporphyrinogen to uroporphyrin occurs in the liver of PCB-induced porphyria. Peroxides formed during the autoxidation of uroporphyrinogen may cause lipid peroxidation. In addition, uroporphyrin thus formed cannot be further metabolized by UROD.

C57BL/6 mice fed 500 ppm of Kanechlor-500 for 3 weeks accumulated an extremely large amount of hepatic uroporphyrin. Under the same treatment, however, only a small amount of uroporphyrin accumulated in ddY mice (FIGURE 1). To determine why porphyria developed in C57BL/6 mice and not in ddY mice, we studied how the onset of porphyria; the induction of cytochrome P-450, mixed function oxidases, and ALAS; and the inhibition of UROD could be related. C57BL/6 mice were highly responsive to PCBs, which induced the drug-metabolizing function, and ddY mice could also be regarded as a responsive strain (FIGURE 2). On the other hand, there was an obvious difference in the inhibition of UROD between the two mice strains (FIGURE 3). These results indicate that the development of porphyria was related to the inhibition of UROD rather than to the induction of the drug-metabolizing function. This conclusion is supported by the findings of Jones and Sweeney.[22] They showed that a dose of TCDD sufficient to cause maximal induction of aryl hydrocarbon hydroxylase activity, even in DBA/2J mice, did not produce any increase of urinary porphyrin excretion in DBA/2J mice. It is noteworthy that the control value for UROD activity in ddY mouse liver was higher than the control value for UROD activity in C57BL/6 mouse liver (about 1.5 times higher when the activities were expressed as per g of liver and about 1.7 times higher when they were expressed as per mg of cytosolic protein) whereas the ALAS activities were similar in the two strains (FIGURE 3). These results suggest that the initial difference between the activities may be attributable to the development of porphyria by toxic agents.

ALAS is a rate-limiting enzyme for heme biosynthesis in the liver. The other enzymes involved in the heme biosynthetic pathway have sufficient biosynthetic capability compared with ALAS. Hence, the decrease in UROD activity does not always lead to the immediate accumulation of uroporphyrinogen. This accumulation should be due to the accompanying increase in ALAS activity. From this point of view, the activities of UROD were compared to those of ALAS.[23] The UROD to ALAS ratio was defined as the UROD activity divided by one-eighth the ALAS activity (eight molecules of ALA convert to one molecule of uroporphyrinogen). When the UROD to ALAS ratio decreases to less than 1.0, hepatic accumulation of uroporphyrinogen should be expected. TABLE 5 shows the ratios calculated for two strains of mice.

The liver of control C57BL/6 mice has approximately 20 times higher UROD activity than normally required to maintain heme biosynthesis. After the administration of Kanechlor-500, the ratio sharply declined, but a small amount of porphyrin accumulated while the ratios exceeded the value of 1.0 (until 2 weeks). When the ratios fell to less than 1.0 (after 3 weeks), marked accumulation of uroporphyrin began. In fact, the amount of uroporphyrin was 2100 times greater than that in the controls. This may explain why the onset of porphyria was suddenly caused between the 2nd and 3rd weeks. On the other hand, because UROD activity was originally higher in ddY mice and not inhibited strongly by Kanechlor-500, the ratios remained between 4.4 and 22.4, and little uroporphyrin accumulated.

It should be clarified whether PCBs can induce porphyria in humans. Fung *et al.*[24] reported that the average amount of urinary uroporphyrin for 69 Taiwanese patients with PCB poisoning was 41.2 ± 24.6 μg/24 hr, a value significantly higher ($p < .01$) than the normal control value, 13.6 ± 11.8 μg/24 hr. Urinary coproporphyrin, however, showed no significant difference. In addition, our data (TABLE 4) showed that urinary uroporphyrin increased in some patients even after 2 years of PCB ingestion. It might be that Yu-Cheng patients who show uroporphyrinuria have lower UROD activities.[25] Finally, it would be worth while to study further whether PCDFs or 2,3,4,5,3′,4′-hexachlorobiphenyl (found in Yusho patients) could cause hepatic porphyria. Kunita *et al.*[26] reported that contaminants such as PCDFs were the main causative agents of the clinical response of Yusho disease (acneiform eruption of skin as well as edema of the lips and eyelids) in cynomolgus monkeys.

TABLE 5. Activities and Ratios of Heme Biosynthetic Pathway Enzymes in the Liver of Mice Treated with Kanechlor-500

Strain	Treatment	ALAS Activity	UROD Activity	UROD/ (ALAS/8)	Uroporphyrin Level
C57BL/6	Control	8.2	21.4	20.9	1
	PCB				
	1 week	12.1	7.4	4.9	9
	2 weeks	47.6	8.5	1.4	16
	3 weeks	123	4.4	0.3	2100
ddY	Control	7.1	32.8	36.9	1
	PCB				
	1 week	10.8	30.3	22.4	3
	3 weeks	42.9	31.8	5.9	3
	6 weeks	38.7	21.1	4.4	14
	10 weeks	31.9	18.4	4.6	13

NOTE: Seven-week-old male mice were treated with a diet containing 500 ppm of Kanechlor-500 and killed after the indicated intervals. Activities of ALAS and UROD in the liver were measured and expressed in nmol ALA formed and nmol coproporphyrinogen III formed, respectively, per hr per g of liver. The ratios are explained in the text. The hepatic uroporphyrin levels are expressed relative to the control values. Each result is a mean of values from four mice.

SUMMARY

PCBs are known to be potent inducers of chemical porphyria. We studied the structure–activity relationship of synthetic PCBs as porphyrin inducers and found that 3,4,3′,4′-tetrachlorobiphenyl, 3,4,3′,4′,5′-pentachlorobiphenyl, and 3,4,5,3′,4′,5′-hexachlorobiphenyl were the most active inducers. The structural requirement for potent porphyrinogenic activity of PCB isomers was the substitution of chlorine atoms at the *para* and *meta* positions. Isomers fulfilling this requirement had more highly conjugated and nearly coplanar conformations. In addition, it was demonstrated that inhibition of UROD by the most active porphrin inducers occurred *in vitro* using purified enzyme. These findings could explain how porphyrinogenic PCBs, by primarily inhibiting UROD and hence depleting heme, ultimately increase ALAS synthesis.

The outbreak of Yusho disease in Japan (which was caused by PCB-contaminated rice oil) and a similar episode of food poisoning in Taiwan (which was also related to PCB contamination) were discussed, particularly with reference to urinary porphyrin levels and clinical response.

REFERENCES

1. KURATSUNE, M., Y. MORIKAWA, T. HIROHATA, M. NISHIZUMI, S. KOHCHI, T. YOSHIMURA, J. MATSUZAKA, A. YAMAGUCHI, N. SARUTA, N. ISHINISHI, E. KUNITAKE, O. SHIMONO, K. TAKIGAWA, K. OKI, M. SONODA, T. UEDA & M. OGAWA. 1969. An epidemiologic study on "Yusho" or chlorobiphenyl poisoning. Fukuoka Acta Med. **60:** 513–532.
2. VOS, J. G. & J. H. KOEMAN. 1970. Comparative toxicologic study with polychlorinated biphenyls in chickens with special reference to porphyria, edema formation, liver necrosis, and tissue residues. Toxicol. Appl. Pharmacol. **17:** 656–668.
3. SANO, S. 1966. 2,4-Bis-(β-hydroxypropionic acid)deutroporphyrinogen IX, a possible intermediate between coproporphyrinogen III and protoporphyrin IX. J. Biol. Chem. **241:** 5276–5283.
4. GRANICK, S. 1966. The induction *in vitro* of the synthesis of δ-aminolevulinic acid synthetase in chemical porphyria: A response to certain drugs, sex hormones, and foreign chemicals. J. Biol. Chem. **241:** 1359–1375.
5. KAWANISHI, S., Y. SEKI, & S. SANO. 1983. Uroporphyrinogen decarboxylase: Purification, properties, and inhibition by polychlorinated biphenyl isomers. J. Biol. Chem. **258:** 4285–4292.
6. SASSA, S., A. KAPPAS, S. E. BERNSTEIN & A. P. ALVARES. 1979. Heme biosynthesis and drug metabilism in mice with hereditary hemolytic anemia: Heme oxygenase induction as an adaptive response for maintaining cytochrome P-450 in chronic hemolysis. J. Biol. Chem. **254:** 729–735.
7. MAUZERALL, D. & S. GRANICK. 1956. The occurrence and determination of δ-aminolevulinic acid and porphobilinogen. J. Biol. Chem. **219:** 435–446.
8. PROUGH, R. A., M. D. BURKE & R. T. MAYER. 1978. Direct fluorometric methods for measuring mixed-function oxidase activity. Methods Enzymol. **52:** 372–377.
9. OMURA, T. & R. SATO. 1964. The carbon monoxide-binding pigment of liver microsomes. I. Evidence for its hemoprotein nature. J. Biol. Chem. **239:** 2370–2377.
10. VEIEROV, D. & N. AHARONSON. 1980. Economic method for analysis of fluid milk for organochlorine residues at the 10 ppb level. J. Assoc. Off. Anal. Chem. **63:** 532–535.
11. LOWRY, O. H., N. G. ROSEBROUGH, A. L. FARR, & R. J. RANDALL. 1951. Protein measurement with the Folin phenol reagent. J. Biol. Chem. **193:** 265–275.
12. KAWANISHI, S., Y. SEKI & S. SANO. 1981. Polychlorobiphenyls that induce δ-aminolevulinic acid synthetase inhibit uroporphyrinogen decarboxylase in cultured chick embryo liver cells. FEBS Lett. **129:** 93–96.
13. GOLDSTEIN, J. A., P. HICKMAN, H. BERGMAN, & J. G. VOS. 1973. Hepatic porphyria induced by 2,3,7,8-tetrachlorodibenzo-*p*-dioxin in the mouse. Res. Commun. Chem. Pathol. Pharmacol. **6:** 919–928.
14. MASUDA, Y. & M. KURATSUNE. 1979. Toxic compounds in the rice oil which caused Yusho. Fukuoka Acta Med. **70:** 229–237.
15. GOTO, M. & K. HIGUCHI. 1969. The symptomatology of Yusho in dermatology. Fukuoka Acta Med. **60:** 409–431.

16. Lan, C. F., P. H. S. Chen, L. L. Shich & Y. H. Chen. 1981. An epidemiological study on polychlorinated biphenyl poisoning in Taichung area. Clin. Med. (Taipei) **7:** 96–100.
17. Kawanishi, S., T. Mizutani & S. Sano 1978. Induction of porphyrin synthesis in chick embryo liver cell cultures by synthetic polychlorobiphenyl isomers. Biochim. Biophys. Acta **540:** 83–92.
18. Sinclair, P. R., W. J. Bement, H. L. Bonkovsky, R. W. Lambrecht, J. E. Frezza, J. F. Sinclair, A. J. Urquhart & G. H. Elder. 1986. Uroporphyrin accumulation produced by halogenated biphenyls in chick embryo hepatocytes. Biochem. J. **237:** 63–71.
19. Verneuil, H. D., S. Sassa & A. Kappas. 1983. Effect of polychlorinated biphenyl compounds, 2,3,7,8-tetrachlorodibenzo-*p*-dioxin, phenobarbital and iron on hepatic uroporphyrinogen decarboxylase. Biochem. J. **214:** 145–151.
20. Swain, M. G., S. B. Follows & G. S. Marks. 1983. Inhibition of uroporphyrinogen decarboxylase by 3,3′,4,4′-tetrachlorobiphenyl in chick embryo liver cell culture. Can. J. Physiol. Pharmacol. **61:** 105–108.
21. Ferioli, A., C. Harvey & F. DeMatteis. 1984. Drug-induced accumulation of uroporphyrin in chicken hepatocyte cultures: Structural requirement for the effect and role of exogenous iron. Biochem. J. **224:** 769–777.
22. Jones, K. G. & G. D. Sweeney. 1980. Dependence on the porphyrinogenic effect of 2,3,7,8-tetrachlorodibenzo-*p*-dioxin upon inheritance of aryl hydrocarbon hydroxylase responsiveness. Toxicol. Appl. Pharmacol. **53:** 42–49.
23. Wood, J. S. & R. M. Kardish. 1983. Development aspects of hepatic heme biosynthetic capability and hematotoxicity. II. Studies on uroporphyrinogen decarboxylase. Biochem. Pharmacol. **32:** 73–78.
24. Fung, J. L., J. C. King, C. L. Su & C. T. Ta. 1980. Studies on patients with polychlorinated biphenyl poisoning: Determination of urinary coproporphyrin, uroporphyrin, δ-aminolevulinic acid, and porphobilinogen. J. Formosan Med. Assoc. **79:** 990–995.
25. Doss, M., H. Sauer, R. Tiepermann & A. M. Colombi. 1984. Development of chronic hepatic porphyria (porphyria cutanea tarda) with inherited uroporphyrinogen decarboxylase deficiency under exposure to dioxin. Int. J. Biochem. **16:** 369–373.
26. Kunita, N., T. Kashimoto, H. Miyata, S. Fukushima, S. Hori & H. Obana. 1984. Causal agents of Yusho. Am. J. Ind. Med. **5:** 45–58.

Regulatory Roles of High-Affinity Metal-Binding Proteins in Mediating Lead Effects on δ-Aminolevulinic Acid Dehydratase

PETER L. GOERING[a]

Division of Life Sciences
Office of Science and Technology
Center for Devices and Radiological Health
Food and Drug Administration
Rockville, Maryland

BRUCE A. FOWLER

National Institute of Environmental Health Sciences
Research Triangle Park, North Carolina

INTRODUCTION

The enzyme δ-aminolevulinic acid dehydratase (ALAD; porphobilinogen synthase; EC 4.2.1.24) is localized in the cell cytoplasm and catalyzes the second step in heme biosynthesis, which involves the condensation of 2 mol of substrate, aminolevulinic acid, to form 1 mol of the monopyrrole, porphobilinogen. ALAD requires Zn as an essential cofactor and can be activated by excess Zn *in vitro* and *in vivo*.[1,2] ALAD is readily inhibited by Pb, a ubiquitous environmental contaminant, in most tissues and is a sensitive indicator of Pb exposure and toxicity.[3–5] In contrast, ALAD in kidney, and to a lesser extent in brain, is known to be relatively resistant to inhibition by Pb.[6–10] The resistance of ALAD to Pb inhibition in these tissues could be explained by the presence of tissue-specific factors mediating the interaction of Pb with the enzyme.

Many essential metals do not simply exist as free ions within the cell but are under complex homeostatic control. One mechanism by which these metals are regulated is via high-affinity metal-binding proteins. For example, Fe metabolism is controlled by ferritin and transferrin,[11] Cu by ceruloplasmin,[12] and Ca by calmodulin[13] and other Ca-binding proteins.[14] Metallothionein (MT), a low molecular weight, cysteine-rich protein, is postulated to serve as a reservoir for the storage of Zn and Cu, as a donor of these ions for the biosynthesis of Cu- and Zn-requiring enzymes, or both.[15] In this regard, ZnMT and CuMT have been shown to reactivate several apoenzymes *in vitro*.[16–18] ZnMT also appears to play a role in the detoxication of metals of environmental health significance, such as Cd and Hg. Following induction of ZnMT synthesis by prior exposure to these metals, this protein sequesters a high percentage of

[a]Address for correspondence: Center for Devices and Radiological Health, Food and Drug Administration (HFZ-112), 12709 Twinbrook Parkway, Rockville, Maryland 20857.

these cations intracellularly into an inert metal–protein complex, making them less available to interact with sensitive organelles or enzyme systems.[19–22]

Tissue-specific, cytosolic Pb-binding proteins (PbBPs), which are not detectable in liver or lung, have recently been discovered in kidney and brain.[23] These endogenous proteins, which are distinct from ZnMT, appear to act as the initial cytosolic ligands for Pb in kidney and brain and are capable of mediating the nuclear translocation of Pb.[24] The dissociation constant (K_d) of these PbBPs for Pb is approximately 10^{-8} M, demonstrating the high affinity of these proteins for Pb.[24]

The following series of experiments were undertaken to examine the role these metal-binding proteins (PbBPs and ZnMT) play in regulating the activity of ALAD, mediating the inhibition of ALAD by Pb, or both. ALAD is an ideal molecule to study these relationships as it is dependent upon Zn as an essential cofactor and is a sensitive indicator of Pb exposure and toxicity.

RESULTS AND DISCUSSION

Previous results from this laboratory have demonstrated that ALAD in rat kidney is relatively resistant to inhibition by Pb after chronic *in vivo* exposure.[8] The relative resistance of rat kidney ALAD was further demonstrated after acute exposure of rats to 10 mg Pb/kg body wt, ip (TABLE 1). Sixteen hours after injection of Pb, liver ALAD activity was inhibited to 54% of the control value, whereas kidney ALAD activity was inhibited to only 75% of the control value. To eliminate the possibility that the differential sensitivity of ALAD to Pb *in vivo* may be due to pharmacokinetic parameters associated with differences in absorption, distribution, or tissue accumulation of Pb, the sensitivity of liver, kidney, and brain ALAD to Pb *in vitro* was studied. In order to determine whether this phenomenon occurred after exposure of ALAD to Pb *in vitro,* the IC_{50} values for the inhibition of ALAD by Pb were estimated by the method of Dixon.[25] The IC_{50} value for Pb inhibition of kidney ALAD was 3 μM Pb (FIGURE 1, top panel), which is 7.5-fold higher than the 0.4 μM Pb value obtained for liver (FIGURE 1, lower panel). In a similar experiment in brain, the IC_{50} value for inhibition of brain ALAD by Pb was estimated to be 0.7 μM Pb (FIGURE 2, top panel), which is more than 2-fold higher than the 0.3 μM Pb value obtained for liver (FIGURE 2, lower panel). These findings confirm tissue differences in sensitivity of ALAD to inhibition by Pb.

In light of the above findings, experiments were initiated to determine whether a tissue-specific factor(s) was responsible for the relative insensitivity of kidney and brain ALAD to inhibition by Pb. The presence of low molecular weight PbBP species in both rat kidney and brain, previously observed by Oskarsson *et al.,*[23] was confirmed

TABLE 1. Effect of Pb on ALAD Activity in Rat Kidney and Liver after Acute Injection

	ALAD Activity (nmol PBG/hr/mg protein)	
	Liver	Kidney
Control	9.33	2.84
Pb	5.03 (54)	2.14 (75)

[a]ALAD activity was determined 16 hr after injection of 10 mg Pb/kg, ip. Values in parentheses are percentages of the control values.

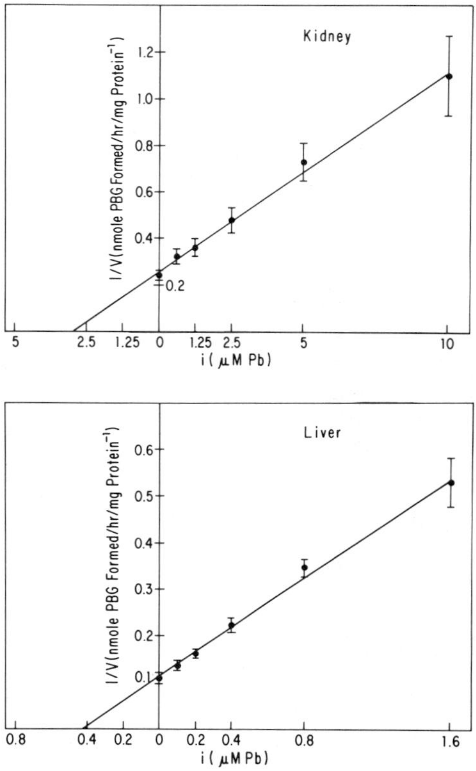

FIGURE 1. Determination of the IC_{50} for inhibition of rat kidney (top panel) and liver (lower panel) ALAD by Pb *in vitro* as described by Dixon.[25] The IC_{50} value is obtained from the intersection of the curve with the abscissa. Each value is a mean ± SEM ($N = 4$). The IC_{50} value for kidney is 3.31 ± 0.62 μM Pb and is significantly higher ($p < .05$) than the IC_{50} value for liver, 0.44 ± 0.08 μM Pb. Sources of ALAD were tissue postmitochondrial supernatants obtained as described in reference 9. PBG: porphobilinogen; i: inhibitor concentration; 1/V: reciprocal velocity. (Used with permission from reference 9, copyright by the American Society of Pharmacology and Experimental Therapeutics.)

using gel filtration chromatography (FIGURE 3). These PbBPs present in kidney and brain eluted from the Sephadex G-75 column with retention coefficients of 2.0 to 2.5 and 1.7 to 2.2, respectively, indicating a higher molecular mass for the protein isolated from brain. The molecular mass for kidney and brain PbBPs estimated by column chromatography was 9000 and 12 000 daltons, respectively. This protein fraction was not a major constituent in liver.

After identifying the PbBPs in kidney and brain, subsequent experiments were performed to determine whether inhibition of hepatic ALAD by Pb could be reversed by addition of kidney or brain PbBP to liver postmitochondrial supernatants. Partially purified kidney and brain PbBPs were isolated for use in an *in vitro* ALAD assay system as described by Goering and Fowler[9,26] and Goering *et al.*[27] The effect of increasing concentrations of kidney and brain PbBP on hepatic ALAD activity in the presence of 0.4 μM Pb is shown in FIGURE 4. In the presence of 0.4 μM Pb alone, ALAD activity was inhibited to 36% of the control value. Inhibition was attenuated in a concentration-related manner with the magnitude of this effect increasing progressively to a point at which activity was inhibited to only 62% of the control value after addition of 1.8 mg of renal PbBP and to 55% of the control value after addition of 0.6 mg of brain PbBP. The different concentrations of renal and brain PbBP required to achieve comparable attenuations of Pb inhibition of ALAD and the different gel filtration elution profiles (FIGURE 3) suggest that the nature and number of Pb-binding

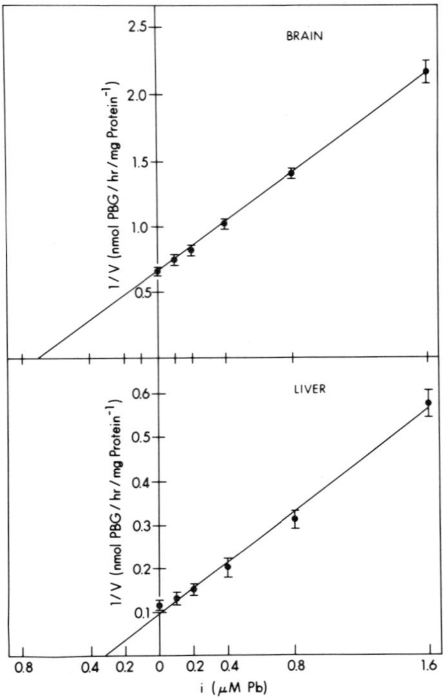

FIGURE 2. Determination of the IC_{50} for inhibition of brain and liver ALAD by Pb *in vitro* as described in FIGURE 1. Each value is a mean ± SEM ($N = 3$). The IC_{50} value for brain is 0.71 ± 0.03 μM Pb and is approximately two times higher ($p < .0005$) than the IC_{50} value of 0.32 ± 0.01 μM Pb for liver. Sources of ALAD were tissue supernatants obtained as described in reference 9. PBG: porphobilinogen; i: inhibitor concentration. (Used with permission from reference 27, copyright by the American Society of Pharmacology and Experimental Therapeutics.)

FIGURE 3. Sephadex G-75 gel filtration elution profiles of ^{203}Pb in rat brain, kidney, and liver cytosol. Tissue cytosols were incubated for 4 hr at 4° C with 0.5 μCi/ml of ^{203}Pb. The V_e/V_o values for the brain and kidney PbBPs that eluted were 1.7 to 2.2 (large arrow) and 2.0 to 2.5 (small arrows), respectively, whereas this protein was not a major constituent in liver. The peaks where V_e/V_o is 3.0 to 3.5 represent Pb bound to VLMW species, amino acids, or free Pb. For experimental details see references 9 and 27. (Used with permission from reference 27, copyright by the American Society of Pharmacology and Experimental Therapeutics.)

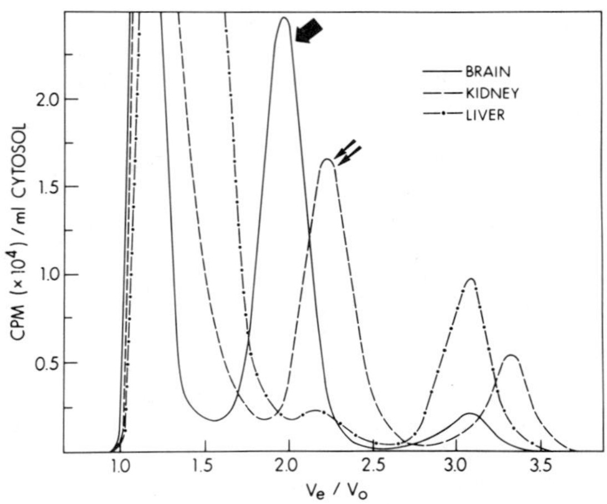

sites on the kidney and brain PbBPs may vary and that the two proteins are separate and distinct moieties.

To determine whether the mechanism of reversal of Pb inhibition of ALAD was related to an altered distribution of Pb within the ALAD reaction mixture resulting from PbBP chelation of Pb into a metal–protein complex, ALAD reaction mixtures plus or minus kidney PbBP were fractionated by gel filtration chromatography on a Sephadex G-150 column. The distribution of Pb in control reaction mixtures and reaction mixtures to which renal PbBP had been added was different as indicated by representative elution profiles (FIGURE 5). In control reaction mixtures, the majority of Pb was associated with high molecular weight (HMW) proteins (V_e/V_o: 1.0 to 1.3) and a fraction that eluted as free Pb or Pb bound to very low molecular weight (VLMW) species (V_e/V_o: 3.3 to 3.7). In contrast, in reaction mixtures to which PbBP had been added, a high percentage of eluted Pb was bound to the PbBP (V_e/V_o: 2.7 to 3.1)

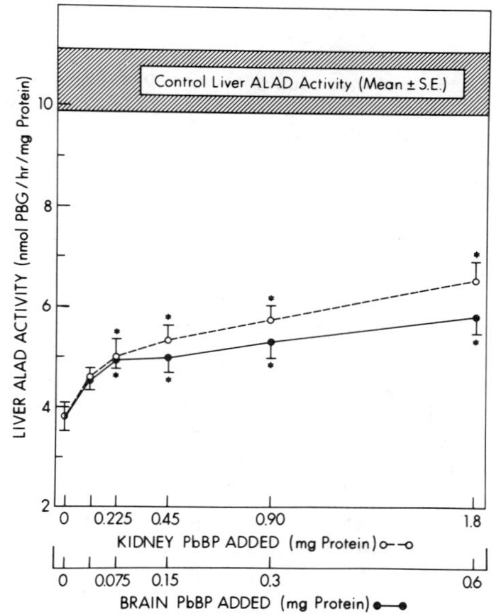

FIGURE 4. Effect of increasing concentrations of rat brain and kidney PbBPs on rat liver ALAD activity in the presence of 0.4 μM Pb. PbBP was added before Pb. Each value is a mean ± SEM ($N = 3$). Asterisks indicate values that are significantly higher ($p < .05$) than the value obtained with 0.4 μM Pb alone (no PbBP). Amount of PbBP added expressed as mg total protein. For details, see references 9 and 27. PBG: porphobilinogen. (Used with permission from reference 27, copyright by the American Society of Experimental Therapeutics.)

whereas binding to HMW proteins and VLMW species was reduced by 5% and 18%, respectively. Because ALAD is included in the HMW peak, one would expect a much greater decrease in binding to HMW proteins if the mechanism of reversal of Pb inhibition of ALAD was solely due to sequestration of Pb by PbBP. It should be noted, however, that not all HMW proteins may be equally affected by the chelation effect. Thus, some proteins such as ALAD may have greatly reduced Pb binding whereas other proteins, which exist in much higher cellular concentrations and with more stable binding sites, would retain Pb to a higher degree and hence mask the reduction in Pb binding to ALAD.

Excess Zn activates ALAD[1,2] and has an antagonistic effect on the inhibition of ALAD by Pb.[4] Therefore, because Zn has been shown to be an endogenous component of both kidney and brain PbBPs[26,27] and addition of PbBP alone to reaction mixtures activated ALAD 20 to 25% above the control value,[9,26,27] the possibility of Zn donation

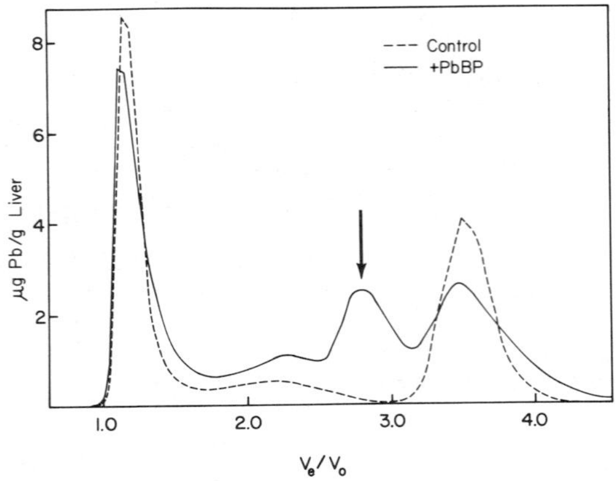

FIGURE 5. Representative Sephadex G-150 gel filtration elution profiles of Pb in hepatic ALAD reaction mixtures containing renal PbBP (solid line) or buffer (broken line). Total high-affinity Pb binding equivalents added was 189 pmol based on Scatchard analysis of the PbBP by Mistry *et al.*[24] The Pb eluting with V_e/V_o values of 1.0 to 1.4, 2.7 to 3.1, and 3.3 to 3.7 was considered bound to HMW proteins, PbBP (see arrow), and VLMW species, respectively. Source of ALAD was tissue postmitochondrial supernatants. (Used with permission from reference 26, copyright by the American Society of Pharmacology and Experimental Therapeutics.)

from PbBP to ALAD as a contributing factor to the reversal of Pb inhibition of ALAD was explored. Fractionation (via gel filtration chromatography) of reaction mixtures containing ^{65}Zn-labeled kidney or brain PbBP resulted in approximately 80% Zn bound to kidney or brain PbBP with 20% eluting in the free region (FIGURE 6). After incubation of ^{65}Zn-labeled kidney or brain PbBP with purified bovine liver ALAD, a change in the distribution of Zn was observed, with the amount of Zn bound to kidney or brain PbBP decreasing to 20% and the amount of Zn bound to purified ALAD being 60%. No change occurred in the amount of Zn eluting in the free region.

The data presented above demonstrate the relative resistance of kidney and brain ALAD to Pb inhibition compared to liver ALAD and that addition of these PbBPs to liver ALAD reaction mixtures results in partial protection of the enzyme from Pb inhibition. Thus, the resistance of kidney and brain ALAD to Pb inhibition probably involves the endogenous PbBPs present in those tissues that are not major constituents of liver. The mechanism of reversal of Pb inhibition of ALAD results from the sequestration of Pb into an inert metal–protein complex and Zn donation with activation of the enzyme.

Based on these data, it was of interest to determine the effect of Zn-induced kidney ZnMT on the activity of ALAD and potential attenuation of Pb inhibition of ALAD. The interaction of ZnMT with Zn-dependent enzymes such as ALAD, the activity of which requires continuous Zn availability, has not been studied. The effects of ZnMT on the intracellular distribution and toxicity of metals such as Pb that are poor inducers of ZnMT[28–30] but are capable of binding to it *in vitro*[31–33] and *in vivo*[31] also have not been reported.

The effect of Zn treatment on inhibition of rat kidney ALAD activity by Pb *in vitro* using postmitochondrial supernatants as the enzyme source is shown in FIGURE 7. In the presence of 3 μm Pb alone, ALAD activity was inhibited to 50% of the control value. Pretreatment of rats with 200 μmol Zn/kg body wt, sc, 48 and 24 hr before assay partially prevented this inhibition; ALAD activity was inhibited to only 85% of the control value. Pretreatment with Zn alone increased enzyme activity such that it was 55% above the control value.

In an attempt to explain the resistance of ALAD to inhibition by Pb in Zn-pretreated rats, the binding patterns of Zn and Pb in kidney cytosol from control and Zn-pretreated rats injected with ^{203}Pb were determined via Sephadex G-75 gel filtration chromatography (data not shown). Induction of ZnMT by pretreatment with Zn resulted in a marked increase in ^{203}Pb and Zn binding to the ZnMT fraction eluting at V_e/V_o: 2.0 to 2.5. It should be noted that the total kidney concentration of Pb was not different in control and Zn-pretreated rats (data not shown).

To more accurately assess the effect of ZnMT on inhibition of ALAD by Pb, we incubated purified rat kidney ZnMT (50 μg of protein containing 2 μM Zn) with purified bovine liver ALAD (2 mg protein) in the presence and absence of Pb (FIGURE 8). Addition of 1 μM Pb to the reaction mixture inhibited ALAD activity to 50% of the control value. In contrast, addition of ZnMT attenuated the inhibition of ALAD by 1 μM Pb; activity was reduced to only 90% of the control value. Although addition of 2 μM Zn as free ion increased enzyme activity such that it was greater than three times the control value, ZnMT alone increased enzyme activity such that it was almost two times the control value.

Because ALAD in liver is activated by Zn,[1,2] the contribution of the Zn donation

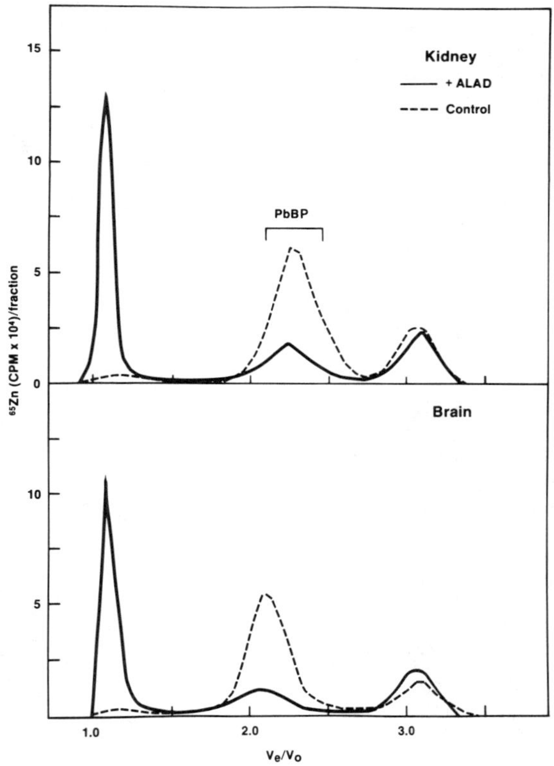

FIGURE 6. Representative Sephadex G-75 gel filtration elution profiles of ^{65}Zn in ALAD reaction mixtures demonstrating Zn donation from ^{65}Zn-labeled kidney and brain PbBPs to purified bovine liver ALAD (Sigma, St. Louis, MO). Reaction mixture contents are described in detail in reference 27. The PbBP-bound ^{65}Zn and free ^{65}Zn were separated before this experiment by gel filtration (Sephadex G-25). The peak where V_e/V_o is 1.0 to 1.2 represents purified bovine liver ALAD whereas kidney PbBP and brain PbBP are represented by peaks where V_e/V_o is 2.0 to 2.3 and 1.8 to 2.1, respectively. Nonprotein-bound Zn eluted where V_e/V_o is 3.0. Profiles are representative of data obtained from three determinations. (Used with permission from reference 27, copyright by the American Society of Pharmacology and Experimental Therapeutics.)

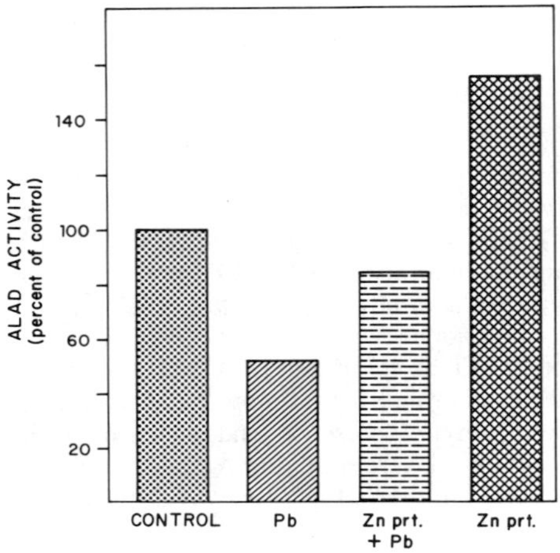

FIGURE 7. Effect of Zn pretreatment on inhibition of rat kidney ALAD activity by Pb *in vitro*. Rats were pretreated with $ZnCl_2$ (200 μmol Zn/kg, sc) 48 and 24 hr before assay, and the concentration of Pb added *in vitro* was 3 μM. Values are expressed as mean percentages of the control value. The mean control activity was 3.3 nmol porphobilinogen/hr/mg protein. Each value represents the mean of six individual determinations, and the SEM for each value was less than 10%. ALAD assayed as described in reference 9.

from ZnMT to ALAD as a factor to explain the attenuation of Pb inhibition of ALAD was assessed. The total amount of metal that eluted within a given fraction after gel filtration (on Sephadex G-75) is illustrated in FIGURE 9. Incubation of control (no ALAD) reaction mixtures containing ^{65}ZnMT resulted in approximately 70% Zn bound to ZnMT with 30% eluting as free ions. In contrast, in the presence of purified bovine liver ALAD, the amount of Zn bound to ZnMT was reduced to 20% whereas 35% was bound to ALAD. The amount of Zn eluting as free ions was increased to 45%. Thus, under the conditions of these experiments, kidney ZnMT appeared to release 3 to 4 μg-atoms Zn/μmole ZnMT; however, net transfer of Zn from ZnMT to ALAD was approximately 1 to 2 μg-atoms Zn/μmole ZnMT. The balance of the released Zn appeared as free ions possibly as a result of Zn dissociating from ALAD. These data indicate that ZnMT has the capacity to donate Zn to a Zn-dependent enzyme, such as ALAD, resulting in activation. That a net transfer of Zn was occurring is supported by

FIGURE 8. Effect of purified rat kidney ZnMT on inhibition of purified bovine liver ALAD activity by Pb. Values expressed as mean percentages of the control value. The mean control activity was 22 nmol porphobilinogen/hr/mg protein. Each value is the mean of three determinations. The concentration of Pb added *in vitro* was 1 μM. The concentration of Zn added as ZnMT was 2 μM; 50 μg protein was added. The chemical form of free Zn (2 μM) was $ZnCl_2$.

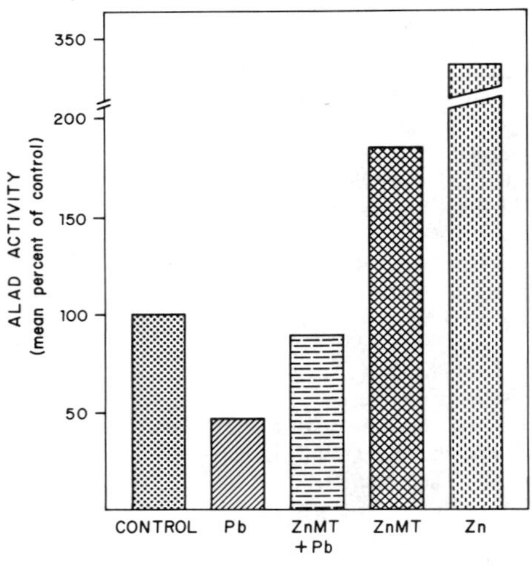

the data demonstrating ALAD activation by addition of ZnMT (FIGURE 8). The present data also indicate that under the conditions of these experiments, rat kidney ZnMT is not capable of releasing all bound Zn when donating ions to metalloenzymes because free Zn, added at the same concentration as Zn bound to ZnMT, increased ALAD activity such that it was 1.8 times greater than the activity with ZnMT. This observation is probably the result of differences in the capacity of ZnMT metal clusters to release Zn. The release of only three or four Zn ions is consistent with the data of Nielson and Winge,[34] who demonstrated that metal binding to MT is ordered and suggested that release of metal ions may also be cooperative, with release occurring at the three-metal cluster B before occurring at the four-metal cluster A.

Li *et al.*[16] have provided evidence suggesting that Zn transfer from ZnMT to another Zn metalloenzyme, carbonic anhydrase, involves a protein–protein interaction

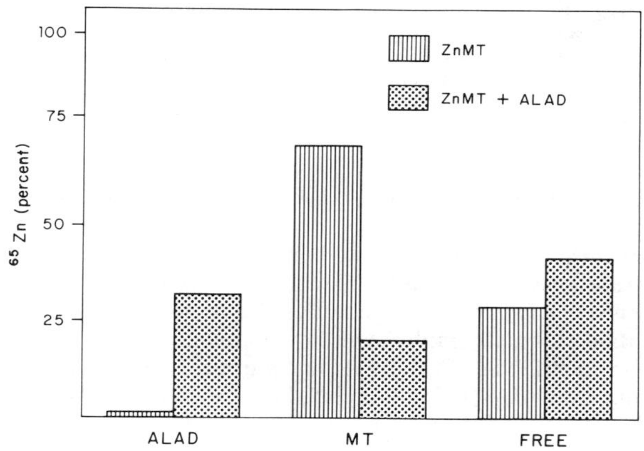

FIGURE 9. Distribution of ^{65}Zn (2 μCi/ml, 0.6 mg Zn/ml) in ALAD reaction mixtures containing purified rat kidney ZnMT (50 μg of protein containing 2 μM Zn) in the absence (vertical shading) or presence (dot shading) of purified bovine liver ALAD (2 mg of protein). Reaction mixtures were fractionated on Sephadex G-75 and each value represents the mean percentage ($N = 3$) of total ^{65}Zn in three major fractions: ALAD, where V_e/V_0 was 1.0 to 1.2; ZnMT, 2.1 to 2.4; nonprotein-bound Zn, 3.0 to 3.3. The ^{65}ZnMT was fractionated on Sephadex G-25 to separate bound and free ^{65}Zn before this experiment. It was estimated that kidney ZnMT released 3 to 4 μg-atoms Zn/μmole ZnMT under the conditions of this experiment; however, net transfer of Zn from ZnMT to ALAD was approximately 1 to 2 μg-atoms/μmole ZnMT.

and that an intermediate dissociation of Zn from ZnMT is unlikely. They suggested that the unique prolate ellipsoid conformation of MT[35] may facilitate the orientation of the protein toward the active site of the enzyme. Although the kinetics of Zn transfer from ZnMT to ALAD were not studied in the present series of experiments, the results suggest that exchange reactions between ZnMT and ALAD can occur; however, it is equivocal as to whether these reactions involve a protein–protein interaction or a free Zn intermediate.

To further examine the role of ZnMT as a regulatory molecule, the effect of rat kidney apothionein on Zn-activated ALAD was examined (FIGURE 10). Purified bovine liver ALAD was fully activated (100% activity) after addition of 50 μM Zn. In the presence of apothionein (150 μg of protein), ALAD activity was reduced to 15% of

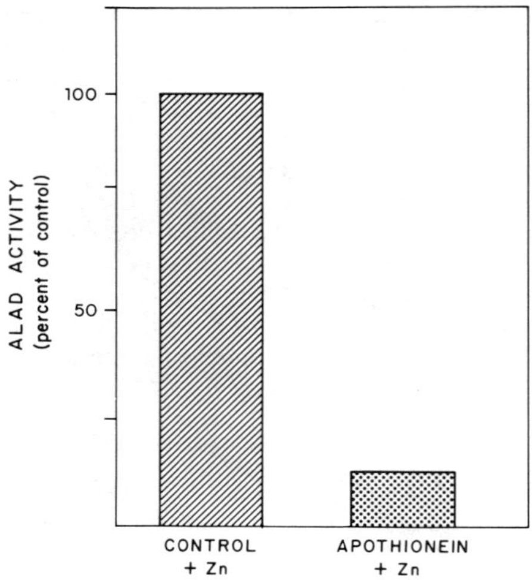

FIGURE 10. Down-regulation of purified bovine liver ALAD (2 mg of protein) activated with Zn (50 μM) by rat kidney apothionein (150 μg). Values expressed as percentages of Zn-activated control activity, which was 44 nmol porphobilinogen/hr/mg protein. Briefly, apothionein was prepared under N_2 atmosphere by adjusting the pH of ZnMT in solution to 1.0, desalting the solution to remove metal, readjusting the solution to pH 7.0, and concentrating to the desired protein concentration.

the control value. These data suggest that apothionein can "down-regulate" the activity of a Zn metalloenzyme.

To determine whether the reversal of Pb inhibition of ALAD was related in part to an altered distribution of Pb within the reaction mixture resulting from the sequestration of Pb by ZnMT, ALAD reaction mixtures plus or minus ZnMT were fractionated using gel filtration chromatography. The total amount of metal that eluted within a given fraction is illustrated in FIGURE 11. In control reaction mixtures, Pb was bound to purified bovine liver ALAD (55%) or eluted as free Pb (45%). In contrast, a marked

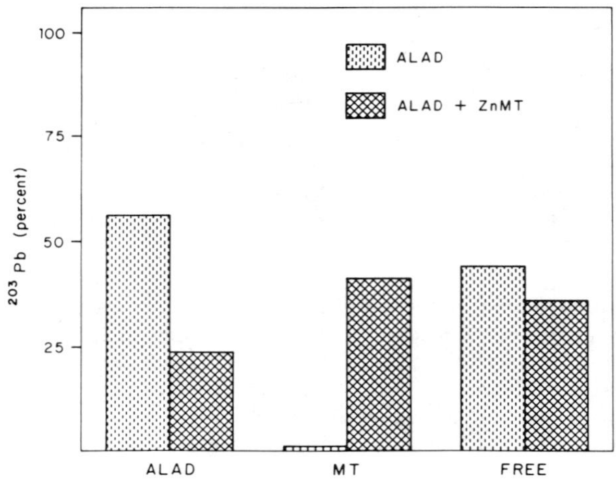

FIGURE 11. Distribution of ^{203}Pb (1 μM Pb) in ALAD reaction mixtures containing purified bovine liver ALAD (2 mg of protein) in the absence (vertical shading) or presence (crosshatched shading) of purified rat kidney ZnMT (50 μg of protein containing 2 μM Zn). Reaction mixtures were fractionated on Sephadex G-75, and each value represents a mean percentage ($N = 3$) of total ^{203}Pb eluted in three major fractions as described in FIGURE 9.

increase in ^{203}Pb binding to ZnMT (40%) was observed in reaction mixtures to which it had been added. This increase was accompanied by a concomitant decrease in Pb binding to ALAD with 25% of Pb being associated with the enzyme. The amount of Pb eluting as free ions was reduced to 35%. Thus, under conditions of these experiments, it was estimated that the amount of Pb chelated by rat kidney ZnMT was approximately 1 μg-atom Pb/μmole ZnMT.

Despite the well-known reactivity of Pb with sulfhydryl groups,[36] the interaction of Pb with MT has received little attention and is probably attributable to the fact that Pb is a poor inducer of ZnMT.[29–30] Ulmer and Vallee[31] demonstrated Pb binding *in vivo* to equine kidney MT and estimated that the association constant of Pb for the protein was between that of Cd and that of Zn. Other *in vitro* studies have also shown that Pb displaces Zn from rat liver ZnMT[32,33] and displaces Zn but not Cd from rat liver Cd-ZnMT.[33]

In order to corroborate gel filtration experiments demonstrating the binding of ^{203}Pb to rat kidney ZnMT, we examined the effect of purified rat kidney apothionein (150 μg of protein) on the inhibition of purified bovine liver ALAD (2 mg of protein)

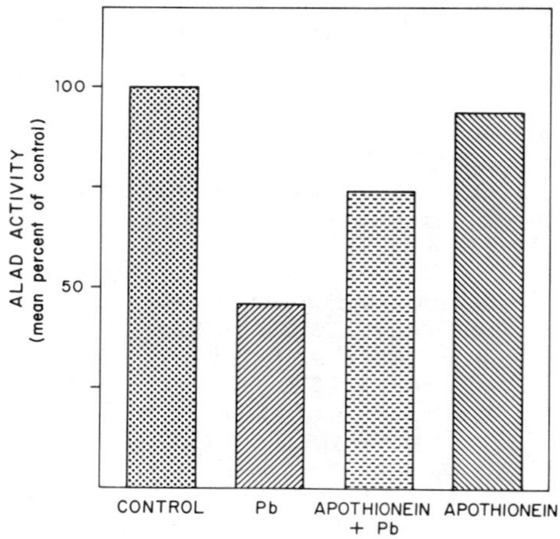

FIGURE 12. Effect of rat kidney apothionein on inhibition of purified bovine liver ALAD activity by Pb. The concentration of Pb added *in vitro* was 1 μM, and 150 μg of apothionein was added. Values are expressed as mean percentages of the control value. The mean control activity was 25 nmol porphobilinogen/hr/mg protein. Each value represents the mean of three determinations. Apothionein was prepared as described in FIGURE 10.

by Pb (FIGURE 12). Addition of 1 μM Pb to the reaction mixture inhibited ALAD activity to 45% of the control value. In the presence of apothionein, inhibition of ALAD by Pb was partially prevented; activity was reduced to only 75% of the control value. Apothionein alone had relatively little effect on basal ALAD activity, suggesting that endogenous Zn ions on ALAD, which in this experiment was not fully activated, are unavailable for chelation by apothionein (in contrast, see FIGURE 10). These data demonstrate that MT appears to be capable of chelating a biologically significant amount of Pb to decrease the inhibitory effects of Pb on a sensitive target enzyme, ALAD.

SUMMARY

The present series of studies demonstrate that several high-affinity metal-binding proteins regulate essential metal availability and play a role in metal detoxication. Rat kidney PbBP and ZnMT were shown to mediate both the interaction of Zn and Pb to a

Zn metalloenzyme, that is, ALAD. The mechanism for this interaction involves the donation of Zn from PbBP and ZnMT to ALAD and chelation of Pb to these proteins. In addition to activating ALAD via donation of Zn from ZnMT to the enzyme, kidney apothionein was shown to deactivate ALAD. Thus, these proteins may provide an intracellular reservoir for essential metals such as Zn and Cu, serving as a homeostatic control mechanism to readily dispense and sequester these cations to meet cellular metabolic requirements, such as the regulation of metalloenzymes, including ALAD. Although Pb does not effectively induce the synthesis of PbBP or ZnMT, these proteins, via binding of Pb, alter the biological activity of Pb toward a highly sensitive target molecule for Pb, that is, ALAD.

Finally, when ALAD is utilized as a biological indicator for Pb exposure, the effect of ZnMT on Pb availability and ALAD activity must be taken into consideration. Tissue-specific differences in these high-affinity, metal-binding proteins and Zn status may alter the expected relationships between total tissue Pb concentrations and inhibition of ALAD.

REFERENCES

1. KOMAI, H. & J. B. NEILANDS. 1968. Arch. Biochem. Biophys. **124:** 456–461.
2. ABDULLA, M. & B. HAEGER-ARONSEN. 1971. Enzyme **12:** 708–710.
3. GRANICK, J. L., S. SASSA, S. GRANICK, R. D. LEVERE & A. KAPPAS. 1973. Biochem. Med. **8:** 149–159.
4. FINELLI, V. N., D. S. KLAUDER, M. A. KARAFFA & H. G. PETERING. 1975. Biochem. Biophys. Res. Commun. **65:** 303–311.
5. BUCHET, J.-P., H. ROELS, G. HUBERMONT & R. LAUWERYS. 1976. Toxicology **6:** 21–34.
6. HUBERMONT, G., J.-P. BUCHET, H. ROELS & R. LAUWERYS. 1976. Toxicology **5:** 379–384.
7. ROELS, H., R. LAUWERYS, J.-P. BUCHET & G. HUBERMONT. 1977. Toxicology **8:** 107–113.
8. FOWLER, B. A., C. A. KIMMEL, J. S. WOODS, E. E. MCCONNELL & L. D. GRANT. 1980. Toxicol. Appl. Pharmacol. **56:** 59–77.
9. GOERING, P. L. & B. A. FOWLER. 1984. J. Pharmacol. Exp. Ther. **231:** 66–71.
10. SILBERGELD, E. K., R. E. HRUSKA, D. BRADLEY, J. M. LAMON & B. C. FRYKHOLM. 1982. Environ. Res. **29:** 459–471.
11. WEST, E. S., W. R. TODD, H. S. MASON & J. T. VAN BRUGGEN. 1966. *In* Textbook of Biochemistry. 4th edit.: 1376–1413. Macmillan. New York, NY.
12. HOLMBERG, C. G. & C.-B. LAURELL. 1948. Acta Chem. Scand. **2:** 550–556.
13. MEANS, A. R. & B. W. O'MALLEY. 1983. Methods Enzymol. **102:** 1–310.
14. WASSERMAN, R. H. & A. N. TAYLOR. 1966. Science **152:** 791–793.
15. KOJIMA, Y. & J. H. R. KAGI. 1978. Trends Biochem. Sci. **3:** 90–93.
16. LI, T. Y., A. J. KRAKER, C. F. SHAW & D. H. PETERING. 1980. Proc. Natl. Acad. Sci. USA **77:** 6334–6338.
17. UDOM, A. O. & F. O. BRADY. 1980. Biochem. J. **187:** 329–335.
18. GELLER, B. L. & D. R. WINGE. 1982. Arch. Biochem. Biophys. **213:** 109–117.
19. PROBST, G. S., W. F. BOUSQUET & T. S. MIYA. 1977. Toxicol. Appl. Pharmacol. **39:** 61–69.
20. ROBERTS, S. A. & R. C. SCHNELL. 1982. Toxicol. Appl. Pharmacol. **64:** 42–51.
21. SQUIBB, K. S., J. B. PRITCHARD & B. A. FOWLER. 1984. J. Pharmacol. Exper. Ther. **229:** 311–321.
22. GOERING, P. L. & C. D. KLAASSEN. 1984. Toxicol. Appl. Pharmacol. **74:** 321–329.
23. OSKARSSON, A., K. S. SQUIBB & B. A. FOWLER. 1982. Biochem. Biophys. Res. Commun. **104:** 290–298.
24. MISTRY, P., G. W. LUCIER & B. A. FOWLER. 1985. J. Pharmacol. Exp. Ther. **232:** 462–469.
25. DIXON, M. 1953. Biochem. J. **55:** 170–171.
26. GOERING, P. L. & B. A. FOWLER. 1985. J. Pharmacol. Exp. Ther. **234:** 365–371.
27. GOERING, P. L., P. MISTRY & B. A. FOWLER. 1986. J. Pharmacol. Exp. Ther. **237:** 220–225.

28. Eaton, D. L., N. H. Stacey, K.-L. Wong & C. D. Klaassen. 1980. Toxicol. Appl. Pharmacol. **55:** 393–402.
29. Waalkes, M. P. & C. D. Klaassen. 1985. Fundam. Appl. Toxicol. **5:** 473–477.
30. Maitani, T., A. Watahiki & K. T. Suzuki. 1986. Toxicol. Appl. Pharmacol. **83:** 211–217.
31. Ulmer, D. D. & B. L. Vallee. 1969. *In* Trace Substances in Environmental Health. Vol. **2:** 7–27. University of Missouri Press. Columbia, MO.
32. Waalkes, M. P., M. J. Harvey & C. D. Klaassen. 1984. Toxicol. Lett. **20:** 33–39.
33. Nielseon, K. B., C. L. Atkin & D. R. Winge. 1985. J. Biol. Chem. **260:** 5342–5350.
34. Nielson, K. B. & D. R. Winge. 1983. J. Biol. Chem. **258:** 13063–13069.
35. Kagi, J. H. R., S. R. Himmelhoch, P. D. Whanger, J. L. Bethune & B. L. Vallee. 1974. J. Biol. Chem. **249:** 3537–3542.
36. Kagi, J. H. R. & H.-J. Hapke. 1984. *In* Changing Metal Cycles and Human Health. J. O. Nriagu, Ed.: 237–250. Springer-Verlag. New York, NY.

Drug-Induced Modulations of Hepatic Heme Metabolism

Neurological Consequences[a]

MARIA ALMIRA CORREIA, DAVID A. LITMAN, AND JENNINE M. LUNETTA

Department of Pharmacology
and
The Liver Center
University of California, San Francisco
San Francisco, California 94143

INTRODUCTION

Administration of the porphyrinogenic agent 3,5-diethoxycarbonyl-2,6-dimethyl-4-ethyl-1,4-dihydropyridine (DDEP) to phenobarbital (PB)-pretreated rats results in acute hepatic heme depletion that progresses into a biochemically discernible porphyric state.[1,2] The observed hepatic heme depletion not only is rapid and dramatic, but is also sustained for over 24 hr, making this particular model of chemically induced hepatic porphyria preferable to others in existence. Three synergistic features of the combined PB/DDEP regimen—marked destruction of cytochrome P-450 heme, inhibition of heme synthesis and heme drainage by PB-induced nascent apocytochromes, and heme-stripped apocytochromes—appear to contribute to the marked depletion of the "free" heme pool observed within 4 hr of treatment (FIGURE 1). As a consequence, the activity of cytosolic holotryptophan pyrrolase, a hepatic hemoprotein in intimate contact with the free heme pool is markedly reduced. This hemoprotein is the rate-limiting enzyme in the metabolism of tryptophan to kynurenine.[3,4] Because the liver plays a predominant role in this metabolism, the activity of the hepatic isozyme critically regulates plasma tryptophan levels. Thus, any changes in such activity would tend to alter tryptophan delivery to various tissues and, consequently, influence tryptophan-dependent pathways and functions (FIGURE 2). Thus, reduction of hepatic tryptophan pyrrolase activity would, on one hand, tend to lower kynurenine required for nicotinic acid biosynthesis. On the other hand, such a reduction would be expected to enhance the conversion of tryptophan to 5-hydroxytryptamine (5-HT) in the central nervous system (CNS) and small intestine with consequent enhancement of their serotonergic tone.

Additionally, because tryptophan even at near physiological levels (0.1 mM) has been shown to regulate hepatic gluconeogenesis by blocking the step catalyzed by phosphoenolpyruvate carboxykinase (PEPCK),[5,6] its impaired degradation in the liver may also be expected to result in accumulation of glucogenic precursors (notably lactate, pyruvate, malate, and aspartate) proximal to the tryptophan-induced PEPCK block. We find it noteworthy that not only tryptophan but 5-hydroxytryptophan and 5-HT also produce hypoglycemia in various mammals, including man.[7,8] This condition

[a]These studies were supported by Grant DK-26506 from the National Institutes of Health. The use of the Liver Center Core Facility in Spectrophotometry was made possible by Grant AM-26743 from the National Institutes of Health.

is apparently reversed by central 5-HT antagonists, methysergide and cyproheptadine.[9,10]

In this report, we provide evidence that acute hepatic heme depletion and impaired tryptophan degradation in the rat liver are associated with alterations of 5-HT-dependent functions in the CNS and the small intestine. In addition, we show that in these rats hepatic gluconeogenesis is also impaired as evidenced by the marked accumulation of the tell-tale glucogenic precursors malate, aspartate, and lactate in the liver. The clinical implications of these experimental findings to the pathognomic signs and symptoms of acute attacks of hepatic porphyria are discussed.

MATERIALS AND METHODS

Methodological procedures for most assays employed have been previously detailed.[1,2] Gastric emptying and small intestinal transit were monitored by the method

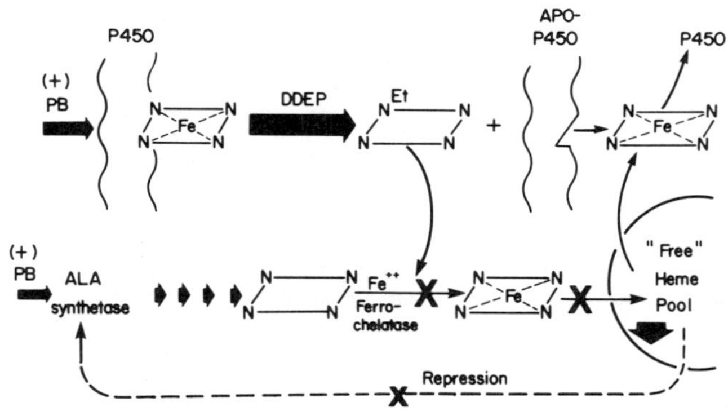

FIGURE 1. Induction of acute hepatic porphyria by administration of DDEP to PB-pretreated rats. Three synergistic features of such combined treatment that lead to marked depletion of the hepatic free heme pool are depicted: 1) DDEP-mediated destruction of cytochrome P-450 heme to yield *N*-ethylporphyrin(s); 2) Inhibition of ferrochelatase, which is the terminal enzyme in the heme synthetic pathway and which inserts iron into the porphyrin core; 3) Further drainage of the free heme pool by newly synthesized and heme-stripped apocytochrome(s) P-450.

of Galligan and Burks.[11] Specifically, gastric emptying was monitored by instilling [^{3}H]polyethylene glycol 900 into a surgically implanted gastric cannula, the radioactivity retained in the stomach being determined 35 min later. Small intestinal transit was monitored by the temporal geometric distribution of radiochromium (^{51}Cr) along the length of the small intestine, the radiochromium having been instilled in the duodenum. Hepatic glucogenic precursors were assayed exactly as described by Ray *et al.*,[5] in liver samples frozen within seconds of sacrificing the donor.

RESULTS AND DISCUSSION

As previously reported,[1,2] DDEP administration to PB-pretreated rats results in acute hepatic heme depletion and impaired degradation of tryptophan in the liver with

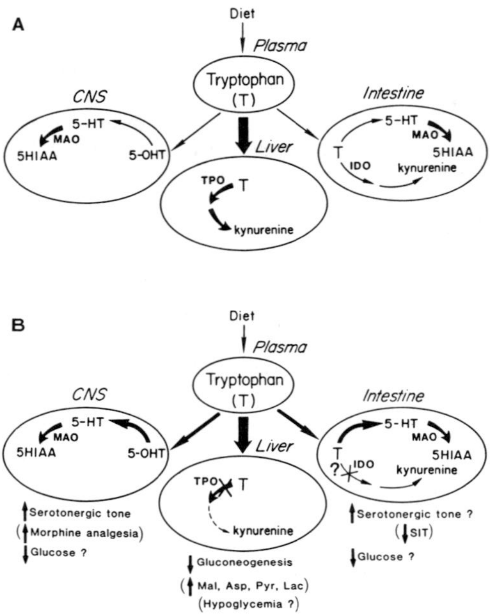

FIGURE 2. Disposition of dietary tryptophan in **(A)** normal and **(B)** heme-deficient livers. TPO: tryptophan pyrrolase; MAO: monoamine oxidase; IDO: indoleamine 2,3-dioxygenase; SIT: small intestinal transit.

TABLE 1. Heme-Mediated Reversal of Impaired Hepatic Tryptophan Degradation and Enhanced Serotonergic Tone in the CNS

	Duration of DDEP Treatment (hrs)/Heme Treatment			
	0/–	6/–	11/–	11/+
Before morphine-induced analgesia				
Tail-flick reaction time (sec)	4.7 ± 0.7	4.6 ± 1.0	5.1 ± 0.2	4.7 ± 0.4
After morphine-induced analgesia[a]				
Tail-flick reaction time (sec)	8.8 ± 2.7 (187)	12.6 ± 3.1 (274)	13.1 ± 3.9 (257)	9.8 ± 3.3 (207)
		Percentage of Basal Values		
After sacrifice				
Tryptophan pyrrolase activity (μmol kynurenine formed/g liver/hr)	2.34 ± 0.76	22	30	86
Tryptophan (μg/g brain)	1.75 ± 0.16	249	193	127
5-HT (ng/g brain)	420.9 ± 9.06	138	110	102
5-HIAA (ng/g brain)	426.5 ± 25.5	160	135	133

NOTE: PB-pretreated rats were given DDEP (125 mg/kg, ip). Six hours after DDEP, half the animals were given heme (11 mg/kg, in divided doses), and the other half received an equivalent volume of saline. Five hours later, morphine (2 mg/kg, sc) was injected, and the tail-flick reaction was monitored 1 hr later in order to assess analgesia. The animals were then sacrificed and their brain tryptophan, 5-HT, and 5-HIAA levels as well as their hepatic tryptophan pyrrolase activities were monitored.

[a]Figures in parentheses expressed as percentages of the values obtained before morphine-induced analgesia.

consequent increase in brain tryptophan levels and 5-HT turnover (TABLE 1). That this increase was associated with significantly elevated serotonergic tone of the CNS was revealed by the potentiation of morphine analgesia, an indirect albeit reliable reporter of central serotonergic tone (TABLE 1). All these events were reversed by administration of heme (TABLE 1). Furthermore, in these rats, exogenous tryptophan potentiated the serotonergic tone of the CNS, whereas *p*-chlorophenylalanine (PCPA), an inhibitor of 5-HT synthesis, abolished its enhancement.[2] To determine whether impaired hepatic tryptophan degradation in the PB–DDEP-treated rats could similarly enhance the serotonergic tone of tissues such as the small intestine, we monitored small intestinal motility, a function believed to be mediated in part by 5-HT.[3,4] Preliminary findings with a relatively crude index, the "charcoal-meal" test, indicated that small intestinal transit was indeed impaired in the PB–DDEP-treated rats.[12] However, because this test assesses both gastric emptying and small intestinal motility, it was necessary to employ the technique described by Galligan and Burks,[11] which permits independent assessment of gastric emptying and small intestinal transit. Using this technique, we have found that, although gastric emptying is unaffected, small intestinal transit is impaired in the porphyric rats (FIGURE 3). These results thus

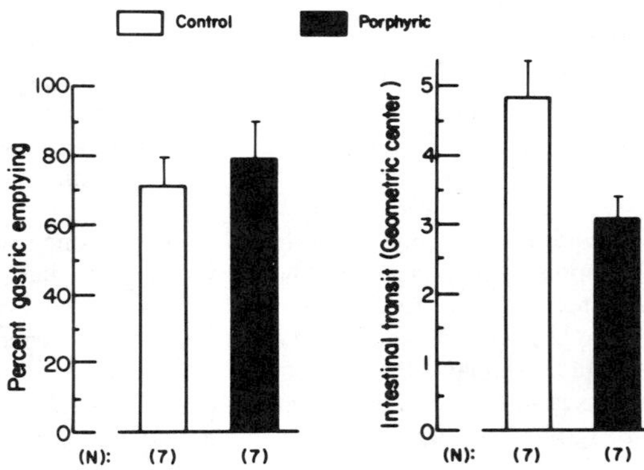

FIGURE 3. Gastric emptying and intestinal transit in control and porphyric rats. PB-pretreated rats were given DDEP (125 mg/kg, ip) in corn oil (porphyric) or an equal volume of corn oil (control). Four hours later, either [^{3}H]polyethylene glycol 900 was instilled into the gastric cannula (gastric emptying) or $Na^{51}CrO_4$ was instilled into the duodenal cannula (small intestinal transit). The animals were sacrificed 35 min later and assessed as described in the text.

confirm previous findings with the cruder charcoal-meal test that small intestinal transit is indeed impaired in the porphyric animals.

Our attempts to link such impairment to enhanced enteric serotonergic tone in the porphyric animals, however, have been deterred because, until very recently, no good enteric 5-HT antagonists had been identified. PCPA, which depletes rat brain 5-HT and reversed the enhanced serotonergic tone in the CNS,[2] proved to be unsuccessful because despite its depletion of intestinal 5-HT in guinea pigs, it fails to deplete rat intestinal 5-HT[13] and hence is a poor probe for such assessments (FIGURE 4). Similarly, a whole host of central serotonergic S_1 and S_2 antagonists failed to reverse the impairment of small intestinal transit in porphyric rats (FIGURE 5). Together, these findings indicate that the impairment of small intestinal transit observed in these rats was independent of their centrally enhanced serotonergic tone, and are fully consistent with the recent report that enteral 5-HT receptors differ from the classical S_1 and S_2 serotonergic receptors in the CNS.[14] Furthermore, the failure of methysergide, an agent also known to antagonize the serotonergic "D" receptors of intestinal smooth muscle,[14] exonerated this particular 5-HT receptor from participation in the observed impairment of small intestinal transit. Whether such impairment is indeed due to

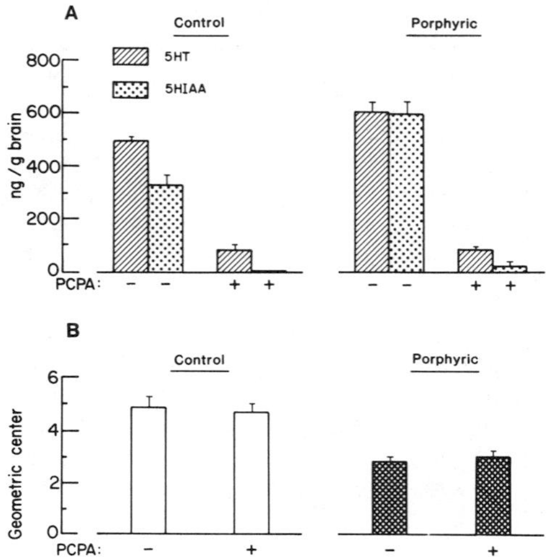

FIGURE 4. Effects of PCPA on (**A**) brain 5-HT turnover and (**B**) intestinal transit in control and porphyric animals. PCPA (300 mg/kg/day, for 3 days) or an equal volume of the vehicle (corn oil) was given to PB-pretreated rats. On the fourth day, the animals received either DDEP (porphyric) or corn oil (control), and small intestinal transit was determined 4 hr later. Each value is a mean ± SEM of at least six rats.

enhanced serotonergic tone of the enteric nervous system awaits studies with two novel synthetic 5-hydroxytryptophan dipeptides, recently described to be the only known specific enteral 5-HT antagonists.[15] Furthermore, studies are currently in progress to determine whether impairment of small intestinal transit in the porphyric rats can be reversed by exogenous heme.

In addition to enhancing the serotonergic tone in nervous tissues, decreased hepatic tryptophan degradation in the porphyric rats was also found to be associated with impaired hepatic gluconeogenesis. Thus, 20 hr after DDEP administration, when tryptophan pyrrolase activity remains 80% depressed, hepatic levels of malate, aspartate, and lactate were found to be increased. Specifically, nine-, five-, and twofold increases in the respective metabolites (TABLE 2) were observed—a result consistent with reduced utilization of glucogenic precursors proximal to the PEPCK block. On the other hand, hepatic levels of fructose diphosphate, a glucogenic precursor distal to the block, were unchanged. The modest but significant elevation of brain tryptophan observed in parallel does attest to the fact that tryptophan levels were indeed increased in these rats. However, the magnitude of this increase and consequently the accumulation of glucogenic precursors might have been even greater had the rats not been fasted and thus deprived of dietary tryptophan.

Collectively, these findings show that acute hepatic heme depletion by impairing tryptophan degradation results in increased tryptophan levels. The elevated tryptophan may not only be shunted into serotonergic pathways, thereby enhancing the serotonergic tone, but may also block hepatic gluconeogenesis and thereby deprive tissues such as the brain and small intestine of their vital energy sources. We find it intriguing that fasting has been reported to elevate plasma levels of lactate and pyruvate in individuals with acute hepatic porphyria.[16] Furthermore, those individuals apparently also exhibited an abnormal glucose tolerance test with relatively greater accumulation of pyruvate and lactate after glucose administration than normal individuals.[17]

Whether such accumulation of the glucogenic precursors in patients with acute hepatic porphyria was indeed a reflection of increased tryptophan availability and consequent block of hepatic gluconeogenesis is unknown. It is reasonable to speculate that impaired hepatic glucogenesis, by depriving neuronal tissues such as the brain of their vital energy source, could contribute to the neurological manifestations of acute attacks of hepatic porphyria observed clinically. This event might explain the dramatic amelioration observed in some patients given glucose—the so-called glucose effect in human hepatic porphyria.[18] Any etiological implications of this sort, however, pivot on the yet to be established demonstration that elevated tryptophan with associated impairment of hepatic glucogenesis is a common clinical feature of acute hepatic porphyria.

Limited clinical studies by us and others reveal that urinary excretion of 5-hydroxyindoleacetic acid (5-HIAA) (the proximal metabolite of 5-HT) is indeed increased in patients with acute attacks of hepatic porphyria[2,12,19] and that heme therapy normalizes this excretion.[2,12] Moreover, in one patient studied recently, during an attack of acute intermittent porphyria, the abdominal symptomatology correlated better with urinary 5-HIAA levels than with urinary excretion of the heme precursors δ-aminolevulinic acid (ALA) and porphobilinogen (PBG) (FIGURE 6). Accordingly, in

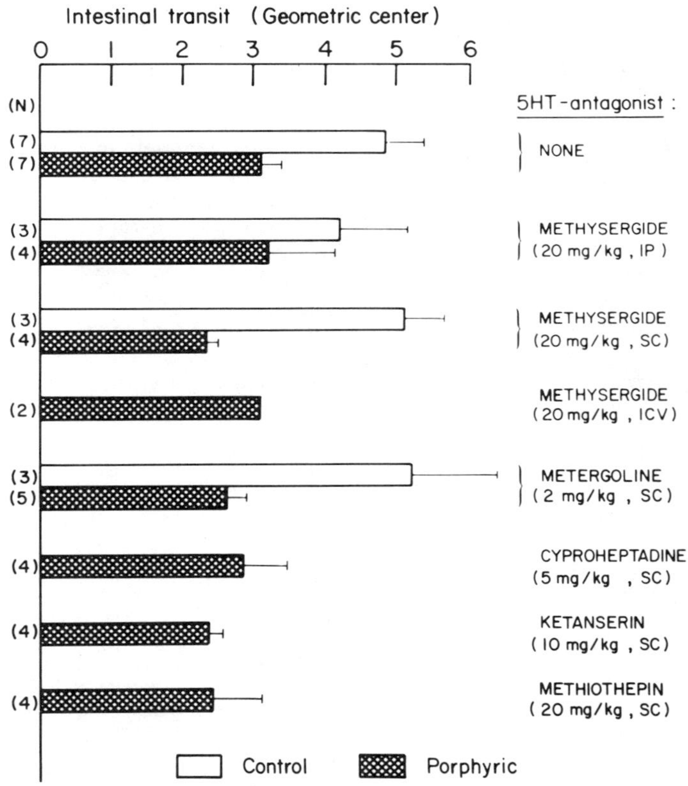

FIGURE 5. Effects of serotonin antagonists on intestinal transit in control and porphyric rats. PB-pretreated rats were administered either DDEP (porphyric) or an equal volume of corn oil (control). Four hours later, the indicated 5-HT antagonists were administered; 30 min later, $Na^{51}CrO_4$ was instilled; 35 min later, small intestinal transit was determined as described in the text. Each value represents the mean ± SEM of *N* individual animals.

TABLE 2. Impaired Tryptophan Degradation and Elevation of Glucogenic Precursors in DDEP-Treated Rat Liver

	Control Liver	DDEP-Treated Liver[a]
Tryptophan pyrrolase[b]	2.84 ± 0.32	0.51 ± 0.17 (18)
Malate[c]	0.09 ± 0.03	0.81 ± 0.42 (900)
Aspartate[c]	0.41 ± 0.07	2.00 ± 1.09 (488)
Lactate[c]	0.35 ± 0.06	0.77 ± 0.13 (220)
Fructose diphosphate[c]	0.02 ± 0.00	0.02 ± 0.01 (100)
Brain tryptophan[d]	1.64 ± 0.22	1.90 ± 0.16 (116)

NOTE: PB-pretreated rats were given DDEP (150 mg/kg, ip, in corn oil) for 20 hr and fasted overnight before sacrifice.

[a]Figures in parentheses are percentages of the control values.

[b]μmol kynurenine formed/g liver/hr.

[c]μmol/g wet wt liver.

[d]μg/g brain.

this patient, although the markedly elevated ALA and PBG excretion was normalized (FIGURE 6), the symptoms persisted in severity in parallel with high 5-HIAA urinary excretion. Parental heme therapy not only dramatically relieved his symptoms but also normalized 5-HIAA excretion. These limited clinical findings thus support the possibility that acute hepatic heme depletion may indeed result in elevated tryptophan levels and enhanced serotonergic tone. They are consistent with our hypothesis that tryptophan might be a common denominator of the biochemical events and the neurological symptomatology of the acute hepatic porphyrias. It is intriguing that many of the neurological manifestions of acute hepatic porphyria observed clinically resemble to a considerable degree some of the side effects observed after administration of tryptophan or 5-HT to humans.[2,20,21] In particular, abdominal pain and constipation are predominantly observed in tryptophan-treated individuals as well as in porphyric individuals during acute attacks. Whether accumulation of lactic acid in the muscle is, as previously suggested,[16,22] responsible for the abdominal pain remains to be determined. We find it intriguing that this metabolite along with other glucogenic precursors should also accumulate in the porphyric animals that we have examined. In

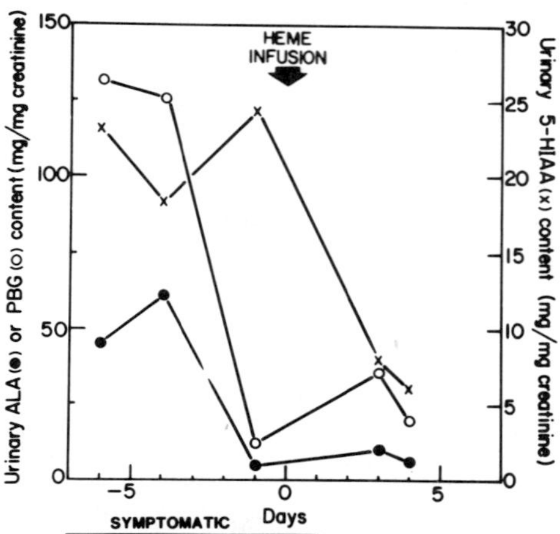

FIGURE 6. Urinary excretion of ALA, PBG, and 5-HIAA before and after heme infusion in a patient with acute hepatic porphyria. Urine samples were obtained at indicated times before and after initiation of heme treatment at day 0.

addition, the finding of impaired small intestinal transit in these porphyric animals provides another direct symptomatological correlate to the constipation observed clinically. Further clinical studies are obviously necessary to fully evaluate our hypothesis that tryptophan may be responsible for many of the neurological symptoms of acute hepatic porphyria.

SUMMARY

Administration of the porphyrinogenic agent DDEP to PB-pretreated rats results in acute hepatic heme depletion, which is a characteristic of acute hepatic porphyria. Such acute heme depletion is associated with impaired hepatic tryptophan degradation and enhanced serotonergic tone in the CNS. We showed that intestinal motility in these rats is also significantly decreased, indicating that the serotonergic tone of the enteric nervous system may also be enhanced. In addition, the marked hepatic accumulation of glucogenic precursors, observed in parallel, indicates that the elevated tryptophan levels may also block hepatic glucogenesis. The clinical implications of these findings to acute heme-deficient states, such as the acute hepatic porphyrias, was discussed.

REFERENCES

1. Litman, D. A. & M. A. Correia. 1983. Science **222:** 1031–1033.
2. Litman, D. A. & M. A. Correia. 1985. J. Pharmacol. Exp. Ther. **232:** 337–345.
3. Gershon, M. D. 1981. Annu. Rev. Neurosci. **4:** 227–272.
4. Gershon, M. D. & S. M. Erde. 1981. Gastroenterology **80:** 1571–1594.
5. Ray, P. D., D. O. Forster & H. A. Lardy. 1966. J. Biol. Chem. **241:** 3904–3908.
6. Veneziale, C. M., P. Walter, N. Kneer & H. A. Lardy. 1967. Biochemistry **6:** 2129–2138.
7. Ajdukiewicz, A. B., P. Keane, J. Pearson, A. E. Read & P. R. Salmon. 1968. Scand. J. Gastroenterol. **3:** 622–624.
8. Mirsky, I. A., G. Perisutti & D. Diengott. 1956. Endocrinology **59:** 369–379.
9. Furman, B. L. 1974. Br. J. Pharmacol. **50:** 575–580.
10. Wilson, G. A. & B. L. Furman. 1982. Eur. J. Pharmacol. **78:** 263–270.
11. Galligan, J. J. & T. E. Burks. 1983. J. Pharmacol. Exp. Ther. **226:** 356–361.
12. Correia, M. A & D. A. Litman. 1986. *In* Porphyrins and Porphyrias. Y. Nordman, Ed. Vol. **134:** 119–126. John Libbey Eurotext. Montrouge.
13. Ternaux, J. P., J. Gonella, C. Legay, M. Faundin, M. C. Barrit & F. Hery. 1981. J. Physiol. (Paris) **77:** 319–326.
14. Branchek, T., M. Kates & M. D. Gershon. 1984. Brain Res. **324:** 107–118.
15. Takaki, M., T. Branchek, H. Tamir & M. D. Gershon. 1985. J. Neurosci. **5:** 1769–1780.
16. Ridley A. 1969. Q. J. Med. **38:** 307–333.
17. Tschudy, D. P. 1968. Ann. N.Y. Acad. Sci. **151:** 850–860.
18. Tschudy, D. P. 1978. *In* Heme and Hemoproteins: Handbook of Experimental Pharmacology. F. De Matteis & W. N. Aldridge, Eds. Vol. **44:** 255–271. Springer-Verlag. New York, NY.
19. Ludwig, G. D. & I. S. Epstein. 1961. Ann. Intern. Med. **55:** 81–93.
20. Baloh, R. W., J. Dietz & J. W. Spooner. 1982. Ann. Neurol. **11:** 95–97.
21. Smith, B. & D. J. Prockop. 1962. N. Engl. J. Med. **276:** 1338–1341.
22. Roman, W., R. Oon, R. F. West & D. P. Reid. 1967. Lancet **ii:** 716.

Hormonal Regulation of the Heme Pathway

RICHARD A. GALBRAITH

The Rockefeller University Hospital
New York, New York 10021

INTRODUCTION

The acute hepatic porphyrias comprise three separate diseases, each predicated on the inheritance of a single defective gene. Deficiency of porphobilinogen deaminase (PBGD) activity is mandatory for clinical expression of acute intermittent porphyria (AIP). Decreased PBGD activity, however, does not necessarily result in clinical expression of AIP, and some authorities estimate that only 1 of every 10 carriers of the defective gene experiences porphyria symptoms.[1] The observations that clinically active AIP is rare both before puberty and after menarche, is more common in women, and may be exacerbated by menses have prompted much inquiry into the relationships between hormones and the pathophysiology of heme pathway regulation. During the 23 years since Granick and Urata[2] first demonstrated the inducibility of δ-aminolevulinic acid synthase, a wealth of basic research has spawned an understanding of the interlocking relationships of the heme pathway and the endocrine system in regulatory terms. Recently, the process has gone full circle with the advent of clinically effective treatment of menstrual AIP exacerbations by medical menopause induced by use of a long-acting (and hence desensitizing) synthetic agonist of luteinizing hormone releasing hormone.[3]

This paper will highlight certain interactions between hormones and the heme pathway in an effort to illustrate key points. No attempt to provide an all-inclusive review has been made. The monograph will focus on three hepatic enzymes: δ-aminolevulinic acid synthase (ALAS), heme oxygenase (HO), and cytochrome P-450 (P-450). The former two enzymes represent the rate-limiting enzymes of, respectively, heme synthesis and heme degradation; the latter reflects heme utilization by the major hepatic hemoprotein.

RESULTS AND DISCUSSION

Effects of Steroid Hormones

The clinical association of AIP symptoms and sex steroids triggered the development of primary cultures of chick embryo hepatocytes and the demonstration that sex steroids and their metabolites induce ALAS activity.[4] A series of experiments by Kappas and Granick revealed that 5β-H steroids containing either 19 or 21 carbons were the most potent steroid inducers of ALAS.[5–8] Similar but less pronounced effects have been reported in rat liver.[9]

AIP patients are known to excrete increased concentrations of steroids into the urine,[10] and they have been shown to be predominately of the 5β-H configuration because of a deficiency of hepatic 5α-reductase.[11–13] Such 5β-H steroids may, as in chick embryo hepatocytes, induce ALAS and predispose AIP patients to porphyria crises. Clinically latent AIP gene carriers, however, are not associated with abnormal steroid metabolism.[14] AIP patients, or women receiving oral contraceptives, have also been shown to produce a factor(s) that is found in the serum and that stimulates porphyrin formation when added to chick embryo hepatocytes.[15] Further studies revealed that human sera could be either stimulatory or inhibitory to chick embryo hepatocyte ALAS activity dependent on the hormonal and metabolic state of the donor (for example, pregnant women and women taking oral steroids had stimulatory sera; women immediately post partum had inhibitory sera).[16]

Effects of Peptide Hormones

A significant development in the chick embryo hepatocyte model was the demonstration that these cells can be maintained in primary culture in a chemically defined medium.[17] This medium enabled more precise examination of the effects of hormones on heme homeostasis without the confounding influence of calf serum and its attendant complement of hormones. Utilizing this model, it was demonstrated that the addition of insulin, hydrocortisone, and 3,5,3′-triiodothyronine (T_3) had essentially no effect on basal ALAS activity, but their presence during allylisopropylacetamide (AIA) treatment produced a 110-fold increase in ALAS activity compared to a 6-fold increase in response to AIA alone.[17] Growth hormone was also found to increase porphyrin synthesis in response to AIA.[18] These hormonal effects are termed "permissive" in that they permit a greater response to an inducer, but do not themselves change basal activities. In contrast, hydrocortisone and insulin decreased basal HO activity but increased HO induction after exposure to an inducing agent, $CoCl_2$.[19] Glucagon was found to inhibit both basal and $CoCl_2$-induced HO activity.[19] Similarly, both basal and 2-propyl-2-isopropylacetamide (PIA) induction of P-450 were stimulated by addition of insulin, hydrocortisone, and T_3.[19] Addition of T_3 only enhanced both the induction of HO by Co and the induction of ALAS by AIA,[20] whereas the addition of insulin only suppressed basal and Co-induced HO activity.[21]

Cytochrome P-450 content and activities have been shown to fall markedly within 24 hr of establishing primary cultures of adult rat hepatocytes.[22] However, inclusion in the medium of various mixtures of insulin, hydrocortisone, thyroxine, estradiol, and testosterone have been reported to restore both P-450 content and activities toward *in vivo* levels.[23–26]

Effects of Surgically Induced Endocrinopathies

Adrenalectomy in adult rats leads to decreased P-450 content and activities[27] and blocks AIA induction of ALAS.[28] This latter blockade is reversed after *in vivo* administration of hydrocortisone[28] and constitutes another permissive effect of hormones. Adrenalectomized rats also display elevations in both basal and Co-induced HO activities; hydrocortisone treatment of adrenalectomized animals abrogated the exaggerated response of HO to Co.[29] Thyroidectomized rats displayed increased basal and induced HO activity with a concurrent depletion of P-450 after replacement with T_3.[30]

Hypophysectomy of adult male rats leads to a 50% reduction in HO activity and a modest, but consistent, elevation of P-450 content; ALAS activity is unaffected (see TABLE 1). Subcutaneous implantation of Alzet osmotic pumps was used to study the effects of replacement of hormones in hypophysectomized rats. Growth hormone administration led to a decrease in ALAS activity but restored HO activity and P-450 content to sham-operated control values. Dexamethasone also inhibited ALAS but failed to restore HO activity and P-450 content, whereas T_3 partially restored HO activity and was without effect on ALAS and P-450 (TABLE 1).

The Hypothalamic–Pituitary–Hepatic Axis

The contribution of the hypothalamic–pituitary–hepatic axis to hepatic heme homeostasis has been the recent focus of intense interest. Several investigators have shown that androgen administration to rats increases many P-450-dependent hepatic enzyme activities including steroid Δ^4-hydrogenase activity and that estrogens produce opposing effects.[31–33] Treatment of hypophysectomized male rats with androgens or estrogens, however, failed to alter hepatic P-450 activities,[34] suggesting an absolute requirement of the pituitary for expression of these steroid effects. Furthermore, it has been shown that hepatic monooxygenase activities are generally higher in postpubertal male rats than in female rats.[31] Castration or hypophysectomy of neonatal male rats leads to subsequently decreased, or "feminized," monooxygenase activities.[35,36] The existence of a pituitary feminizing hormone under hypothalamic feminostatin control was postulated.[36] A series of studies have identified this feminizing hormone to be growth hormone that is controlled positively by hypothalamic growth hormone releasing hormone and negatively by hypothalamic somatostatin.[37–39] In males, growth hormone is secreted episodically whereas in females secretion is more continuous; each pattern of growth hormone secretion is regulated in part by gonadal steroids.[40] Thus the working hypothesis is that the pattern of growth hormone release is largely responsible for sexual dimorphisms, which include hepatic monooxygenase activities and the subsequent metabolic sequelae of exposure to both endogenous and exogenous chemicals.

TABLE 1. Effect of Hypophysectomy and Hormonal Replacement on Hepatic Heme Homeostasis

	δ-Aminolevulinic Acid Synthase (nmol ALA/mg protein/hr)	Heme Oxygenase (nmol bilirubin/mg protein/hr)	Cytochrome P-450 (nmol/mg protein)
Shams	0.115 ± 0.009	1.79 ± 0.17	0.78 ± 0.05
Hypox	0.121 ± 0.005	0.90 ± 0.14[a]	0.90 ± 0.09
Hypox + growth hormone	0.077 ± 0.016	1.74 ± 0.22	0.72 ± 0.06
Hypox + dexamethasone	0.043 ± 0.002	0.72 ± 0.09	0.34 ± 0.05
Hypox + T_3	0.109 ± 0.007	1.21 ± 0.10	0.92 ± 0.10

NOTE: Hormonal replacements to hypophysectomized (hypox) rats were made via subcutaneously implanted Alzet Osmotic Minipumps which, in each 24-hr period, were loaded to deliver 120 μg of growth hormone, 280 mg of 3,5,3′-triiodothyronine (T_3), and 0.1 mg of dexamethasone. After 6 days, animals were sacrificed and assays performed as described in reference 28. Each value is a mean ± SEM of three determinations.

[a]Significantly different from the sham and the hypox + growth hormone values ($p < .05$).

CONCLUSIONS

The multitude of hormonal effects described above serve to indicate the pervasive effects of the endocrine system on the heme pathway and heme utilization in liver.

Additionally, it should be understood that only small portions of the huge literature pertaining to this topic have been discussed. The effects of single pathologic endocrinopathies are seen to be relatively easy to test and document, whereas the subtler changes in hormone concentrations that are permissive for an induced response are rather more difficult to dissect. The hardest task, however, is to understand the integration of the entire endocrine status in modifying the response to a specific chemical stimulus.

The concept of a genetic disease (AIP), the expression of which is dependent on environmental factors and hormonal status, was introduced at the beginning of this report. Perhaps this is also the most appropriate way to visualize chemically induced porphyrinopathies. Thus, the hepatic response to a chemical is dependent not only on the genetically determined activities of heme pathway and heme-dependent enzymes, but also on the metabolic milieu of the liver, which is largely orchestrated by the integrated effects of many components of the endocrine system.

ACKNOWLEDGMENTS

The author is grateful to Dr. A. Kappas for his encouragement, to Dr. G. S. Drummond for performing the assays depicted in TABLE 1, and to Dr. S. Sassa for helpful discussions.

REFERENCES

1. KAPPAS, A., S. SASSA & K. E. ANDERSON. 1983. *In* The Metabolic Basis of Inherited Disease. J. B. Stanbury, J. B. Wyngaarden, D. S. Frederickson, J. L. Goldstein & M. S. Brown, Eds.: 1301–1384. McGraw-Hill. New York, NY.
2. GRANICK, S. & G. URATA. 1963. J. Biol. Chem. **238:** 821–827.
3. ANDERSON, K. E., I. M. SPITZ, S. SASSA, W. BARDIN & A. KAPPAS. 1984. N. Engl. J. Med. **311:** 643–645.
4. GRANICK, S. 1966. J. Biol. Chem. **241:** 1359–1375.
5. GRANICK, S. & A. KAPPAS. 1967. Proc. Natl. Acad. Sci. USA **57:** 1463–1467.
6. GRANICK, S. & A. KAPPAS. 1967. J. Biol. Chem. **242:** 4587–4593.
7. KAPPAS, A. & S. GRANICK. 1968. Ann. N. Y. Acad. Sci. **151:** 842–849.
8. KAPPAS, A. & S. GRANICK. 1968. J. Biol. Chem. **243:** 346–351.
9. MOORE, M. R., J. W. PAXTON, A. D. BEATTIE & A. GOLDBERG. 1973. Enzyme **16:** 314–325.
10. GOLDBERG, A., M. R. MOORE, A. D. BEATTIE, P. E. HALE, J. MCCALLUN & J. K. GRANT. 1969. Lancet **ii:** 115–118.
11. KAPPAS, A., H. L. BRADLOW, P. N. GILLETTE & T. F. GALLAGHER. 1971. Ann. N. Y. Acad. Sci. **179:** 611–624.
12. KAPPAS, A., H. L. BRADLOW, P. N. GILLETTE & T. F. GALLAGHER. 1972. J. Exp. Med. **136:** 1043–1053.
13. BRADLOW, H. L., P. N. GILLETTE, T. F. GALLAGHER & A. KAPPAS. 1973. J. Exp. Med. **138:** 754–763.
14. ANDERSON, K. E., H. L. BRADLOW, S. SASSA & A. KAPPAS. 1979. Am. J. Med. **66:** 644–650.
15. KAPPAS, A., C. S. SONG, S. SASSA, R. D. LEVERE & S. GRANICK. 1969. Proc. Natl. Acad. Sci. USA **64:** 557–564.
16. RIFKIND, A. B., S. SASSA, I. R. MERKATZ, R. WINCHESTER, L. HARBER & A. KAPPAS. 1974. J. Clin. Invest. **53:** 1167–1177.
17. SASSA, S. & A. KAPPAS. 1977. J. Biol. Chem. **252:** 2428–2436.
18. SASSA, S., D. W. GOLDE & C. HAO. 1981. J. Biol. Chem. **256:** 8882–8884.
19. SASSA, S., M. K. SARDANA & A. KAPPAS. 1980. *In* Biochemistry, Biophysics and Regulation of Cytochrome P-450. J. A. Gustafsson, Ed.: 187–190. Elsevier. Amsterdam.

20. SARDANA, M. K., S. SASSA & A. KAPPAS. 1983. *In* Isolation, Characterization and Use of Hepatocytes. R. Harris & N. Cornell, Eds.: 111–1167. Elsevier. Amsterdam.
21. SARDANA, M. K., S. SASSA & A. KAPPAS. 1985. Biochem. Pharmacol. **34:** 2937–2944.
22. GUZELIAN, P. S., D. M. BISSEL & U. A. MEYER. 1977. Gastroenterology **72:** 1232–1239.
23. MICHALOPOULOS, G., G. L. SATTLER & H. C. PITOT. 1976. Life Sci. **18:** 1139–1144.
24. DECAD, G. M., D. P. H. HSIEH & J. L. BYARD. 1977. Biochem. Biophys. Res. Commun. **78:** 279–287.
25. PAINE, A. J. & R. F. LEGG. 1978. Biochem. Biophys. Res. Commun. **81:** 672–679.
26. DICKINS, M. & R. E. PETERSON. 1980. Biochem. Pharmacol. **29:** 1231–1237.
27. CASTRO, J. A., F. E. GREEN, P. GIGON, H. SASAME & J. R. GILLETTE. 1970. Biochem. Pharmacol. **19:** 2461–2467.
28. MARVER, H. S., A. COLLINS & D. P. TSCHUDY. 1966. Biochem. J. **99:** 31C–33C.
29. SARDANA, M. K., S. SASSA & A. KAPPAS. 1980. J. Biol. Chem. **255:** 11320–11323.
30. SMITH, T. J., G. S. DRUMMOND, I. A. KOURIDES & A. KAPPAS. 1982. Proc. Natl. Acad. Sci. USA **79:** 7537–7541.
31. EL DEFRAWY EL MASRY & G. J. MANNERING. 1974. Drug Metab. Dispos. **2:** 279–284.
32. EINERSSON, K., J. A. GUSTAFSSON, J. SJOVALL & E. ZIETZ. 1975. Acta Endocrinol. **78:** 54–64.
33. KRAMER, R. E., J. W. GREINER, R. C. RUMBAUGH, T. D. SWEENEY & H. D. COLBY. 1978. J. Pharmacol. Exp. Ther. **204:** 247–254.
34. KRAMER, R. E., J. W. GREINER, R. C. RUMBAUGH, T. D. SWEENEY & H. D. COLBY. 1979. J. Pharmacol. Exp. Ther. **208:** 19–23.
35. DEMOOR, P. & C. DENEF. 1968. Endocrinology **82:** 480–492.
36. GUSTAFSSON, J. A., P. ENEROTH, T. HÖKFELT, A. MODE, G. NORSTEDT & P. SKETT. 1981. Environ. Health Perspect. **38:** 129–141.
37. SKEET, P. & C. YOUNY. 1982. Acta Endocrinol. **100:** 421–426.
38. MODE, A., G. NORSTEDT, P. ENEROTH & J. A. GUSTAFSSON. 1983. Endocrinology **113:** 1250–1260.
39. GUSTAFSSON, J. A., S. EDEN, P. ENEROTH, T. HÖKFELT, O. ISAKSSON, J. JANSSON, A. MODE & G. NORSTEDT. 1983. J. Steroid Biochem. **19:** 691–698.
40. JANSSON, J., E. STAFFEN & O. ISAKSSON. 1983. Endocr. Rev. **6:** 128–149.

PART VI. PORPHYRINOPATHIES IN CHEMICAL-INDUCED DISEASE STATES INVOLVING SPECIFIC ORGAN SYSTEMS

The Dermatologic Manifestations of Human Porphyria

DAVID R. BICKERS

Department of Dermatology
Case Western Reserve University
University Hospitals of Cleveland
Cleveland, Ohio 44106

INTRODUCTION

The skin is a major target tissue for the injurious effects of environmental agents, some of which are directly noxious, others of which exert their toxicity in concert with metabolic derangements in the body. The porphyrias are a group of diseases that have in common the aberrant synthesis of heme, a cyclic iron-chelated tetrapyrrole. Heme is crucial for a variety of fundamental biochemical activities, including gas exchange and electron transport, that are essential for homeostasis.

Heme synthesis occurs in all cells but is found predominantly in the liver and the bone marrow and is dependent on a complex series of chemical reactions catalyzed by both membrane-bound and soluble enzymes. Altered activity of these enzymes can result from either inherited or acquired abnormalities, the consequences of which include neuropsychiatric and dermatologic symptom complexes that are thought to relate directly to the excessive accumulation of porphyrins or their precursors in the body.

HISTORY

Awareness of the possible relationship between porphyrins and cutaneous photosensitivity can be found at least as early as 1898 when T. M'Call Anderson gave an account of two brothers who were fishermen from Scotland. They had suffered for many years from strikingly similar recurrent skin eruptions limited to sun-exposed body areas and occurring in the summer months with spontaneous improvement in the winter.[1] This was associated with darkening of the urine to a "light burgundy-red colour." Spectral examination of the urinary pigment revealed it to possess absorption characteristics similar to those of a porphyrin rather than to those of hemoglobin.

Subsequent experimental studies by Haussman showed that hematoporphyrin, a synthetic porphyrin prepared by acid hydrolysis of whole blood, was a potent photosensitizer in red blood cells and in paramecia, and that the reaction was "photodynamic" and required oxygen.[2] Despite these observations, some controversy persisted regarding the capacity of porphyrins to produce photosensitivity in human skin. This was permanently laid to rest by the heroic "Selbstversuch" (self-experiment) by Meyer-Betz who injected himself with hematoporphyrin (200 mg) intravenously and then exposed his face and arms to sunlight.[3] He observed there was an immediate burning discomfort of his skin limited to the sun-exposed sites, and that the burning sensation was followed by edema but no blisters. His cutaneous hyperresponsiveness to sunlight persisted for more than 2 months as a result of the slow release of the porphyrin from the skin.

Subsequent studies by Harold Blum and his colleagues showed that the action spectrum for porphyrin photosensitization of human skin agrees closely with the absorption spectrum of the compound as determined by the use of light from the sun, carbon arcs, and mercury arcs.[4] Intracutaneous injection of hematoporphyrin hydrochloride (0.01%) followed by exposure to light with spectral irradiance between 380 and 410 nm evoked an itching wheal with surrounding flare within minutes. Similar though less intense reactions occurred with exposure to wavelengths greater than 650 nm.

In the 1960s, studies to define the mechanism of porphyrin photosensitization were carried out by Slater and Riley[5] and by Allison *et al.*[6] They speculated that since porphyrin photosensitization requires oxygen, an essential part of the process may be the formation of oxygen-containing radicals of the peroxy type. Lipid peroxidation, a process involving such radicals, was thought to be capable of damaging lipid-rich subcellular structures such as membrane-bound organelles and endoplasmic reticulum. In their studies, Slater and Riley showed that phylloerythrin and light was capable of damaging lysosomes as evidenced by the enhanced release of lysosomal enzymes. Allison *et al.* showed that uroporphyrin and light was also capable of damaging lysosomes, the damage resulting in increased permeability of lysosomes of endothelial cells with release of one or more chemical mediators.

In 1967, Rimington *et al.* reported an exhaustive review of porphyria and cutaneous photosensitization.[7] Employing a high-output irradiation monochromator, they studied 29 patients with porphyria. These investigators were able to demonstrate abnormal cutaneous responses to radiation of 400 nm and, to a lesser extent, to radiation between 500 and 600 nm in porphyric patients. They emphasized that in some patients these abnormal responses were subjective, with burning, pricking, or scalding sensations occurring during or immediately after irradiation; in other patients, an objective response— 1) immediate erythema, 2) delayed erythema, 3) wheal and flare, 4) delayed edema, or 5) bullous lesions—could be observed.

MECHANISM OF PORPHYRIN PHOTOSENSITIZATION

Evidence for the role of lipid peroxidation as a mechanism of porphyrin photosensitization came from *in vitro* studies with red blood cells that demonstrated selective loss of unsaturated fatty acids, along with malonaldehyde and diene bond formation following irradiation of protoporphyrin-enriched red blood cells.[8] Similar results were obtained in protoporphyrin-treated human skin fibroblasts that were shown to undergo a porphyrin concentration-dependent, light-dependent, and time-dependent cytolysis during exposure to radiation of 400 nm.[9] Pulse-chase experiments with radiolabeled proline demonstrated precipitation–denaturation of cellular proteins in protoporphyrin-photosensitized cells whereas no evidence for release of lysosomal enzymes could be identified. These latter findings were confirmed by Sandberg and Romslo, who showed that liver mitochondrial membranes in mice were much more susceptible to protoporphyrin-photolyzed damage than were lysosomal membranes.[10] These studies further emphasized that porphyrin-induced photodamage is largely dependent upon the solubility properties of the porphyrins as well as their target structures. Additional studies in human neutrophils also confirmed that porphyrin photodamage occurs earlier in the plasma membrane and in mitochondria as compared to lysosomes.[11] Finally, irradiation of cultured human skin fibroblasts that had been induced to synthesize increased amounts of protoporphyrin by incubation with the porphyrin precursor δ-aminolevulinic acid was shown to produce photodamage. The earliest indications of this response were rarification of the mitochondrial matrix with dilation of cristae, dilatation of endoplasmic reticulum, and loss of plasma membrane integrity suggesting generalized cellular injury.[12]

Studies with epidermal microsomes *in vitro* to which hematoporphyrin derivative was added showed that exposure to radiation of 400 nm resulted in destruction of cytochrome P-450, loss of heme content, and inhibition of the catalytic activity that is dependent upon this hemoprotein.[13] Use of inhibitors of singlet oxygen and the hydroxyl radical diminished photoxic injury to the microsomes whereas scavengers of superoxide anion and of hydrogen peroxide were ineffective in this regard.

The role of singlet oxygen in porphyrin photosensitization has been verified by direct detection using a method of measurement of its luminescence at 1270 nm in a system employing hematoporphyrin.[14]

Although the evidence for the role of reactive oxygen species in porphyrin photosensitization seems incontrovertible, the precise steps involved in producing skin injury remain unclear. One possibility that has been extensively explored by Gigli, Lim, and their associates is the role of complement activation.[15–18] In a series of elegant studies in animals and in human subjects with erythropoietic protoporphyria and porphyria cutanea tarda, these workers have shown that irradiation of serum-containing porphyrins or irradiation of the skin of patients with these disorders results in complement activation with release of proinflammatory fragments including C5a.[15–18] One effect of this photosensitization reaction is enhanced chemotactic activity, which may contribute to the cutaneous inflammation that occurs in these disorders. In further studies, it has been shown that mast cells may also play an important role in the cutaneous photosensitization reaction, particularly in the delayed phase that occurs in guinea pig skin.[19,20]

PORPHYRIA CUTANEA TARDA

Incidence

This porphyria of hepatic origin is perhaps the commonest of porphyric disorders with cutaneous manifestations. It occurs in approximately one in 25 000 individuals in the United States. Historically, this disease was believed to primarily affect middle-aged males who were heavy imbibers of alcohol, but more recent reviews indicate that with the widespread use of oral contraceptive hormones, the occurrence in females has increased steadily to the point where the sex incidence is approximately equal.[21]

Dermatologic Manifestations

A major cutaneous feature of this disease is increased fragility of light-exposed skin, particularly areas of repeated minor trauma such as the hands, forearms, face, and ears. This is often the major complaint that brings the patient to the physician. Associated with the enhanced fragility are fluid-containing vesicles and bullae that often develop after minimal trauma. It should be emphasized that most patients are unaware of any role of sunlight in the pathogenesis of their abnormal skin findings. The vesicles and bullae usually rupture leaving an exudative hemorrhagic base that becomes crusted and then reepithelializes slowly leaving a scar, milia formation, or both.

Hypertrichosis is a second major cutaneous manifestation of porphyria cutanea tarda. This usually affects the face and begins insidiously with increased growth of fine vellus hairs that gradually thicken and darken. Typically, these are most prominent in the periorbital areas and on the forehead between the eyebrows and the scalp hairline. In selected individuals there may be extensive hair growth on the trunk and extremities as well. The mechanism of this is unknown.

Mottled melasma-like hyperpigmentation often occurs, particularly in sun-exposed areas. This may be a postinflammatory phenomenon, but the precise mechanism has not been defined.

Yet another dermatologic manifestation of porphyria cutanea tarda is the development of indurated ivory-colored plaques on either sun-exposed or protected areas, particularly the upper portions of the trunk. These bear a strong clinical resemblance to morphea or scleroderma and have been called sclerodermoid changes. Studies on the mechanism of this feature of porphyria cutanea tarda have shown that uroporphyrin I, which is the major abnormal porphyrin present in skin and blood of these patients, stimulates collagen synthesis up to threefold in cultured human fibroblasts.[22] This effect is dependent upon the dose of uroporphyrin I added to the cells but is independent of light exposure. Thus, porphyrin enhancement of collagen synthesis in dermal fibroblasts may explain the sclerodermoid skin changes in this disease.

Other less common skin findings in porphyria cutanea tarda include dystrophic calcification, alopecia, and onycholysis.

Histopathology

Vesicles and bullae are subepidermal and characteristically demonstrate little or no inflammatory cell infiltrate. Some thickening of the epidermis and of blood vessel walls in the upper dermis also is seen. Direct immunofluorescence studies of lesional skin has demonstrated that immunoglobulin is present in a patchy distribution around blood vessels and to a lesser extent at the dermal–epidermal junction.

Differential Diagnosis

The combination of increased skin fragility and vesicles and bullae occurs in a limited number of disorders including epidermolysis bullosa acquisita and phototoxicity associated with drugs such as tetracycline, nalidixic acid, and naprosyn.[23] In addition, megadoses of pyridoxine have been associated with similar cutaneous findings.[24]

Patients with chronic renal failure undergoing hemodialysis may develop bullous skin lesions that are indistinguishable from those of porphyria cutanea tarda.[25] Although careful assessment may reveal that some of these individuals suffer from porphyria cutanea tarda, the majority do not; furthermore, the pathogenesis of the skin changes is unknown.

Variegate porphyria and hepatoerythropoietic porphyria are the major types of porphyria with dermatologic manifestations that are indistinguishable from those of porphyria cutanea tarda. The cutaneous manifestations in variegate porphyria, however, typically have their onset between puberty and the third or fourth decade of life whereas in hepatoerythropoietic porphyria photosensitivity most often begins in early childhood. Each has unique porphyrin excretion abnormalities that readily permit appropriate classification. In porphyria cutanea tarda, the major abnormality is elevated excretion of urinary uroporphyrin and coproporphyrin (ratio of uroporphyrin to coproporphyrin $>$ 3:1) and elevated isocoproporphyrin excretion in the feces. In variegate porphyria, the major abnormality is increased fecal excretion of protoporphyrin and coproporphyrin (protoporphyrin > coproporphyrin) and increased urinary excretion of the porphyrin precursors δ-aminolevulinic acid and porphobilinogen. In hepatoerythropoietic porphyria, the excretion pattern is similar to that in porphyria cutanea tarda. There is increased urinary excretion of uroporphyrin; 7-, 6-, and 5-carboxylporphyrins; and coproporphyrin. Excretion of isocoproporphyrin in the feces is increased, and the level of protoporphyrin in red blood cells is increased.

Treatment

The treatment of the dermatologic manifestations of porphyria cutanea tarda is based upon reducing the excessive hepatic porphyrin production because there are no known ways to provide effective photoprotection of the skin. The difficulty in protecting the skin is due to the absorption of the porphyrins within the visible portion of the solar spectrum.

Avoidance of porphyrinogenic agents such as ethanol, estrogens, and chlorinated hydrocarbons may be helpful in some cases. The depletion of increased hepatic iron stores by phlebotomy and the chronic intermittent administration of low doses of antimalarials (chloroquine and hydroxychloroquine) are the major therapeutic approaches.[26] The dermatologic manifestations improve at different rates. Typically, the enhanced fragility and vesiculobullous lesions resolve first followed by the hypertrichosis. The pigmentary and sclerodermoid changes improve very slowly.

ERYTHROPOIETIC PROTOPORPHYRIA

Incidence

This porphyria of bone marrow origin is probably second only to porphyria cutanea tarda in frequency. Many cases may go undiagnosed because of the relatively banal cutaneous photosensitivity that occurs in some affected individuals. On the other hand, certain patients are incapacitated by the severity of their symptoms. The disease typically begins in early childhood, affects both sexes equally, and is inherited as an autosomal dominant trait with highly variable expressivity.

Dermatologic Manifestations

In contrast to porphyria cutanea tarda, in which there is chronic photosensitivity, patients with erythropoietic protoporphyria usually manifest acute symptoms and signs, including subjective burning and stinging during or shortly after even brief periods of sun exposure. These symptoms are occasionally severe, and could, for example, cause a child to be brought to the emergency room because of intractable discomfort. Because these symptoms are initially unassociated with any visible cutaneous change, early recognition of the disease is difficult. Erythema, edema, and eczematous changes may ensue with repeated episodes. Purpura may be visible in some patients. The painful purpuric changes gradually subside over a period of several days. Chronically affected skin sites begin to appear thickened and leather-like. Acral areas such as the nose and knuckles are particularly vulnerable to these changes. Fine linear scars become apparent and may have a wax-like translucence. Linear furrows at the corners of the mouth ("pseudorhagades") may also be seen.

Histopathology

Typical features include epidermal acanthosis and thickening of the walls of capillaries in the upper dermis. The latter is due to the deposition of dense hyaline-like material in and around the walls of these vessels. This material is PAS-positive and diastase-negative, indicating the presence of glycosaminoglycans. Ultrastructural studies have shown extensive reduplication of the basal laminae of affected vessels.

Differential Diagnosis

The acute photosensitivity that occurs in erythropoietic protoporphyria is rather distinctive, and once this porphyria is suspected, its diagnosis can readily be made by measurement of red blood cell protoporphyrin or by enzyme assay for heme synthase (ferrochelatase) in fibroblasts or leukocytes. Only two of the other porphyrias, congenital erythropoietic porphyria (Gunther's disease) and hepatoerythropoietic porphyria, usually begin in childhood. In congenital erythropoietic porphyria, the photosensitivity is much more profound. The urinary uroporphyrin and coproporphyrin excretion is increased (uroporphyrin:coproporphyrin ratio $>$ 5:1), and is often sufficient to produce a red-pink staining of the diapers. In erythropoietic protoporphyria, urine porphyrin excretion is normal.

In patients with hepatoerythropoietic porphyria, the typical excretion profile includes increased urinary excretion of uroporphyrin; 7-, 6-, and 5-carboxylporphyrins; and coproporphyrin. There is also increased fecal excretion of isocoproporphyrin.

Three other disorders that must be considered in the differential diagnosis of erythropoietic protoporphyria are polymorphous light eruption, lipoid proteinosis, and psychoneurosis.

Polymorphous light eruption may manifest acute photosensitivity, but the onset of signs and symptoms is typically delayed by several hours. Pruritus, rather than burning and stinging, is the characteristic symptom in affected individuals. Scarring changes are also much less likely to occur in polymorphous light eruption. Measurement of red blood cell protoporphyrin readily differentiates these two disorders.

Lipoid proteinosis is a disorder also known as hyalinosis cutis et mucosae and is characterized by wax-like thickening of the skin with warty excrescences. Hoarseness may occur because of laryngeal involvement. The mucosal involvement of lipoid proteinosis is not a feature of erythropoietic protoporphyria nor is the disorder limited to light-exposed areas. Periodic acid–Schiff stains of skin biopsies in lipoid proteinosis reveal deposition of material throughout the dermis. This disorder affects not only blood vessels, but appendages as well.

Psychoneurosis may be considered in severely symptomatic children with erythropoietic protoporphyria, but the history of acute photosensitivity and measurement of erythrocyte protoporphyrin can confirm the diagnosis of the latter.

Treatment

The therapeutic management of the dermatologic manifestations of erythropoietic protoporphyria is based upon reducing cutaneous photosensitivity either by reducing the phototoxic effect of the excessive porphyrin in the skin or by diminishing its production in the bone marrow.

The simple expedient of avoiding sunlight is effective but is impractical for active children. Extensive experience with oral administration of the yellow-orange vitamin A doublet, β-carotene, has clearly shown that this drug does afford substantial photoprotection in many, though not all, affected individuals.[27–29] The drug is administered in doses ranging from 30 to 180 mg daily to maintain serum carotene levels above 400–500 μg/dl. The drug appears to be remarkably nontoxic. The major side effect is yellow-orange discoloration of the skin that is slowly reversible when the drug is discontinued.

The mechanism of action of β-carotene is not fully understood, but many believe that it is related to the ability of the drug to quench reactive oxygen species and to scavenge free radicals.

One other therapeutic approach has been the use of transfusions of packed red blood cells in an effort to diminish bone marrow heme synthesis. The risk:benefit ratio of this modality must be considered carefully in light of the potential hazards of iron overload, transfusion reactions, and blood-borne infectious diseases.

REFERENCES

1. M'Call Anderson, T. 1898. Br. J. Dermatol. **10:** 1–4.
2. Hausmann, W. 1914. Biochem. Z. **67:** 309–321.
3. Meyer-Betz, F. 1913. Dtsch. Arch. Klin. Med. **112:** 476–488.
4. Blum, H. F. 1941. Photodynamic Action and Diseases Caused by Light. Reinhold. New York, NY.
5. Slater, T. F. & P. A. Riley. 1966. Nature **209:** 151–154.
6. Allison, A. C., I. A. Magnus & M. R. Young. 1966. Nature **209:** 874–878.
7. Rimington, C., I. A. Magnus, E. A. Ryan & D. J. Cripps. 1967. Q. J. Med. **36:** 29–57.
8. Goldstein, B. D. & L. C. Harber. 1972. J. Clin. Invest. **51:** 892–901.
9. Wakulchik, S. D., J. R. Schiltz & D. R. Bickers. 1980. J. Lab. Clin. Med. **96:** 158–167.
10. Sandberg, S. & I. Romslo. 1981. Biochem. J. **198:** 67–74.
11. Sandberg, S., J. Glette, G. Hopen, C. O. Solberg & I. Romslo. 1981. Photochem. Photobiol. **34:** 471–475.
12. Latham, P. S. & J. R. Bloomer. 1983. Photochem. Photobiol. **37:** 553–557.
13. Bickers, D. R., R. Dixit & H. Mukhtar. 1982. Biochem. Biophys. Res. Commun. **108:** 1032–1039.
14. Keene, J. P., D. Kessel, E. J. Land, R. W. Redmond & T. G. Truscott. 1986. Photochem. Photobiol. **43:** 117–120.
15. Gigli, I., A. A. Schothorst, N. A. Soter & M. A. Pathak. 1980. J. Clin. Invest. **66:** 517–522.
16. Lim, H. W., H. D. Perez, M. B. Poh-Fitzpatrick, I. M. Goldstein & I. Gigli. 1981. N. Engl. J. Med. **304:** 212–216.
17. Lim, H. W. & I. Gigli. 1981. J. Invest. Dermatol. **76:** 4–9.
18. Lim, H. W., M. B. Poh-Fitzpatrick & I. Gigli. 1984. J. Clin. Invest. **74:** 1961–1965.
19. Kamide, R., I. Gigli & H. W. Lim 1984. J. Invest. Dermatol. **82:** 485–490.
20. Lim, H. W., L. Young, M. Hagan & I. Gigli. 1985. J. Invest. Dermatol. **84:** 114–117.
21. Grossman, M. E., D. R. Bickers, M. B. Poh-Fitzpatrick & L. C. Harber. 1979. Am. J. Med. **67:** 277–286.
22. Varigos, G., J. R. Schlitz & D. R. Bickers. 1981. J. Clin. Invest. **69:** 129–135.
23. Ramsay, C. A. & E. Obreshkova. 1974. Br. J. Dermatol. **91:** 523–528.
24. Baer, R. L. & M. A. Stillman. 1984. J. Am. Acad. Dermatol. **10:** 527.
25. Poh-Fitzpatrick, M. B., N. Bellet, V. A. DeLeo, M. E. Grossman & D. R. Bickers. 1977. N. Engl. J. Med. **299:** 292–294.
26. Bickers, D. R. & H. Merk. 1986. Semin. Dermatol. **5:** 186–197.
27. Mathews-Roth, M. M. 1978. J. Infect. Dis. **138:** 924–927.
28. Mathews-Roth, M. M. 1984. Photochem. Photobiol. **40:** 63–67.
29. Mathews-Roth, M. M. 1984. Photodermatology **1:** 310–321.

Alterations in Renal Heme Biosynthesis during Metal Nephrotoxicity

AGNETA OSKARSSON

National Food Administration
Toxicology Laboratory
Uppsala, Sweden

BRUCE A. FOWLER

National Institute of Environmental Health Sciences
Research Triangle Park, North Carolina 27709

INTRODUCTION

The kidney is a major target organ for many trace metals, and it has been shown that many of these metals have the ability to alter heme metabolism in this organ. The effects on heme metabolism may represent an initial response to the metal-induced toxicity and may be of importance in understanding the development of the tissue damage. The effects of metals on heme metabolism can occur at various steps in the heme biosynthetic pathway, and in this respect lead has been extensively studied.

Much of the research on the effects of metals on the regulation of heme metabolism has been conducted in hepatic and erythropoietic tissues. There is evidence that the regulatory mechanisms for the renal activity of δ-aminolevulinic acid synthetase (ALAS), which catalyzes the first and rate-limiting step in heme biosynthesis, differ from those in the liver.[1] The normal activities of enzymes of heme biosynthesis and the contents of heme have been compared in kidney and liver of rats and found to be lower in kidney than in liver.[2] The concentration of porphyrins in the kidney, however, exceeded that in the liver severalfold.

Some metals regulate the cellular content of heme in the kidney by interacting with the degradation of heme via stimulation of the enzyme heme oxygenase (such metals include Ni^{2+}, Sb^{3+}, Sn^{2+}, Co^{2+}, Pt^{4+}, Pb^{2+}, Hg^{2+}, and Cu^{2+}).[3–5] The following paper is mainly concerned with the effects of these metals on heme biosynthesis.

LEAD

Lead has been shown to interfere with heme biosynthesis at several steps, and the results of these interactions in the erythroid tissues can be used for biological monitoring of lead exposure.[6,7] In the kidney, however, the heme biosynthetic pathway enzymes show different responses to lead in comparison with other tissues. In chronic lead toxicity the most prominent feature of morphologic response in the kidney is the formation of lead-containing intranuclear inclusion bodies.[8,9] Goyer *et al.*[8] have suggested that the formation of nuclear inclusions may act to protect sensitive biochemical systems, such as the mitochondria, from lead by reducing the availability of lead in the cytoplasm and mitochondria of the cell.

The effects of lead on renal heme biosynthesis have recently been reviewed.[10] TABLE 1 is modified from this review. Gibson and Goldberg[11] did not find any significant effects on ALAS and ferrochelatase activities in the kidneys of lead-treated rabbits (injected sc daily with doses of 10, 30, 150, and 200 mg/kg body wt of lead acetate/week for up to 24 weeks) compared to controls. Renal δ-aminolevulinic acid dehydratase (ALAD) activity was reduced in the lead-treated rabbits by a factor of 50%. The inhibition of ALAD could be reversed by addition of glutathione to the enzyme assay system, indicating that lead inhibits the enzyme through its active sulfhydryl groups. A single peroral dose of lead at 100, 250, 500, 1500, or 3000 μg to rats (100–110 g body wt) produced no effects on ALAS activity in the kidney.[12] ALAD activity in the kidney was inhibited to the highest degree at 6 hr after the administration of lead when the activity was about 20% of control levels in the two highest dose groups. Although there was a tendency for recovery of ALAD activity within the 6-day period after exposure, recovery was not complete in the two highest dose groups. Buchet *et al.*[13] provided rats with drinking water containing lead at concentrations from 10 to 5000 ppm. They concluded that the kidney ALAD was the most susceptible parameter to lead because reduced activity was observed in all treated groups after 3 weeks of exposure. The inhibition of kidney ALAD, however, appeared transitory because after 6 weeks of exposure it was only observed in the 5000 ppm group. When

TABLE 1. Effects of Lead on Enzymes Regulating Heme Biosynthesis in the Kidney

		Enzyme Affected			
Species	Dose	ALAS	ALAD	FC	References
Rabbit	10, 30, 150, 200 mg/kg per week sc	NC	↓	NC	Gibson & Goldberg, 1970[11]
Rat	100, 250, 500, 1500, 3000 μg po	NC	↓		Kao & Forbes, 1973[12]
Rat	10, 100, 1000, 5000 ppm in dw for 3 and 6 weeks		↓[a]		Buchet *et al.*, 1976[13]
Rat	0.1, 1, 10 ppm in dw for 9 weeks pre- and postnatally		NC		Hubermont *et al.*, 1976[14]
Rat	1, 10, 100 ppm in dw for 3 weeks postnatally		↓[b]		Roels *et al.*, 1977[15]
Rat	0.5, 5, 25, 50, 250 ppm in dw for 6 and 9 months	↓[c]	NC	↓[c]	Fowler *et al.*, 1980[9]
Rat	5, 25, 50, 100 mg/kg ip		NC		Woods & Fowler, 1982[16]
Rat	10 000 ppm in dw for 1.5–2 months	NC	↓		Silbergeld *et al.*, 1982[17]
Rat	2 mg/kg ip for 3 days	↑	↓		Silbergeld *et al.*, 1982[17]
Rat	500, 1000 ppm in dw for 3 and 6 months	NC	↑	↓	Oskarsson & Fowler, 1985[18]
Rat	50 mg/kg ip	↑	NC	↑	This study

NOTE: Adapted from reference 10. FC: ferrochelatase; NC: no change; ↓: decreased activity; ↑: increased activity; dw: drinking water.

[a]Three weeks.

[b]Sucklings.

[c]Nine months.

drinking water containing lead was administered to female rats for 3 weeks before mating, during pregnancy, and during 3 weeks after delivery at concentrations from 0.1 to 10 ppm, no change in ALAD activity was seen, in either the dams or the newborns.[14] In the newborns, however, there was an increase in the level of "free tissue" porphyrins in the kidney in the 10 ppm group. The conclusion was that the developing organism is more susceptible to the biological action of lead than adults. Higher doses of lead, up to 100 ppm, were administered in drinking water to lactating rats for 3 weeks after delivery.[15] A slight reduction in ALAD activity and a marked increase in free tissue porphyrins in the kidney were found in the suckling rats, whereas in the mothers there was no change in these parameters. Fowler *et al.*[9] studied rats exposed to lead in drinking water in doses of 0.5 to 250 ppm for 6 and 9 months. There were no significant differences in enzyme activities in the renal heme biosynthetic pathway after 6 months of exposure. After 9 months, effects were observed on several of these enzymes. The activity of ALAS was significantly decreased at the 50 and 250 ppm dose levels to 64 and 43% of control levels, respectively, and ferrochelatase was significantly inhibited at 5, 50, and 250 ppm with a decrease to 63% of control levels at the highest dose level. In contrast, ALAD activity was not altered in the lead-treated rats at any exposure levels.[9] Acute treatment of rats with lead in doses up to 100 mg/kg ip 16 hr before sacrifice did not result in any significant effect on renal ALAD activity.[16] Silbergeld *et al.*[17] studied the effects of lead on the heme biosynthesis both after chronic exposure (10 mg/ml of drinking water, that is, 10 000 ppm; first indirectly, until 30 days of age, via milk from dams provided the drinking water, and then directly, via the drinking water, until 40–60 days of age) and after acute exposure (10 μmol/kg/day, that is, 2 mg/kg/day; received ip for 3 days). After acute exposure there was an increase in ALAS activity in the kidney, but this was not seen after chronic exposure. Renal ALAD activity was decreased to 55% of control levels after chronic lead treatment and to 40% of control levels after acute treatment. Renal tissue content of δ-aminolevulinic acid (ALA) was markedly increased both after chronic and acute treatments, 3- and 10-fold increases, respectively, compared to control levels.[17] Small effects on enzymes in the heme biosynthetic pathway in kidney were observed in rats exposed to lead in drinking water in doses of 500 and 1000 ppm for 3 and 6 months.[18] Renal activity of ALAS was not significantly changed in the lead-treated groups. There was a slight, but significant, increase in renal ALAD activity after 6 months to approximately 110% of control values. An increase in ALAD activity in liver up to 250% of control values was reported by Kao and Forbes.[12] They explained the increased activity as resulting from a release of inhibition or as an adaptive response to elevated ALA levels. Ferrochelatase activity was significantly reduced after 3 and 6 months of treatment with lead.[18] In the highest dose group the activity was about 80% of controls at 6 months. The effects on enzymes in the heme biosynthetic pathway in the rat kidney were also studied after acute treatment with lead, which is known to result in the formation of nuclear inclusion bodies in renal proximal tubule cells.

EFFECTS ON RENAL HEME BIOSYNTHESIS IN RATS AFTER ACUTE TREATMENT WITH LEAD

It has been shown that in the kidney there are cytosolic lead-binding proteins,[19] which are capable of greatly reducing the inhibitory effects of lead on ALAD activity.[20,21] Formation of lead inclusion bodies, which may act to protect the kidney from toxic effects from lead, has also been shown in rats after acute treatment with lead.[22] Pretreatment of rats with lead, resulting in formation of inclusion bodies in the renal proximal tubule cells, decreases the binding of lead in whole kidney homogenate and in mitochondrial and supernatant fractions.[23] The present study was performed to investigate the impact of inclusion bodies on lead toxicity to the renal heme biosynthetic pathway after acute treatment of rats with lead.

MATERIALS AND METHODS

Male Sprague–Dawley rats (CD strain, obtained from Charles River Laboratories, Boston, MA) weighing 286 ± 3 g were randomly divided into two groups of 18 rats each. The rats were fed NIH 31 Diet (Ziegler Brothers, Gardner, PA) and tap water *ad libitum*. Formation of lead intranuclear inclusion bodies was initiated in kidney proximal tubule cells by a single ip injection of lead acetate (lead acetate trihydrate was dissolved in 0.25 ml of distilled water so that the dose of lead would be 50 mg/kg body wt) using the regimen of Choie and Richter.[22] In a previous study[23] the optimal time period for inclusion body formation was found to be 6 days as monitored by electron microscopy. At day 1 after lead treatment there was no evidence of inclusion bodies in proximal tubule cells. After 3 days fibrillar cytoplasmic inclusions were found; after 6 days the cytoplasmic inclusions were less prominent, and intranuclear inclusions were observed.

At 1, 3, and 7 days after treatment, 6 rats per group were decapitated. Kidneys were rapidly excised, the left kidney was weighed, and both kidneys were homogenized in nine volumes of 0.25 M sucrose in 0.05 M Tris-HCl buffer, pH 7.5, in a Potter–Elvehjem homogenizer fitted with a Teflon pestle. Homogenates were centrifuged at 600 g for 10 min. Mitochondria were isolated from the supernatant by centrifugation at 9000 g for 15 min. The mitochondria were washed twice and resuspended in 0.05 M Tris-HCl buffer to give a final protein concentration of approximately 4 mg/ml. All steps were performed at about +4 °C.

ALAD and uroporphyrinogen I synthetase (UROS) activities were measured in the 9000 g supernatant fraction. The ALAD activity was measured by a modification of the method of Gibson *et al.*[24] as described by Baron and Tephly[25]; the UROS activity, after heating the supernatant fraction to 65 °C for 15 min by the method of Levin and Coleman[26] as described by Woods.[27] Activities of ALAS and ferrochelatase were determined in the mitochondrial fraction. The ALAS activity was measured by a modification of a method of Scholnick *et al.*[28] as described by Woods[29]; the ferrochelatase activity, by a modification of the method of Porra[30] as described by Woods and Carver.[31] Protein determinations were carried out by a method of Lowry *et al.*[32] Analyses of significance of differences between the control and lead-treated groups were performed by means of Student's t test. The level of significance was chosen as $p \leq .05$.

RESULTS AND DISCUSSION

Lead treatment did not cause any significant changes in body and kidney weights within 7 days after treatment (TABLE 2). Lead-treated rats were somewhat reduced in weight resulting in a small, but significant increase in the kidney weight:whole body weight ratio at 3 days after treatment.

Activities of the heme enzymes ALAS, ALAD, UROS, and ferrochelatase after acute treatment of rats with lead are shown in TABLE 3. Renal ALAD and UROS activities were not significantly affected by lead treatment at any of the studied time

TABLE 2. Body Weight, Weight of Left Kidney, and Kidney Weight:Body Weight Ratio in Rats after Acute Treatment with Lead

Group	Time after Treatment (days)	Body Weight (g)	Kidney Weight (g)	Kidney Weight:Body Weight ($\times 10^3$)
Control	1	286 ± 3	1.13 ± 0.03	3.95
	3	282 ± 9	1.06 ± 0.04	3.76
	7	328 ± 11	1.25 ± 0.07	3.81
Lead	1	286 ± 3	1.22 ± 0.07	4.27
	3	266 ± 8	1.23 ± 0.06	4.62[a]
	7	300 ± 10	1.27 ± 0.07	4.23

NOTE: Each value for weight is a mean ± SEM ($N = 6$).
[a]Significantly different from the control group ($p \leq .05$).

points. At 3 days after treatment there was an increase in ALAS activity in the lead-treated group to about 120% of the control value and an increase in ferrochelatase activity to about 140% of the control value. Ferrochelatase activity was not determined at 1 day after treatment. Possibly, the stimulated rather than inhibited activity seen at 3 days could be explained by an enzyme activation following release of a preceding inhibition of the enzyme activity. At 7 days, when intranuclear inclusion bodies were present, these enzymes were not significantly affected by lead treatment.

Inhibition of ALAD activity in the red blood cells is generally recognized as the most sensitive biological indicator of the individual's exposure to lead. When depressed in red blood cells, the activity of this enzyme is also depressed in other tissues (see reviews by Chisolm[6] and Sassa[7]). It is evident from the present results and from the review in TABLE 1 that the renal heme biosynthetic pathway is relatively mildly affected by lead treatment compared to other tissues. An explanation to this can be the intracellular complexation of lead with high-affinity renal lead-binding proteins[20,21] and the donation of zinc to this zinc-dependent enzyme. Formation of intranuclear inclusion bodies[8] also appears to protect the heme biosynthetic pathway in kidney from lead effects.

TABLE 3. Effects of Enzymes of the Renal Heme Biosynthetic Pathway in the Rat after Acute Treatment with Lead

		Enzyme Activity			
Group	Time after Treatment (days)	ALAS (nmol ALA/mg protein/hr)	ALAD (nmol PBG/ mg protein/hr)	UROS (nmol/PBG/ mg protein/hr)	FC (nmol mesoheme/ mg protein/hr)
Control	1	0.89 ± 0.08	4.60 ± 0.21	5.94 ± 0.10	ND
	3	0.53 ± 0.02	4.56 ± 0.34	5.13 ± 0.28	1.78 ± 0.05
	7	0.52 ± 0.04	3.43 ± 0.08	3.58 ± 0.70	1.44 ± 0.09
Lead	1	0.69 ± 0.11	4.44 ± 0.31	6.12 ± 0.24	ND
	3	0.65 ± 0.07[a]	4.00 ± 0.16	6.50 ± 0.43	2.45 ± 0.17[a]
	7	0.66 ± 0.10	3.04 ± 0.26	3.82 ± 0.74	1.67 ± 0.01

NOTE: Each value is a mean ± SEM of three determinations from pooled kidneys (two rats/determination). FC: ferrochelatase; PBG: porphobilinogen; ND: not determined.
[a]Significantly different from the control group ($p \leq .05$).

Methyl Mercury

The effects of methyl mercury on the heme biosynthetic pathway in the kidney has been studied by Woods and Fowler[33] and by Fowler and Woods.[34] Rats were administered 3, 5, and 10 ppm mercury as methyl mercury hydroxide for 6 weeks, and the activities of several heme biosynthetic pathway enzymes both in liver and kidney were investigated. In the liver the enzyme activities were not significantly altered by treatment with methyl mercury hydroxide whereas in the kidney effects at several steps were observed. ALAS was significantly increased at all dose levels tested, and a maximum increase to more than 250% of the control value was observed at 5 ppm. The activity of ALAD was not altered at any dose level tested. Both UROS and ferrochelatase were decreased in activity, and as with ALAS, maximal effects were seen at 5 ppm, where the activity of UROS was 58% of control levels and the activity of ferrochelatase was approximately 70% of control levels. Urinary levels of heme precursors were also measured, and a slight elevation of ALA and porphobilinogen excretion was observed. In contrast, levels of urinary excretion of uroporphyrin and coproporphyrin were increased, 12.5- and 21-fold, respectively, in the highest dose group.

Also the effects of acute exposure to methyl mercury hydroxide on activities of renal heme biosynthetic pathway enzymes were studied by Woods and Fowler.[33] In rats treated ip with 0.7 or 1.4 mg/kg methyl mercury hydroxide at 48 and 24 hr before sacrifice, ALAS and ferrochelatase were inhibited to 83 and 90% of their control values, respectively, at the high dose level. Activities of ALAD and UROS were not affected by the treatment.

The authors suggested that chronic exposure to methyl mercury hydroxide resulting in renal porphyria is mediated via an initial inhibition of ferrochelatase and a subsequent inhibition of UROS. Secondarily to these effects, induction of ALAS in the kidney occurs. The unique response to exposure to methyl mercury hydroxide in urinary excretion of porphyrins was suggested for use as a biological indicator of pretoxic exposure to methyl mercury in human populations.[33]

Indium

Indium is retained in the kidney and produces renal failure in high doses.[35] Woods and Fowler[16] reported a tissue-specific inhibition of ALAD in rat kidney after ip treatment with indium chloride in doses of 10, 20, and 40 mg/kg body wt 16 hr before sacrifice. At the highest dose level ALAD activity was inhibited to 41% of control levels. Treatment with indium produced no significant changes in the renal activities of ALAS, UROS, or ferrochelatase. The inhibited ALAD activity did not result in any changes in the urinary excretion of ALA or porphobilinogen or changes in the renal microsomal heme content.

Nickel

Effects of nickel on enzymes of heme metabolism in the kidney have been investigated by Maines and Kappas.[3–5] Nickel and platinum were chosen as examples of metals that do not enzymatically form heme complexes and thereby would not inhibit ALAS by mimicking the action of heme. Rats were treated sc with nickel dichloride at a dose of 250 μmol/kg body wt. The time courses of the enzyme effects were followed in the kidney. There was a marked increase in the activity of renal heme oxygenase, an enzyme involved in heme degradation. ALAS activity remained unchanged for 12 hr, after which the activity was increased and reached about 200% of control levels at 48 hr. Treatment with nickel caused a significant inhibition of ALAD activity, and at 24 hr the activity was inhibited to 50% of control levels. UROS and ferrochelatase activities were slightly decreased during the first 24 hr. Total porphyrin content was decreased at 3 hr after treatment with nickel but had returned to normal at 16 hr. In contrast, microsomal content of heme did not change during the first 3 hr but had decreased to about 60% of control levels at 16 hr, whereafter the content increased to control levels at 48 hr. At concentrations of 125 and 250 μM *in vitro,* nickel did not change the renal activity of ALAS and slightly decreased the activity of ALAD.

Platinum

The investigation of platinum, which was concerned with the effects of platinum on renal heme metabolism, was similar to the investigation of nickel by Maines and Kappas.[3] Rats were treated sc with platinic potassium chloride at a dose of 125 μmol/kg body wt. The activity of ALAS was decreased during the first 12 hr after treatment with platinum, after which there was a rapid increase to approximately 250% of control levels. Platinum was, as was nickel, a potent inducer of heme oxygenase in the kidney. ALAD activity was inhibited to 34% of control levels at 24 hr. The activities of UROS and ferrochelatase were reduced to about 50 and 90% of their respective control levels at 24 hr. The total porphyrin content of kidney was decreased markedly at 24 hr. The microsomal content of heme was decreased from 3 hr after treatment, and the effect was sustained for 24 hr. *In vitro,* platinum markedly reduced the activity of ALAD in the kidney; at 250 μM the activity was 56% of the control value.

Tin

Tin has been shown to greatly enhance heme degradation in the kidney by inducing heme oxygenase.[36,37] Rats were treated sc with tin dichloride at a dose of 250 μmol/kg body wt. Animals were sacrificed after 16 hr, and parameters of heme metabolism in the kidney were examined. Heme oxygenase activity was increased to a value 23-fold higher than that in the control kidneys. Concomitantly, the microsomal content of cytochrome P-450 in kidney was decreased by an average of 50%. Kappas and Maines[38] found no alteration in ALAS activity. However, a 3–4-fold increase in renal ALAS activity was reported at 16 hr after treatment with tin by Drummond and Kappas.[37] The latter authors also found that the effects of tin and those of nickel on heme metabolism could be inhibited with simultaneous or prior treatment of the rats with manganese or zinc.

Antimony

Trivalent antimony, but not pentavalent, is a potent inducer of heme oxygenase in the kidney.[39] Rats were injected sc with trivalent antimony in doses of up to 250 μmol/kg body wt. Heme oxygenase activity increased to a value approximately 11-fold higher than the control level. ALAS activity decreased shortly after antimony administration and then increased and reached a level almost three times higher than the control level at 24 hr. Microsomal content of cytochrome P-450 in kidney decreased to 70% of the control level at 16 hr after treatment. *In vitro,* 0.1 mM antimony increased renal heme oxygenase to 160% of the control level, and ALAS activity was inhibited to 50% of the control level at 1.0 mM of antimony.[39]

Cobalt

Cobalt is a potent inducer of hepatic heme oxygenase in the liver. Its effects on heme biosynthesis and degradation in other tissues were studied by Maines and Kappas.[36] Rats were treated sc daily with 250 μmol/kg body wt for 3 days. The activity of heme oxygenase was increased 5-fold in kidney cortex, and the microsomal heme content was decreased. ALAS activity was first decreased and then returned to the control level.

Bismuth

The effects of acute bismuth administration on the renal heme biosynthetic pathway have been reported by Woods and Fowler.[40] These studies demonstrated dose-related decreases in the renal activities of ALAS and of ALAD but not of ferrochelatase at 16 hr after subcutaneous injection of basic bismuth nitrate ($BiONO_3$) at doses of 0, 20, 40, or 80 mg Bi/kg body wt. Similar effects were observed for this bismuth salt *in vitro* over a concentration range of 0–100 μg Bi/ml.

Gallium

Intratracheal instillation studies with particulate gallium arsenide[41,42] have shown a moderate inhibitory effect on renal ALAD with increased urinary excretion of ALA. *In vitro* studies[42] demonstrated that ALAD was highly sensitive to gallium in this organ.

SUMMARY

The regulation of the heme biosynthetic pathway in the kidney by various metals has been reviewed. In addition, a study on the effects of lead on renal heme biosynthesis after acute treatment of rats has been reported. Chronic low-level lead exposure in rats results in relatively small effects on renal heme biosynthetic pathway enzymes. After acute treatment of rats with lead, no effects on ALAD or UROS and mild, transitory effects on ALAS and ferrochelatase are observed. The intracellular binding of lead within intranuclear inclusion bodies in the proximal tubule cells and to high-affinity cytosolic lead-binding proteins probably protects sensitive subcellular systems, such as the heme pathway, from lead toxicity. Chronic exposure to methyl mercury results in increased urinary excretion of uro- and coproporphyrins in rats, mediated via inhibition of ferrochelatase and UROS and stimulation of ALAS. A tissue-specific inhibition of ALAD occurs in the kidney after treatment of rats with indium. Acute treatment of rats with nickel, platinum, tin, antimony, bismuth, and cobalt results in induction of heme oxygenase, followed by decreased microsomal heme content and ALAS stimulation in the kidney.

REFERENCES

1. MAINES, M. D. 1984. New developments in the regulation of heme metabolism and their implications. CRC Crit. Rev. Toxicol. **12:** 241–314.
2. MAINES, M. D., J. M. TOMIO & V. JANOUSEK. 1978. Differential patterns of heme porphyrin biosynthesis in liver and kidney: Metal ion blockade of δ-aminolevulinate synthetase introduction by porphyrinogenic agents. *In* Diagnosis and Therapy of Porphyrias and Lead Intoxication. M. Doss, Ed.: 8–14. Springer-Verlag. New York, NY.
3. MAINES, M. D. & A. KAPPAS. 1977. Regulation of heme pathway enzymes and cellular glutathione content by metals that do not chelate with tetrapyrroles: Blockade of metal effects by thiols. Proc. Natl. Acad. Sci. USA **74:** 1875–1878.
4. MAINES, M. D. & A. KAPPAS. 1977. Regulation of cytochrome P-450-dependent microsomal drug-metabolizing enzymes by nickel, cobalt and iron. Clin. Pharmacol. Ther. **22:** 780–790.
5. MAINES, M. D. & A. KAPPAS. 1977. Nickel-mediated alterations in the activity of hepatic and renal enzymes of heme metabolism and heme-dependent cellular activities. *In* Clinical Chemistry and Chemical Toxicology of Metals. S. S. Brown, Ed.: 75–81. Elsevier/North-Holland. Amsterdam.
6. CHISOLM, J. J., JR. 1964. Disturbances in the biosynthesis of heme in lead intoxication. J Pediatr. **64:** 174–187.
7. SASSA, S. 1978. Toxic effects of lead, with particular reference to porphyrin and heme metabolism. *In* Handbook of Experimental Pharmacology. Vol. **44:** 333–371. Springer-Verlag. Berlin.
8. GOYER, R. A., D. L. LEONARD, J. F. MOORE, B. RHYNE & M. R. KRIGMAN. 1970. Lead dosage and the role of the intranuclear inclusion body. Arch. Environ. Health **20:** 705–711.
9. FOWLER, B. A., C. A. KIMMEL, J. S. WOODS, E. E. MCCONNELL & L. D. GRANT. 1980. Chronic low-level lead toxicity in the rat. III. An integrated assessment of long-term toxicity with special reference to the kidney. Toxicol. Appl. Pharmacol. **56:** 59–77.
10. Environmental Protection Agency. 1986. Air Quality Criteria for Lead. 12-164-167. U.S. Environmental Protection Agency. Washington, D.C.
11. GIBSON, S. L. M. & A. GOLDBERG. 1970. Defects in haem synthesis in mammalian tissues in experimental lead poisoning and experimental porphyria. Clin. Sci. **38:** 63–72.
12. KAO, R. L. C. & R. M. FORBES. 1973. Effects of lead on heme-synthesizing enzymes and urinary δ-aminolevulinic acid in the rat. Proc. Soc. Exp. Biol. Med. **143:** 234–237.
13. BUCHET J.-P., H. ROELS, G. HUBERMONT & R. LAUWERYS. 1976. Effect of lead on some parameters of the heme biosynthetic pathway in rat tissues *in vivo.* Toxicology **6:** 21–34.
14. HUBERMONT, G., J.-P. BUCHET, H. ROELS & R. LAUWERYS. 1976. Effect of short-term administration of lead to pregnant rats. Toxicology **5:** 379–384.
15. ROELS, H., R. LAUWERYS, J.-P. BUCHET & G. HUBERMONT. 1977. Effects of lead on lactating rats and their sucklings. Toxicology **8:** 107–113.
16. WOODS, J. S. & B. A. FOWLER. 1982. Selective inhibition of δ-aminolevulinic acid dehydratase by indium chloride in rat kidney: Biochemical and ultrastructural studies. Exp. Mol. Pathol. **36:** 306–315.
17. SILBERGELD, E. K., R. E. HRUSKA, D. BRADLEY, J. M. LAMON & B. C. FRYKHOLM. 1982. Neurotoxic aspects of porphyrinopathies: Lead and succinylacetone. Environ. Res. **29:** 459–471.
18. OSKARSSON, A., & B. A. FOWLER. 1985. Effects of lead on the heme biosynthetic pathway in rat kidney. Exp. Mol. Pathol. **43:** 409–417.
19. OSKARSSON, A. K. S. SQUIBB & B. A. FOWLER. 1982. Intracellular binding of lead in the kidney: The partial isolation and characterization of postmitochondrial lead-binding components. Biochem. Biophys. Res. Commun. **104:** 290–298.
20. GOERING, P. L. & B. A. FOWLER. 1984. Regulation of lead inhibition of δ-aminolevulinic acid dehydratase by a low molecular weight, high affinity renal lead-binding protein. J. Pharmacol. Exp. Ther. **231:** 66–71.
21. GOERING, P. L. & B. A. FOWLER. 1985. Mechanism of renal lead-binding protection against lead inhibition of δ-aminolevulinic acid dehydratase. J. Pharmacol. Exp. Ther. **234:** 365–371.

22. CHOIE, D. D. & G. W. RICHTER. 1972. Lead poisoning: Rapid formation of intranuclear inclusions. Science (Washington, DC) **177:** 1194–1195.
23. OSKARSSON, A. & B. A. FOWLER. 1985. Effects of lead inclusion bodies on subcellular distribution of lead in rat kidney: The relationship to mitochondrial function. Exp. Mol. Pathol. **43:** 397–408.
24. GIBSON, K. D., A. NEUBERGER, & J. J. SCOTT. 1955. The purification and properties of δ-aminolevulinic acid dehydrase. Biochem. J. **61:** 618–629.
25. BARON, J. & T. R. TEPHLY. 1969. Effect of 3-amino-1,2,4-triazole on the induction of rat hepatic microsomal oxidases, cytochrome P-450, and NADPH-cytochrome reductase by phenobarbital. Mol. Pharmacol. **5:** 10–20.
26. LEVIN, E. Y. & D. L. COLEMAN. 1967. The enzymatic conversion of porphobilinogen to uroporphyrinogen catalyzed by extracts of hematopoietic mouse spleen. J. Biol. Chem. **242:** 4248–4253.
27. WOODS, J. S. 1976. Developmental aspects of hepatic heme biosynthetic capability and hematotoxicity. Biochem. Pharmacol. **25:** 2147–2152.
28. SCHOLNICK, P. L., L. E. HAMMAKER & H. S. MARVER. 1972. Soluble δ-aminolevulinic acid synthetase of rat liver. I. Some properties of the partially purified enzyme. J. Biol. Chem. **247:** 4126–4131.
29. WOODS, J. S. 1974. Studies on the role of heme in the regulation of δ-aminolevulinic acid synthetase during fetal hepatic development. Mol. Pharmacol. **10:** 389–397.
30. PORRA, R. J. 1975. A rapid spectrophotometric assay for ferrochelatase activity in preparations containing much endogenous hemoglobin and its application to soybean root-nodule preparations. Anal. Biochem. **68:** 289–298.
31. WOODS, J. S. & G. T. CARVER. 1977. Action of cobalt chloride on the biosynthesis, degradation and utilization of heme in fetal rat liver. Drug Metab. Dispos. **5:** 487–492.
32. LOWRY, O. H., N. J. ROSEBROUGH, A. L. FARR & R. L. RANDALL. 1951. Protein measurement with the Folin phenol reagent. J. Biol. Chem. **193:** 265–275.
33. WOODS, J. S. & B. A. FOWLER. 1977. Renal porphyrinuria during chronic methyl mercury exposure. J. Lab. Clin. Med. **90:** 266–272.
34. FOWLER, B. A. & J. S. WOODS. 1977. Ultrastructural and biochemical changes in renal mitochondria during chronic oral methyl mercury exposure: The relationship to renal function. Exp. Mol. Pathol. **27:** 403–412.
35. FOWLER, B. A. 1986. Indium. *In* Handbook on the Toxicology of Metals. 2nd edit. L. Friberg, G. F. Nordberg & V. Vouk, Eds.: 267–275. Elsevier. Amsterdam.
36. MAINES, M. D. & A. KAPPAS. 1976. The induction of heme oxidation in various tissues by trace metals: Evidence for the catabolism of endogenous heme by hepatic heme oxygenase. Ann. Clin. Res. **8**(Suppl. 17): 39–46.
37. DRUMMOND, G. S. & A. KAPPAS. 1980. Metal ion interaction in the control of haem oxygenase induction in liver and kidney. Biochem. J. **192:** 637–648.
38. KAPPAS, A. & M. D. MAINES. 1976. Tin: A potent inducer of heme oxygenase in kidney. Science **192:** 60–62.
39. DRUMMOND, G. S. & A. KAPPAS. 1981. Potent heme-degrading action of antimony and antimony-containing parasiticidal agents. J. Exp. Med. **153:** 245–256.
40. WOODS, J. S. & B. A. FOWLER. 1983. Studies on the ultrastructural and hematotoxic effects of bismuth in rat liver and kidney. Toxicologist **3:** 135.
41. GOERING, P. L. & B. A. FOWLER. 1986. Effect of intratracheal gallium arsenide (GaAs) administration on δ-aminolevulinic acid dehydratase (ALAD) in rats. Pharmacologist **28:** 229.
42. FOWLER, B. A. & P. L. GOERING. 1986. Mechanism of gallium arsenide (GaAs) inhibition of δ-aminolevulinic acid dehydratase (ALAD). Pharmacologist **28:** 229.

Lead-Induced Abnormalities of Porphyrin Metabolism

The Relationship with Iron Deficiency[a]

SERGIO PIOMELLI, CAROL SEAMAN,
AND SHIV KAPOOR

Columbia University
College of Physicians and Surgeons
Division of Pediatric Hematology
New York, New York 10032

INTRODUCTION

Lead poisoning remains a serious public health problem in this country, particularly for underprivileged urban children. The major source of Pb is household dust,[1] containing not only Pb from peeling paint in old buildings but also the deposits of airborne Pb.[2]

Epidemiological studies from our laboratory have provided evidence that low-level exposure to Pb results in biochemical toxicity to children.[3] These studies, and the demonstration that neurological damage is possible at a low level of exposure,[4] have been instrumental in inspiring the Environmental Protection Agency to legislate the progressive reduction of Pb in gasoline. There has been a substantial decrease in the average blood Pb (BPb) in recent years, associated with a reduction in the use of Pb in gasoline.[5–8] This has greatly reduced the average BPb of all Americans and the frequency of clinically severe childhood Pb poisoning. The situation is greatly improved from the deplorable state that existed in the 1960s. Many children, however, still have an excessive body burden of Pb.[9] The HANES study[10] indicates that 4% of American children 6 months to 5 years old (that is, over 650 000 children) have an excessive body burden of Pb; the percentage rises to 18% among poor urban black children.

A simple and accurate fluorometric technique for measurement of erythrocyte protoporphyrin (EP) from a finger puncture was developed by us[11] in 1972. This method (the EP Test) has become the standard for Pb-poisoning screening[12] and has opened the way to a nationwide screening program for childhood Pb poisoning; this program has tested several millions of children and detected hundreds of thousands of unsuspected cases. The initial observations were rapidly confirmed.[13–16] It was soon apparent that elevation of EP has greater clinical relevance to childhood Pb poisoning and to neurological toxicity than the BPb level itself.[17–19] The Center for Disease Control[12] recommended that children with BPb >30 μg/dl be given medical attention, if they also had an elevated EP. These recommendations have been recently updated to include children with BPb >25 μg/dl, and with elevated EP.[20]

As our knowledge of the toxic effects of Pb increases, the acceptable body burden decreases. In 1965, Patterson—on purely geophysical considerations—postulated a

[a]This work was supported by Grants ES02343 and HL28381 from the National Institutes of Health and by Grant CU50184601 from the New York Community Trust.

negligible "natural" body burden of Pb.[21] His hypothesis was shown to be correct by studies of ancient skeletons,[22] and of remote populations, both unacculturated, the Yanomama,[23] and acculturated, the Himalayans studied by us[24] and others.[25] He also postulated, on the basis of the extreme toxicity of Pb, that adverse health effects would be detected in normal populations, if studied with sensitive enough techniques.[21] Patterson's second hypothesis was initially confirmed in 1970 by the demonstration of a progressive exponential decline of δ-aminolevulinic acid dehydratase (ALAD) with an increase in BPb.[26]

We demonstrated that EP increases exponentially with BPb[27] and showed with a large epidemiological study that most children with BPb >40 μg/dl have markedly elevated EP values.[28] These studies demonstrated clearly that there was evidence of severe adverse effects on the heme synthetic pathway at BPb levels well below those associated with increased frequency of severe clinical neurotoxicity (at that time a child was considered a case of Pb intoxication only if the BPb exceeded 60 μg/dl). Our more recent studies[3] have shown that the effect on EP starts at BPb levels between 14 and 17 μg/dl, levels that are well below the currently accepted "normal" BPb level. Thus, well before clinical symptoms become obvious, profound alterations of the heme biosynthetic pathway occur. A number of studies have shown significant biochemical and functional alterations of various organ systems at low levels of Pb exposure.[29–31] Neurobehavioral abnormalities in normal children have been correlated with a body burden, albeit a small one, of Pb.[4]

Several studies[32,33] suggest that abnormalities of heme synthesis may be the biochemical mediators of neurological toxicity of Pb.[34] The neurotoxicity of byproducts of deranged heme synthesis is obvious in the porphyrias[35]; the symptoms of Pb poisoning are similar to those of acute intermittent porphyria.[36,37] In organotypic cultures of neural tissue, Pb induces both accumulation of porphyrins and demyelination, the latter of which is reversible by heme.[38,39] In Pb poisoning, alterations in heme synthesis appear pivotal not only as mediators of neurological toxicity, but also as sensitive tools for diagnosis.

It is clear that Pb is a nonessential, toxic metal present in the body as a result of environmental pollution; that Pb exerts toxic effects even at extremely low concentrations; and that the effects of heme synthesis are crucial, both as a yardstick of biological toxicity and as key to the biochemical basis of neurotoxicity.

In this report, the interaction between Pb poisoning and Fe deficiency and its implications for heme synthesis are considered at two levels: 1) at the biochemical level (the mechanism of the effect on ferrochelatase) and 2) at the population level (through an epidemiological study of children).

BIOCHEMICAL MECHANISMS OF THE Pb EFFECT ON FERROCHELATASE

In children, at BPb 16 μg/dl (0.8 μM) the elevation of EP starts; at BPb 50 μg/dl (2.5 μM) all children have EP > mean + 2σ; and at BPb 100 μg/dl (5 μM) all children have EP > 20 times normal and anemia. Despite this obvious dose–effect relationship, our current understanding of the stoichiometry of the Pb effect on ferrochelatase is poor. Old studies by Tephly[40] on rat liver mitochondria suggested 50% inhibition at 30 μM Pb and 13% inhibition at 20 μM Pb using an insensitive pyridine-hemochromogen spectrophotometric assay.[41] With this method, Fe must be held at optimal concentration (100 μM) to obtain measurable optical density (OD), and 13% inhibition represents an insignificant OD change. These studies do not allow the assessment of the interaction of Pb poisoning and Fe deficiency.

Using a sensitive ^{59}Fe assay,[42] we have investigated the hypothesis that Pb is a competitive inhibitor of Fe with regard to ferrochelatase. Competitive inhibition implies that the effect of Pb is highest when the concentration of Fe is lowest: thus the greater sensitivity to Pb of children with Fe deficiency.

We have undertaken the study of human ferrochelatase from erythropoietic tissue. For this study, we have utilized the reticulocyte-rich blood of patients with sickle-cell diseases, removed at time of therapeutic exchange transfusions. Our division includes the Comprehensive Sickle Cell Center of Columbia University, which monitors the care of over 200 patients with sickle-cell syndromes at Presbyterian Hospital. Elective exchange transfusions are performed, on the average, in one or two patients each month. The blood removed, which would be otherwise discarded, provides a large, excellent source of mitochondria (reticulocytes usually constitute over 15% of the red blood cells). This represents a unique source of human reticulocyte mitochondria and places our laboratory in an extraordinary position to study for the first time human erythropoietic ferrochelatase in detail.

Isolation of Mitochondria from Reticulocytes

After the clot was broken, leukocytes and platelets were removed from the blood by filtering it through Whatman #2 paper. The blood was then washed with buffered saline with glucose. The red cells were lysed by osmotic shock, and their osmolarity was then reconstituted with concentrated NaCl. The mitochondria were then separated by differential centrifugation according to Pedersen and Flatmark.[43] Their functional integrity was tested by measuring the respiratory control ratio with adenosine diphosphate, with glutamate as substrate.[44] Preparations of ruptured mitochondria were obtained by sonication; the completeness of disruption was verified by the release in the supernatant of malate dehydrogenase, a matrix enzyme.[45]

Assays of Ferrochelatase

We performed the assays according to the ^{59}Fe incorporation technique described by Williams *et al.*[42] We used mesoporphyrin as a substrate and introduced some modifications. More reproducible results and better control of the pH can be obtained by hydrolyzing the dimethyl ester of the porphyrins in HCl, then drying and dissolving the resulting dihydrochloride in NH_4OH, rather than by utilizing the ethanolic–KOH preparation suggested by Williams *et al.*[42]; similar observations have also been reported recently by Bloomer and Morton.[46] We have been able to obtain satisfactory measurements of ferrochelatase even at an Fe concentration of 1 μM by using ^{59}Fe of greater specific activity and by appropriate changes in the volumes in the incubation phase of the assay.

Results

First, we measured the K_m for Fe and the K_m for mesoporphyrin with standard biochemical methods. Next, we determined the kinetics of the effect of Pb on ferrochelatase. The preparations were exposed to varying concentrations of Pb; the concentration of one of the two substrates was kept constant at three times K_m, and the concentration of the other was varied above and below K_m. This type of three-tier titration is necessary to study with precision the kinetics of the inhibition; although very time consuming, it yields the most precise results.

The kinetics of the inhibition were analyzed by drawing Lineweaver–Burk plots in which the reciprocals of activity were plotted against the concentrations of the substrates. The K_i values were determined by replotting the intercepts and slopes from the Lineweaver–Burk plots against the inhibitor concentrations, as well as by using the methods of Dixon and Hunter-Downs.[47,48] With all the techniques of analysis, it became obvious that the effect of Pb was one of competitive inhibition with regard to Fe.

The competitive inhibition by Pb on ferrochelatase does not stem from an affinity for the Fe pocket of porphyrin; in fact, Pb would not physically fit within the ring. On the contrary, these kinetics suggest that Pb binds to ferrochelatase and modifies the conformation of its active site; however, if Fe is already bound before exposure to Pb, the structure of the active site is stabilized, and it cannot be altered by Pb.

The competitive nature of the inhibition of ferrochelatase by Pb explains the extreme sensitivity of children with Fe deficiency to the effect of Pb. Since Pb increases the K_m for Fe, the efficiency of ferrochelatase is further decreased in these children already hampered by the scarcity of Fe.

Our data show that at a low concentration of Fe, ferrochelatase becomes inhibited by concentrations of Pb that would not affect it at a higher Fe concentration. These findings have important therapeutic implications and indicate that Fe supplementation may decrease the toxicity of Pb exposure.

RELATIONSHIP BETWEEN Pb INTOXICATION AND Fe DEFICIENCY AT THE POPULATION LEVEL

Previous work from our laboratory has demonstrated an exponential correlation between BPb and EP in children with markedly elevated BPb (>30 μg/dl).[11] By confining the analysis to children with BPb in the normal range (5 to 30 μg/dl), we have shown that there is a threshold BPb between 14 and 17 μg/dl for the EP effect.[49] These studies have clearly demonstrated that biochemical evidence of damage by Pb to key biological systems is evident at BPb levels well below those currently accepted as being "normal."

It has been shown by us and others[50–52] that Fe deficiency potentiates the adverse effects of Pb in children. It has also been shown that both Pb poisoning and Fe deficiency occur at greatest frequency in the youngest and most socially deprived children[10]; that the absorption of Pb and Fe are interrelated with each other[52,53] and with other nutritional factors[54]; that both Pb poisoning[27] and Fe deficiency[51] raise EP; and that the elevation of EP with BPb is greater in women and children.[2,15,55]

It appears important to clarify whether Fe deficiency aggravates the biochemical effect of Pb on ferrochelatase and, in that case, whether the threshold BPb is lowered or the slope of elevation is different. It is also important to clarify whether the greater elevation of EP observed in women and children at equal levels of BPb reflects a different biological sensitivity or just a greater incidence of Fe deficiency.

A larger epidemiological study was directed to assess whether age and concomitant Fe deficiency influences the threshold BPb for elevation of EP in normal children. Questions of this type can only be answered by large-scale epidemiological analysis, as they obviously cannot be addressed prospectively. Iron deficiency may 1) render children more sensitive to the adverse effects of Pb by facilitating Pb absorption, in which case Fe-deficient children would have higher BPb; 2) render them more sensitive to the effects of lower doses, in which case the threshold BPb for the EP effect would be lower; or 3) enhance the effect of equal doses, in which case the slope of correlation of Pb with log EP would be steeper. Of course, more than one of these mechanisms may be operating simultaneously.

The assessment of the interactions of Fe and Pb in various groups is valuable not only for its implication of biological effect, but also because it is important to define the *most sensitive group of individuals* with respect to the effect of Pb on heme synthesis, which is considered to be a pivotal health effect. In fact, if the threshold BPb for the EP effect is lower for Fe-deficient children, it will be necessary to reduce the amount of air Pb further, to assure that all the individuals in the population maintain their BPb below the level that results in that adverse effect. In other words, these epidemiological studies have important implications to public health, particularly for those children of low socioeconomic class, who are most likely to be exposed to excess Pb and to suffer from Fe deficiency at the same time.

Source of Samples

At Babies Hospital–Presbyterian Hospital we have established an active program for childhood Pb-poisoning control. This includes both a screening program and a Pb-poisoning follow-up clinic.

Under the screening program, a total of about 5000 children are screened each year. This total includes children from two groups: 1) Approximately 3000–3500 children/year are screened through a voluntary screening program in the Pediatric Clinic. A hematofluorometer is used, and children are screened while they wait in the clinic. 2) An additional 1800–2000 children/year are screened through a program in which venous samples are collected for complete blood counts. When blood is drawn for the complete blood count, an additional small sample (~1 ml) is set aside for Pb-poisoning screening.

TABLE 1A. Iron Deficiency in Children: Criteria to Diagnoses

	Diagnosis of Fe Deficiency		
	SF $< \mu - 2\sigma$	SF $\geq \mu - 2\sigma$ and $\leq \mu - \sigma$	SF $> \mu - \sigma$
MCV			
$<\mu - \sigma$	Certainly	Most likely	Unclassified
$>\mu - \sigma$	Most likely	Unclassified	Certainly not

TABLE 1B. Iron Deficiency in Children: Diagnoses to Criteria

Diagnosis of Fe Deficiency	Criteria: SF	Criteria: MCV
Certainly	$<\mu - 2\sigma$	$<\mu - \sigma$
Most likely	$<\mu - 2\sigma$	$>\mu - \sigma$
	$\geq\mu - 2\sigma$ and $\leq\mu - \sigma$	$<\mu - \sigma$
Unclassified	$\geq\mu - 2\sigma$ and $\leq\mu - \sigma$	$>\mu - \sigma$
	$>\mu - \sigma$	$<\mu - \sigma$
Certainly not	$>\mu - \sigma$	$>\mu - \sigma$

TABLE 2A. Diagnosis of Fe Deficiency in Children Aged 12 to 23 Months: Criteria to Diagnoses

	Number of Children			
	SF < 8 ng/ml	SF ≥ 8 ng/ml and ≤ 14 ng ml	SF > 14 ng/ml	Total
MCV				
<74 fl	37	31	125	193
>74 fl	48	115	669	832
Total	85	146	794	1025

TABLE 2B. Diagnosis of Fe Deficiency in Children Aged 12 to 23 Months: Diagnoses to Criteria

	Criteria		
Diagnosis of Fe Deficiency	SF (ng/ml)	MCV (fl)	Number of Children
Certainly	<8	<74	37
Most likely	<8	>74	31
	≥8 and ≤14	<74	48
Total			116[a]
Unclassified	≥8 and ≤14	>74	115
	>14	<74	125
Certainly not	>14	>74	669
Total			1025

NOTE: For SF in this age group, $\mu - 2\sigma = 8$ ng/ml and $\mu - \sigma = 14$ ng/ml; for MCV in this age group, $\mu - \sigma = 74$ fl.

[a]This value represents the total number of Fe-deficient children in this age group.

The latter group of children was the subject of the current study. For the purpose of this study, in all samples collected, whether EP positive or negative, the following information was obtained: a complete blood count (the purpose of the collection); the BPb level, by graphite furnace atomic absorption spectrophotometry[56]; the EP level, by extraction; the ALAD activity, with and without SH reactivation[57]; and the serum ferritin (SF) level, by radioimmunoassay.

All data were stored in the Pediatric Division's VAX 11-750 utilizing the SIR relational database software. The CBC data and the sex, date of birth, and basic diagnosis for each child were entered via a modem from the HP3000 hospital computer (where this information is stored on line) utilizing software developed by us. Statistical analysis and graphic documentation were obtained utilizing the SAS software. Only data from well children were analyzed.

It is of paramount importance in this type of epidemiological study to define with the greatest possible precision all three conditions involved: 1) level of exposure to Pb (BPb level), 2) degree of effect (EP level), and 3) Fe status (SF level, mean corpuscular volume (MCV), and hemoglobin level). The first two conditions are relatively easy to define: the BPb and EP levels can be utilized as continuous variables. For the purposes of our analysis, however, the third condition would be better defined if all individuals were separated into two categories: Fe deficient and Fe sufficient.

Diagnosis of Fe Deficiency

The diagnosis of Fe deficiency in children aged 1 to 6 years is particularly difficult because the normal values for both SF and MCV are different than in adults, are themselves subject to developmental changes, and vary with age. Hence it was necessary to use our own data for each age group to determine the normal range. In order to exclude the influence of those children with Fe deficiency, we first determined, for each age group, a cutoff MCV value, obtained only from children with SF values greater than 30 ng/ml. The MCV cutoff point was defined as being $\mu - \sigma$. Then, for the same age group, we determined the normal μ and σ values for SF, this time analyzing only the children with MCV values greater than the cutoff point. To classify Fe deficiency, we then utilized a combination of the results of both SF and MCV, considering for each group its specific range of normal values (TABLE 1). A certain percentage of children (approximately 20%) yielded equivocal results and remained unclassified. This substantial exclusion should be accepted to assure a reliable classification of all the data points ultimately to be used for threshold measurements.

TABLE 2 shows, as an example, the results obtained for the group aged 12 to 23 months and the corresponding diagnostic classification. Our clinic serves a lower-middle-class population, and this population is served by programs of Fe supplementation, such as the WIC program,[58] that are designed to reduce Fe deficiency. The

TABLE 3. Frequency of Fe Deficiency by Age

	Age (months)					
	12–18	19–24	25–36	37–48	≥49	Total
Percentage of Fe-deficient children	10.91	12.17	12.83	6.78	6.76	10.09

relatively high frequency of low MCV with high SF reflects the fact that our cutoff point is only $\mu - \sigma$ and also the presence of α-thalassemia in our clinic population. The frequency of Fe deficiency is strongly age related (TABLE 3).

Data Analysis

First, all duplicate samples were excluded: only data from the first encounter were used for the analysis. Next, each medical record was verified, and children with any illnesses (including Pb poisoning or any minor acute illness that could have raised the level of SF, an acute-phase reactant) were excluded. The remaining samples were all from normal children.

At this writing, we have analyzed the results for 3950 samples. Duplicate samples and any samples from children that had a specific illness or could not be classified as being either Fe deficient or Fe sufficient were excluded from the study. We have, at this writing, 2299 samples obtained from healthy children, precisely classified by Fe status, and tested exclusively for the purpose of pediatric follow-up and screening.

The correlation between EP and BPb was analyzed using the hockey-stick technique of Hasselblad[59] and the segmented-line technique of Hudson[60] to ascertain the threshold BPb for EP elevation. First, we analyzed the results for the entire population, without any subdivision by Fe status. The observed threshold, 17.8 μg/dl, is in agreement with our previous findings[3] on a similar undivided group of children. Next we analyzed the Fe-sufficient children and the Fe-deficient children.

Although preliminary and of marginal statistical significance, the results of this study appear to be of great importance, as they clearly show the effect of Fe deficiency on the threshold BPb for the elevation of EP. The threshold BPb for the EP effect appears lower for Fe-deficient than for Fe-sufficient children; the slope of elevation is the same. The baseline EP is higher in the Fe-deficient than in the Fe-sufficient children; these differences cannot be attributed to age.

The data indicate that Fe deficiency makes the child more sensitive to the effect of Pb, not only because the EP level is already higher, but also because the elevation of EP due to Pb starts at a low level of exposure, that is, at a level that would be of no consequence to an Fe-sufficient child. This finding is the natural corollary to the biochemical mechanism of competitive inhibition that we have shown. It is clear that the Fe-deficient child is most sensitive to the effect of Pb; hence it is clear that Fe supplementation may reduce the Pb toxicity. This conclusion suggests it is important to maintain Fe supplementation programs for children. Such a program, that is, the WIC program, has been demonstrated to be successful in reducing the frequency of Fe deficiency.[58] The Fe-deficient child does appear to be sensitive to doses of Pb that would not noticeably affect the Fe-sufficient child.

These results have implications that transcend purely scientific interests. This study, by defining the category of children most sensitive to the effect of Pb on heme synthesis, is relevant to any attempts to define acceptable levels of Pb in the air. Air quality standards for Pb must, by law, aim at avoiding health effects in the most sensitive individuals. Our preliminary findings make it clear that Fe-deficient children below 2½ years of age are the most sensitive individuals with regard to Pb. Air quality standards—if they are to protect everyone—should be low enough to assure protection to this group of children.

ACKNOWLEDGMENTS

The authors are grateful to Ms. D. Ciprian, Ms. J. Mathis, and Mr. B. Tilton for their expert technical assistance; to Ms. D. Bouras for accurate computer data entry; and to Ms. V. Sadoff for help in the statistical analysis. We extend special thanks to Ms. I. Carter for her cooperation and perseverance in obtaining, from thousands of pediatric clinic patients, the blood samples required for the epidemiological study.

REFERENCES

1. Sayre, J. W., E. Charney, J. Vostal & D. B. Pless. 1974. House and hand dust as a potential source of childhood lead exposure. Am. J. Dis. Child. **127:** 167–170.
2. U.S. Environmental Protection Agency. 1977. Air Quality Criteria for Lead. Publication 600/8-77-017. U.S. Environmental Protection Agency. Washington, DC.
3. Piomelli, S., C. Seaman, D. Zullow, A. Curran & B. Davidow. 1982. The threshold for lead damage to heme synthesis in urban children. Proc. Natl. Acad. Sci. USA **79:** 3335–3339.
4. Needleman, H. L., C. Gunnoe, A. Leviton, R. Reed, H. Peresie, C. Maher, & P. Barrett. 1979. Deficits in psychologic and classroom performance of children with elevated dentine lead levels. N. Engl. J. Med. **300:** 689–695.
5. Annest, J. L., J. L. Pirkle, D. Makuc, J. W. Neese, D. D. Bayse & M. G. Kovar. 1983. Chronological trend in blood lead levels between 1976 and 1980. N. Engl. J. Med. **308:** 1373–1377.

6. BILLICK, I. H., A. S. CURRAN & D. R. SHIER. 1979. Analysis of pediatric blood lead levels in New York City for 1970–1976. Environ. Health Perspect. **31:** 183–190.
7. BILLICK, I., A. CURRAN & D. SHIER. 1977. Presentation to the U.S. EPA Lead Subcommittee of the Science Advisory Board, October 7, 1977. Available from Docket ECAO, Central Docket Section, U.S. Environmental Protection Agency, Washington, DC.
8. Center for Disease Control. 1982. Blood Lead Levels in U.S. Population. Morbid. Mortal. Wkly. Rep. **31:** 132–134.
9. LIN-FU, J. S. 1982. Children and lead: New findings and concerns. N. Engl. J. Med. **307:** 615–616.
10. MAHAFFEY, K. R., J. L. ANNEST, J. ROBERTS & R. S. MURPHY. 1982. National estimates of blood lead levels: United States 1976–1980. Association with selected demographic and socioeconomic factors. N. Eng. J. Med. **307:** 573–579.
11. PIOMELLI, S. & B. DAVIDOW. 1972. Free erythrocyte protoporphyrin concentration: A promising screening test for lead poisoning. Pediatr. Res. **6:** 366.
12. Center for Disease Control. 1978. Preventing Lead Poisoning in Young Children. J. Pediatr. **93:** 709–720.
13. ALESSIO, L., P. A. BARTAZZI, O. MORELLI & U. FOA. 1976. Free erythrocyte protoporphyrin as an indicator of the biological effects of lead in adult males. II. Comparison between free erythrocyte protoporphyrin and other indicators of effect. Int. Arch. Occup. Environ. Health **37:** 89–105.
14. GRANICK, J. L., S. SASSA, S. GRANICK, R. D. LEVERE & A. KAPPAS. 1973. Studies in lead poisoning. II. Correlation between the ratio of activated to inactivated δ-aminolevulinic acid dehydratase of whole blood and the blood lead level. Biochem. Med. **8:** 149–159.
15. LAMOLA, A. A., M. JOSELOW & T. YAMANE. 1975. Zinc protoporphyrin (ZPP): A simple sensitive fluorometric screening test for lead poisoning. Clin. Chem. **21:** 93–97.
16. SASSA, S., J. L. GRANICK, A. KAPPAS & R. D. LEVERE. 1973. Studies in lead poisoning. I. Microanalysis of erythrocyte protoporphyrin levels by spectrofluorometry in the detection of chronic lead intoxication in the subclinical range. Biochem. Med. **8:** 135–148.
17. LEVI, D. T., A. S. CURRAN, S. D. ALEXANDER, B. DAVIDOW & S. PIOMELLI. 1976. Physiological significance of erythrocyte protoporphyrin (EP) in moderate body lead burden. Pediatr. Res. **10:** 378.
18. REIGERT, J. R. & N. H. WHITLOCK. 1976. Longitudinal observations of the relationships between free erythrocyte porphyrins and whole blood lead. Pediatrics **57:** 54–59.
19. VALCIUKAS, J. A., R. LILIS, A. FISCHBEIN, I. J. SELIKOFF, J. EISINGER & W. E. BLUMBERG. 1978. Central nervous system dysfunction due to lead exposure. Science **201:** 465–467.
20. Center for Disease Control. 1985. Preventing Lead Poisoning in Young Children: A statement by the Center for Disease Control. Publication 99–2230. U.S. Department of Health, Education, and Welfare. Altanta, GA.
21. PATTERSON, C. C. 1965. Contaminated and natural lead environments of man. Arch. Environ. Health **11:** 344–360.
22. GRANDJEAN, P., O. V. NIELSEN & I. M. SHAPIRO. 1978. Lead retention in ancient Nubian and contemporary populations. J. Environ. Pathol. Toxicol. **2:** 781–787.
23. HECKER, L., H. E. ALLEN, D. D. DINMAN & J. V. NEEL. 1974. Heavy metal levels in acculturated and unacculturated populations. Arch. Environ. Health **29:** 181–185.
24. PIOMELLI, S., L. CORASH, M. D. CORASH, C. SEAMAN, P. MUSHAK, B. GLOVER & R. PADGETT. 1980. Blood lead concentrations in a remote Himalayan population. Science **210:** 1135–1137.
25. DAVIDSON, C. I., T. C. GRIMM & M. A. NASTA. 1981. Airborne lead and other elements derived from local fires in the Himalayas. Science **214:** 1344–1346.
26. HERNBERG, S. & J. NIKKANEN. 1970. Enzyme inhibition by lead under normal urban conditions. Lancet **i:** 63–64.
27. PIOMELLI, S. 1973. A micromethod for free erythrocyte porphyrins: The FEP test. J. Lab. Clin. Med. **81:** 932–940.
28. PIOMELLI, S., B. DAVIDOW, V. F. GUINEE, P. YOUNG & G. GAY. 1973. The FEP (free erythrocyte protoporphyrin) test: A screening micromethod for lead poisoning. Pediatrics **51:** 254–259.
29. ANGLE, C. R. & M. S. MCINTIRE. 1978. Low-level lead and inhibition of erythrocyte pyrimidine nucleotidase. Environ. Res. **17:** 296–302.

30. ANGLE, C. R., M. S. MCINTIRE, M. S. SWANSON & S. J. STOHS. 1982. Erythrocyte nucleotidase in children: Increased blood lead and cytidine triphosphate. Pediatr Res. **16:** 331–334.

31. ROSEN, J. F., R. W. CHESNEY, A. HAMSTRA, H. F. DELUCA & K. R. MAHAFFEY. 1980. Reduction in 1,25-dihydroxyvitamin D in children with increased lead absorption. N. Engl. J. Med. **302:** 1128–1131.

32. BRENNAN, G. D. & R. C. CANTRILL. 1979. δ-Aminolaevulinic acid is a potent agonist for GABA autoreceptors. Nature (London) **280:** 514–515.

33. SILBERGELD, E. K., R. E. HRUSKA, D. BRADLEY, J. M. LAMON & B. C. FRYKHOLM. 1982. Neurotoxic aspects of porphyrinopathies: Lead and succinylacetone. Environ. Res. **29:** 459–471.

34. SILBERGELD, E. K. & J. M. LAMON. 1980. Role of altered heme synthesis in lead neurotoxicity. J. Occup. Med. **22:** 680–684.

35. MOORE, M. R. & P. A. MEREDITH. 1976. The association of δ-aminolevulinic acid with the neurological and behavioral effects of lead exposure. *In* Trace Substances in Environmental Health. D. D. Hemphill, Ed.: 363–371. University of Missouri. Columbia, MO.

36. DAGG, J. H., A. GOLDBERG, A. LOCHHEAD & J. A. MITH. 1965. The relationship of lead poisoning to acute intermittent porphyria. Q. J. Med. **34:** 163–175.

37. GIBSON, S. L. M. & A. GOLDBERG. 1970. Defects in haem synthesis in mammalian tissues in experimental lead poisoning and experimental porphyria. Clin. Sci. **38:** 63–72.

38. SASSA, S., W. O. WHETSELL & K. KAPPAS. 1979. Studies on porphyrin-heme biosynthesis in organotypic cultures of chick dorsal root ganglia. II. The effect of lead. Environ Res. **19:** 415–426.

39. WHETSELL, W. O., S. SASSA & A. KAPPAS. 1984. Porphyrin-heme biosynthesis in organotypic cultures of mouse dorsal root ganglia: Effects of heme and lead on porphyrin synthesis and peripheral myelin. J. Clin. Invest. **74:** 600–607.

40. TEPHLY, T. R., J. HASEGAWA & J. BARON. 1971. Effect of drugs on heme synthesis in the liver. Metabolism **20:** 200–214.

41. PORRA, R. J. & O. T. C. JONES. 1963. Studies on ferrochelatase. I. Assay and properties of ferrochelatase from pig-liver mitochondria. Biochem. J. **87:** 181–185.

42. WILLIAMS, H. L., D. J. JOHNSON, M. J. HAUT, S. SLATER, R. G. BRENNAN & L. B. ALLSTATT. 1980. Improved radiochemical method for measuring ferrochelatase activity. Clin. Chem. **26:** 153–156.

43. PEDERSEN, J. I. & T. FLATMARK. 1972. Studies on the energy state of isolated brown adipose tissue mitochondria: The cytochrome *b* complex as a probe of the energy state of the mitochondrial inner membrane. Biochim. Biophys. Acta **275:** 135–147.

44. ESTABROOK, R. W. 1967. Mitochondrial respiratory control and the polarographic measurements of ADP:O ratios. *In* Methods in Enzymology. Volume X. Oxidation and Phosphorylation. R. W. Estabrook & M.E. Pullman, Eds.: 41–47. Academic Press. New York, NY.

45. BERGMEYER, H. U. & E. BERNT. 1963. Malic dehydrogenase. *In* Methods of Enzymatic Analysis. H. U. Bergmeyer, Ed.: 759–760. Academic Press. New York, NY.

46. BLOOMER, J. R. & K. O. MORTON. 1982. A radiochemical assay for heme synthase activity. Enzyme **28:** 220–232.

47. DIXON, M. & E. C. WEBB. 1964. Enzymes: 327–331. Academic Press. New York, NY.

48. SEGEL, I. H. 1975. Enzyme Kinetics. John Wiley & Sons. New York, NY.

49. PIOMELLI, S., A. A. LAMOLA, M. B. POH-FITZPATRICK, C. SEAMAN & L. HARBER. 1975. Erythropoietic protoporphyria and Pb intoxication: The molecular basis for difference in cutaneous sensitivity. I. Different rates of diffusion of protoporphyrin from the erythrocytes, both *in vivo* and *in vitro*. J. Clin. Invest. **56:** 1519–1527.

50. KAPOOR, S., C. SEAMAN, D. HURST, S. MATOS & S. PIOMELLI. 1984. The biochemical basis of the clinical interaction of Fe deficiency and Pb intoxication. Pediatr. Res. **18:** 242.

51. PIOMELLI, S., A. BRICKMAN & E. CARLOS. 1976. Rapid diagnosis of Fe deficiency by the measurement of free erythrocyte porphyrins (FEP) and hemoglobin: The FEP/hemoglobin ratio. Pediatrics **57:** 136–141.

52. YIP, R., T. N. NORRIS & A. S. ANDERSON. 1981. Iron status of children with elevated blood lead concentrations. J. Pediatr. **98:** 922–955.

53. HAMILTON, D. L. 1978. Interrelationship of lead and iron retention in iron-deficient mice. Toxicol. Appl. Pharmacol. **46:** 651–661.
54. MAHAFFEY, K. R. 1981. Nutritional factors in lead poisoning. Nutr. Rev. **39:** 353–362.
55. ROELS, H. A., R. R. LAUWERYS, J. P. BUCHET & M. T. URELUST. 1975. Response of free erythrocyte protoporphyrin and urinary δ-aminolevulinic acid in men and women moderately exposed to lead. Int. Arch. Arbeitsmed. **34:** 97–108.
56. FERNANDEZ, F. J. 1975. Micromethod for lead determination in whole blood by atomic absorption, with the use of the graphite furnace. Clin. Chim. Acta **21:** 558–562.
57. SASSA, S. 1982. δ-Aminolevulinic acid dehydratase assay. Enzyme **28:** 133–145.
58. MILLER, V., S. SWANEY & A. DEINARD. 1985. Impact of the WIC program on the iron status of infants. Pediatrics **75:** 100–105.
59. HASSELBLAD, V., J. P. CREASON & W. C. NELSON. 1976. Regression using "hockey stick" functions. Publication EPA-600/1-76-024. U.S. Environmental Protection Agency. Research Triangle Park, NC.
60. HUDSON, D. J. 1966, Fitting segmented curves whose joint points have to be estimated. J. Am. Stat. Assoc. **61:** 1097–1129.

Porphyrins and Hepatotoxicity[a]

RENATE D. KIMBROUGH[b]

Center for Environmental Health
Centers for Disease Control
Public Health Service
U.S. Department of Health and Human Services
Atlanta, Georgia 30333

INTRODUCTION

In humans, porphyrias are rare inborn disorders of porphyrin metabolism, characterized by an increased urinary excretion of porphyrins or their precursors. The two most common types of hepatic porphyria are acute intermittent porphyria and porphyria cutanea tarda. Patients with acute intermittent porphyria have predominantly psychiatric or neurologic symptoms, characterized by severe abdominal pain, constipation, hypertension, and motor disturbances that may result in respiratory paralysis. The disease is transmitted as Mendelian autosomal dominant and rarely appears before the age of 20. Patients with porphyria cutanea tarda have increased skin sensitivity to solar radiation and trauma, resulting in bullae, hirsutism, and local hyperpigmentation. Neurologic manifestations and abdominal pain are absent. The disease usually appears during middle age. Only porphyria cutanea tarda and porphyrinuria will be discussed here.

CLINICAL TYPES OF PORPHYRIA CUTANEA TARDA

Two types of porphyria cutanea tarda are recognized: the "constitutional type," which can occur spontaneously, can follow the ingestion of drugs and alcohol, or can be associated with hepatic cirrhosis, lymphoma, and various malignant neoplasias, and the "acquired type," which has been observed in thousands of patients with hexachlorobenzene poisoning.[1] A similar disease, usually termed experimental porphyria or hepatic porphyria, can also be produced in a variety of experimental animals. The experimental porphyrias provide excellent systems for the study of the complex control mechanism of the porphyrin metabolism in the mammalian liver.

Disturbances of porphyrin metabolism can be further divided into the porphyrias and the porphyrinurias. The porphyrias can be considered primary and are usually caused by hereditary abnormalities of porphyrin or porphyrin precursor metabolism. In contrast, porphyrinurias are secondary to a number of conditions that are accompanied by increased urinary excretion of coproporphyrin or on rare occasion, of uroporphyrin or porphyrin precursors. Such diseases include certain malignant

[a]The use of trade names in this paper is for identification only and does not constitute endorsement by the Public Health Service or the U.S. Department of Health and Human Services.

[b]Present address: U.S. Environmental Protection Agency, Office of Regional Operations, 401 M Street S.W., Washington, DC 20460.

diseases, liver diseases, a number of hematological diseases, heavy metal intoxication, and other disorders.[2]

Several drugs and other conditions have been associated with inducing constitutional porphyria cutanea tarda. These include Sufonal[3] and other sulfonamides.[4] Other drugs are sedormid and its more soluble congener, allylisopropylacetatamide, and griseofulvin. Apparently, the type of porphyria induced by drugs such as allylisopropylacetamide and griseofulvin is different from the type induced by hexachlorobenzene, and the mechanisms by which these different types of porphyria are induced also appear to be different.[2] Other chemicals that have been associated with the induction of increased urinary uroporphyrin excretion in humans are ethyl alcohol, the steroids, and an iron overload of the liver, as well as phenobarbital overload.[1,5,6] Women who use birth control pills may have elevated urinary excretion of uroporphyrins.[7]

Hepatic porphyria can be produced in various species of laboratory animals with a number of environmental chemicals. These include 2,3,7,8-tetrachlorodibenzodioxin, hexachlorobenzene, and the polychlorinated and polybrominated biphenyls. This is reviewed in detail by Strik *et al.*[8]

The constitutional type of porphyria cutanea tarda seems to occur in people who are deficient in the enzyme uroporphyrinogen decarboxylase in the liver. Elder *et al.*[9,10] showed that uroporphyrinogen decarboxylase activity in the liver may be decreased to half the normal values in such patients. Most, but not all, will have a family history of porphyria cutanea tarda. Some patients with familial porphyria cutanea tarda, however, may have normal, or close to normal, uroporphyrinogen decarboxylase activities. The activity of this enzyme may also be decreased in erythrocytes[9] but the activity of uroporphyrinogen I synthase remains normal. This enzyme defect seems to be a prerequisite for the development of porphyria cutanea tarda in response to conditions such as alcoholic liver disease with siderosis. At present, it is not clear what the overall incidence of this condition is in the general population nor is it clear whether the susceptibility to porphyrinogenic agents varies. If iron stores in the liver continue to increase over the lifetime in these individuals, perhaps a critical iron level in the liver must be reached before overt disease develops. This would explain the onset of disease in middle age. The distribution of iron among the various cell types that make up the liver may play a role as well.

In the acquired form of porphyria cutanea tarda, uroporphyrinogen decarboxylase is inhibited by the porphyrinogenic agent. Secondarily, the inhibition of uroporphyrinogen decarboxylase may cause other enzymes, such as δ-aminolevulinic acid synthetase, to become elevated in the liver. Furthermore, iron potentiates the porphyrinogenic effect of the different agents.[6,11] A decrease in uroporphyrinogen decarboxylase has been demonstrated in experimental animals that were dosed with porphyrino-

TABLE 1. Morphological Changes in the Liver of Humans with Porphyria Cutanea Tarda as Seen with Light and Electron Microscopy

	Morphological Changes
Light microscopy	Portal fibrosis
	Fatty change
	Focal necrosis
	Increased iron in hepatocytes and Kupffer cells
	Round cell inhibition of portal triads
Electron microscopy	Mitochondrial changes, bizarre shapes with disorganization of cristae, and large electron dense granules
	Increase in smooth endoplasmic reticulum and dilatation

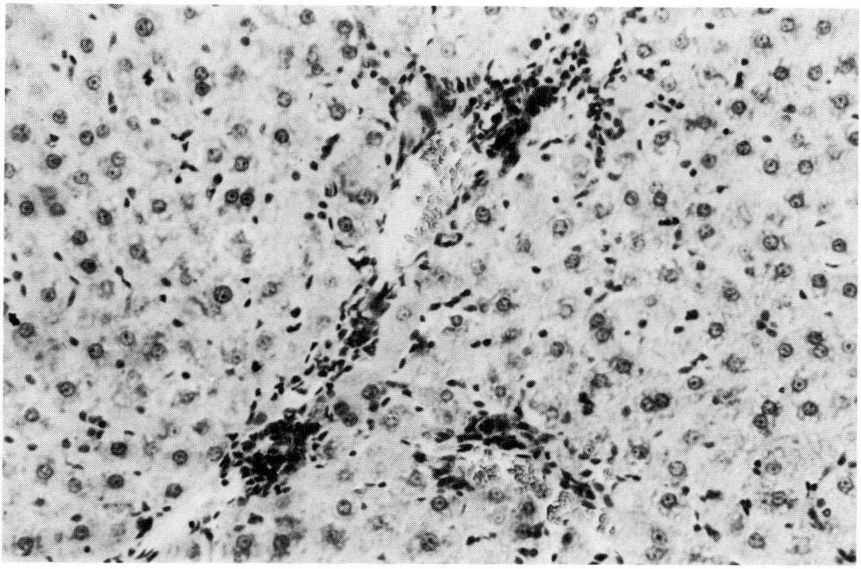

FIGURE 1. Section of liver from rat given technical pentachlorophenol. This liver shows minor changes, such as focal round cell infiltrates and binucleation. Stained with hematoxylin and eosin. Magnification: ×125.

genic agents.[12] Kushner *et al.*[13] demonstrated the inhibition of uroporphyrinogen decarboxylase by ferrous iron *in vitro* and pointed out that this, together with the inhibitory effect of iron on uroporphyrinogen III cosynthetase, may explain the clinical association of porphyria cutanea tarda with a deranged iron metabolism.

The role of iron has particularly been emphasized by authors from South Africa.[14] Many persons in South Africa have an oral iron overload,[15] and Lamont and Hathorn[14] have described increased plasma iron and liver pathology in Africans with porphyria. Apparently, the increase in iron overload in these patients is caused by the increased iron absorption caused by drinking the locally brewed beer, which is made of maize and sorghum, and which has a high iron content.[16]

MORPHOLOGICAL CHANGES

The morphological changes associated with both the constitutional and the acquired porphyria cutanea tarda in humans and the hepatic porphyria in animals vary a great deal.

Experimental Animals

Morphological changes associated with hepatic porphyria in experimental animals have been described previously.[17] The morphological appearance can be quite varied. At the light microscopic level, there is usually some lipid accumulation within the hepatocytes. The Kupffer cells are quite prominent and contain a brown pigment that stains positive, in part, for iron. In addition, there may be necrobiosis of hepatocytes and aggregates of inflammatory cells. The hepatocytes themselves may be enlarged and may contain eosinophil inclusions within their cytoplasm. Livers fluoresce pink under ultraviolet light.

In Sherman rats, it is easier to produce hepatic porphyria in females than in males.[18,19] The light microscopic appearance of the liver in the two sexes may also vary. If these laboratory animals are dosed with polychlorinated or polybrominated biphenyls or similar compounds, the male rat usually develops a much larger liver, which may be seen to contain a great deal more fat, whereas in the female, more necrobiosis and pigment are noted. Thus, lipid peroxidation and the presence of iron in the liver apparently affect the development of hepatic porphyria in a given animal.[20] In addition, the polybrominated, as well as the polychlorinated biphenyls, produce hepatocellular carcinomas and neoplastic nodules in the livers of rats and mice. Some animals with such tumors will have hepatic porphyria.[20]

The various light microscopic lesions that may be associated with hepatic porphyria in rodents are illustrated in FIGURES 1–3. Similarly, at the ultrastructural level, the changes that are noted in hepatocytes vary greatly with the dose, length of exposure, and the type of compound to which the laboratory animals are exposed. Usually, there is an increase in smooth endoplasmic reticulum, and the mitochondria may be of an abnormal size and may have abnormal shapes. Lipid vacuoles surrounded by concentric rings of membranous material may be noted. Furthermore, the mitochondria may contain crystalline inclusions. The Kupffer cells and the hepatocytes contain a dark pigment. Some of this pigment is thought to represent porphyrin. Some, however, may

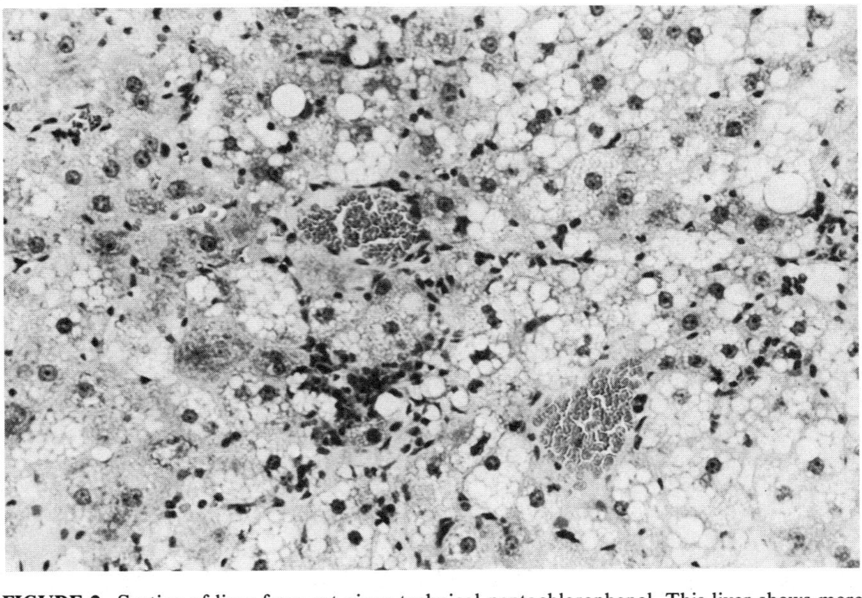

FIGURE 2. Section of liver from rat given technical pentachlorophenol. This liver shows more pronounced changes. The hepatocytes are enlarged, highly vacuolated, and pleomorphic. Infiltration by inflammatory cells is also noted. Stained with hematoxylin and eosin. Magnification: ×125.

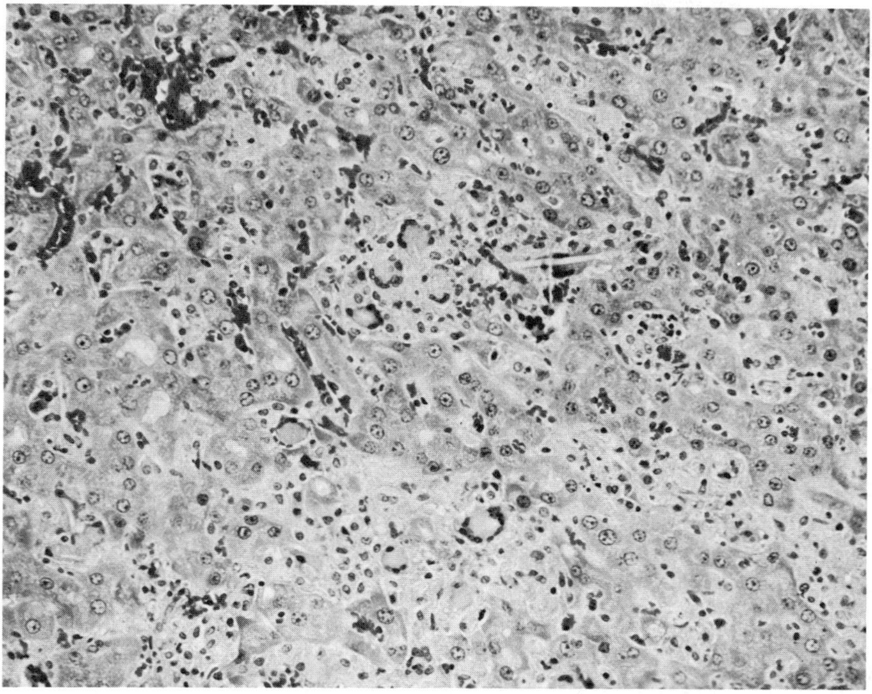

FIGURE 3. Section of liver from rat given technical hexachlorobenzene. Necrobiosis of hepatocytes, infiltration of sinuosids by inflammatory cells, multinucleated cells, and proliferation of fibrocytes are noted. Stained with hematoxylin and eosin. Magnification: ×125.

be ceroid pigment. In addition, in studies with rats, technical pentachlorophenol resulted in livers that were totally black. Whether the livers became black because of an accumulation of melanin is not clear—the cause of the change was not specifically identified.[18] The technical pentachlorophenol was contaminated with chlorinated dibenzofurans and dibenzodioxins. In a poisoning episode in Japan as well as in Taiwan,[21,22] dark discoloration of the skin was noted in the patients, particularly in neonates. This hyperpigmentation was caused by increased melanin deposits in the skin.

Humans

Lamont and Hathorn[14] describe morphological changes in the livers of their patients with iron overload and porphyria cutanea tarda. These changes varied from minor alterations, such as a slight increase in reticulum fibers, to gross distortion of the lobular architecture and liver cirrhosis. In addition, there was much iron in Kupffer cells and widespread fatty change in the hepatocytes.

Similar changes, however, were also described by Bothwell *et al.*[15] in patients who did not suffer from porphyria cutanea tarda. Thus, the role of iron in porphyria cutanea tarda remains controversial. Iron overload is a characteristic finding, but many patients with diseases of iron overload, such as hemochromatosis and hemosiderosis, do not develop porphyria cutanea tarda. The morphological changes observed in the liver appear to be related to the increased iron in the liver rather than the porphyria cutanea tarda.

In the United States, Grossman *et al.*[5] reviewed the clinical and laboratory findings of 40 patients with porphyria cutanea tarda. Ethanol and use of estrogens were the most frequently associated etiological factors. Two-thirds of the patients had abnormal liver function tests and elevated serum iron levels. The abnormalities of liver function varied with the etiological agent. Patients ingesting estrogens had abnormal bromosulfothalein retention, more often while patients in whom alcohol ingestion was a factor had elevated aspartate aminotransferase. Elevated serum iron levels occurred most frequently in the idiopathic group. Observations made in liver biopsies were similar to those seen in experimental animals. Siderosis was the most frequent abnormality. In addition, periportal inflammation, focal necrosis, fatty infiltration, and mild fibrosis were seen in more than half of the biopsy specimens. Needle-like inclusion bodies in hepatocytes were seen in the livers of several patients.

TABLE 2. Liver Changes Associated with Alcoholism in Humans

Fatty change	Intracellular clumps of hyaline (Mallory's bodies)
Lipogranulomas	Sclerosis around central veins
Swollen liver cells	Hemosiderosis
Hepatocellular necrosis	Cirrhosis
Inflammatory reaction with polymorphonuclear leucocytes	Hepatoma

CONCLUSION

Porphyria cutanea tarda in humans should be subdivided into two diseases. The constitutional type of porphyria cutanea tarda occurs sporadically in the general population because some individuals have a deficiency in uroporphyrinogen decarboxylase. This disease expresses itself upon environmental exposure to alcohol, certain medications, or some industrial chemicals.[6] It seems to be more prevalent in certain subpopulations. Then there is acquired porphyria cutanea tarda, first noted in a population in Turkey after exposure to hexachlorobenzene. In this situation, porphyria cutanea tarda developed in almost all of the exposed persons.[23]

In experimental animals, the sporadic type of hepatic porphyria has thus far not been described. Hepatic porphyria can, however, be induced in laboratory animals by a variety of chemicals. Some animals, such as birds, seem to be more susceptible than rodents.

The morphological changes noted in the liver seem to be dependent on the type of chemical that produces the porphyria in humans or the hepatic porphyria in animals. Porphyria cutanea tarda may be associated with very little or no liver pathology or with very pronounced liver pathology, depending on the agent responsible for the porphyria cutanea tarda and the length of time the condition has existed. The same type of changes may also be present in livers without a deranged porphyrin metabolism, after exposure to chemicals at concentrations that damage the liver. Morphological changes do not appear to be more severe in humans or animals that have, in addition to liver toxicity, porphyria cutanea tarda.

SUMMARY

Human porphyrias are rare inborn disorders of porphyrin metabolism. Porphyria cutanea tarda is one such disorder, which may occur spontaneously after the ingestion of drugs and alcohol. This is the "constitutional" type and is associated with a deficiency of uroporphyrinogen decarboxylase in the liver. This constitutional type must be differentiated from the "acquired" type, which occurred in hundreds of people poisoned by hexachlorobenzene in Turkey. Morphological changes observed in the livers of animals and humans vary greatly. Porphyria cutanea tarda may be associated with very little liver pathology or with very pronounced changes, depending on the agent causing the porphyria. Usually, the liver has an increased iron load, and mitochondria contain a crystalline material.

REFERENCES

1. BIEMPICA, L., N. KOSOWER, M. H. MA & S. GOLDFISCHER. 1974. Hepatic porphyrias: Cytochemical and ultrastructural studies of liver in acute intermittent porphyria and porphyria cutanea tarda. Arch. Pathol. **98:** 336–343.
2. TSCHUDY, D. P. & H. L. BONKOWSKY. 1972. Experimental porphyria. Fed. Proc. **31:** 147–159.
3. STOKVIS, B. J. 1889. Ned. Tijdschr. Geneeskd. **25:** 409.
4. RIMMINGTON, C. & A. W. HEMMINGS. 1938. Lancet **i:** 770.
5. GROSSMAN, M. F., D. R. BICKERS, M. B. POH-FITZPATRICK, V. A. DELEO & L. C. HARBER. 1979. Porphyria cutanea tarda: Clinical features and laboratory findings in 40 patients. Am. J. Med. **67:** 277–286.
6. DOSS, M. O. 1985. Alcohol and porphyrin metabolism. *In* Alcohol-Related Disease in Gastroenterology. H. K. Seitz & B. Kommerell, Eds.: 232–252. Springer-Verlag. Berlin.
7. KOSKELO, P., A. EISALO & I. TOIVONEN. 1966. Urinary excretion of porphyrin precursors and coproporphyrin in healthy females on oral contraceptives. Br. Med. J. **1:** 652–654.
8. STRIK, J. J. T. W. A., F. M. H. DEBETS & G. KOSS. 1980. Chemical porphyria. *In* Topics in Environmental Health. R. D. Kimbrough, Ed. Vol. 4: 191–239. Elsevier/North-Holland Biomedical Press. Amsterdam.
9. ELDER, G. H., D. M. SHEPPARD, R. E. SALAMANCA & A. OLMOS. 1980. Identification of two types of porphyria cutanea tarda by measurement of erythrocyte uroporphyrinogen decarboxylase. Clin. Sci. **58:** 477–484.
10. DOSS, M. O., H. SAUER, R. VON TIEPERMANN & A. M. COLOMBI. 1984. Development of chronic hepatic porphyria (porphyria cutanea tarda) with inherited uroporphyrinogen decarboxylase deficiency under exposure to dioxin. Int. J. Biochem. **16**(4): 369–373.
11. FELSHER, B. F., N. M. CRPIO, D. W. ENGLEKING & A. T. NUNN. 1982. Decreased hepatic uroporphyrinogen decarboxylase activity in porphyria cutanea tarda. N Engl. J. Med. **306:** 766–769.
12. TALJAARD, J. J. F., B. C. SHANLEY & S. M. JOUBERT. 1971. Decreased uroporphyrinogen decarboxylase activity in experimental symptomatic porphyria. Life Sci. **10:** 887–893.
13. KUSHNER, J. P., D. P. STEINMULLER & G. R. LEE. 1975. The role of iron in the pathogenesis of porphyria cutanea tarda. II. Inhibition of uroporphyrinogen decarboxylase. J. Clin. Invest. **56:** 661–667.
14. LAMONT, N. M. & M. HATHORN. 1960. Increased plasma iron and liver pathology in Africans with porphyria. S. Afr. Med. J. **34:** 279.
15. BOTHWELL, T. H., R. W. CHARLTON & H. C. SEFTEL. 1965. Oral iron overload. S. Afr. Med. J. **39:** 892.
16. DERMAN, D. P., T. H. BOTHWELL, J. D. TORRANCE, W. R. BEZWODA, A. P. MACPHAIL, M. C. KEW, M. H. SAYERS, P. B. DISLER & R. W. CHARLTON. 1980. Iron absorption from maize (*Zea mays*) and sorghum (*Sorghum vulgare*) beer. Br. J. Nutr. **43:** 271.
17. KIMBROUGH, R. D. 1979. The carcinogenic and other chronic effects of persistent halogenated organic compounds. Ann. N.Y. Acad. Sci. **320:** 415–418.

18. KIMBROUGH, R. D. & R. E. LINDER. 1978. The effect of technical and purified pentachlorophenol on the rat liver. Toxicol. Appl. Pharmacol. **46:** 151–162.
19. GROCE, D. F. & R. D. KIMBROUGH. 1984. Stunted growth, increased mortality and liver tumors in offspring of polybrominated biphenyl (PBB)-dosed Sherman rats. J. Toxicol. Environ. Health **14:** 695–706.
20. KIMBROUGH, R. D., D. F. GROCE, M. P. KORVER & V. W. BURSE. 1981. Induction of liver tumors in female Sherman strain rats by polybrominated biphenyls. J. Natl. Cancer Inst. **66:** 535–542.
21. YAMASHITA, F. & M. HAYASHI. 1985. Fetal PCB syndrome: Clinical features, intrauterine growth retardation and possible alteration in calcium metabolism. Environ. Health Perspect. **59:** 41–51.
22. LU, Y.-C. & Y.-C. WU. 1985. Clinical findings and immunological abnormalities in Yu-Cheng patients. Environ. Health Perspect. **59:** 17–29.
23. CAM, S. & G. NIGOGOSYAN. 1963. Acquired toxic porphyria cutanea tarda due to hexachlorobenzene: Report of 348 cases caused by this fungicide. J. Am. Med. Assoc. **183:** 88–91.

Role of Altered Heme Synthesis in Chemical Injury to the Nervous System[a]

ELLEN K. SILBERGELD

Toxic Chemicals Program
Environmental Defense Fund
Washington, DC 20036

INTRODUCTION

The clinical similarities in the neuropsychiatric signs and symptoms of the acquired and inherited porphyrinopathies have long been noted.[1] Because of these parallels, it has been hypothesized that common mechanisms of neurotoxicity, involving alterations in heme synthesis, are important to both sets of diseases. It should be noted, however, that in the case of toxicant-induced porphyrinopathies, some of the agents themselves (such as heavy metals and certain halogenated hydrocarbons) have direct effects on the nervous system, which play a role in their neurotoxicity. The most extensive studies of neurotoxicology in inherited porphyria have concerned acute intermittent porphyria (AIP); in acquired porphyrinopathy, they have been done in connection with lead exposure. Although, as noted above, the direct neurotoxic effects of lead must be included in a full discussion of etiology and mechanisms, this paper will be limited in focus to discussing the role of altered porphyrin synthesis in neurotoxicity.

The neuropsychiatric manifestations of inherited and acquired porphyrinopathies include dysfunctions of both the central nervous system (CNS) and the peripheral nervous system (PNS).[2] Descriptions of PNS toxicity include stomach pain, parasthenia, fatigue, weakness, muscle and joint pain, and slowed conduction velocity of both sensory and motor nerves.[3,4] Gait disturbance may be found. Bradykinesia developing into quadriplegia can occur in cases of prolonged disease, particularly AIP. CNS toxicity includes mild and severe psychiatric disturbances of mood and affect. These can be incapacitating and diagnosed as psychosis or dementia, as in the classic examples described in Wetterberg's famous studies of institutionalized patients in Sweden who were demonstrated to have AIP.[5] Seizures (*grand mal*) have also been reported in cases of AIP such that in some instances, differential diagnosis has been a problem.[6] Learning ability and intelligence (as measured psychometrically) are impaired in the acquired porphyrinopathies, specifically those associated with heavy metal intoxication.[7,8]

NEUROTOXIC MANIFESTATIONS OF INHERITED AND ACQUIRED PORPHYRINOPATHIES

Experimental models of porphyrinopathy have focused primarily on toxicant-induced states. The drug-induced porphyrinopathies do not lend themselves to investigations of chronic neurotoxicity because many of these agents markedly affect

[a]This paper is dedicated to the late M. R. Krigman, MD.

the nervous system themselves and most cannot be administered chronically in the doses required to alter heme synthesis. Succinylacetone—a suicide inhibitor of δ-aminolevulinic acid dehydrase (ALAD) found in persons with hereditary tyrosinemia[9]—may be an exception: it does not possess marked neuroactive properties, and it can be administered chronically.[10] Only a few studies, however, have been done of neurobiologic function in animals treated with this or other porphyrinopathic agents (see below).

Neuropsychiatric symptoms are also reported in persons exposed to polychlorobiphenyls (PCBs), polybromobiphenyls (PBBs), hexachlorobenzene, and 2,3,7,8-tetrachlorodibenzo-*p*-dioxin (TCDD).[11–14] These are generally expressed as so-called "soft" signs of neurological dysfunction. Only rarely have objective signs or clinically defined neuropathy been described.[15,16] Altered personality (including depression and sociopathy, as measured on standardized personality profiles) have been reported in occupational groups exposed to PCBs and TCDD.[15,17] The neurotoxicity of hexachlorobenzene and hexachlorophene intoxication, involving morphological damage to the CNS, contributes to their toxicity.[13] Porphyrinopathy may also be an important factor, however, particularly in the case of hexachlorophene, which has been shown to be a potent inducer of hepatic porphyrinopathy.[18] The presence of dioxins in these compounds[19] may also have contributed to both their neurotoxicity and their porphyrinopathy.

Morphologically, the most consistent finding in porphyrinopathic conditions has been CNS and PNS demyelination.[4] Wallerian-type degeneration of motoneuron fibers is also observed,[20,21] which would indicate that the primary lesion is axonal. In the spinal cord, chromatolysis and vacuolization of the anterior and lateral horn cells, the nuclei medullaris, and the dorsal nuclei vagalis have been reported. In lead poisoning, both CNS and PNS demyelination, axonopathy, and—at high doses—cerebellar hemorrhage and edema have been described.[22]

Electrophysiological studies of patients with porphyrinopathy have focused on PNS abnormalities. Decreased nerve conduction has been reported in both AIP[23] and in lead toxicity, even at relatively low levels of lead exposure.[24] Slowed nerve conduction velocity has also been reported in persons exposed to dibenzofurans, dioxins, and hexachlorobenzene[12,25] but not PCBs (Fischbein, personal communication). A range of electroencephalographic (EEG) abnormalities have been reported in lead-intoxicated persons, including altered evolved response.[26] EEG analyses of a patient with AIP found focal discharges consistent with the diagnosis of *grand mal* epilepsy.[6]

Neurochemical studies in humans with inherited or acquired porphyrinopathies have been rare because of the problems in sampling and interpreting biochemical parameters related to the nervous system. Neurochemical metabolites have been studied in lead-exposed humans (TABLE 1). Increased urinary levels of 5-hydroxyindoleacetic acid (5-HIAA), a metabolite of the neurotransmitter 5-hydroxytryptamine (5-HT), were found in adult workers.[27] Lead-exposed children excreted increased levels of the aminergic metabolites vanillylmandelic acid (VMA) and homovanillic acid (HVA) in urine (TABLE 1).[28] Noradrenergic function may be altered in porphyrinopathy, as suggested by the therapeutic efficacy in clinical and experimental porphyrinopathies of the β-adrenergic blocker propranolol.[29] The cardiovascular problems found in porphyrinopathies also suggest dysfunctions in peripheral noradrenergic neurotransmission.

In animal models of lead poisoning, a range of neurochemical alterations have been reported involving cholinergic, aminergic, GABA-ergic, and peptidergic pathways.[30,31] Because of the functional interactions among these neurotransmitter pathways, it is difficult to determine which of these pathways are primarily affected in toxicant-induced neurotoxicity and which are altered in response to a single site or set of primary sites of action. In lead intoxication, not all the neurochemical alterations found in rodents after *in vivo* treatment can be produced by adding lead *in vitro* to preparations of neural tissue. This is particularly true for the γ-aminobutyric acid (GABA) pathway.[32] *In vivo* lead treatment decreased neuronal release of GABA and increased GABA receptor sensitivity in cerebellum,[32,33] whereas lead *in vitro* affected GABA-ergic function only in specific media.[32,34] These apparent inconsistencies may be due to the disruption *in vitro* of interconnections among neural circuits that exist *in vivo* or to nonneurochemical actions of lead, which affect the brain and are expressed in the intact animal. Porphyrinopathy has been proposed as one of the latter mechanisms.[32]

HEME SYNTHESIS IN THE NERVOUS SYSTEM

Cells in the nervous system contain porphyrins and the enzymes of the heme synthetic pathway.[36–39] Brain also has heme-containing microsomal monooxygenases,

TABLE 1. Neurotransmitter Metabolites in Lead-Exposed Children and Adults[a]

Group	Urinary Aminergic Metabolite HVA ($mg/m^2/day$)	VMA ($mg/m^2/day$)	5-HIAA (mg/day)
Children			
Exposed to lead[b]	14.07	27.54	
Controls	2.55	4.52	
Adults			
Exposed to lead[c]			12.84
Controls			2.36

[a]Data from Dugandzic *et al.*[27] and Silbergeld and Chisolm.[28]
[b]Blood lead: 59–68 μg/100 ml.
[c]Urinary lead: 0.213 mg/l.

which metabolize benzo[*a*]pyrene[40,41] and which can be induced by manganese.[42] Brain tissue and PC12 cells also contain cytosolic receptors for polycyclic aromatic hydrocarbons (PAHs) and the so-called *Ah* or dioxin receptor, which regulates activity of microsomal monooxygenases.[43,44]

The role of heme synthesis in the nervous system, apart from cellular metabolism, is not clear. Heme is a necessary constituent for maintaining nerve cells in culture. Heme synthesis may be involved in the complex functional interrelationship that exists between neurons and supporting cells. *In vivo,* heme or heme precursors may be supplied by glia and Schwann cells, which can convert δ-aminolevulinic acid (ALA) to porphyrins.[36,38,39]

Relatively few studies of porphyrin biochemistry have been done in the CNS or PNS of animals exposed to porphyrinopathic agents. As shown in TABLE 2, there appear to be differences with respect to the effects of metals on heme synthesis in brain as compared to liver, kidney, or spleen.[45,46] As noted above, treatment with benzo[*a*]pyrene can induce brain arylhydrocarbon hydroxylases. In chronic lead toxicity, levels of the heme precursors ALA were elevated, and activity of ALA dehydrase was inhibited.

δ-Aminolevulinic acid synthase (ALAS) activity was depressed after acute lead exposure, but was not changed after chronic exposure. In manganese-treated rats, brain ALAS activity was decreased.[42] Unfortunately, levels of heme or activity of other enzymes in brain were not examined.

PORPHYRINOPATHY AND NEUROTOXICITY

The discussion above provides general evidence that altered heme synthesis and porphyrinopathy may play a role in the neurotoxicity common to inherited and acquired disorders. Five hypotheses can be advanced for a role of this biochemical lesion in neurotoxicity:

TABLE 2. Heme Metabolism in Lead-Exposed Rats[a]

	Acute		Chronic	
	Control	Exposed	Control	Exposed
Brain				
ALA	58	49	74	98
ALAD	134	99	95	94
ALAS	4.4	2.8	2.2	1.8
Kidney				
ALA	135	409	118	1086
ALAD	652	264	656	360
ALDS	10	22	7.4	9.3
Liver				
ALA	33	32	20	24
ALAD	1696	1032	1822	1521
ALAS	28.2	19.6	24.2	13.9
Spleen				
ALA	150	239	130	1937
ALAD	607	234	404	320
ALAS	44.8	39.7	24.9	71.4

NOTE: The ALA level is expressed as nmol/g wet wt; the ALAS activity, as nmol porphobilinogen/g wt wt/hr; and the ALAD activity, as nmol ALA/g wt wt/hr.

[a]Data from Silbergeld and Lamon.[10]

1. Decreased available heme for cytochromes selectively toxic to nerve cells because of their high energy requirements and rate of metabolism.
2. Impacts upon the availability of certain amino acids (tryptophan and glycine), which have important roles in neurochemical pathways.
3. Increased tissue levels of neuroactive heme precursors and products.
4. Impaired functional relationship between neurons and supporting cells in the nervous system.
5. Changes in nerve cell membrane receptors as a consequence of the binding of certain halogenated organics to the *Ah* receptor with pleiotropic effects regulated at the DNA level.

Heme Depletion

Porphyrinopathies commonly result in depletion of cellular pools of heme, the end product of the pathway.[47] Enzyme inducers, such as the PAHs and barbiturates, exert a demand for heme as a prosthetic group in microsomal oxidative pathways; the total cellular concentrations of heme may increase under these conditions, but "free" heme and heme available for other functions may decrease.[48] Aryl hydrocarbon hydroxylase (AHH) induction by PAHs may deplete the amount of heme available for oxidative phosphorylation, but such depletions have not been measured. In lead poisoning, brain and liver concentrations of cytochromes are reduced.[49–51] Amounts of heme-containing monooxygenases are also decreased in lead poisoning, as demonstrated functionally by increased time for antipyrene clearance in humans[52] and prolonged hexobarbital sleeping time in rats.[10,51]

Depletion of functional heme pools may be selectively toxic to the brain because of the high oxygen demands and metabolic activity of neurons in the CNS.[53] The relative sensitivity of the brain (particularly in early development) to even transient anoxia is well known. Indirect evidence for the hypothesized heme depletion comes from observations of the efficacy of exogenous heme to suppress some neurotoxic effects in AIP and in lead-exposed humans and animals.[10]

Neurotransmitter Metabolism

Altered heme synthesis affects the peripheral metabolism of glycine and tryptophan, two amino acids that have roles in neurotransmission. Glycine, the amino acid precursor of ALA, is a major inhibitory neurotransmitter, with pathways in the CNS and the spinal cord.[54] Glycine is transported into neurons by high-affinity uptake processes, which are specific markers for this pathway. Neuronal activity in glycinergic pathways depends at least in part upon circulating levels of glycine in order to supply releasable stores.

GNS glycine status has not been investigated in porphyrinopathies. A provocative test involving glycine loading was proposed by Richards and Scott[55] and used by Barltrop[56] for identification of latent AIP. Glycine loading in homozygotes for AIP caused an increased excretion of ALA in urine. Glycine produced no expression of neurological symptoms, however, even in patients with AIP. This argues against any supersensitivity of glycinergic receptors in clinical disease. Glycine response has not been examined, however, in persons with acquired porphyrinopathy.

Tryptophan metabolism has also been reported to be significantly altered in clinical porphyria.[57] Tryptophan is the amino acid precursor of the neurotransmitter 5-HT (or serotonin). In an animal model of acute porphyria produced by inhibiting hepatic cytochrome P-450, Litman and Correia[58] and Correia *et al.*[59] reported inhibition of hepatic tryptophan pyrrolase, increased brain levels of tryptophan, and increased turnover of 5-HT. As demonstrated by Wurtman,[60] increasing plasma tryptophan can increase impulse flow through serotonergic pathways in the CNS. These results suggest that serotonergic neurotransmission may be altered in conditions where hepatic heme is depleted. Moreover, the finding of increased urinary 5-HIAA excretion in patients occupationally exposed to lead[27] is consistent with increased 5-HT turnover. This, however, may occur only in peripheral serotonergic pathways; serotonergic metabolism in the brain was not consistently changed in two studies of experimental lead exposure.[31,61]

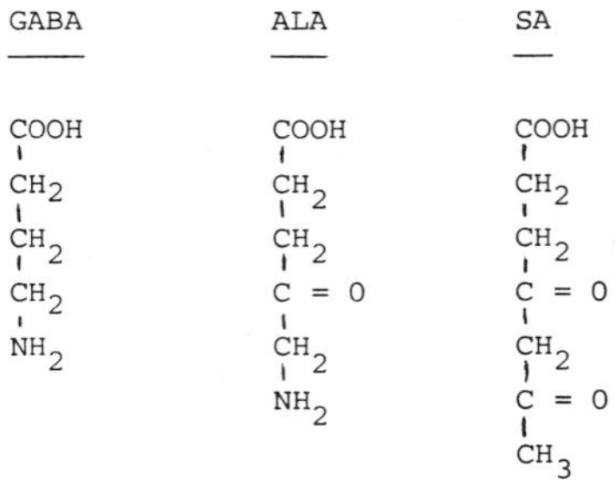

FIGURE 1. Structure of GABA, ALA, and succinylacetone (SA).

Neurotoxic Heme Precursors and Products

One mechanism of neurotoxicity that has received much attention is the overproduction of certain heme precursors, particularly ALA, that may have neuroactive properties. ALA has been shown to inhibit transmission at cholinergic and GABA-ergic synapses.[62–66] Moreover, in some reports, intracerebrovascular injection of ALA produces behavioral excitation and running seizures in rodents.[67,68] Because of the structural resemblance between ALA and the inhibitory amino acid transmitter GABA (see FIG. 1), work has focused on possible action of ALA at GABA-ergic synapses in the CNS and the spinal cord. We and others have shown that ALA competitively inhibits specific high-affinity [^{3}H]GABA binding to synaptic membranes from rat cerebellum.[35] ALA also inhibits sodium-dependent high-affinity uptake and release of GABA.[63,64] ALA did not affect the release of dopamine by rat brain synaptosomes (Silbergeld, unpublished observations) or of norepinephrine by rat heart tissue or human tissue.[69]

Other investigators, however, have found ALA be inactive in neural cell cultures.[70] ALA administered peripherally did not affect the behavior of rats injected once or continuously infused with ALA.[68,71] Only at high doses (1 g/kg) did ALA decrease nerve conduction velocity acutely in mice.[72] This contradictory evidence, especially when considered in light of the relatively high concentrations needed for *in vitro* activity (in the range of 0.1 to 10 mM), tends to argue against ALA as a mechanism in porphyrinopathy. Relatively high concentrations of ALA, however, can be found in the plasma of animals that have porphyria or are intoxicated. Studies with radiolabeled tracers demonstrate that intravenous ALA can enter the brain; for example, in nephrectomized rats that had been perfused and sacrificed by focused microwave irradiation to remove blood from brain and to prevent postmortem metabolism, as

TABLE 3. Comparison between the Effects of Succinylacetone and the Effects of Lead on Hepatic Heme Synthesis[a]

	Parameter		
Treatment	ALA (μg/24 hr in urine)	ALAS (nmol ALA/g wet wt)	ALAD (nmol PBG/g wet wt/hr)
Control	9	24	1822
Lead (Chronic)	385	24	1521
Control	16	19	1366
Succinylacetone	1745	34	42

[a]Data from Silbergeld and Lamon.[10]

much as 10% of an administered dose was found in brain tissue (Silbergeld and Lamon, unpublished observations). ALA levels in the CNS are increased in two experimental porphyrinopathic states: lead toxicity and succinylacetone treatment (TABLE 3). This property is consistent with the *in vitro* inhibition by lead of the conversion of ALA into porphyrins by chick and mouse dorsal root ganglia.[37,38]

Other evidence of ALA-mediated effects on GABA-ergic neurochemistry is only

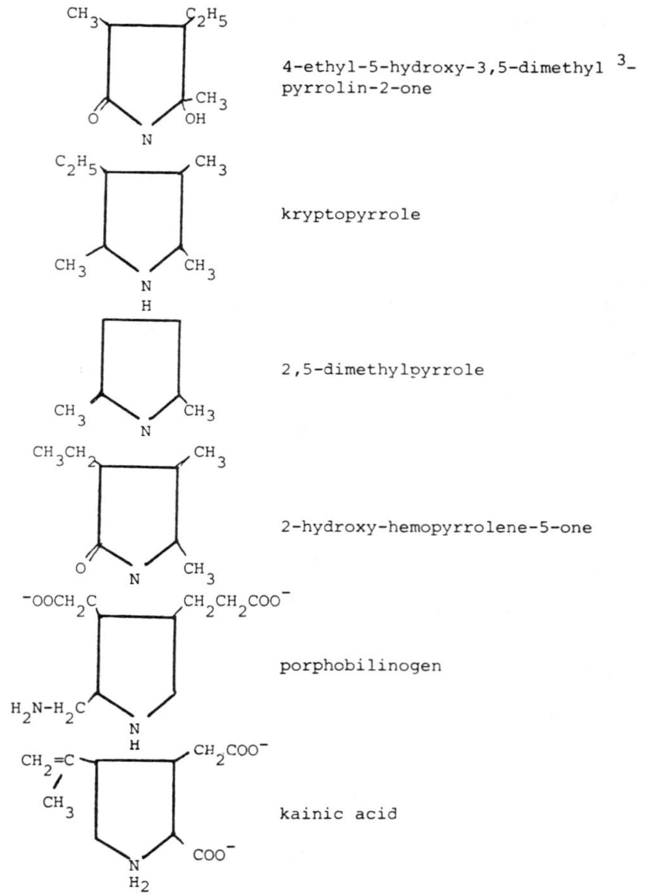

FIGURE 2. Some biologically active pyrroles.

indirect. Both lead- and succinylacetone-treated rats are supersensitive to convulsants that antagonize GABA-ergic neurotransmission; in both groups, intravenous administration of hematin reduced their hyperresponsiveness to picrotoxin.[10,73]

Other heme-related compounds and metabolic products may be neuroactive. Among these are the synthetic porphyrin tetraphenylporphine sulfate (TPPS), the acid derivatives of hematoporphyrin (HP), and certain pyrroles. *In vivo,* TPPS and HP were acutely toxic in high doses and caused a rapid decrease in motor nerve conduction velocity as well as body weight loss. A more chronic peripheral neuropathy was also observed with TPPS.[72] *In vitro,* TPPS and HP inhibited neurite outgrowth and killed nerve cells.[70]

Pyrroles have a somewhat controversial history in neuropsychiatric research. The structures for some pyrroles hypothesized to be neurotoxic are shown in FIGURE 2. Kryptopyrrole and 3-ethyl-5-hydroxy-4,5-dimethyl(Δ^3)pyrrolin-2-one are reported to be elevated in certain conditions of porphyrinopathy, including AIP and lead poisoning.[74] The neurobiological properties and etiologic roles of these compounds in neuropsychiatric disease, however, are not certain.[75–77] Recently, there has been an elegant demonstration of a mechanistic role for another pyrrole, 2,5-dimethylpyrrole, in the neuropathy of the 2,5-substituted hexacarbon solvents.[78,79] This pyrrole irreversibly cross-links neurotubules, resulting in inhibition of axonal transport and, over time, the morphological manifestations of focal axonopathy that characterize hexacarbon neurotoxicity. Whether other pyrroles could have similar neurotoxic effects is not known, although the structural requirements for such activity can be predicted. Another highly neurotoxic pyrrole is kainate.[80] As shown in TABLE 4, intrastriatal injection of kainate causes major changes in cortical neurochemistry.[81] These findings in models of nonporphyrinopathic neurotoxic disease suggest that reexamination of the pyrrole hypothesis may be appropriate, particularly because of some of the neurobehavioral and pathologic similarities among hexacarbon neuropathy, kainate neurotoxicity, the inherited porphyrias, and lead toxicity.

Porphyrins and Neuroglial Interactions

Porphyrins appear to participate in some way in cell–cell communication between neurons and myelinating cells of the CNS and PNS. The essential role of heme in

TABLE 4. Effects of the Pyrrole Kainate on CNS Neurochemistry

Region	Parameter	Percentage of Control
Striatum	[^{3}H]QNB binding	59
	CAT activity	16
	Chlorine uptake	30
	Dopamine uptake	99
	Glutamate uptake	94
	GAD activity	18
Globus pallidus	CAT activity	26
	GAD activity	27
Cortex	[^{3}H]QNB binding	92
	CAT activity	65
	Choline uptake	58
	Dopamine uptake	162
	GABA uptake	95

NOTE: Rats were unilaterally injected with 1 μg of kainate into the stratum; results are expressed as percentages of the value from the sham-injected side. QNB: quinuclidinyl benzilate; CAT: choline acetyltransferase; GAD: glutamine acid decarboxylase; GABA: γ-aminobutyric acid.

[a]Data from Silbergeld and Hruska.[18]

TABLE 5. Correlations between Neurochemical and Porphyrin Markers and Lead Poisoning

	PbB	PbU	ALA	EP	HVA
Blood Lead (PbB)	1.00	0.13	0.47	0.31	0.73
Urinary lead (PbU)		1.00	0.37	0.01	0.13
Urinary ALA (ALA)			1.00	0.03	0.19
Erythrocytic Protoporphyrin (EP)				1.00	−0.20
Homovanillic acid (HVA)					1.00

[a]Data from Chisolm and Silbergeld.[90]

maintaining organotypic cultures of CNS cells is well known, although its mechanism is not fully understood. *In vitro* inhibition of glial cell porphyrin synthesis by lead is associated with segmental demyelination; this inhibition can be prevented by adding hematin in the presence of lead.[37,38] Porphyrin mediation of neuroglial interrelationships may explain the findings of damage to both myelin and axons in inherited porphyrinopathy.[20] Inhibition of heme synthesis, which may occur primarily in glia and Schwann cells (where the pathway is most active), not only impairs the fundamental biochemical functions of these cells, but also deprives neurons of the extracellular source of heme and other factors that may result in cytotoxicity.

Ah *Receptor Mediated Neurotoxicity*

Certain PAHs, such as TCDD and the polyhalogenated biphenyls, exert their porphyrinopathic effects through inhibition of uroporphyrinogen decarboxylase and enzyme induction.[82] This induction is functionally associated with PAH binding to a steriod-like cytosolic receptor, known as the *Ah* or TCDD receptor.[48,83] The highly porphyrinopathic PAH hexachlorobenzene also binds to the *Ah* receptor.[84] The receptor–PAH complex acts at the level of DNA, increasing transcription of specific structural genes resulting in increased concentrations of mRNA that directs increased synthesis of heme-containing P-450 enzymes. Other cellular effects are also associated with receptor binding, including down-regulation of a number of membrane-bound receptors.[85] Effects on neurotransmitter receptors have not been studied, although PCBs are reported to affect dopamine receptors.[86]

Some of these receptor-mediated effects may be important to the neurotoxic and neuromyotoxic effects of these agents. On the other hand, it may be the case that altered porphyrin synthesis is parallel rather than serial to these aspects of PAH-induced neurotoxicity. In this case, altered heme synthesis would provide markers of exposure and effect related to neurotoxicity.

SUMMARY AND DISCUSSION

Neuropsychiatric symptoms are prominent clinical features of many inherited and acquired porphyrinopathies. Indeed, the remarkable similarities in these symptoms first suggested a role of altered heme synthesis as a common etiology, with particular focus on lead poisoning and AIP. Also striking are the observations that reestablishment of the regulation of the porphyrin pathway through the administration of hematin, the end product of the pathway, is beneficial in reducing neurotoxic problems in both inherited and acquired porphyrinopathy—even in the latter cases, where no reduction in internal exposure occurred.

Although it is tempting to conclude that altered heme synthesis can cause neurotoxicity, it is difficult to find instances where agents that produce porphyrinopathy do not also have direct neurotoxic properties (for example, lead and hexachlorobenzene). Succinylacetone may be the best model for investigating neurotoxicity, and further studies should be undertaken with this compound. Also, more research is needed on heme biochemistry in nervous tissue during intoxication states.

Investigations of porphyrin-related mechanisms of neurotoxicity have focused on four major areas: depletion of cell heme; changes in amino acid availability; neuroactivity of heme precursors and products; and interdependence of porphyrin and myelin synthesis in the nervous system. At present, the evidence for any of these mechanisms providing a common basis for the similar neurotoxic manifestations of inherited and acquired porphyrinopathy is incomplete. Moreover, our understanding of the neuropathology of the porphyrinopathies is still unresolved. Uncertainty remains over such basic issues as whether nerve or myelin is the primary site in PNS neurotoxicity, and over which neurotransmitter pathways in the CNS are altered even in such well-studied conditions as lead intoxication.

In addition to their possible etiologic roles, alterations in heme synthesis induced by toxicants have provided biological markers useful for estimating dose and outcome, as well as responsiveness to treatment. This has been most clearly demonstrated for lead intoxication, where levels of red cell protoporphyrin are widely used as a marker for lead exposure in individuals and for purposes of mass screening.[87] In some cases, changes in the porphyrin markers appear to correlate better than blood lead concentrations with neurotoxic end points.[88,89] Porphyrin markers, however, are not well correlated with urinary HVA, a neurochemical marker (TABLE 5).[90] Although biological monitoring—the measurement of the toxicants themselves in biological compartments—can be done for some toxicants (for example, blood lead, PCBs, or TCDD in lipids), measurement of markers such as porphyrins provides additional information on concentrations at sites of cellular action. As our understanding of mechanisms increases, so will our ability to interpret the significance of these signals, and their predictive relationship to organism-level neurotoxicity.

[NOTE ADDED IN PROOF: S. H. Snyder, A. Verma, and R. R. Trifiletti (*FASEB J.* **1:** 282–284, 1987) have reported that porphyrins can bind with high affinity to the peripheral benzodiazepine receptor.]

REFERENCES

1. DAGG, J. H., A. GOLDBERG, A. LOCHHEAD & J. A. SMITH. 1965. Q. J. Med. **134:** 163–168.
2. KAPPAS, A., S. SASSA & K. E. ANDERSON. 1982. *In* The Metabolic Basis of Inherited Disease. McGraw-Hill. New York, NY.
3. NAGLER, W. 1971. Arch. Phys. Med. Rehabil. 426–431.
4. BECKER, D. M. & S. KRAMER. 1977. Medicine **56:** 411–423.
5. WETTERBERG, L. 1967. A Neuropsychiatric and Genetic Investigation of Acute Intermittent Porphyria. Svenska Bokforlaget. Stockholm.
6. MAGNUSSEN, C. R., J. M. DOHERTY, R. A. HESS & D. P. TSCHUDY. 1975. Neurology **25:** 1211–1125.
7. SILBERGELD, E. K. 1985. *In* Neurotoxicology. K. Blum & L. Manzo, Eds.: 299–322. Marcel Dekker. New York, NY.
8. NEEDLEMAN, H. L. & P. J. LANDRIGAN. 1981. Annu. Rev. Public Health **2:** 277–298.

9. EBERT, P. S., R. A. HESS, B. C. FRYKHOLM & D. P. TSCHUDY. 1979. Biochem. Biophys. Res. Commun. **88:** 1382–1390.
10. SILBERGELD, E. K. & J. M. LAMON. 1982. Neurobehav. Toxicol. Teratol. **4:** 635–642.
11. SUSKIND, R. & V. HERZBERG. 1984. J. Am. Med. Assoc. **251:** 2372–2380.
12. CHEN, R. C., S. Y. TANG, H. MIYATA, T. KASHIMOTO, Y. C. CHANG, K. J. CHANG & T. C. TUNG. 1985. Environ. Res. **37:** 340–348.
13. CRIPPS, D. J., H. A. PETERS, A. GOCMEN & I. DOGRAMICI. 1984. Br. J. Dermatol. **111:** 413–422.
14. REGGIANI, G. 1983. Chemosphere **12:** 463–475.
15. ANDERSON, H. A., R. LILIS & I. J. SELIKIOFF. 1978. Environ. Health Perspect. **23:** 217–226.
16. FILIPPINI, G., B. BORDO, P. CRENNA, N. MASETTO, M. MUSICCO & R. BOERI. 1981. Scand. J. Work Environ. Health **7:** 257–262.
17. LATHROP, G. D., W. H. WOLFE, R. A. ALBANESE & P. M. MOYNAHAN. 1984. U.S. Air Force Project. Ranch Hand II. Brooks Air Force Base. San Antonio, TX.
18. JONES, R. E. & M. CHESLEY. 1986. Arch. Environ. Health **41:** 100–103.
19. Environmental Protection Agency. 1980. Dioxins. U.S. Environmental Protection Agency. Cincinnati, OH.
20. CAVANAGH, J. B. & A. R. RIDLEY. 1967. Lancet **ii:** 1023–1024.
21. RIDLEY, A., R. HIERONS & J. B. CAVANAGH. 1968. Lancet **ii:** 708–711.
22. KRIGMAN, M. R., T. W. BOULDIN & P. MUSHAK. 1980. *In* Experimental and Clinical Neurotoxicology. P. S. Spencer & H. H. Schaumberg, Eds.: 490–507. Williams & Wilkins. Baltimore, MD.
23. MUSTAJOKI, P. & A. M. SEPPALAINEN. 1975. Br. Med. J. **20:** 310–314.
24. COOPER, G. P. & C. D. SIGWART. 1980. *In* Lead Toxicity. R. L. Singhal & J. A. Thomas, Eds.: 401–423. Urban and Schwarzenberg. Baltimore, MD.
25. SINGER, R., M. MOSES, J. VALCIUKAS, R. LILAS & I. J. SELIKOFF. 1982. Environ. Res. **29:** 297–311.
26. OTTO, D., G. ROBINSON, S. BAUMANN, S. SCHROEDER, P. MUSHAK, D. KLEINBAUM & L. BOONE. 1985. Environ. Res. **20:** 168–186.
27. DUGANDZIC, M., B. STANKOVIC, L. MILOVANOVIC & Z. KORICANAC. 1973. Arch. Hig. Rada **24:** 37–44.
28. SILBERGELD, E. K. & J. J. CHISOLM. 1976. Science **192:** 153–155.
29. BLUM, I., N. SCHOENFELD & A. ATSMON. 1973. Biochim. Biophys. Acta **320:** 242–248.
30. SILBERGELD, E. K. & R. E. HRUSKA. 1980. *In* Low-Level Lead Exposure. H. L. Needleman, Ed.: 135–157. Raven Press. New York, NY.
31. WINDER, C. & I. KITCHEN. 1984. Prog. Neurobiol. **22:** 159–87.
32. SILBERGELD, E. K., R. E. HRUSKA, L. P. MILLER & N. ENG. 1980. J. Neurochem. **34:** 1712–1718.
33. MEMO, M., L. LUCCHI, P. F. SPANO & M. TABUCCHI. 1980. Toxicol. Lett. **6:** 427–432.
34. SPENCE, I., C. DREW, G. A. R. JOHNSTON & D. LODGE. 1985. Brain Res. **333:** 103–109.
35. SILBERGELD, E. K. & J. M. LAMON. 1980. J. Occup. Med. **22:** 680–684.
36. WHETSELL, W. O., S. SASSA, D. R. BICKCERS & A. KAPPAS. 1978. J. Neuropathol. Exp. Neurol. **37:** 497–507.
37. WHETSELL, W. O., S. SASSA & A. KAPPAS. 1984. J. Clin. Invest. **74:** 600–607.
38. SASSA, S., W. O. WHETSELL & A. KAPPAS. 1979. Environ. Res. **19:** 415–426.
39. DURKO, I. & A. JUHASZ. 1986. Neurochem. Res. **11:** 607–615.
40. DAS, M., P. K. SETH & H. MUKHTAR. 1981. J. Pharmacol. Exp. Ther. **216:** 156–161.
41. DAS, M., P. K. SETH & H. MUKHTAR. 1985. Bull. Environ. Contam. Toxicol. **35:** 500–504.
42. QATO, M. K. & M. D. MAINES. 1985. Biochem. Biophys. Res. Commun. **128:** 18–24.
43. GASIEWICZ, T. A. & G. RUCCI. 1984. Mol. Pharmacol. **26:** 90–98.
44. MAX, S. R. & E. K. SILBERGELD. 1987. Toxicol. Appl. Pharmacol. **87:** 523–527.
45. SUNDERMAN, F. W. 1987. Ann. N.Y. Acad. Sci. This volume.
46. WOODS, J. S. & B. A. FOWLER. 1987. Ann. N.Y. Acad. Sci. This volume.
47. MARKS, G. S. 1985. CRC Crit. Rev. Toxicol. **15:** 151–179.
48. EISEN, H., H. R. HANNAH, C. LEGRAVERAND, A. B. OKEY & D. NEBERT. 1973. Biochem. Act. Horm. **x:** 277–280.
49. HOLTZMAN, D. & J. SHEN HSU. 1976. Pediatr. Res. **10:** 70–75.

50. BULL, R. J., S. D. LUTKENHOFF, G. E. MCCARTY & R. G. MILLER. 1979. Neuropharmacology **18:** 83–92.
51. HJELLE, J. J., M. KEMAN & A. POKLIS. 1978. Toxicol. Lett. **2:** 361–364.
52. ALVARES, A. P., S. KAPNER, S. SASSA & A. KAPPAS. 1973. J. Exp. Med. **135:** 1406–1411.
53. CHIEN, S. 1981. *In* Principles of Neural Science. E. Kandel & J. Schwartz, Eds.: 660–660. Elsevier. New York, NY.
54. MCGEER, P. L. & E. MCGEER. 1981. *In* Basic Neurochemistry. G. Siegel, R. W. Albers, B. W. Agranoff & R. Katzman, Eds.: 233–254. Little, Brown. Boston, MA.
55. RICHARDS, F. F. & J. J. SCOTT. 1961. Clin. Sci. **20:** 387–400.
56. BALTROP, D. 1964. Pediatrics **34:** 696–697.
57. PRICE, J. M., R. R. BROWN & F. C. LARSON. 1957. J. Clin. Med. **36:** 1600–1605.
58. LITMAN, D. A. & M. A. CORREIA. 1985. J. Pharmacol. Exp. Ther. **232:** 337–345.
59. CORREIA, M. A., D. A. LITMAN & J. M. LUNETTA. 1987. Ann. N.Y. Acad. Sci. This volume.
60. WURTMAN, R. J. 1980. *In* Biochemical and Medical Aspects of Tryptophan Metabolism. O. Hayaishi, Y Ishimura & R. Kideo, Eds.: 31–46. Elsevier. New York, NY.
61. LASLEY, S. M., R. D. GREENLAND, D. J. MINNEMA & I. A. MICHAELSON. 1984. Neurochem. Res. **9:** 1675–1688.
62. DICHTER, H. D., L. TADDEINI, S. LIN & G. F. AYALA. 1977. Brain Res. **126:** 189–195.
63. BRENNAN, M. J. W. & R. C. CANTRILL. 1979. J. Neurochem. **32:** 1781–1786.
64. BRENNAN, M. J. W. & R. C. CANTRILL. 1979. Nature **280:** 514–575.
65. BORNSTEIN, J. C., J. B. PICKETT & I. DIAMOND. 1978. Ann. Neurol. **5:** 94–96.
66. CUTLER, M. G., J. M. DICK & M. R. MOORE. 1978. Life Sci. **23:** 2233–2238.
67. MOORE, M. R., F. B. MCGILLION & A. GOLDBERG. 1975. *In* Porphyrins in Human Diseases. M. Doss, Ed.: 148–154. Karger. Basel.
68. SHANELY, B. C., V. A. PERCY & A. C. NEETHLING. 1976. *In* Porphyrins in Human Diseases. M. Doss, Ed.: 155–162. Karger. Basel.
69. BEAL, M. F., N. O. ATUK, T. C. WESTFALL & S. M. TURNER. 1977. J. Clin. Invest. **60:** 1141–1148.
70. RIOPELLE, R. J. & J. C. KENNEDY. 1982. Can. J. Physiol. Pharmacol. **60:** 707–714.
71. EDWARDS, S., D. JACKSON, J. REYNOLDSON & B. SHANLEY. 1984. Neurosci. Lett. **50:** 169–173.
72. SIMA, A. A. F., J. C. KENNEDY, O. BLAKESLEE & D. M. ROBERTSON. 1981. Can. J. Neurol. Sci. **8:** 105–114.
73. SILBERGELD, E. K., R. E. HRUSKA & D. BRADLEY. 1982. Environ. Res. **29:** 459–471.
74. MOORE, M. R. & D. J. M. GRAHAM. 1980. Int. J. Biochem. **12:** 827–832.
75. WALKER, J. L. 1975. Pharmacol. Biochem. Behav. **3:** 243–250.
76. IRVINE, D. G. 1978. Life Sci. **23:** 983–990.
77. GORCHEIN, A., L. D. MITCHELL & A. T. ROGERS. 1979. Br. J. Pharmacol. **65:** 23–27.
78. DECAPRIO, A. P. & E. A. O'NEILL. 1985. Toxicol. Appl. Pharmacol. **78:** 235–247.
79. GRAHAM, D. G., D. C. ANTHONY, K. BOEKELHERDE, N. A. MASCHMANN, R. G. RICHARDS, J. W. WOLFRAM & B. R. SHAW. 1982. Toxicol. Appl. Pharmacol. **64:** 415–422.
80. COYLE, J. T., R. SCHWARCZ, J. P. BENNETT & P. CAMPOCHIARO. 1977. Prog. Neuropsychopharmacol. **1:** 13–30.
81. SILBERGELD, E. K. & R. E. HRUSKA. 1979. *In* Advances in Neurology. Vol. 25. T. N. Chase, E. Roberts & N. Wexler, Eds.: 633–644. Rover Press. New York, NY.
82. SASSA, S., H. FUJITA & O. SUGITA. 1987. Ann. N.Y. Acad. Sci. This volume.
83. SAFE, S. H. 1986. Annu. Rev. Pharmacol. Toxicol. **26:** 371–399.
84. HAHN, M. E., T. A. GASIEWICZ, P. LINKO & J. A. GOLDSTEIN. 1987. Ann. N.Y. Acad. Sci. This volume.
85. MATSUMURA, F., D. W. BREWSTER, B. V. MADHUKER & D. W. BOMBICK. 1984. Arch. Environ. Contam. Toxicol. **13:** 509–515.
86. AGRARWAL, A. K., H. A. TILSON & S. C. BONDY. 1981. Toxicol. Lett. **7:** 417–424.
87. PIOMELLI, S., C. SEAMAN & S. KAPOOR. 1987. Ann. N.Y. Acad. Sci. This volume.
88. BAKER, E. L. *et al.* 1985. Br. J. Ind. Med. **42:** 507–516.
89. VALCIUKAS, J. A., R. LILIS, A. FISCHBEIN, J. J. SELIKOFF, J. EISINGER & W. E. BLUMBERG. 1978. Science **201:** 465–467.
90. CHISOLM, J. J. & E. K. SILBERGELD. 1981. Proceedings of the International Symposium on Heavy Metals: 565–568. CEP Association. Edinburgh.

Mechanisms of Toxicity of Photoactivated Artificial Porphyrins

Role of Porphyrin–Protein Interactions

ANN SMITH

Department of Biochemistry and Molecular Biology
Louisiana State University Medical Center
New Orleans, Louisiana 70112

INTRODUCTION

This topic has two sides: one is the generation of toxic oxygen species by light-activated porphyrins; the other concerns the mechanism of uptake and intracellular localization of porphyrins. The porphyrins may be derived exogenously as in photodynamic therapy of cancer or generated endogenously as a consequence of disturbances of heme metabolism, which can be genetic, chemically induced, or a combination of both. Importantly, the binding of porphyrins by proteins determines not only sites of cellular damage upon illumination but also has biochemical consequences to cells, including inhibition of enzymic activities. The physicochemical state of porphyrins and their interaction with proteins in the circulation and with cellular components are being investigated in this laboratory.

Although there are differences between the chemistry of porphyrins and the chemistry of their derivatives, metalloporphyrins, it is well established that the chemical reactivity of heme (iron-porphyrin) is profoundly influenced by its physical state and by the nature of the proteins with which it associates.[1] Porphyrins, like heme, aggregate in solution generally at concentrations higher than about 1 μM.[2,3] Porphyrins and heme are found associated with proteins, and exist in an unbound state only transiently, for example, during intracellular synthesis. Heme aggregates, such as dimers, are much less chemically active than the monomeric form,[1] and only monomeric heme is known to be accommodated by the binding sites of hemoproteins. Although much of this is due to the iron atom of the heme, certain similarities in effects on porphyrins are expected. For example, porphyrin dimers and aggregates do not fluoresce,[4] and binding to a protein, such as albumin, prevents porphyrin transport into cells (see below).[5] In this context, the following should be noted: although the active component of the mixture used for photodynamic therapy is currently suggested to be a dimeric dihematoporphyrin ether (DHE)[6] or ester,[7] monomeric porphyrins are considered to be the effective intracellular photosensitizers.[8,9]

The localization of porphyrins by binding to cell constituents, principally proteins, will play a major role in determining the effects upon cells of exogenous porphyrins or those produced endogenously in excessive amounts because of perturbations of heme synthesis in certain porphyrias. The effects exerted can be specific or nonspecific. Specific effects result from inhibition of enzyme activity due to binding of porphyrins to proteins even in the absence of light, as demonstrated by some of the work presented here. Nonspecific effects are produced in porphyrin-loaded cells upon illumination resulting in intense fluorescence. Consequently, highly active oxygen species, which are cytotoxic, are produced.[10] By diffusion, these reactive molecules produce cross-linking of macromolecules, including protein, lipid, and DNA molecules.

Knowledge of the interaction of porphyrins with proteins is thus necessary to understand the pharmacokinetic processes that occur during metabolism of artificial and naturally occurring porphyrins. In photodynamic therapy, artificial porphyrins injected intravenously first encounter serum proteins together with membrane proteins and lipids of circulating cells before reaching the membrane proteins and lipids of the various tissues where uptake occurs. After transport, the intracellular locations of the porphyrins are determined by interaction with intracellular components. Although porphyrin–protein interactions predominate, porphyrins also have the potential to reside in the lipid phase of membranes. However, the distribution between cytosolic protein and membrane *in vivo* is not known. Endogenous porphyrins (as porphyrinogens) first interact with their site of generation, the soluble proteins involved in heme biosynthesis, before being compartmentalized in the mitochondria. In certain porphyrias, other cellular organelles are involved, and excretion of porphyrins into the circulation and bile can occur.[11]

There are thus three principal areas where significant porphyrin–protein interactions can be expected to occur: in the circulation, at the cell membrane, and intracellularly. Detrimental effects to a variety of cells can be expected even in the absence of light in addition to the well established photocytotoxic effects of porphyrins.[12] Each of these areas will be addressed in turn, emphasizing known as well as putative porphyrin–protein interactions. The relevance of these interactions on processes likely to be involved in porphyrin clearance by cells, and the metabolic–biochemical effects on the cell will be stressed.

GENERAL APPROACH AND QUESTIONS

We took an approach using purified proteins and porphyrins to examine specific porphyrin–protein interactions under well-defined conditions *in vitro*. The results from such work are needed to provide basic information both on porphyrin–protein binding and on potential biochemical consequences to the cell, which in turn will provide a rational basis for improving porphyrin photodynamic therapy of cancers and aid understanding of the pathology of the porphyrias and porphyrinopathies.

In the circulation, questions of interest include the following: What forms of porphyrin are bound by which plasma components (for example, albumin,[5,13–16] hemopexin,[5,13,14,16] lipoproteins,[17] fibrin,[18] or atherosclerotic plaques[19])? Are the associations specific or nonspecific? What are the stoichiometry and affinity of binding? Which form of protein-bound porphyrin is likely to predominate in serum? What factors affect the distribution of porphyrins among plasma components?

At the cell membrane, other questions arise: Is the porphyrin specifically bound to a protein or dissolved in the membrane by partitioning in lipid? Does transport of porphyrin occur by diffusion, by facilitated diffusion, or by active transport? Is pinocytosis or endocytosis (perhaps via the low-density lipoprotein receptor) involved? What features of the porphyrin molecule affect transport? Finally, does the transport mechanism vary with cell type? If transport is not specific, then the damage produced upon photoactivation will be "incidental" and hard to predict, let alone control.

With regard to the intracellular pathway, questions of note include the following: Do exogenous and endogenous porphyrins take the same intracellular route? What is this intracellular pathway? Do the porphyrins interact with cytosolic proteins? Are they transported into mitochondria and converted by ferrochelatase to heme or an analogue? Is this followed by transfer, again perhaps via a cytosolic intermediate, to heme oxygenase in the endoplasmic reticulum? If transport to the mitochondria does not occur, are other organelles (for example, endosomes, lysosomes) involved? If heme analogues are formed from exogenous porphyrins, are abnormal hemoproteins produced? How do these abnormal hemoproteins affect cell function?

PORPHYRIN–PROTEIN INTERACTIONS IN THE CIRCULATION

Two major porphyrin-binding proteins, albumin and hemopexin, are known to exist in human serum with concentrations near 30 and 0.8 g/l, respectively.[13] Their physicochemical properties were reviewed in another volume of the *Annals of the New York Academy of Sciences* in 1975.[13] In collaboration with Drs. Morgan and Koskelo, we addressed the protein-bound state in human serum of protoporphyrin IX (PP IX), zinc protoporphyrin IX (ZnPP IX), coproporphyrin I (CP I), uroporphyrin I (UP I), and uroporphyrinogen III (UP III) in order to understand more fully the plasma clearance of porphyrins in patients with porphyria and lead intoxication.[14]

TABLE 1. Stoichiometry and Affinity of Exogenous and Endogenous Porphyrins for Human Serum Albumin and Human Hemopexin

Protein	Porphyrin	Apparent K_d (μM)	Porphyrin/Protein (mol of porphyrin/ mol of protein)
Human albumin	Hp	1.9	1.1
	HpAc	1.5	0.9
	PP IX	0.75	1.85
	ZnPP IX	1.2	1.6
	CP I	7.7	0.9
	CP III	6.0	1.5
Human hemopexin	Hp	2.8	1.2
	HpAc	4.0	1.05
	PP IX	1.75	0.95
	ZnPP IX	0.65	1.9
	CP I	0.4	1.05
	CP III	0.4	1.05

NOTE: These data are summarized from quenching of intrinsic fluorescence and equilibrium experiments from published sources.[5,12] Uncertainties based on the observed experimental range are approximately ±50% for K_d values and ±0.1 for molar ratios. The data for albumin are estimated from the single apparent high-affinity site.

The binding properties of CP I, CP III, UP I, PP IX, and ZnPP IX with human serum albumin and human hemopexin were examined quantitatively using equilibrium dialysis and porphyrin fluorescence spectroscopy.[14] Albumin bound the porphyrins with a high affinity, the dissociation constants (K_d) being in the μM range, with the exception of UP I, which was only weakly bound, the K_d being on the order of 100 μM. PP IX and ZnPP IX were more strongly bound (K_d: 0.75 and 1.2 μM, respectively) than CP I and III (K_d: 6–7 μM) (TABLE 1). Analysis of the quenching of intrinsic protein fluorescence reveals that human serum albumin, known to have one high-affinity site for heme (K_d: ~10 nM) and several sites of lower affinity,[20] has several binding sites for each of the three porphyrins. The porphyrin and heme binding sites are the same because the amounts of CP bound are considerably decreased in the presence of heme.[14]

Human hemopexin also forms complexes with CP I and CP III, the K_d being near

0.5 μM; with PP IX, the K_d being near 2 μM; and with ZnPP IX, the K_d being near 0.5 μM.[14] All porphyrins are bound by hemopexin with an apparent stoichiometry of 1:1, except ZnPP IX, 2 mol of which appear to be bound per mol of protein (TABLE 1). As with albumin, only a weak interaction with UP I is evident, the K_d being near 50 μM.

Competition experiments were conducted to directly test the relative affinities of the two proteins for these porphyrins. In these experiments, the fluorescence spectra of mixtures varying in the molar ratio of porphyrin:albumin:hemopexin were used to determine and identify which complex predominated (FIGURE 1). Albumin clearly has a higher affinity than hemopexin for both PP IX and ZnPP IX, and hemopexin is able to compete effectively with albumin for CP I or CP III only at low molar ratios of the two proteins, not at the molar ratios (about 1:40) of hemopexin and albumin found in human serum. Because albumin and hemopexin are the major porphyrin-binding proteins in serum,[13] the protein-bound state of a porphyrin in serum can also be inferred from spectral data. For example, addition of PP IX (0.5 μM) to an equimolar mixture of albumin and hemopexin (2.5 μM) rapidly produces a spectrum characteristic of PP IX–albumin (FIGURE 1). Similar results were found with all the other porphyrins examined.[5,14]

A second series of investigations[5] was carried out to determine whether albumin or hemopexin is the predominant binding agent in human plasma for exogenous porphyrins encountered by patients during porphyrin photodynamic therapy. Hematoporphyrin (Hp) and *O,O'*-diacetylhematoporphyrin (HpAc) were chosen as representative phototherapeutic porphyrins. Titrations of the quenching of the intrinsic protein fluorescence for albumin and hemopexin with both Hp and HpAc were similar, with stoichiometries of 1 mol of porphyrin bound per mol of protein and apparent K_d values in the range 1–2 μM for albumin and 3–4 μM for hemopexin. Again, in view of the molar excess of albumin to hemopexin in serum,[14] Hp and HpAc will be preferentially bound to albumin (FIGURE 2). Although endogenous substances associated with albumin in serum, such as fatty acids, may alter the environment where the porphyrin binds, the spectra of the porphyrins in hemopexin-free serum closely resemble those of

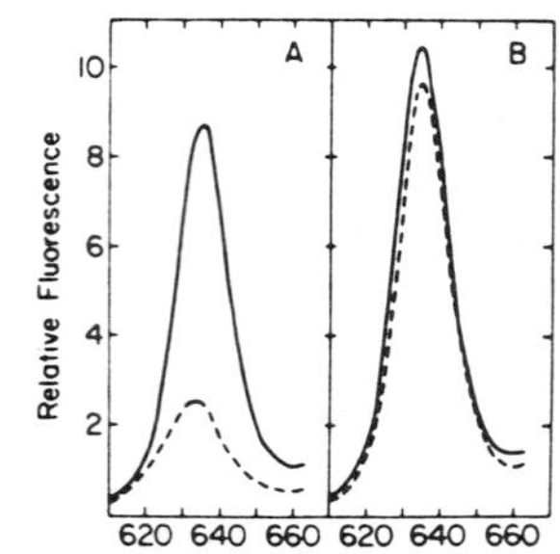

FIGURE 1. Fluorescence emission spectra of protoporphyrin in human albumin–hemopexin mixtures and in human serum. (**A**) The emission spectra of protoporphyrin (0.5 μM) added to a mixture of human hemopexin (2.5 μM) and serum albumin (2.5 μM) immediately after mixing (-----) and 15 minutes later (——). The excitation wavelength was 410 nm. (**B**) The emission spectra of protoporphyrin–albumin (——) and protoporphyrin–serum (-----) mixtures are depicted. The concentration of porphyrin was 0.5 μM, and that of albumin was 2.5 μM. The serum was diluted 1:100 with phosphate buffer giving an albumin concentration of 4.5 μM and a hemopexin concentration of 0.15 μM. Mixtures were incubated in the dark at ambient temperature for 1 hr to allow ample time for complex formation. The excitation wavelength was 410 nm. Reprinted from Morgan *et al.*[14]; permission granted by Elsevier Science Publishers.

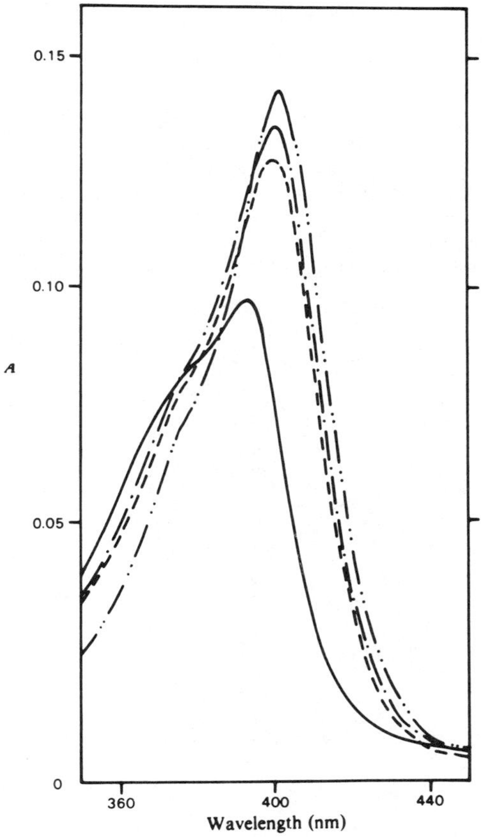

FIGURE 2. The absorbance spectra of free and protein-bound *O,O'*-diacetylhematoporphyrin. The absorbance of HpAc (1 μM) was determined in each of the following conditions: in a buffer including 25 mM HEPES and 0.15 M sodium chloride, pH 7.2 (——); in 100 μM human serum albumin (– · –); in human serum diluted 1:10 with buffer (– · · –); and in hemopexin-free serum diluted 1:10 with buffer (-----). Similar spectra were generated for Hp except all were shifted 1–2 nm toward the blue end of the spectrum.[5] For example, the λ_{max} for the Hp-albumin complex was 400 nm compared with 402 nm for the HpAc-albumin complex.[5] Reprinted from Smith and Neuschatz[5]; permission granted by the Biochemical Society.

the porphyrin–albumin complexes.[14] Whether, by binding to serum albumin, drugs or endogenous compounds affect the amount of porphyrin associated with albumin and with which serum components, dimers, and aggregates of porphyrins are associated in patients undergoing photodynamic therapy is not yet established.

Information concerning the nature of the porphyrin-binding site can be obtained from the intensity and wavelength maxima of the fluorescence spectrum of the porphyrin. For example, the CP–albumin spectra (λ_{max}: 618 nm), like porphyrins in detergent, showed shifted emission maxima (λ_{max}: 623 nm) compared with porphyrins in buffer (λ_{max}: 612 nm). In contrast, the hemopexin spectra resemble those of porphyrins in buffer, suggesting that the ligand is in a polar environment and may be largely exposed to solvent. That the binding site for heme on hemopexin is exposed has been suggested further by solvent pertubation spectroscopy.[21]

PORPHYRIN INTERACTIONS AT THE CELL MEMBRANE

The affinity that normal and cancerous cells have for porphyrins can be ascribed to binding sites on the plasma membrane (lipid, protein, or both) or to intracellular porphyrin-binding proteins. In this section, the form of porphyrin that interacts with cells; the effects of porphyrin structure and cell type on transport; and the mechanism of porphyrin transport in Morris hepatoma tissue culture (HTC 7288) cells are addressed.

There is a direct, but complex, relationship between free porphyrin concentration in the media and the amount of porphyrin associated with the cells (FIGURE 3). Moreover, the same data plotted against albumin concentration (inset in FIGURE 3) show a decrease in cell-associated porphyrin with increasing albumin concentration. This provides evidence that free porphyrin, not porphyrin–albumin complexes, is transported into the cells.[5] Similar results were obtained with Hp; over the same range of free porphyrin concentrations the amount of cell-associated Hp was about fivefold lower with HpAc. The concentration of free porphyrin presented to the cells was

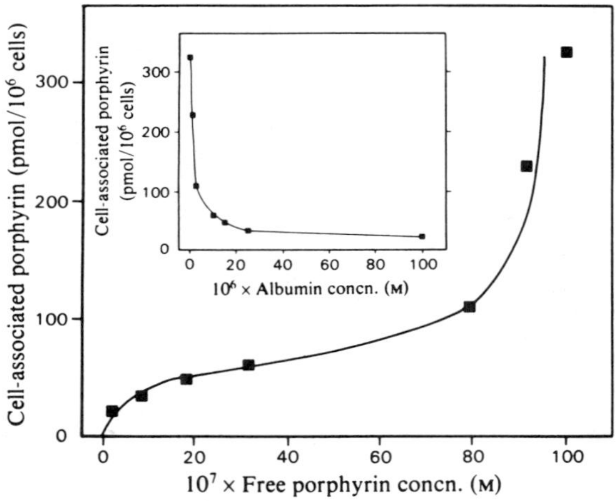

FIGURE 3. The effect of free porphyrin concentration on cell-associated porphyrin. The data represent the difference between porphyrin associated with the cells at 30 min and 30 sec in HTC cells after incubation with preformed complexes of HpAc and albumin. HpAc (10 μM) was added to albumin (0–100 μM) to produce a range of free porphyrin concentrations calculated from the experimentally determined K_d values. In addition, the cell-associated porphyrin is shown versus the albumin concentration (see inset). Reprinted from Smith and Neuschatz[5]; permission granted by the Biochemical Society.

carefully controlled by adjusting the molar proportions of albumin and porphyrin, and the amount of free porphyrin was calculated using the experimentally determined K_d values.[5,14] The range of free porphyrin concentrations studied reflect the estimated intravascular free porphyrin (0–2 μM) in patients undergoing photodynamic therapy. Current dose ranges are 2–3 mg/kg body wt for Hp derivative and 1–2 mg/kg for DHE,[22] and 8 μM Hp in serum has been recorded 1 hr after Hp derivative infusion in patients.[23] Although most of the porphyrin in serum will circulate bound to albumin, which is present at 500 μM in serum, binding is a dynamic process and ligand constantly dissociates and becomes available to interact with cells. Similar porphyrin–serum protein cell dynamics are likely to occur in several porphyrias and porphyrinopathies.

Hp and HpAc (0.2 μM free porphyrin) rapidly associate with HTC 7288 cells, and the amount of porphyrin that accumulates is time- and dose-dependent.[5] The time course of association of porphyrin with cells shows a rapid porphyrin accumulation for the first 15 min followed by a lower rate over the next 30 min. HpAc is accumulated at a clearly higher rate (>2pmol of HpAc/10^6 cells/min) than is Hp.[5]

Two classes of porphyrin association with HTC cells are observed that are saturable and can be related to events occurring at the membrane and after transport. The first occurs within 30 sec at 37° C under conditions of low levels of free, predominately monomeric porphyrin (less than 2 μM). Porphyrin is likely to be essentially membrane-associated at this early stage. Eadie-Hofstee analyses were consistent with a high-affinity saturable process with a K_m of approximately 10 nM for both Hp and HpAc. The maximum porphyrin association was 20 pmol/10^6 cells, which by using Avogadro's number, suggests there are 4×10^9 porphyrin binding sites/HTC cell. These sites may reflect association with a protein such as the heme-binding protein of the hemopexin system[24] or the heme receptor of duodenal mucosal[25] and erythroleukemic cells.[26] Although heme is bound by the latter with a K_d of 30 nM, the affinity for PP IX is lower.[26]

After this rapid initial association a slower but also saturable process of higher capacity is evident (FIGURE 3), a process likely to be related to the saturation of intracellular binding sites after transport across the membrane. The difference in cell-associated porphyrin at 30 min and 30 sec is plotted to reflect the major saturable process observed. This process is half-saturated at 1.0 μM free porphyrin and reaches a maximum of 60 pmol of HpAc/10^6 cells. A nonsaturable association of porphyrin with HTC cells occurs at free porphyrin levels above 2 μM, perhaps because of porphyrin aggregates at the surface or uptake via pinocytosis of incubation medium. Investigations on the mechanism of cell–porphyrin interactions conducted under conditions likely to produce free porphyrin concentrations greater than 2 μM should be interpreted with caution.

The effect of cell type on porphyrin accumulation by cells was addressed using S49.1 lymphoma cells[5] and the same protocols described for the HTC cells. In side-by-side incubations, the lymphoma cells accumulated both Hp and HpAc but at a significantly lower rate than the HTC cells, even after taking into account the smaller size and smaller surface area of the lymphoma cells. Again, more HpAc than Hp was taken up (0.5 pmol of HpAc/10^6 lymphoma cells/min compared with 1.5 pmol/10^6 HTC cells/min).[5] These and other published studies[27–29] demonstrate the importance both of porphyrin structure and cell type in porphyrin–cell interactions. The role of additional factors in affecting the rate of uptake, including cell cycle (A. Smith, unpublished observations) and cell density (M. Kreimer-Birnbaum, unpublished observations) are being determined.

The final group of experiments described in this section addressed the mechanism of porphyrin transport by cells. The porphyrin associated with HTC cells represents an accumulation of porphyrin over the initial concentration of free porphyrin in the incubation medium. For example, taking a value of cell-associated HpAc of 45 pmol/10^6 cells and using a volume of 1.9×10^{-12} liter/HTC cell,[30] the concentration of porphyrin in the cell pellet is 24 μM, compared with an initial free porphyrin level of 1 μM. Moreover, preincubation of the HTC cells with potassium cyanide (10 mM for 2 hr at 37° C) reduced the amount of cell-associated porphyrin at 30 min by 50%. Other metabolic inhibitors, including sodium fluoride, ouabain, and oligomycin, also decreased the amount of cell-associated porphyrin uptake but were less effective (20–35% inhibition). The association of porphyrin with the HTC cells was also temperature-dependent. By 1 hr at 4° C, cell-associated porphyrin was only 50% of that observed at 37° C.[5] These results strongly suggest that at least part of the process involved in the interaction of Hp and HpAc with HTC cells involves active transport.[5]

PORPHYRIN INTERACTION WITH INTRACELLULAR PROTEINS

GSH Sulfotransferase

The aim of this investigation was to begin to assess the importance of protein binding for the retention of porphyrins by cells, to assess the effects of this binding on cell metabolism, and to identify potential sites of damage to cells upon illumination.

Glutathione (GSH) sulfotransferases (EC 2.5.1.18) were considered to be likely to be involved in intracellular porphyrin binding for several reasons. They are abundant cytoplasmic proteins, particularly in liver, which is a major site of porphyrin metabolism.[11] Ligandin (one GSH sulfotransferase isoenzyme) binds hematin,[31] protoporphyrin IX, and Hp.[32,33] Several other transferases also bind nonsubstrate ligands, including hematin and bilirubin.[34] GSH sulfotransferases are found in both normal and cancerous tissues.[35–37] Furthermore, PP IX, Hp, and other porphyrins present in the mixture of hematoporphyrin derivatives used in photodynamic therapy share similarities in structure, suggesting that certain exogenous porphyrins would interact with these proteins. GSH sulfotransferases represent an important group of inducible, multifunctional enzymes. They are involved in the detoxification of various pharmacologically active compounds[38] and in the metabolism of endogenous compounds.[39,40] Porphyrin binding to these enzymes might cause inhibition in a manner analogous to other hydrophobic ligands known to inhibit certain GSH sulfotransferases *in vitro,*[34,41] and thus be detrimental to cells. A role in the intracellular transport of heme has been suggested for these enzymes.[42]

In general, all naturally occurring and artificial porphyrins investigated bind to the GSH sulfotransferases with an affinity in the range 1–2 μM.[44] This binding inhibits enzymic activity noncompetitively, and K_{ii} equals K_{is}, except for form 2-2 (FIGURE 4). Porphyrin structure is an important factor in determining the affinity of binding, as shown by the relative inhibitory effectiveness. Of the dicarboxylic porphyrins examined (present in some mixtures used clinically in photodynamic therapy), HpAc and monohydroxyethylmonovinyldeuteroporphyrin (HVD) are more effective inhibitors than Hp and PP IX. Of the naturally occurring porphyrins, the order of effectiveness is PP IX (dicarboxylic) > CP (tetracarboxylic) > UP (octacarboxylic), with the type I isomers of both CP and UP being more effective than the type III isomers. Of particular interest is the greater effective inhibition by the synthetic tetra-mesophenylporphine-tetra-sulfonate (TMPPTS) (apparent K_i: 250 nM) compared with CP, which contains a comparable number of negative charges. TMPPTS contains four negatively charged phenylsulfonic acid groups in the meso positions of the porphyrin macrocycle. This suggests that the porphyrin binding site on the transferases contains positive charges that are in a particular orientation and that facilitate binding.[43] Importantly, all these porphyrins were shown to inhibit transferase activity in the presence of other proteins in cytosol (see below).[43]

Not only do many artificial and naturally occurring porphyrins inhibit GSH sulfotransferase isoenzymes,[43] but there is tissue-specific expression of these enzymes in a variety of normal tissues,[44] and there are differences in the type and amount of these isoenzymes in normal and cancerous rat liver[43] and human tumors (see below). Thus, these enzymes appear to be potential cytosolic binding sites for naturally occurring porphyrins in the porphyrias and for porphyrins used in cancer photodynamic therapy. Other porphyrins that are taken up by tumors in animals are also effective inhibitors of GSH sulfotransferase. These include UP I, which is taken up by the KHJJ mouse mammary carcinoma,[45] and TMPPTS, which is taken up by the Walker 256 carcinoma,[46] by tumors induced by the murine sarcoma virus,[47] and by HEp-2 cells.[48] It seems likely that these transferases also provide binding sites, albeit weak ones, for UP I and TMPPTS.

Because heme and heme analogues are probably metabolic products of many of the artificial porphyrins (see below),[48] and because hematin inhibits ligandin (for example, form 1-1[31]), we examined the effect of the protoheme analogue mesoheme (containing ethyl groups on positions 2 and 4 of the porphyrin) on the GSH sulfotransferases. Iron-porphyrin chelates have quite different properties from those of the free porphyrins and are bound more tightly than are the free porphyrins to serum proteins[14] and to a membrane heme-binding protein in erythroleukemic cells.[26] Mesoheme is a potent noncompetitive inhibitor of these transferases: 50% inhibition is produced in rat liver cytosol by 150 nM mesoheme, and the K_{ii} is 70 nM for form 3-3.[43]

Of interest too is the apparent altered expression of the GSH sulfotransferases in the liver caused by an extrahepatic tumor.[43] This is suggested by the differences we found in relative activities toward specific substrates of these enzymes in cytosol from

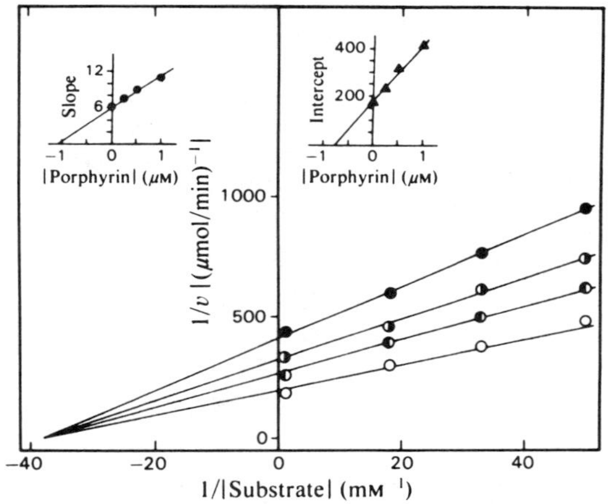

FIGURE 4. Noncompetitive inhibition of purified GSH sulfotransferases by porphyrins. Assays were carried out at 30° C in 0.1 M potassium phosphate buffer, pH 6.5, with various 1-chloro-2,4,-dinitrobenzene concentrations from 0.2 to 1.0 mM in the presence of a saturating concentration of GSH (10 mM). *O,O'*-Diacetyl hematoporphyrin was added at 0 μM (○), 0.25 μM (◐), 0.5 μM (◑), and 1.0 μM (●) final concentration. Insets show the intercept and slope replots of the data for form 3-3. All data, fitted by computer using published procedures, showed noncompetitive inhibition with K_{is} equal to K_{ii}.[5] Reprinted from Smith *et al.*[43]; permission granted by the Biochemical Society.

normal rat liver and from the liver of rats bearing the Morris 7288C hepatoma.[43] The presence of tumors in animals also alters the activity of heme-synthesizing enzymes[50] and causes induction of heme oxygenase.[51] If similar changes occur in the liver of patients with tumors, alterations in the type and amount of the GSH sulfotransferase isoenzymes could affect not only the metabolism of drugs but also that of exogenous porphyrins.

Ferrochelatase

We postulated that the retention of exogenous porphyrins by tumors occurs because although both normal and cancerous tissues contain porphyrin-binding molecules, cancerous cells have a decreased ability to metabolize these porphyrins to heme analogues that can then be catabolized.[49] A variety of porphyrins are substrates for the mitochondrial enzyme ferrochelatase (EC 4.99.1.1). Several representative porphyrins present in the porphyrin mixture administered to patients were found to be substrates for the purified ferrochelatase. The K_m values for PP IX, Hp, and HVD are 11, 22, and 23 μM, respectively.[49] HpAc, however, is an effective competitive inhibitor with a K_i of 11 μM.[49] This is not unexpected: porphyrins with bulky substituents in the 2,4 positions, such as 2,4-disulfonicdeuteroporphyrin, and 2,4-bisglycoldeuteroporphyrin,[52] and 2,4-bisacetyldeuteroporphyrin, are competitive inhibitors (K_i: 13–70 μM) with respect to deuteroporphyrin.[49]

Tumors, although the extent of their de-differentiation varies, commonly use glycolysis more extensively than oxidative phosphorylation (especially rapidly growing tumors[53]) and contain lower activities of mitochondrial cytochrome oxidase.[54] We therefore thought it likely that tumors in general, and the Morris hepatoma 7288C cells in particular, would be deficient in ferrochelatase. Hepatic ferrochelatase specific activity is 5.9 and 5.5 nmol of heme/hr per mg protein in normal Buffalo rats and in those bearing the extrahepatic Morris 7288C hepatoma, respectively, but is only 0.13 nmol/hr per mg in the hepatomas.[49] Therefore low ferrochelatase activity in cancerous cells may provide one means whereby some porphyrins accumulate in tumors, and the ability of certain porphyrins to act as ferrochelatase inhibitors may provide another.[49] Thus, ferrochelatase may provide one porphyrin binding site in mitochondria, where exogenous porphyrins have been shown to localize.[55,56]

Porphyrins that are inhibitors and not substrates for ferrochelatase would not be expected to cause major problems in normal hepatic tissues, owing to the relatively high ferrochelatase activity. In cancerous cells (or other tissues), however, where ferrochelatase activity is low, small amounts of these competitive inhibitors may profoundly affect the cell's heme metabolism. A de-repression the initial rate-limiting step in the heme biosynthetic pathway, a step catalyzed by 5-aminolevulinate

TABLE 2. Comparison of Relative Activities of GSH Sulfotransferases from Rat and Human Tissues

	Initial Rate (μmol/min/mg protein)			
Enzyme Source	1-Chloro-2,4-dinitrobenzene	Cumene Hydroperoxide	Ethacrynic Acid	*trans*-4-Phenylbut-3-en-2-one
Normal rat liver	1.13 ± 0.34, 6	0.501 ± 0.014, 4	0.047 ± 0.002, 2	0.022 ± 0.009, 5
Morris 7288C hepatoma	0.29 ± 0.13, 6	0.048 ± 0.008, 4	0.042 ± 0.01, 2	0.020 ± 0.008, 5
Human squamous cell carcinoma				
Cytosol 1	0.816 ± 0.041	0.306 ± 0.194	0.0010	0.008
Cytosol 2	0.206 ± 0.006	ND	ND	ND
Human basal cell carcinoma				
Cytosol 1	0.414 ± 0.075	ND	ND	0.021
Cytosol 2	ND	0.163	0.0017 ± 0.0007	ND

NOTE: Samples of human tumor were frozen immediately after excision in liquid nitrogen and stored at −70° C. Cytosol was prepared from a 100 000 × g supernatant of a 33% homogenate of the tissue prepared in 100 mM potassium phosphate buffer, pH 6.5. The initial rates of transferase activity in cytosol samples were measured with each substrate using published procedures.[43] Where applicable, the initial rate has been expressed as a mean ± SD, the number of samples following a comma. ND: not determined.

TABLE 3. Inhibition by Porphyrins of GSH Sulfotransferase Activity in Cytosol from Human Squamous Cell Carcinoma

Porphyrin	Concentration (μM)	GSH Sulfotransferase Activity (percentage of control)	
		Cytosol 1	Cytosol 2
Hp	10	37	27
HpAc	1	64	45
UP I	20	52	65

NOTE: The inhibition of GSH sulfotransferase activity was determined in cytosol prepared from individual tumors (taken from two males, of 54 and 65 years, respectively). The initial rates were 0.82 and 0.21 μmol/min/mg protein for cytosols 1 and 2, respectively.

Porphyrins (at the final concentrations indicated above) were incubated in cytosol samples for 5 min at the appropriate incubation temperature before initiation of the enzyme reaction by addition of the electrophilic substrate. These porphyrin concentrations were chosen because they produced 50% inhibition of transferase activity in normal rat liver cytosol.[43] The extent of porphyrin inhibition varied no more than $\pm$5% in duplicate determinations. Each "percentage of control" is a ratio of the enzyme activity measured in the presence of a porphyrin to that measured in the absence of a porphyrin. Initial rates were determined after subtraction of nonenzymatic rates.

synthase,[57] results in the consequent accumulation of endogenous porphyrins that may then contribute to the observed photosensitivity of these cells. Accumulation of endogenous porphyrins might also contribute to the cutaneous photosensitivity observed in some patients[23,58] and to the sensitivity of certain normal tissues.[59]

Porphyrin-Binding and Metabolizing Enzymes in Human Tissues

We chose to investigate cancers of the skin because not only do these epithelial cancers represent the most common type of tumor but also because cutaneous and subcutaneous lesions constitute 40% of the cases treated by Hp derivative photodynamic therapy. In addition, these tumors respond extremely well to this photodynamic therapy; for example, 75% of basal cell carcinomas have shown complete remission.[24] The data in TABLE 2 summarizes the GSH sulfotransferase activities in squamous and basal cell carcinomas for four defined substrates. The human tumors showed some variation in total activity (initial rates of 0.21–0.81 μmol/min/mg protein). The activities were measured with 1-chloro-2,4-dinitrobenzene, which is active with all isozymes, as substrate. In contrast to the Morris 7288C hepatoma, transferase activity in the human carcinomas was high with respect to cumene hydroperoxide but low with respect to ethacrynic acid. These results strongly suggest that considerable differences in the level of expression of these enzymes occur in various tumors, even if they are of the same origin in separate patients. Such differences may be due to variation in pretranslational expression, considered to be one mechanism for normal tissue specificity[44] or to differential induction caused by drug exposure. Affinity purification of GSH sulfotransferases from cytosol of the squamous cell carcinoma revealed negligible levels of the Yc subunit, which has an apparent M_r of 28 000 on sodium dodecyl sulfate–polyacrylamide gel electrophoresis.

There is already evidence for considerable variation in the affinity for heme in transferase isozymes isolated from human liver.[60] We investigated the potential for the human GSH sulfotransferases isolated from these tumors to act as porphyrin binding sites by inhibition assays. The results, summarized in TABLE 3, clearly show that two artificial porphyrins, Hp and HpAc, and the naturally occurring UP I were all inhibitors. This is interesting because Hp is considered to be the active intracellular, monomeric porphyrin in Hp derivative photodynamic therapy.

Consistent with our earlier work on the Morris hepatoma, all human squamous and basal cell carcinoma samples examined so far have negligible ferrochelatase activity. Six normal human skin samples were examined using a sensitive, radiochemical assay adapted for use with tissue homogenates and found to have extremely low ferrochelatase activity. The values ranged from 0.46 to 0.20 $\pm$ 0.07 nmol heme/hr/mg protein, about 8% of the activity in normal rat liver mitochondria. If a major determinant in the clearance of exogenous porphyrins by cells is their ability to be metabolized by ferrochelatase and heme oxygenase, then these low ferrochelatase levels in skin provide an explanation of why cutaneous photosensitivity is so persistent (up to 3–4 weeks) in patients after phototherapy. Likewise, this may relate to the photosensitivity of porphyric patients and other conditions.

CONCLUSIONS

The apparent preferential accumulation of artificial porphyrins by cancerous cells must be the result of differences in the clearance of porphyrins by these cells compared with normal tissue. Clearance involves many processes, including binding by serum proteins, cell uptake (which may involve interaction with one or more membrane components), intracellular binding, and elimination. Albumin is shown to be the principal porphyrin-binding protein, in serum, whereas nonprotein bound monomeric porphyrin is the kinetically active species in the cell transport process. Both porphyrin structure and cell type are important determinants of the rate and extent to which porphyrin associates with cells.

Cellular protein binding of porphyrins, as described here for GSH sulfotransferases and ferrochelatase, may play a role in determining the intracellular sites of photodamage after photodynamic therapy, that is, after administration of a mixture of hematoporphyrin derivatives, and in certain porphyrias. Although the cytosol from the Morris 7288C hepatoma and human squamous cell carcinomas contains lower amounts of the GSH sulfotransferases, several isoenzymes that interact with porphyrins are present. These enzymes have also been located in the endoplasmic reticulum,[61] mitochondria,[62] and other organelles.[63] This variety of binding sites could readily account for the extreme sensitivity that is observed in certain normal tissues, particularly liver (a major site of porphyrin metabolism), and that currently hampers treatment of certain tumors, including hepatic metastases.

In addition, because GSH sulfotransferases are important detoxifying enzymes and because porphyrin and heme binding inhibits the enzyme activity of the transferases, metabolic consequences to the cells of the administration of artificial porphyrins are likely, and may also occur in porphyrias and porphyrinopathies.

Many artificial porphyrins, likely to be found intracellularly, are substrates for ferrochelatase, and this enzyme activity is negligible in the rat and human tumors examined. PP IX and deuteroporphyrin are transported into intact mitochondria *in vitro,* and iron is incorporated there.[64,65] Also, artificial porphyrins are transported into Morris 7288 hepatoma cells *in vitro.*[5] We therefore propose that, in normal liver and perhaps other tissues, clearance of porphyrins, including PP IX, Hp, and HVD, is facilitated by their metabolism to their corresponding iron chelates. Further metabolism by heme oxygenase is likely because the heme analogue formed will be similar in structure to hematoheme, which is known to be a substrate for this enzyme *in vitro.*[66]

Conversely, tissues with low ferrochelatase activity—including certain tumors (such as the Morris hepatoma and the human squamous and basal cell carcinomas) and normal human skin—would retain porphyrins.

ACKNOWLEDGMENTS

We are indebted to Dr. H. A. Dailey for developing and carrying out the ferrochelatase determinations in the human tissue samples and to Dr. R. Mora for his help in obtaining the human tissues.

REFERENCES

1. BROWN, S. B., H. HATZIKONSTANTINOU & D. G. HERRIES. 1980. Int. J. Biochem. **12:** 701–707.
2. FALK, J. E. 1964. Porphyrins and Metalloporphyrins. Elsevier. Amsterdam.
3. BROWN, S. B., M. SHILLCOCK & P. JONES. 1976. Biochem. J. **153:** 279–285.
4. MARGALIT, R., N. SHAKLAI & S. COHEN. 1983. Biochem. J. **209:** 547–552.
5. SMITH, A. & T. NEUSCHATZ. 1983. Biochem. J. **214:** 503–509.
6. DOUGHERTY, T. J., W. R. POTTER & K. R. WEISHAUPT. 1984. *In* Porphyrin Localization and Treatment of Tumors. D. R. Doiron & C. J. Gomer, Eds.: 301–314. Allan R. Liss. New York, NY.
7. KESSEL, D. 1986. Cancer Res. **46:** 2248–2251.
8. DOUGHERTY, T. J. 1984. CRC Crit. Rev. Oncol./Hemat. **2:** 83–116.
9. KESSEL, D. 1984. Photochem. Photobiol. **39:** 851–859.
10. WEISHAUPT, K. R., C. J. GOMER & T. J. DOUGHERTY. 1976. Cancer Res. **36:** 2326–2329.
11. KAPPAS, A., S. SASSA & K. E. ANDERSON. 1983. *In* The Metabolic Basis of Inherited Diseases. 5th edit. D. S. Frederickson, J. B. Stanbury, J. B. Wyngaarden, J. L. Goldstein & M. S. Brown, Eds.: 1301–1384. McGraw-Hill. New York, NY.
12. SPIKES, J. D. 1984. *In* Porphyrin Localization and Treatment of Tumors. D. R. Doiron & C. J. Gomer, Eds.: 19–39. Allan R. Liss. New York, NY.
13. MULLER-EBERHARD, U. & W. T. MORGAN. 1975. Ann. N.Y. Acad. Sci. **244:** 624–650.
14. MORGAN, W. T., A. SMITH & P. KOSKELO. 1980. Biochim. Biophys. Acta **624:** 271–285.
15. REDDI, E., F. RICCHELLI & G. JORI. 1981. Int. J. Peptide Res. **18:** 402–408.
16. LAMOLA, A. A., I. ASHER, U. MULLER-EBERHARD & M. POH-FITZPATRICK. 1981. Biochem. J. **196:** 693–698.
17. JORI, G., M. BELTRAMINI, E. REDDI, B. SALVATO, A. PAGNAN, L. ZIRON, L. TOMIO & T. TSANOV. 1984. Cancer Lett. **24:** 291–297.
18. MUSSER, D. A., J. M. WAGNER, F. J. WEBER & N. DATTA-GUPTA. 1980. Res. Commun. Chem. Pathol. Pharmacol. **28:** 505–525.
19. SPEARS, J. R., J. SERUR, D. SHROPSHIRE & S. PAULIN. 1983. J. Clin. Invest. **71:** 395–399.
20. BEAVEN, G. H., S.-H. CHEN, A. D'ALBIS & W. B. GRATZER. 1974. Eur. J. Biochem. **41:** 539–546.
21. MORGAN, W. T. 1976. Ann. Clin. Res. **8**(Suppl. 17): 223–232.
22. DOIRON, D. R. & G. S. KELLER. 1986. Curr. Probl. Dermatol. **15:** 85–93.
23. ZALAR, G. L., M. POH-FITZPATRICK, D. L. KROHN, R. JACOBS & L. C. HARVER. 1977. Arch. Dermatol. **113:** 1392–1397.
24. SMITH, A. & W. T. MORGAN. 1985. J. Biol. Chem. **260:** 8325–8329.
25. GRASBECK, R., R. MAJURI, I. KOUVONEN & R. TENHUNEN. 1982. Biochim. Biophys. Acta **700:** 137–142.
26. GALBRAITH, R. A., S. SASSA & A. KAPPAS. 1985. J. Biol. Chem. **260:** 12198–12202.
27. MOAN, J. & T. CHRISTENSEN. 1981. Cancer Lett. **11:** 207–214.
28. DUBBELMAN, T. M. A. R. & T. VAN STEVENINCK, 1984. Biochim. Biophys. Acta **771:** 201–207.
29. KESSEL, D. 1981. Cancer Res. **41:** 1318–1323.

30. MAPES, J. P. & K. S. AULTMAN. 1982. Arch. Biochem. Biophys. **214:** 782–791.
31. MEUWISSEN, J. A. T. P., B. KETTERER & B. B. E. MERTENS. 1972. Digestion **6:** 293–294.
32. TIPPING, E., B. KETTERER & P. KOSKELO. 1978. Biochem. J. **169:** 509–516.
33. KETTERER, B., T. CARNE & E. TIPPING. 1978. *In* Transport by Proteins. G. Blauer & H. Sund, Eds.: 79–94. W. de Gruyter. New York, NY.
34. KETLEY, J. N., W. H. HABIG & W. B. JAKOBY. 1975. J. Biol. Chem. **250:** 8670–8673.
35. ARIAS, I. M. 1979. Med. Biol. **57:** 328–334.
36. ARIAS, I. M., N. OHMI & M. BHARGARVA. 1979. Trans. Assoc. Am. Physicians **92:** 113–120.
37. SCULLY, N. C. & T. J. MANTLE. 1981. Biochem. J. **193:** 367–370.
38. KETTERER, B., D. BEALE, J. B. TAYLOR & D. J. MEYER. 1983. Biochem. Soc. Trans. **11:** 466–467.
39. BOOTH, I., E. BOYLAND & P. SIMS. 1961. Biochem. J. **79:** 516–524.
40. BOYLAND, E. & L. F. CHASSEAUD. 1969. Adv. Enzymol. Relat. Areas Mol. Biol. **32:** 172–219.
41. VANDER JAGT, D. L., S. P. WILSON, V. L. DEAN & P. C. SIMONS. 1982. J. Biol. Chem. **257:** 1997–2001.
42. SENJO, M., T. ISHIBASHI & Y. IMAI. 1985. J. Biol. Chem. **260:** 9191–9196.
43. SMITH, A., I. NUIRY & Y. C. AWASTHI. 1985. Biochem. J. **229:** 823–831.
44. TU, C.-P. D., M. J. WEISS, N.-Q. LI & C. C. REDDY. 1983. J. Biol. Chem. **258:** 4659–4662.
45. EL-FAR, M. A. & N. R. PIMSTONE. 1983. Cell Biochem. Funct. **1:** 156–160.
46. WINKLEMAN, J., G. SLATER & J. GROSSMAN. 1967. Cancer Res. **27:** 2060–2064.
47. CARRANO, C. C., M. TSUTSUI & S. MCCONNEL. 1977. Cancer Treat. Rep. **61:** 1297–1300.
48. CARRANO, C. C., M. TSUTSUI & S. MCCONNELL. 1978. Chem.-Biol. Interact. **21:** 233–248.
49. SMITH, A. & H. A. DAILEY. 1984. Biochem. J. **223:** 441–445.
50. WOODS, J. S. 1978. Mol. Pharmacol. **14:** 900–910.
51. SCHACTER, B. A. & P. KURZ. 1984. Biochem. Pharmacol. **33:** 815–820.
52. DAILEY, H. A. & J. E. FLEMING. 1983. J. Biol. Chem. **258:** 11453–11459.
53. FRIEDKIN, M. 1973. Fed. Proc. Fed. Am. Soc. Exp. Biol. **32:** 2148–2153.
54. WEINHOUSE, S. 1960. *In* Amino Acids, Proteins and Cancer Biochemistry. J. T. Edsall, Ed.: 109–119. Academic Press. New York. NY.
55. COZZANI, I., G. JORI, E. REDDI, A. FORTUNATO, B. GRANATI, M. FELICE, L. TOMIO & P. ZORAT. 1981. Chem.-Biol. Interact. **37:** 67–75.
56. BERNS, M. W., A. DAHLMAN, F. M. JOHNSON, R. BURNS, D. SPERLING, M. GUILTINAN, A. SIEMENS, R. WALTER, W. WRIGHT, M. HAMMER-WILSON & A. WILE. 1982. Cancer Res. **42:** 2325–2329.
57. GRANICK, S. & S. I. BEALE. 1978. Adv. Enzymol. Relat. Areas Mol. Biol. **46:** 33–203.
58. DOUGHERTY, T. J. 1981. J. Invest. Dermatol. **77:** 122–124.
59. DOUGLAS, H. O., H. R. NAVA, K. R. WEISHAUPT, D. BOYLE, M. G. SUGERMAN, E. HALPERN & T. J. DOUGHERTY. 1983. Adv. Exp. Med. Biol. **160:** 15–22.
60. VANDER JAGT, D. L., L. A. HUNSAKER, K. B. GARCIA & R. E. ROYER. 1985. J. Biol. Chem. **260:** 11603–11610.
61. MORGANSTERN, R., C. GUTHENBERG & J. W. DEPIERRE. 1982. Eur. J. Biochem. **128:** 243–248.
62. KRAUS, P. 1980. Hoppe-Seyler's Z. Physiol. Chem. **361:** 9–15.
63. FRIEDBERG, T., P. BENTLEY, P. STASIECKI, H. R. GLATT, D. RAPHAEL & F. OESCH. 1979. J. Biol. Chem. **254:** 12028–12033.
64. ROMSLO, I. & P. HUSBY. 1980. Int. J. Biochem. **12:** 709–712.
65. KOLLER, M. E. & I. ROMSLO. 1980. Biochem. J. **188:** 329–335.
66. FRYDMAN, R. B., J. L. TOMARO, G. BULDAIN, J. AWRUCH, L. DIAZ & B. FRYDMAN. 1981. Biochemistry **20:** 5177–5182.

Principles for Application of Porphyrin Markers in Epidemiological Studies[a]

PHILIP J. LANDRIGAN

Division of Environmental and Occupational Medicine
Department of Community Medicine
Mount Sinai School of Medicine
New York, New York 10029

INTRODUCTION

The dominant theme of this volume has been exploration of the biochemical and toxicological mechanisms by which environmental toxins alter porphyrin metabolism. Metals—such as lead—and halogenated aromatic hydrocarbons—including 2,3,7,8-tetrachlorodibenzo-*p*-dioxin (TCDD) and hexachlorobenzene—are among the materials whose effects on porphyrin metabolism have been considered.

In this paper, I shall discuss the potential application of porphyrin toxicology to epidemiology. In particular, I shall consider approaches to using the metabolic alterations characteristic of porphyrin toxicity as "biological markers" in epidemiological studies.

BIOLOGICAL MARKERS

Biological markers are biochemical, physiological, and cytogenetical changes induced in cells by exposures to environmental chemicals.[1] Such changes are signals of an encounter between organism and environment. Some biological markers, such as the blood lead level or the concentration of polychlorinated biphenyls (PCBs) in breast milk, are indicators of exposure or absorption. Others, such as slowing of reaction time in painters exposed to solvents, reduction in forced expiratory volume in textile spinners inhaling cotton dust, or formation of DNA:benz[*a*]pyrene adducts in roofers exposed to coal tar pitch, are indicators of effect.

Biological markers have for many years been used occasionally in epidemiological studies. Alterations in heme biosystems, for example, have been utilized in the screening and clinical assessment of children exposed to lead. At present, however, recognition of the utility of biological markers has become much more widespread. An explosion of interest surrounds the possibility of systematically applying markers to a broad range of epidemiological studies. Expectation exists that successful application of biological markers to epidemiology may bring about a revolution in preventive medicine as profound as that which resulted fifty years ago from the joining of epidemiology with microbiology.

[a]This paper was originally presented as part of a panel discussion entitled Chemical-Induced Porphyrinopathies: Future Directions.

THE NATURE AND LIMITATIONS OF EPIDEMIOLOGY

Epidemiology is the study of diseases in populations.[2] A principal goal of epidemiological research is to establish disease causation. Through description of patterns of occurrence of illness and through analysis of relationships between exposure and illness, epidemiologists seek to identify etiologies of occupational and environmental disease, to identify populations at greatest risk, and ultimately to provide a rational foundation for prevention of these diseases.

Epidemiology is rooted historically in the study of the great infectious diseases. It has enjoyed remarkable successes in eradication of epidemics of those illnesses. More recently, epidemiology has begun to examine chronic, chemically induced illness. Also, epidemiology has moved in the past two decades from a purely descriptive science to a much more highly analytical discipline whose focus has shifted from the qualitative to the quantitative.

Despite its past successes, epidemiology has not escaped criticism. Critics of chronic disease epidemiology have argued that epidemiology can never prove anything—that the processes of inductive reasoning in epidemiology are too tenuous to be believable, and the data too "soft" ever to permit drawing valid conclusions as to causality.

TABLE 1. Criteria for Determining Causality in Epidemiological Studies

Major Criteria
Experimental data (Does there exist evidence from true experiments in humans?)
Consistency (reproducibility)
Temporal direction
Strength of association
Dose–response relationship
Minor Criteria
Does the association make sense epidemiologically?
Does the association make sense biologically?
Is the association specific?

Particularly ingenious criticisms of epidemiology have been developed by apologists for the tobacco industry:

> Epidemiology, like astronomy, is largely an observational rather than an experimental science, and hence the scope for planned and controlled intervention by the investigator is usually either limited or absent. Precautions that are routine in properly conducted experiments are unattainable in epidemiology. Inbred strains of the human species are not available, and strict standardization of environmental conditions throughout the life span is inconceivable. Even randomization—which can often be used to mitigate differences between experimental and control groups of animals—is seldom possible when investigating the causes of human disease.[3]

These criticisms have been a powerful goad to epidemiologists. Their refutation has stimulated elegant rebuttals, defending the ability of epidemiology to establish causality. More importantly, this debate has fostered the development of well-standardized criteria for examining data pertaining to causality (TABLE 1).[4] The two most important of these criteria are the reproducibility of an observed association and the documentation of a positive dose–response relationship.

Strict application of these criteria coupled with development of powerful statistical instruments has greatly enhanced the ability of epidemiology to study relationships between toxic environmental exposures and human disease.

Classical epidemiology, however, has been limited in the study of diseases caused by chemical toxins. In evaluation of these often chronic conditions, epidemiology is hampered by its need to wait for death or for the appearance of clinically obvious symptoms as indicators of biological damage. Also, because these diseases are often rare events, analysis of their links with toxic exposures requires evaluation of very large populations. Such approaches are inherently cumbersome and, therefore, inefficient in disease prevention. A further limitation on classical epidemiology is the difficulty it faces in reconstructing information on past exposures; such reconstructions are difficult at best and may, if not properly conducted, lead to serious misclassification and subsequent blurring of calculated dose–response relationships.

APPLICATION OF BIOLOGICAL MARKERS TO EPIDEMIOLOGY

Many of these limitations on epidemiology could be overcome by applying to epidemiological studies biological markers of exposure and effect. Gains resulting from this combination of disciplines would include the following:

1. Earlier detection than heretofore of environmentally induced dysfunction and disease.
2. Detection of dysfunction at low levels of exposure.
3. Identification of large numbers of affected persons, because markers permit identification of early, subtle dysfunction.
4. Precise assessment of individual exposure.

Successful integration of biological markers into epidemiological research, however, will require close scrutiny of proposed markers. Their sensitivity and specificity must be assessed. Their quantitative relationships to chemical exposures must be defined, possibly through combination of biological monitoring with measurement of environmental exposures. Their pharmacokinetics must be understood, particularly their temporal relationships to exposure. Finally, and most importantly, the ability of markers to predict the future occurrence of disease must be delineated. Often, this last step will require prospective, longitudinal study.

Additionally, it is most important that biological markers be applied only in well-designed epidemiological studies. Experimentalists and epidemiologists must recognize that relationships of biological markers to environmental exposures and to health outcomes are probabilistic. Epidemiological studies utilizing markers, therefore, must take cognizance of the statistical nature of these relationships. Several epidemiologic principles must be heeded:

1. Populations included in studies of biological markers must be large enough to permit analyses of adequate statistical power. Studies on small populations tend typically to produce only inconclusive results.[5]
2. Populations to be studied must have had sufficient exposure to a suspect environmental toxin to produce measurable biological alteration. This objective is frequently achieved by studying populations with a wide range of exposures, including persons with quite intense exposures such as occur in industrial settings.
3. In prospective, longitudinal studies to assess the predictive accuracy of markers in regard to future occurrence of disease, follow-up must be long enough to

permit adequate accrual of cases. No absolute guidelines exist here, but investigators must recall that the latent period between environmental exposures and malignancy may be as long as three to four decades.

CONCLUDING COMMENTS

Epidemiology and porphyrin biology are disciplines poised at a moment of critical opportunity. Epidemiologists have developed powerful methodological instruments that can define more accurately than ever before causal relationships between toxic exposures and human disease. "Epidemiology today is no more restricted to the study of striking outbreaks of disease than meteorology is to the study of hurricanes or astronomy to eclipses of the sun."[6]

Porphyrin biologists, likewise, have reached unanticipated heights of molecular sophistication. The integration of these two disciplines will provide techniques that can burst through methodological limitations that have hindered for years the application of epidemiologic techniques to the study of environmentally induced disease. In our enthusiasm to consummate this integration, however, we must not forget the need for proper validation of markers, and the need for proper design of epidemiological studies.

REFERENCES

1. PERERA, F. P., R. M. SANTELLA & M. C. POIRIER. 1984. Potential methods to monitor human population exposure to carcinogens: Carcinogen-DNA binding as an example. *In* Risk Quantitation and Regulatory Policy. Banbury Report 19. D. G. Hoel, R. A. Merrill & F. P. Perera, Eds. Cold Spring Harbor Laboratory. Cold Spring Harbor, NY.
2. LILIENFELD, A. M. & D. E. LILIENFELD. 1984. Foundations of Epidemiology. 2nd edit. Oxford University Press. New York, NY.
3. BURCH, P. R. J. 1978. Smoking and lung cancer: The problem of inferring cause. J. R. Stat. Soc. **141:** 437–477.
4. HILL, A. B., SR. 1962. Statistical Methods in Clinical and Preventive Medicine. Oxford University Press. New York, NY.
5. HERNBERG, S. 1982. Fact and fiction in occupational epidemiology. J. R. Coll. Physicians London **17:** 139–143.
6. MACMAHON, B. & T. F. PUGH. 1970. Epidemiology: Principles and Methods. Little, Brown. Boston, MA.

Comments on Porphyrin[a]

NORTON NELSON

New York University Medical Center
Institute of Environmental Medicine
New York, New York 10016

I would suggest that while continuing the very important study of abnormal porphyrin metabolism, that is, of its mechanisms and the relationship of its manifestations to disease, that a simultaneous course be followed that would be a deliberate and planned attempt to exploit the potentialities of the system to detect external injuries, stresses, and toxicants of a variety of kinds. It is already apparent that a variety of agents can produce aberrations in the course of porphyrin metabolism that may have, as a minimum, important diagnostic contributions to make. Among these, of course, is the relationship to lead poisoning and to a whole range of organochlorine compounds. Thus, it appears we have a complex system that is quite sensitive to external stresses, that might be systematically exploited in a variety of fields, and that has a high degree of sensitivity and, possibly, specificity.

The system involves a set of compounds that are unique within the organism and for which, I believe, there already exist highly sensitive and highly specific analytical techniques for detection both in blood and in urine. This system, as noted, has also demonstrated that it sensitively responds to a variety of external agents. Accordingly, one could imagine a systematic search for the range of responses of the porphyrin system to external chemical stresses as an addition to our armamentarium for diagnostic detection of a chemical injury. This would involve examining a range of compounds that would test the validity of this diagnostic test for a wider range of applications than has so far been considered. Certainly, at some point these could aid in epidemiological studies as Dr. Landrigan suggests[b]; however, until we are more certain about the range of chemical susceptibilities, there may be a limited number of applications of tests for the porphyrin metabolites as markers of chemical injury. Nevertheless, in what I propose, epidemiological studies could proceed in parallel; moreover, studies such as I propose would not only replace but should actually be in parallel with continued studies of the basic mechanisms of action that lead to the production of aberrant porphyrin metabolites.

In view of the genetic control of the course of porphyrin metabolism, susceptibilities both of genetic origin as well as those based on prior or current disease might be tested through a systematic application of the techniques for detection of abnormal metabolites in the body.

[a]This paper was originally presented as part of a panel discussion entitled Chemical-Induced Porphyrinopathies: Future Directions.

[b]See the paper in this volume by P. J. Landrigan.

Hematoporphyrin Derivative Uptake and Tumor Cell Density[a]

GRETA M. GARBO,[b,c] RICK W. KECK,[c]
STEVEN H. SELMAN,[c] AND MARTHA
KREIMER-BIRNBAUM[b-d]

[b]*St. Vincent Medical Center*
Toledo, Ohio 43608

[c]*Medical College of Ohio*
Toledo, Ohio 43699

Photodynamic therapy of solid malignancies is an evolving modality of cancer treatment.[1] This form of therapy depends upon the retention of systemically administered photosensitizers which accumulate in tumors. Subsequent tissue destruction occurs as a result of activation of the sensitizer with light. The most widely studied tumor photosensitizer is hematoporphyrin derivative (HpD). HpD is a complex mixture of dicarboxylic porphyrins, in addition to dimers (dihematoporphyrin ether[2] or dihematoporphyrin ester[3]) and aggregates of various molecular sizes.[2] Because the uptake mechanisms of porphyrins are not clearly understood, conclusions both on uptake and phototoxicity have been reached often without due regard to the different experimental conditions. One such condition, cell density, and its effect on porphyrin uptake, is the subject of this communication. AY-27 rat bladder tumor cells, derived from an *N*-(4-(5-nitro-2-furyl)-2-thiazolyl)formamide (FANFT)-induced transitional cell carcinoma, were cultured as described.[3] Stock cultures were maintained, and, for individual experiments, cells were trypsinized and viable aliquots were plated on 60 mm × 15 mm plastic dishes. Culture medium was removed from the plastic dishes just prior to the incubation with HpD, and cells were washed three times with PBS. Incubations with HpD[4] were run for 1 hr, at a concentration of 10 μg/ml in PBS. After the incubations were completed, medium was removed, and the cells washed three times with PBS to remove any free or loosely adhered porphyrins. This procedure did not cause cell losses. Porphyrins were extracted from the cells with Triton X:HEPES and quantitated as described previously.[4]

FIGURE 1 depicts HpD uptake by the tumor cells as a function of the number of cells per plate. After a very rapid increase, the uptake decreases markedly as the number of cells per plate increases. Dense cultures (over 7 million cells per plate) reduce the uptake by close to two orders of magnitude when compared with cultures of about 1 million cells. Protein content per cell remained constant and independent of cell density. Therefore, if the cellular protein content is constant, it might not be responsible for the observed kinetic profile of porphyrin uptake. Other variables affecting porphyrin uptake by cultured cells could be 1) position in the cell cycle, 2) morphological changes of the cell surface with increased cell density, and 3) membrane

[a]This work was supported by grants from the F. M. Douglass Foundation and the St. Vincent Medical Center Foundation and by Grant R23 CA38754–01 from the National Institutes of Health.

[d]Address for correspondence and reprint requests: St. Vincent Medical Center, 2213 Cherry Street, Toledo, Ohio 43608.

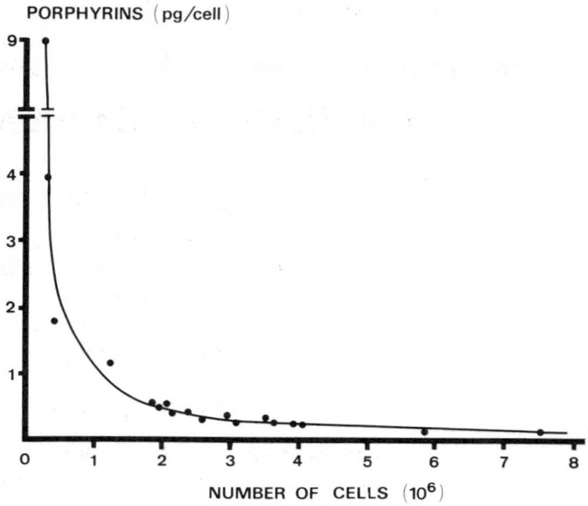

FIGURE 1. HpD uptake by AY-27 rat bladder tumor cells expressed in picograms per cell as a function of the number of cells per plate.

fluidity. Uptake of porphyrins from a protein-free medium can be mediated by processes such as passive or facilitated diffusion, non-receptor-mediated endocytosis, or binding to a specific protein on the membrane. We have shown a clear effect on the uptake of dicarboxylic porphyrins in HpD by cell density. Which one of the possible uptake mechanisms is down-regulated by cell density remains to be further explored.

REFERENCES

1. Moan, J. 1986. Porphyrin photosensitization and phototherapy. Photochem. Photobiol. **43:** 681–690.
2. Dougherty, T. J. 1985. Photodynamic therapy. Clin. Chest Med. **6:** 219–236.
3. Kreimer-Birnbaum, M., J. L. Baumann, J. E. Klaunig, R. Keck, P. J. Goldblatt & S. H. Selman. 1984. Chemical studies with hematoporphyrin derivative in bladder cell lines. *In* Porphyrin Localization and Treatment of Tumors. D. R. Doiron & C. J. Gomer, Eds.: 335–350. Alan R. Liss. New York, NY.
4. Garbo, G. M., J. B. Kramer, R. W. Keck, S. H. Selman & M. Kreimer-Birnbaum. 1985. Solid-phase purification and analysis of dicarboxylic porphyrins extracted from cultured tumor cells. Anal. Biochem. **151:** 70–81.

Mechanism of Urinary Excretion of δ-Aminolevulinic Acid after Intratracheal Instillation of Gallium Arsenide

PETER L. GOERING

Division of Life Sciences
Office of Science and Technology
Center for Devices and Radiological Health
Food and Drug Administration
Rockville, Maryland 20857

BRUCE A. FOWLER

National Institute of Environmental Health Sciences
Research Triangle Park, North Carolina 27709

Gallium arsenide (GaAs) has superior electron-conducting properties compared to other materials, such as silicon, and is one of several compounds revolutionizing semiconductor microcircuit technology.[1] Exposure to airborne particulates of GaAs, however, is a potential health hazard to workers in the semiconductor industry.[2] Assessment of risk from GaAs exposure is complicated by the lack of toxicity data on this compound and of biological indicators for evaluating early exposure or toxicity. At present, GaAs is regulated on the basis of As_2O_3 toxicity data. Recent studies have demonstrated altered urinary porphyrin patterns in rats exposed to GaAs by the intratracheal and oral routes.[3] Therefore, the present study was undertaken to further evaluate the effects of acute intratracheal instillation of GaAs suspensions on the heme biosynthetic pathway. Specifically, efforts were focused on the effects of GaAs on blood δ-aminolevulinic acid dehydratase (ALAD), which catalyzes the second step in heme biosynthesis, and resultant changes in aminolevulinic acid (ALA) in urine.

The effect of intratracheal instillation of GaAs on blood ALAD activity and urinary concentrations of ALA in adult male rats is shown in FIGURE 1. A dose-dependent decrease in ALAD activity was observed 6 days after the intratracheal instillation of 50, 100, and 200 mg GaAs/kg body wt (FIG. 1, upper left), with activity 5% of control at the highest dose. A concomitant dose-dependent increase in the urinary concentration of ALA was also observed (FIG. 1, lower left). A 60% decrease in ALAD activity occurred 3 days after administration of 100 mg GaAs/kg body wt, remained at this level at 6 days after exposure, and progressively recovered toward control values at 18 days after exposure (FIG. 1, upper right). A parallel, transient increase in the urinary concentration of ALA was also observed 3 to 6 days after exposure with levels returning to control values 12 to 18 days after exposure (FIG. 1, lower right). Neither the mean urine volume for the final 24-hr period nor the mean hematocrit for each treatment group was markedly different from controls (data not shown).

Dissolution of GaAs *in vivo* into Ga and As components has been demonstrated after intratracheal and oral administration.[3] Therefore, in order to determine the mechanism of GaAs inhibition of ALAD *in vivo,* the following were determined: the relative sensitivity of blood ALAD to Ga^{3+} and AsO_2^- (arsenite) *in vitro* and whether these cations exerted their effects on ALAD by competing for Zn^{2+} (TABLE 1).

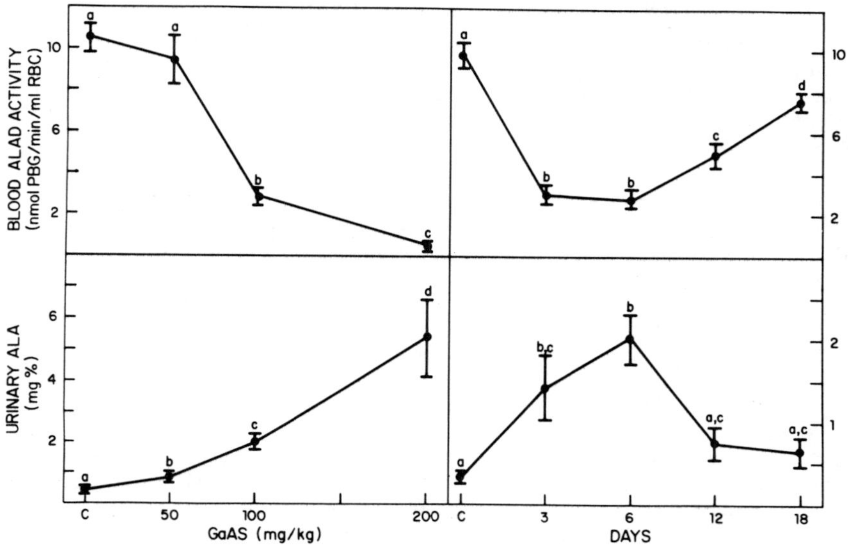

FIGURE 1. Blood ALAD activity and urinary ALA concentrations in rats 6 days after intratracheal instillation of 50, 100, and 200 mg GaAs/kg body wt (left panels) and at various times after administration of 100 mg GaAs/kg body wt (right panels). Controls received 0.9% filter-sterilized saline (1 ml/kg body wt, it). Gallium arsenide (suspensions of 1-μm particles), obtained from Alfa Products (Danvers, MA), was instilled into the lungs of anesthetized rats (100 mg ketamine/kg body wt and 3.2 mg xylazine/kg body wt, im) via an 18-gauge, 6.5-cm-long Teflon catheter. Blood ALAD was assayed by the method of Mitchell *et al.*,[4] and urine ALA by the method of Davis and Andelman[5] using a Bio-Rad kit. Each value represents a mean ± SEM (N = 4–9). Values with no letters in common are significantly different ($p < .05$).

TABLE 1. Effect of Zn^{2+} on Inhibition of Blood ALAD by Ga^{3+} and AsO_2^- *in Vitro*

	ALAD Activity (Mean Percentage of Control)
Control	100
$ZnCl_2$ (36 μM)	120
$Ga(NO_3)_3$ (75 μM)	40
Zn + Ga	72
$NaAsO_2$ (15 mM)	54
Zn + As	60

NOTE: Each value is a mean of three individual determinations and is expressed as a percentage of the control activity, which was 9 nmol porphobilinogen/min/ml RBC. The salts were added at the indicated concentrations of metal immediately before the incubation period.

Addition of ZnCl (36 μM) alone activated blood ALAD activity 20% higher than control. Addition of $Ga(NO_3)_3$ (75 μM) decreased ALAD activity to 40% of control. Addition of $NaAsO_2$ (15 mM) decreased ALAD activity to 54% of control. Inhibition of ALAD by Ga^{3+} and AsO_2^- was partially prevented in the presence of Zn^{2+}; enzyme activity was reduced to only 72% and 60% of controls, respectively. No discernable inhibition of ALAD occurred at 10^{-3} M concentrations of $H_2AsO_4^-$ (arsenate) (data not shown).

The results of this study demonstrate that acute exposure to GaAs via intratracheal instillation results in a dose- and time-dependent inhibition of ALAD in blood. Inhibition of this enzyme results in elevated urinary excretion of ALA. Results also showed that the concentration of Ga^{3+} required to achieve a half-maximal inhibition of blood ALAD *in vitro* is approximately 200-fold less than the concentration of AsO_2^- required to achieve a similar degree of inhibition. These data suggest that Ga^{3+} is the primary inhibitor of ALAD in blood after *in vivo* dissolution of an intratracheally instilled dose of GaAs. Inhibition of ALAD by Ga^{3+} and AsO_2^- *in vitro* was attenuated by Zn^{2+}, suggesting that competition for or displacement of Zn^{2+} from the enzyme active site by these ions may be the primary mechanism of inhibition of ALAD after exposure to GaAs. Finally, measurement of heme precursors such as ALA in urine along with the determination of red blood cell ALAD activity may be of value as early biological indicators of GaAs exposure and toxicity.

REFERENCES

1. ROBINSON, A. L. 1983. Science **219:** 275–277.
2. BOENIGER, M. & T. BRIGGS. 1979. *In* Health Applications of New Energy Technologies. W. N. Rom & V. E. Archer, Eds.: 593–606. Ann Arbor Science Publishers. Ann Arbor, MI.
3. WEBB, D. R., I. G. SIPES & D. E. CARTER. 1984. Toxicol. Appl. Pharmacol. **76:** 96–104.
4. MITCHELL, R. A., J. E. DRAKE, L. A. WITTLIN & T. A. REJENT. 1977. Clin. Chem. **23:** 105–111.
5. DAVIS, J. R. & S. L. ANDELMAN. 1967. Arch. Environ. Health **15:** 53–59.

Studies on the Role of the *Ah* Receptor in Hexachlorobenzene-Induced Porphyria

MARK E. HAHN AND THOMAS A. GASIEWICZ[a]

Division of Toxicology
Department of Biophysics
University of Rochester School of Medicine and Dentistry
Rochester, New York 14642

PATRICIA LINKO AND JOYCE A. GOLDSTEIN

National Institute of Environmental Health Sciences
Research Triangle Park, North Carolina 27709

The halogenated aromatic hydrocarbons hexachlorobenzene (HCB) and 2,3,7,8-tetrachlorodibenzo-*p*-dioxin (TCDD) produce similar hepatic porphyrias and induce the cytochrome P-450 isozymes P_1-450 and P_3-450 in mice.[1-3] Because many of the effects of TCDD—including porphyria—are thought to be mediated by its binding to the *Ah* receptor (a regulatory gene product of the *Ah* locus),[3] we are examining the possible involvement of this receptor in HCB-induced porphyria as well. Previous investigators, in studying the role of the *Ah* receptor in TCDD- or HCB-induced porphyria, have taken advantage of differences in the properties of this receptor among inbred strains of mice. Both TCDD and HCB produce porphyria in C57BL/6 mice, in which a high-affinity hepatic *Ah* receptor may be found, but not in DBA/2 mice, in which the *Ah* receptor is undetectable.[1,4] Studies in other strains, however, have shown an incomplete correlation between the *Ah* phenotype and susceptibility to porphyria.[1,5] Interpretation of these experiments is complicated by the many other genetic differences that exist between these strains of mice.[6] In order to eliminate these other differences, we studied the porphyrogenic effect of HCB in two congenic strains of C57BL/6 mice that differ only at the *Ah* locus (and a limited number of loci closely linked to it).[6]

Female C57BL/6J mice of both the B6-Ah^b and B6-Ah^d strains were pretreated with iron (500 mg/kg body wt) and then fed a diet containing HCB (200 ppm).[1] Mice from the two strains consumed similar amounts of HCB throughout the 15-week-long study. Urinary excretion of porphyrins increased after 7 weeks of HCB treatment in the B6-Ah^b mice, and after 15 weeks was over 200 times greater than that of mice given only iron. In the B6-Ah^d mice, porphyrin excretion did not increase until after 13 weeks; after 15 weeks the levels were only 6 times greater than those of the iron-only controls (TABLE 1). Similarly, hepatic porphyrin concentrations were increased in both strains, but to a much greater extent in the B6-Ah^b mice. Measurement of *Ah* receptor levels confirmed its presence in B6-Ah^b mice and the absence of detectable levels of this protein in B6-Ah^d mice (TABLE 1).

It has been suggested that the induction of one or more isozymes of cytochrome P-450 may play a role in the pathogenesis of HCB-induced porphyria.[7] We therefore measured the amounts of P_1-450 and P_3-450 in these mice, using immunoblotting with antisera raised against the homologous rat isozymes.[2,8] HCB induced relatively large

[a]To whom correspondence should be addressed.

TABLE 1. *Ah* Receptor Levels and Porphyrin Accumulation in Congenic C57BL/6 Mice Treated with HCB

Strain	Number of Mice	Treatment	*Ah* Receptor (fmol/mg protein)	Urine Porphyrins ($nmol \cdot day^{-1} \cdot mouse^{-1}$)	Hepatic Porphyrins (nmol/g liver)
B6-Ah^b	5	Iron	32.6 ± 15.1	1.29[a]	0.20[a]
	4	Iron and HCB	35.6 ± 15.2	265.[a]	1110 ± 393
B6-Ah^d	4	Iron	1.4 ± 3.4[b]	0.97[a]	0.24[a]
	3	Iron and HCB	2.7 ± 1.3[b]	5.76[a]	17.6 ± 14.5

NOTE: Mice were treated with iron or with iron and HCB as described in reference 1. After 15 weeks on either the control or HCB-containing diet, mice were killed and liver samples taken for determination of cytosolic *Ah* receptor levels (by the hydroxylapatite adsorption method[9]) and porphyrin concentrations. Urinary and hepatic porphyrins are the sum of uroporphyrin; hepta-, hexa-, and pentacarboxyporphyrins; and coproporphyrin. These porphyrins were determined by reverse-phase HPLC (Hahn and Gasiewicz, in preparation). Values are means ± standard deviations.

[a]Samples were pooled.

[b]Not significantly different from zero ($p > .05$).

amounts of a protein recognized by anti-P-450d (P_3-450) in both strains, although induction was greater in the B6-Ah^b mice. HCB also induced a small amount of a protein recognized by anti-P-450c (P_1-450) in B6-Ah^b mice, but this protein was not increased in B6-Ah^d mice.

These data suggest that the *Ah* locus is involved in HCB-induced porphyria in mice. Our data are also consistent with the possibility that the *Ah* locus influences susceptibility to porphyria through its control of the induction of P_3-450 or P_1-450 or both. Additional work is needed, however, to determine whether these are causally related.

REFERENCES

1. SMITH, A. G. & J. E. FRANCIS. 1983. Biochem. J. **214:** 909–913.
2. LINKO, P., H. N. YEOWELL, T. A. GASIEWICZ & J. A. GOLDSTEIN. 1986. J. Biochem. Toxicol. **1**(2): 95–107.
3. POLAND, A. & J. C. KNUTSON. 1982. Annu. Rev. Pharmacol. Toxicol. **22:** 517–554.
4. JONES, K. G. & G. D. SWEENEY. 1980. Toxicol. Appl. Pharmacol. **53:** 42–49.
5. GREIG, J. B., J. E. FRANCIS, S. J. E. KAY, D. P. LOVELL & A. G. SMITH. 1984. Toxicol. Appl. Pharmacol. **74:** 17–25.
6. BIRNBAUM, L. S. 1986. Drug Metab. Dispos. **14:** 34–40.
7. SMITH, A. G., J. E. FRANCIS, S. J. E. KAY, J. B. GREIG & F. P. STEWART. 1986. Biochem. J. **238:** 871–878.
8. LINKO, P., T. A. GASIEWICZ, H. N. YEOWELL & J. A. GOLDSTEIN. 1986. Toxicologist **6:** 275.
9. GASIEWICZ, T. A. & R. A. NEAL. 1982. Anal. Biochem. **124:** 1–11.

Porphyria Cutanea Tarda

Manifestation and Therapy

VÁCLAV KORDAČ,[a] PAVEL MARTÁSEK,
VLASTIMIL ZOUBEK, AND MILAN KALÁB

*Charles University School of Medicine
Prague, Czechoslovakia*

The frequency of porphyria cutanea tarda (PCT) has been steadily increasing. In Czechoslovakia, an extensive epidemiologic study covering consecutive urine examinations of 35 000 randomly selected persons (one-third of the total population of two districts) was performed; this study revealed PCT in the age group of 15–70 years in about 1 patient per 1000 inhabitants. The same preliminary results were obtained in a similar study performed in Mongolia.

In the last 35 years, 1002 PCT patients were diagnosed in our department. Among these, 929 were patients with sporadic PCT (797 men, 132 women), and 73 were patients with familial PCT (38 men, 35 women). Liver biopsies were performed in 279 patients. Histological findings were not diagnostically valuable. Quantitative determinations of porphyrins in liver tissues[1] were performed in 97 patients (11 women) whose average age at the time of diagnosis was 46 years. The last group mentioned (97 patients) was studied in detail. Before the initiation of therapy, levels of total porphyrins in 72 patients reached 274 μg/g wet liver tissue (42–1029 μg). In 10 patients before the cessation of therapy (chloroquine in 125-mg doses given twice a week),[2] they decreased to 26 μg (17–38 μg); in 14 patients after the cessation of the therapy, they fell to 7.4 μg (1.1–16.0 μg). Levels for the control group (hepatitis and hyperbilirubinemia) reached 10.2 μg (0.09–29 μg). Alcohol abuse was proved in 58 patients, and anamnestic findings of lues were made in 2 patients. Silicosis was observed in 3 patients; tuberculosis, in 4 patients; diabetes mellitus, in 13 patients; liver

TABLE 1. Urinary Porphyrin Excretion in PCT Patients before and after Chloroquine Therapy

	Before Therapy	After Therapy
Uroporphyrin (μg/24 hr)[a]	3356.5 (590–16 743)	55.4 (6–99)
Coproporphyrin (μg/24 hr)[b]	487.4 (172–1112)	97.6 (21–160)

NOTE: Each value is a mean. Ranges are shown in parentheses.
[a]Normal level: <25 μg/24 hr.
[b]Normal level: <160 μg/24 hr.

[a]Address for correspondence: First Department of Internal Medicine, Charles University School of Medicine, U nemocnice 2, 128 08 Prague 2, Czechoslovakia.

illness, in 21 patients before PCT manifestation; hormonal provocation, in 8 patients (4 men, 4 women); the familial form of the disease, in 6 patients. The most frequent occupation (10 patients) was driver and garageman. Chloroquine proved to be therapeutically effective in both forms of PCT, considerably reducing the deposit of liver porphyrins and providing the prevention of further complications.

The results obtained from two groups of patients were compared. The first group comprised PCT patients observed between 1954 and 1969; the second, PCT patients observed between 1969 and 1983. Only patients in the second group received chloroquine: 367 patients (306 men, 61 women) received 250-mg doses perorally two times a week (TABLE 1). In 96% of the patients in this group, the remission of the disease was achieved, on the average, after 14.2 months of the therapy (FIG. 1). In 11 patients the treatment had to be prolonged to 5 years in order to achieve the remission, and 4 patients were resistant to chloroquine. Primary liver carcinoma in cirrhosis developed only in three patients. Of the patients in the first group, which comprised

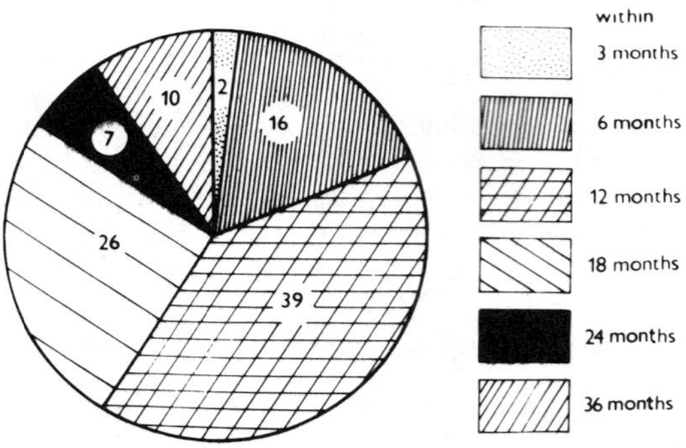

FIGURE 1. Withdrawal of the clinical signs and normalization of laboratory findings in PCT patients during chloroquine therapy.

342 patients (301 men, 41 women), a total of 17 patients died of primary liver carcinoma in cirrhosis.[3]

We conclude that treatment with low doses of chloroquine represents an effective and safe method of PCT therapy that can prevent the development of cirrhosis and hinder the development of primary liver carcinoma in cirrhosis in PCT patients.

REFERENCES

1. MARTÁSEK, P., M. JIRSA & V. KORDAČ. 1982. The effect of chloroquine treatment on the level of the porphyrins in the liver tissue of patients having symptomatic hepatic porphyria (in Czech). Cesk. Gastroenterol. Vyz. **36:** 198–205.
2. KORDAČ, V. & M. SEMRÁDOVÁ. 1974. Treatment of porphyria cutanea tarda with chloroquine. Br. J. Dermatol. **90:** 95–100.
3. KORDAČ, V. 1972. Frequency of occurrence of hepatocellular carcinoma in patients with porphyria cutanea tarda in long-term follow-up. Neoplasma **19:** 135–139.

Comparison of Uroporphyrinogen Decarboxylase Activities and Uroporphyrin Accumulation in Mice, Japanese Quail, and Cultured Chick Embryo Hepatocytes Treated with Polyhalogenated Aromatic Compounds[a]

R. LAMBRECHT,[b] P. SINCLAIR,[b,c] W. J. BEMENT,[b]
J. SINCLAIR,[b,c] AND H. CARPENTER[d]

[b]*Veterans Administration Medical Center*
White River Junction, Vermont 05001

[c]*Dartmouth Medical School*
Hanover, New Hampshire 03755

[d]*Oregon State University*
Corvallis, Oregon 97331

The purpose of these studies was to compare the activities of uroporphyrinogen decarboxylase (UROD) in several experimental models that have been used to investigate the mechanism by which certain polyhalogenated aromatic (PHA) compounds cause hepatic porphyria. Under these conditions, mice developed porphyria in weeks, quail in days, and chick embryo hepatocytes in hours after exposure to PHA compounds as indicated by elevated levels of uroporphyrin and heptacarboxylporphyrin. UROD activity was assayed in 10 000 × g supernatants using pentacarboxylporphyrinogen III as substrate (5 μM substrate, 0.3 mg protein/ml, 30 min, 37°C).

Cultures of chick embryo hepatocytes have been previously reported to have decreased UROD activity when treated with either 3,4,3′,4′-tetrachlorobiphenyl (TCB)[1] or 2,3,7,8-tetrachlorodibenzo-*p*-dioxin (TCDD).[2] Under the conditions of our study, however, no differences in UROD activity were detected for cells treated with 0, 10, 100, or 1000 ng TCB/ml medium, although marked increases in uroporphyrin accumulation with increasing TCB concentration were observed. Cells treated with TCDD (100 or 300 ng/ml) also accumulated uroporphyrin without a detectable decrease in UROD activity from control values.

UROD activity and porphyrin accumulation in liver were analyzed in Japanese quail treated with hexachlorobenzene (HCB). In two separate experiments, uroporphyrin and heptacarboxylporphyrin accumulated in the HCB-treated quail without decreased UROD activity relative to controls.

Male mice (C57BL/6) were treated with iron and HCB, and killed 70 days after initial exposure to HCB. Highly porphyric mice were found to have decreased UROD

[a]This work was supported by the Veterans Administration and by Grant CA25012 from the National Institutes of Health.

activities, ranging from 13% to 63% of the control values, depending on the degree of uroporphyrin accumulation.

In these experimental models, PHA compounds cause uroporphyrin and heptacarboxylporphyrin to accumulate to similar hepatic concentrations; however, the results suggest that this accumulation (in chick embryo hepatocytes and Japanese quail) is not caused by a decrease in UROD activity.

REFERENCES

1. Swain, M. J., S. B. Follows & G. S. Marks. 1983. Can. J. Physiol. Pharmacol. **61:** 105–108.
2. de Verneuil, H., S. Sassa & A. Kappas. 1983. Biochem. J. **214:** 145–151.

Mathematical Models for Effects of Xenobiotics on Heme Biosynthesis

Application to Lead

ALLAN H. MARCUS[a]

Washington State University
Pullman, Washington 99163

The disturbance of heme biosynthesis by a variety of toxic substances can be detected at levels of exposure well below those levels at which overt toxic symptoms occur. The heme biosynthetic pathway is influenced by the trace metals, hydrogen sulfide, methylmercaptan, polyhalogenated aromatic hydrocarbons such as hexachlorobenzene, and some pesticides. The accumulation of the heme precursor protoporphyrin IX in the red blood cells, that is, the erythrocytic protoporphyrin IX (EP) level, is now

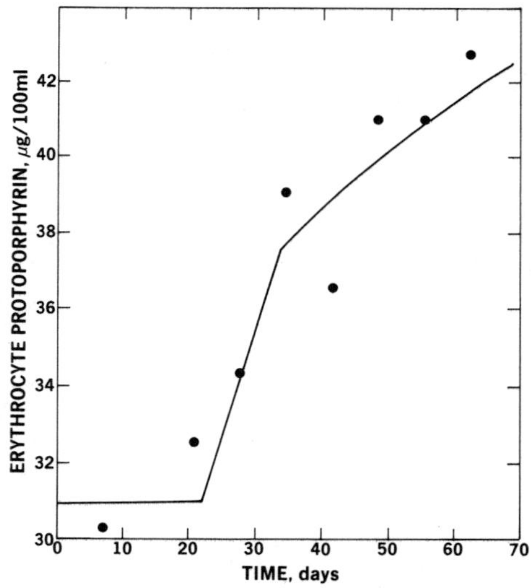

FIGURE 1. Observed EP levels for a healthy young male volunteer ingesting lead from days 14 to 63. Adapted from Cools *et al.*[1] and Marcus and Schwartz.[2]

recommended as a screening indicator of childhood lead poisoning. After we developed a compartmental model for the absorption and distribution of lead in humans, we coupled this model to a heme biosynthesis compartmental model by assuming that lead inhibits certain enzymes controlling the process rates, particularly δ-aminolevulinic acid dehydratase (ALAD) and ferrochelatase in the bone marrow. First-order kinetics were used except where important nonlinear effects occurred: 1) Limited lead-binding

[a]Present address: Battelle Columbus Division, P.O. Box 13758, Research Triangle Park, North Carolina 27709.

capacity of a shallow or fast-moving erythrocytic pool; 2) Michaelis–Menten (MM) kinetics for enzyme function and for decreased MM velocity from lead exposure. The validity of the kinetic model was tested by fitting curves for individual subjects in a number of data sets in which blood lead concentration (PbB) and EP and erythrocytic ALAD levels changed in response to experimental lead exposures. The general adequacy of the toxicokinetic model is shown in FIGURE 1, where we exhibit the fitted EP-versus-time curve for a lead-exposed subject.[1] We also estimated kinetic parameters for ALAD activity and for plasmatic, erythrocytic, and urinary concentrations of lead and zinc. These estimates were based on results taken from another experiment, in which we used a 1-hr infusion of EDTA.

Theoretical dose–effect curves—plots of EP versus PbB—were fitted to data from several population studies. One such study was the Second U.S. National Health and Nutrition Examination Survey, which included data on children. A model parameter

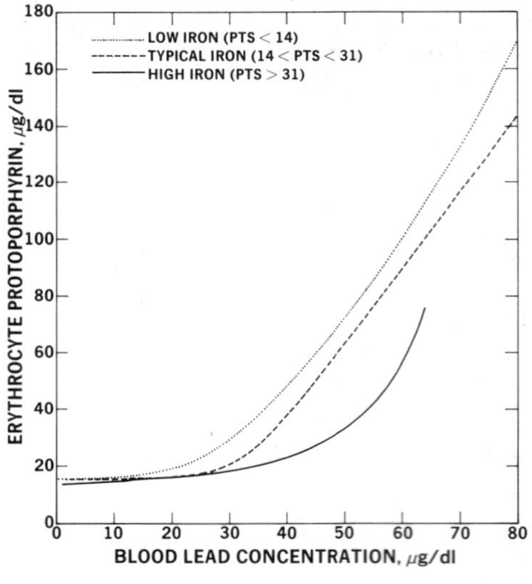

FIGURE 2. Relationship between the EP level and PbB as a function of PTS for U.S. children aged 2–10 years. Dotted line: PTS < 14%. Dashed line: 14% ≤ PTS ≤ 31%. Solid line: PTS > 31%. Adjusted for red cell count and surface area. Analyses by J. Schwartz of the U.S. Environmental Protection Agency; data from the Second U.S. National Health and Nutrition Examination Survey.

that was proportional to the maximum erythrocytic lead-holding capacity (adjusted for red cell surface area) increased systematically with increasing iron status for children, as measured by percent transferrin saturation (PTS). FIGURE 2 shows plots of EP versus PbB fitted by J. Schwartz of the U.S. Environmental Protection Agency to data from children with poor iron status (PTS < 14%), typical iron status (14% ≤ PTS ≤ 31%), and high iron status (PTS > 31%). Not only do children with poor iron status have the highest EP response for a given PbB, but the increased response to PbB also appears to occur at much lower levels (PbB < 20 μg/dl) than for children with better iron status.

Polyhalogenated aromatic hydrocarbons affect heme biosynthesis by inhibiting uroporphyrinogen decarboxylase. A simulation model for hexachlorobenzene combines a physiologically based pharmacokinetic model with a compartmental model for the heme pathway, through inhibition of uroporphyrinogen decarboxylase by hexachlorobenzene in the liver.

REFERENCES

1. Cools, A., H. J. A. Salla, M. Verberk & R. L. Zielhuis. 1976. Biochemical response of male volunteers ingesting lead for 49 days. Int. Arch. Occup. Environ. Health **38:** 129–139.
2. Marcus, A. H. & J. Schwartz. 1987. Dose-response curves for erythrocyte protoporphyrin vs. blood lead: Effects of iron status. Environ. Res. **44:** in press.

Induction of Acute Attack in Acute Hepatic Porphyrias

Experiences with Diagnosis and Therapy over Thirty Years

PAVEL MARTÁSEK,[a] VÁCLAV KORDAČ,
MICHAELA KOZÁKOVÁ, MILAN JIRSA, PETR KOTAL,
AND JIŘÍ HORÁK

Charles University School of Medicine
Prague, Czechoslovakia

The long-term follow-up study performed at our department involved 115 patients (40 men, 75 women) with acute intermittent porphyria (AIP), 51 patients (12 men, 39 women) with variegate porphyria (VP), and 30 patients (13 men, 17 women) with hereditary coproporphyria (HC). We tried to analyze the factors that induced 89 acute attacks in 61 patients. The first attack occurred within 27.9 years, on the average. The

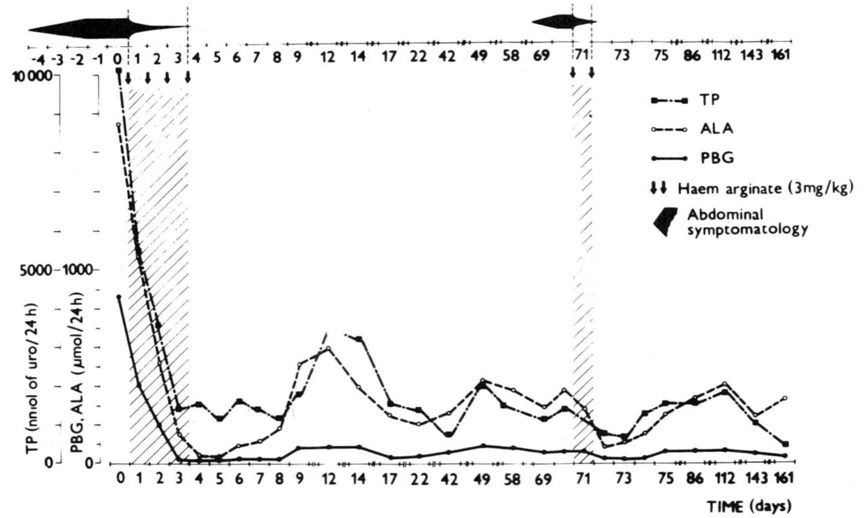

FIGURE 1. Changes in the urinary excretion of total porphyrin (TP), δ-aminolevulinic acid (ALA), and porphobilinogen (PBG) during and after heme arginate treatment in a 55-year-old patient that had acute attacks of AIP and prevailing abdominal symptomatology.

[a]Address for correspondence: First Department of Internal Medicine, Charles University School of Medicine, U nemocnice 2, 128 08 Prague 2, Czechoslovakia.

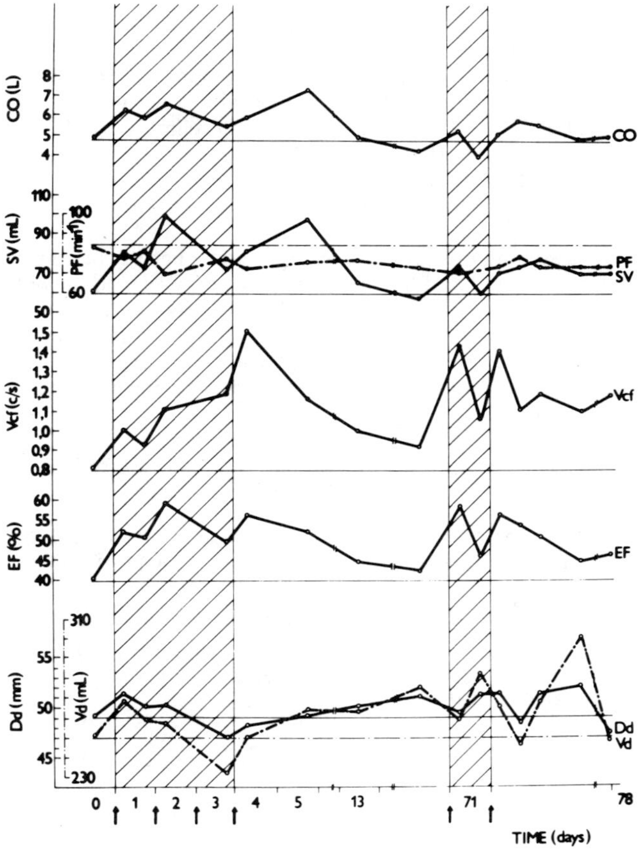

FIGURE 2. The effect of heme arginate administration on left ventricle function in the same patient described in FIGURE 1. Two-dimensional and Doppler echocardiography were used. Dd: end-diastolic diameter (normal below 60 mm); Vd: end-diastolic volume (below 200 ml); EF: ejection fraction (above 60%); Vcf: velocity of shortening of the fiber of the left ventricle (above 1.3 c/s); SV: stroke volume (between 40 and 70 ml/m^2); PF: pulse frequency; CO: cardiac output (between 2.8 and 4 l/m^2).

causes of the attacks, in order of decreasing frequency, were as follows: the use of barbiturates, total anesthesia, stressful family situations, gravidity (manifestations seen in the first trimester), menses, viral infections, tonsilitis, meprobamate, nitrofurantoine, pyrazolidine derivatives, oxytetracycline, sulfathiazole, and tilidine hydrochloride. Up to 1983, we used glucose overload or chlorpromazine and β-blockers in high doses or both for therapy of the acute attack. Generally, porphyrias can be understood as diseases resulting from autointoxication. Therefore, we decided to try hemoperfusion using coated-resin Amberlite XAD-2 for the treatment of three patients with acute attacks of AIP.[1] Excellent clinical effects were achieved: prompt retreat of abdominal spasms in all three patients, rapid regression of paresis in one patient, and marked improvement of psychiatric symptoms in another. Hematin has been used for therapy of acute attacks for more than 15 years. The goal of therapy with heme derivatives such as hematin is to replenish cellular heme and to influence the enzyme limiting heme synthesis by a feedback mechanism.[2] In 1985, an excellent effect and tolerance of heme arginate were reported.[3] When we used heme arginate (Normosang, Medica, Finland), we observed prompt clinical and laboratory effects (FIGURE 1) for four patients that had acute attacks and for seven patients in the subacute phase of the disease. Heme arginate at present seems to be the most reliable preparation in the treatment of acute porphyrias, provided the timely initiation of its application.

Additionally, assuming myocardial damage in acute hepatic porphyria (especially in the acute attack) was induced by a metabolic defect similar to the one observed in the liver, we made detailed noninvasive cardiologic observations of patients receiving heme arginate. In patients with acute attacks, a remarkable improvement of the left ventricle function was observed after the application of heme arginate (FIGURE 2). This clinical observation may give evidence for the replenishment of decreased cellular heme in myocardial tissue.

REFERENCES

1. MARTÁSEK, P., V. KORDAČ, P. KOTAL, J. VACEK, M. JIRSA, J. HORÁK & R. TOMÁŠEK. 1986. Recovery from a severe attack of acute intermittent porphyria during coated-resin hemoperfusion. Int. J. Artif. Organs **9:** 117–118.
2. GOETSCH, C. A. & D. M. BISSELL. 1986. Instability of hematin used in the treatment of acute hepatic porphyria. N. Engl. J. Med. **315:** 235–238.
3. TENHUNEN, R., O. TOKOLA, I. B. LINDÉN & G. GOTHONI. 1985. Heme arginate: A new compound for treatment of porphyrias. Paper presented at the conference Porphyrins and Porphyrias June 19–22. Paris.

Overcoming the Limitations of Hematofluorometry for Assaying Zinc Protoporphyrin

REBECCA L. RETTMER,[a] ELAINE W. GUNTER,[b] AND ROBERT F. LABBÉ[a]

[a]*Department of Laboratory Medicine*
University of Washington
Seattle, Washington 98195

[b]*Nutritional Biochemistry Branch*
Centers for Disease Control
Atlanta, Georgia 30333

Zinc protoporphyrin (ZPP) in whole blood can be directly measured using hematofluorometry. ZPP has been recognized as an acceptable test for defining chronic lead exposure and iron deficiency. This test, as applied during the past decade (largely in lead surveillance programs), proved quite adequate. As the acceptable levels of blood lead have been gradually decreased, however, the limitations of hematofluorometry have become apparent. Accurate, reproducible results for ZPP measurement by hematofluorometry usually require careful and time-consuming oxygenation of the hemoglobin. Deoxygenation causes a spectral shift, resulting in low readings (false negatives), a problem particular to blood that is not fresh. This limitation can be overcome by use of a recently introduced reagent (ProtoFluor, Helena Labs, Beaumont, TX) that is mixed with the specimen before making a reading. The ProtoFluor reagent reacts virtually instantaneously, derivatizing the hemoglobin and eliminating the need for blood aeration. ProtoFluor reagent does not appreciably alter the hemoglobin absorption spectrum of fully oxygenated blood (peak absorbance Hgb = 415 nm; peak absorbance Hgb with ProtoFluor = 416 nm) nor does the reagent alter the spectral properties of ZPP (peak absorbance with and without ProtoFluor = 595 nm). The reagent resolves the limited oxygenation of the commonly practiced stirring method and compares favorably with the complete oxygenation (maximum reading) obtainable with extended aeration. Simple stirring for reoxygenation gave in one series ($N = 98$) a mean of 38.8 μg ZPP/dl blood, whereas ProtoFluor reagent increased the mean to 53.1 μg ZPP/dl blood ($r = .84$). By contrast, meticulous aeration gave virtually identical results when compared with the reagent use ($N = 46$; aeration $\bar{x} =$ 80.8 μmol ZPP/mol heme and ProtoFluor $\bar{x} = 81.0$ μmol ZPP/mol heme; $r = .97$). The reagent improves precision within runs compared to the stirring method (CV = 4% versus 6%). Stable standards are necessary for evaluating daily precision and linearity of hematofluorometers in the field. The precision of the results obtained with ProtoFluor standards, which are stable for 1 year, varied little within any given day as well as from day to day (CV ≤ 5%).

The calibration of existing instruments may be inadequate in some instances. The hematocrit values that are used in the internal calculation of μg ZPP/dl blood are assumed. Errors attributable to faulty calibration, however, may be eliminated by use of a metabolite ratio, that is, the μmol ZPP:mol heme (ZPP:H) ratio. The ZPP:H ratio represents a functional measure of the availability of iron for heme synthesis and reflects relative iron-deficient erythropoiesis. Expressing results as the metabolite ratio offers several advantages. The ratio, in light of the conversion to Système International (SI) units, is likely to become the expression of choice. Furthermore, clinical evaluations (>1000 subjects) revealed that levels of the ZPP:H ratio were unaffected by sex, age (1–80 years), or hematocrit. Multiple regression analysis of the ZPP:H ratio, the serum ferritin concentration, and the saturation of transferrin reveals that the ZPP:H ratio is the best overall predictor of iron deficiency anemia ($N = 206$). These studies suggest that the ZPP:H ratio is useful as an initial screen for iron status as well as a means of monitoring therapy.

Introduction of these improvements into standard hematofluorometry yields accurate and precise results. With these modifications, hematofluorometry effectively identifies marginal or preanemic iron deficiency. These innovations contribute to the elimination of false-negative results and make high-quality results achievable. Thus assaying ZPP:H by hematofluorometry is a highly reliable diagnostic technique that may be applied to blood banking, sports medicine, general pediatrics, obstetrics, and lead surveillance programs.

Evidence for 3,4,3′,4′-Tetrachlorobiphenyl-Mediated Oxidation of Uroporphyrinogen by Chick Embryo Liver Supernatants

PETER R. SINCLAIR[a–c] AND RICHARD W. LAMBRECHT[c]

[b]*Veterans Administration Medical Center*
White River Junction, Vermont 05001

[c]*Department of Biochemistry*
Dartmouth Medical School
Hanover, New Hampshire 03756

Chick embryo hepatocytes in culture have been used in our laboratory for investigations into the mechanism by which polyhalogenated aromatic hydrocarbons cause hepatic uroporphyrin (URO) accumulation.[1–3] These investigations are reviewed in this volume.[4]

From various studies, we have concluded that the mechanism requires some action of a form of cytochrome P-450 induced by 20-methylcholanthrene (MC), which causes oxidation of uroporphyrinogen I (UROgen I) and heptacarboxylporphyrinogen to the corresponding porphyrins. It is difficult, however, to study such oxidation in intact cells. Here we report that postmitochondrial supernatants or microsomes from liver of chicken embryos treated with MC will catalyze the oxidation of UROgen I in the presence of NADPH and 3,4,3′,4′-tetrachlorobiphenyl (TCB).

Chick embryos after a 17–18-day gestation were injected with 0.2 ml DMSO alone or with 0.5 mg MC or 6 mg glutethimide in DMSO. After 48 hr, liver homogenates (20%, w/v) were prepared in 0.25 M sucrose–0.05 M HEPES–1 mM EDTA buffer (pH 7.6) and centrifuged at 12 000 g for 20 min. The supernatants were used for reactions or were centrifuged for 1 hr at 100 000 g to prepare microsomes that were washed with 0.15 M KCl. UROgen I in 5 mM NaOH and in the presence of 60 mM dithiothreitol was reduced with fresh Na–Hg amalgam and transferred to the sucrose–HEPES–EDTA buffer containing dithiothreitol (final concentration: 35 mM). This solution remained stable for several hours with essentially no autooxidation to porphyrins. The oxidation reaction was followed in a Hitachi 512 spectrofluorometer. The cuvette contained 1.5 ml sucrose–HEPES–EDTA buffer, 0.1 ml liver supernatant or microsomal suspension, an NADPH-generating system (isocitrate, isocitrate dehydrogenase, NADPH, $MgCl_2$, and nicotinamide), and freshly reduced UROgen I (approximately 1 μM). Under these conditions the rate of porphyrin formation was slow. Addition of 10 μl DMSO caused a small increase in the rate of porphyrin formation whereas addition of 20 μg TCB in DMSO resulted in linear oxidation rates 5–10-fold greater than with DMSO alone. Nonplanar 2,4,2′,4′-tetrachlorobiphenyl was ineffective in stimulating porphyrin formation and did not cause URO accumulation in the intact hepatocytes. Deletion of the NADPH-

[a]Address for correspondence: Veterans Administration Medical Center (151), White River Junction, Vermont 05001.

generating system greatly decreased the TCB-stimulated rate. The oxidation was inhibited by CO (1/3rd saturation) and in the presence of piperonyl butoxide. The latter compound, an inhibitor of cytochrome P-450, prevents and reverses the accumulation of URO in the chick hepatocyte culture[2,3] and decreases uroporphyria in hexachlorobenzene-treated mice.[5]

The involvement of a particular MC-inducible form of cytochrome P-450 was shown by the following findings: 1) Liver extracts of untreated or glutethimide-treated chick embryo liver did not catalyze the oxidation. 2) Antisera against the 60K MC-induced cytochrome P-450 strongly inhibited the reaction, whereas antisera against the 50K glutethimide-cytochrome P-450 only weakly inhibited the oxidation catalyzed by liver extracts from MC-treated embryos. These forms of chick liver cytochrome P-450s were recently isolated in our laboratory.[6] Superoxide dismutase, catalase, and desferrioxamine were added to the reaction to test for activated forms of oxygen that might be produced by cytochrome P-450 or by iron-catalyzed reactions. None of these additions were inhibitory, suggesting that if oxidation was mediated by active oxygen species these must be sequestered in membranes inaccessible to the enzymes. The reaction remained linear for at least 30 min, a result inconsistent with the oxidation being due to URO-catalyzed autoxidation.

The results presented here with a cell-free system suggest that a form of cytochrome P-450 can mediate oxidation of UROgen I in the presence of polyhalogenated aromatic hydrocarbons known to cause the uroporphyria in cultured hepatocytes. Such oxidation may be the primary action in development of the porphyria rather than the primary action being inhibition of UROgen decarboxylase. The finding that this reaction occurs in cell-free systems will allow further investigations to determine the mechanism of the reaction.

REFERENCES

1. SINCLAIR, P. & S. GRANICK. 1974. Biochem. Biophys. Res. Commun. **61:** 124–133.
2. SINCLAIR, P., W. BEMENT, H. BONKOVSKY & J. SINCLAIR. 1984. Biochem. J. **222:** 737–748.
3. SINCLAIR, P., W. BEMENT, H. BONKOVSKY, R. LAMBRECHT, J. FREZZA, J. SINCLAIR, A. URQUHART & G. ELDER. 1986. Biochem. J. **237:** 63–71.
4. BONKOVSKY, H., P. SINCLAIR, W. BEMENT, R. LAMBRECHT & J. SINCLAIR. 1987. Ann. N. Y. Acad. Sci. This volume.
5. SMITH, A., J. FRANCIS, S. KAY, J. GREIG & F. STEWART. 1986. Biochem. J. **238:** 871–878.
6. SINCLAIR, P., J. FREZZA, J. SINCLAIR, J. BEMENT, J. HEALEY & H. BONKOVSKY. 1986. Fed. Proc. **45:** 1663.

An Association between Zinc Protoporphyrin Elevation and Recovery from Acute Hematopoietic Damage[a]

THOMAS L. WALDEN, JR.,[b] HUDA M. AL-ANSARI, AND WALTER R. FARKAS

Department of Environmental Practice
University of Tennessee College of Veterinary Medicine
Knoxville, Tennessee 37901

We recently reported that whole-body irradiation of mice induces an elevation of zinc protoporphyrin (ZPP) in the red blood cells. The elevation was associated with hematopoietic recovery, but was not the result of an iron deficiency or of ferrochelatase inhibition.[1] These observations have been extended to show that cytotoxic drugs also elevate ZPP in association with hematopoietic recovery (FIGURE 1). ZPP blood levels were determined using a dedicated hematofluorometer (Aviv Associates, Lakewood, NJ). The validity of this technique for measuring the ZPP level in mouse blood and the ZPP elevation that occurs after whole-body irradiation have been previously confirmed by HPLC.[1] The ZPP levels of mice receiving hydroxyurea (1 mg/g body wt, daily for

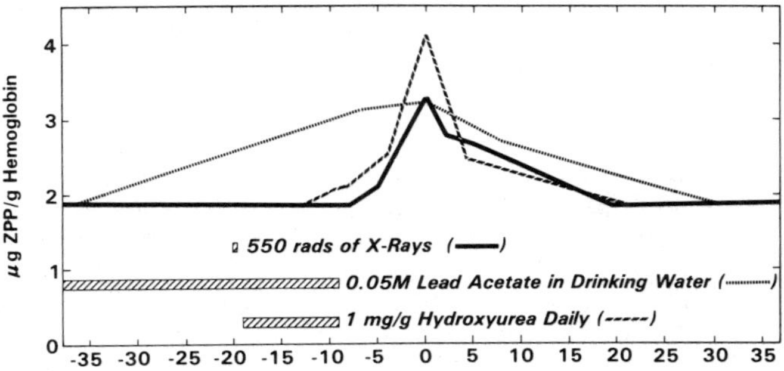

FIGURE 1. Comparison of ZPP elevations in mice observed after lead poisoning, X-irradiation, or treatment with hydroxyurea. The peak ZPP concentration (μg ZPP/g hemoglobin) for each treatment has been adjusted to day 0 in figure. Dotted line: ZPP for mice exposed to lead-contaminated drinking water for 35 days. After the 35-day period, the mice were given normal water. Solid line: ZPP for mice that received 550 rads of whole-body X-irradiation on day −20. Dashed line: ZPP for mice that received 1 mg hydroxyurea/g body wt daily for 10 days (days −19 to −9).

[a]This work was supported by Grants HD14062 and ES01323 from the National Institutes of Health.

[b]Present address: Department of Radiation Biochemistry, Armed Forces Radiobiology Research Institute, Bethesda, Maryland 20814-5145.

10–13 days) increased from control levels (about 2.0 μg ZPP/g hemoglobin) beginning around the 15th day after the first injection, and reached a peak of 3.3 μg ZPP/g hemoglobin on day 19 (represented as day 0 in FIGURE 1). Over the next 4 days, the ZPP levels decreased at a rate similar to the previous rate of increase, and returned close to base levels. The rapid decrease in ZPP resembles the pattern observed after irradiation (reference 1 and FIGURE 1) but differs from the ZPP elevation seen in lead-poisoned mice (FIGURE 1)—an elevation that persisted for over 1 month after removal of the lead source. Increases in ZPP levels were also produced by combining a subthreshold dose of radiation (400 rads of X-irradiation) with subthreshold doses of either cyclophosphamide (0.25 mg/g body wt) or cytarabine (1 mg/g body wt). ZPP elevation was not the result of increased erthyropoiesis per se because 1) it was not elevated by hypoxia, 2) it was not elevated during recovery from megaloblastic anemia, and 3) it could be prevented by a postirradiation bone marrow transplant. The ZPP

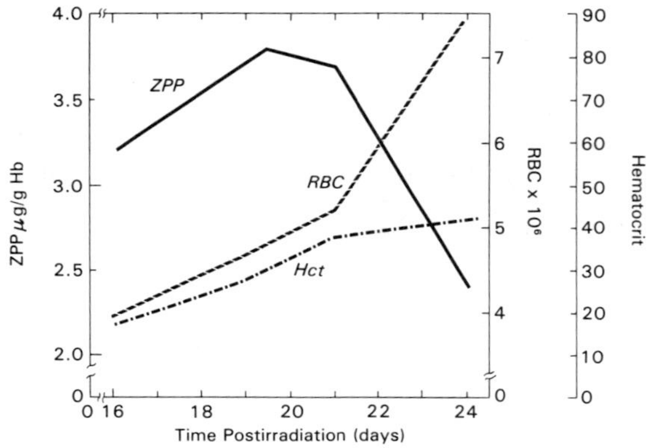

FIGURE 2. ZPP elevation following X-irradiation is associated with an early phase of recovery from acute hematopoietic damage. Mice received 550 rads of whole-body X-irradiation, and their blood levels were measured for hematocrit, ZPP, and red blood cell counts on postirradiation days 16, 19, 21, and 24. Animals for each data point were only bled once.

elevation that follows whole-body irradiation is associated with an early phase of the hematopoietic recovery period (FIGURE 2) because red blood cell numbers are still increasing between postirradiation days 19 and 21, although ZPP levels are decreasing. The rapid decrease of ZPP, which occurs after peak concentrations of ZPP are reached, may be due to a decreased production of ZPP in the new red blood cells during this period, coupled with these cells diluting out the earlier cells, which contained elevations of ZPP. It is of interest that we have also observed an elevation of ZPP in mice with erythroleukemias induced by Friend leukemia virus.

REFERENCE

1. WALDEN, T. L., JR., P. S. DRAGANAC & W. R. FARKAS. 1984. The elevation of blood levels of zinc protoporphyrin in mice following whole-body irradiation. Blood **63:** 1159–1167.

Index of Contributors

Al-Ansari, H. M., 349–350
Astrin, K. H., 23–29

Bement, W. J., 96–112, 337–338
Bernard, A., 41–47
Bickers, D. R., 261–267
Bishop, D. F., 23–29
Bonkovsky, H. L., 96–112
Bryan, G., 183–190

Cantoni, L., 128–140
Carpenter, H., 337–338
Carugo, C., 128–140
Christenson, W. R., 48–54
Correia, M. A., 248–255
Cripps, D., 183–190

Dailey, H. A., 81–86
Davidow, B., 23–29
De Matteis, F., 30–40
Desnick, R. J., 23–29
Doss, M. O., 204–218
Drummond, G. S., 87–95

Elder, G. H., 141–147
Ertürk, E., 183–190

Farkas, W. R., 349–350
Fowler, B. A., 55–64, 172–182, 235–247, 268–277, 330–332
Fujita, H., 15–22

Galbraith, R. A., 256–260
Garattini, S., 128–140
Garbo, G. M., 328–329
Gasiewicz, T. A., 333–334
Gibbs, A. H., 30–40
Göcmen, A., 183–190
Goering, P. L., 235–247, 330–332
Goldberg, A., 191–203
Goldstein, J. A., 333–334
Graziani, A., 128–140
Gunter, E. W., 345–346

Hahn, M. E., 333–334
Holley, A. E., 30–40
Horák, J., 342–344

Jirsa, M., 342–344

Kaláb, M., 335–336
Kapoor, S., 278–288
Kappas, A., 1–6
Kaul, B., 23–29
Kawanishi, S., 222–234
Keck, R. W., 328–329
Kimbrough, R. D., 289–296
Koransky, W., 148–159
Kordač, V., 335–336, 342–344
Koss, G., 148–159
Kotal, P., 342–344
Kozáková, M., 342–344
Kreimer-Birnbaum, M., 328–329

Labbé, R. F., 7–14, 345–346
Lambrecht, R. W., 96–112, 337–338, 347–348
Landrigan, P. J., 323–326
Lauwerys, R., 41–47
Linko, P., 333–334
Litman, D. A., 248–255
Losekam, M., 148–159
Lunetta, J. M., 248–255
Lyon, M., 113–127

McCluskey, S., 113–127
Marcus, A. H., 339–341
Marks, G. S., 113–127
Martásek, P., 335–336, 342–344
Moore, M. R., 191–203
Morris, C., 183–190

Nelson, N., 327
Nonaka, S., 160–169

Oskarsson, A., 172–182, 268–277

Peters, H., 183–190
Piomelli, S., 278–288
Piper, W. N., 48–54
Powles, J., 113–127

Rettmer, R. L., 7–14, 345–346
Rizzardini, M., 128–140
Roberts, A. G., 141–147

Sano, S., 222–234
Sassa, S., 15–22
Seaman, C., 278–288
Seidel, J., 148–159
Seki, Y., 222–234
Selman, S. H., 328–329
Shah, A. G., 7–14
Shimoyama, T., 160–169

Silbergeld, E. K., 170–171, 297–308
Sinclair, J. F., 96–112, 337–338
Sinclair, P. R., 96–112, 337–338, 347–348
Smith, A., 309–322
Sugita, O., 15–22
Steinbach, K., 148–159
Strik, J. J. T. W. A., 219–221
Sunderman, F. W., Jr., 65–80
Sutherland, E., 113–127

Turnlund, J. R., 7–14

Urquhart, A. J., 141–147

Walden, T. L., Jr., 349–350
Wetmur, J. G., 23–29
Woods, J. S., 55–64, 172–182

Yeung-Laiwah, A. A. C., 191–203

Zelt, D., 113–127
Zoubek, V., 335–336